Methods in Enzymology

Volume 112
DRUG AND ENZYME TARGETING
Part A

METHODS IN ENZYMOLOGY

EDITORS-IN-CHIEF

Sidney P. Colowick   Nathan O. Kaplan

Methods in Enzymology

Volume 112

Drug and Enzyme Targeting

Part A

EDITED BY

Kenneth J. Widder

MOLECULAR BIOSYSTEMS, INC.
AND
DEPARTMENT OF PATHOLOGY
SCHOOL OF MEDICINE
UNIVERSITY OF CALIFORNIA
SAN DIEGO, CALIFORNIA

Ralph Green

DEPARTMENT OF LABORATORY HEMATOLOGY
CLEVELAND CLINIC FOUNDATION
CLEVELAND, OHIO

1985

ACADEMIC PRESS, INC.

(Harcourt Brace Jovanovich, Publishers)

Orlando   San Diego   New York   London
Toronto   Montreal   Sydney   Tokyo

ACADEMIC PRESS, INC.
Orlando, Florida 32887

United Kingdom Edition published by
ACADEMIC PRESS INC. (LONDON) LTD.
24–28 Oval Road, London NW1 7DX

LIBRARY OF CONGRESS CATALOG CARD NUMBER: 54-9110
ISBN 0-12-182012-2

PRINTED IN THE UNITED STATES OF AMERICA

85 86 87 88   9 8 7 6 5 4 3 2 1

Table of Contents

Section I. Microencapsulation Techniques

v

Section IV. Polymer Systems

Contributors to Volume 112

Article numbers are in parentheses following the names of contributors.
Affiliations listed are current.

GORDON L. AMIDON (28), *College of Pharmacy, The University of Michigan, Ann Arbor, Michigan 48109*

LYLE J. ARNOLD, JR. (21), *University of California, School of Medicine, La Jolla, California 92093, and Molecular Biosystems, Inc., 11180 Roselle Street, La Jolla, California 92121*

PERRY J. BLACKSHEAR (39, 40), *Department of Medicine, Duke University Medical Center, Durham, North Carolina 27710*

JUDITH L. BODMER (23), *Ciba-Geigy, Basel CH-4010, Switzerland*

NICHOLAS BODOR (29), *Department of Medicinal Chemistry, College of Pharmacy, and J. Hillis Miller Health Center, University of Florida, Gainesville, Florida 32610*

LARRY BROWN (30), *Whitaker College of Health Sciences, Technology and Management, and Department of Nutrition and Food Science, Massachusetts Institute of Technology, Cambridge, Massachusetts 02139, and Department of Surgery, Children's Hospital, Boston, Massachusetts 02115*

HANS BUNDGAARD (27), *Department of Pharmaceutical Chemistry, Royal Danish School of Pharmacy, DK-2100 Copenhagen, Denmark*

J. J. BURGER (3, 4), *Netherlands Cancer Institute, 1066 CX Amsterdam, The Netherlands*

M. CHANG (12), *Jet Propulsion Laboratory, California Institute of Technology, Pasadena, California 91109*

THOMAS MING SWI CHANG (15), *Artificial Cells and Organs Research Centre, McGill University, Montreal, Quebec H36 1Y6, Canada*

M. CHARTON (25), *School of Liberal Arts and Sciences, Pratt Institute, Brooklyn, New York 11205*

YIE W. CHIEN (34), *Controlled Drug Delivery Research Center, College of Pharmacy, Rutgers University, Piscataway, New Jersey 08854*

ALFRED M. COHEN (40), *Surgical Oncology Unit, Massachusetts General Hospital, Boston, Massachusetts 02114*

DONALD R. COWSAR (8), *Applied Sciences Department, Southern Research Institute, Birmingham, Alabama 35255*

GEORGE W. CUFF (7), *Pharmaceutical Research and Development, Eli Lilly and Company, Indianapolis, Indiana 46285*

A. J. CUMBER (16), *Biology Section, Institute of Cancer Research, Royal Cancer Hospital, Chester Beatty Laboratories, London SW3 6JB, England*

ROGER T. DEAN (23), *Cell Biology Research Group, Brunel University, Uxbridge, Middlesex UB8 3PH, England*

DERRICK L. DOMINGO (18), *Department of Cancer Biology, The Salk Institute for Biological Studies, San Diego, California 92138*

CHARLES F. DRISCOLL (5), *Department of Physics, University of California, San Diego, La Jolla, California 92093*

ELAZER EDELMAN (30), *Whitaker College of Health Sciences, Technology and Management, and Department of Nutrition and Food Science, Massachusetts Institute of Technology, Cambridge, Massachusetts 02139, and Department of Surgery, Children's Hospital, Boston, Massachusetts 02115*

JAMES P. ENGLISH (8), *Applied Sciences Department, Southern Research Institute, Birmingham, Alabama 35255*

JOHN W. FARA (35), *ALZA Corporation, 950 Page Mill Road, Palo Alto, California 94304*

HEINZ FAULSTICH (17), *Max-Planck-Institut für Medizinische Forschung, D-6900 Heidelberg 1, Federal Republic of Germany*

LUIGI FIUME (17), *Istituto di Patologia Generale, Universita degli Studi di Bologna, Bologna 40126, Italy*

DAVID FLEISHER (28), *College of Pharmacy, The University of Michigan, Ann Arbor, Michigan 48109*

J. A. FORRESTER (16), *Biology Section, Institute of Cancer Research, Royal Cancer Hospital, Chester Beatty Laboratories, London SW3 6JB, England*

B. M. J. FOXWELL (16), *Cellular Pharmacology Laboratory, Imperial Cancer Research Fund, Lincoln's Inn Fields, London WC2A 3PX, England*

RICHARD M. GILLEY (8), *Applied Sciences Department, Southern Research Institute, Birmingham, Alabama 35255*

EUGENE P. GOLDBERG (2), *Department of Materials Science and Engineering, University of Florida, Gainesville, Florida 32611*

AKIO GOTO (11), *Department of Pharmacy, Akita University Hospital, Akita 010, Japan*

J. HELLER (31), *Department of Polymer Sciences, SRI International, 333 Ravenswood Avenue, Menlo Park, California 94025*

K. J. HIMMELSTEIN (31), *Interx Research Corporation, Merck Sharpe & Dohme Research Laboratories, 2201 West 21st Street, Lawrence, Kansas 66044*

T. A. HORBETT (36), *Department of Chemical Engineering, and Center for Bioengineering, University of Washington, Seattle, Washington 98195*

LISBETH ILLUM (6), *Department of Pharmaceutical Chemistry, Royal Danish School of Pharmacy, DK-2100 Copenhagen, Denmark*

PHILIP D. E. JONES (6), *Department of Pharmacy, University of Nottingham, Nottingham NG7 2RD, England*

TETSURO KATO (11), *Department of Urology, Akita University School of Medicine, Akita 010, Japan*

JUDITH P. KITCHELL (32), *Dynatech Research and Development Company, 99 Erie Street, Cambridge, Massachusetts 02139*

JÖRG KREUTER (10), *Institut für Pharmazeutische Technologie, Johann Wolfgang Goethe Universität, D-6000 Frankfurt am Main, Federal Republic of Germany*

ROBERT LANGER (30), *Whitaker College of Health Sciences, Technology and Management, and Department of Nutrition and Food Science, Massachusetts Institute of Technology, Cambridge, Massachusetts 02139, and Department of Surgery, Children's Hospital, Boston, Massachusets 02115*

WILLIAM E. LONGO (2), *Department of Materials Science and Engineering, University of Florida, Gainesville, Florida 32611*

H. K. LONSDALE (37), *Bend Research, Inc., 64550 Research Road, Bend, Oregon 97701*

SHLOMO MARGEL (13), *Department of Materials Research, The Weizmann Institute of Science, Rehovot 76100, Israel*

JAMES W. MCGINITY (7), *College of Pharmacy, University of Texas, Austin, Texas 78712*

LUIGI MOLTENI (22), *MEDEA Research, Via Pisacane 34/A, Milan 20129, Italy*

SIDNEY D. NELSON (26), *Department of Medicinal Chemistry, University of Washington, Seattle, Washington 98195*

ROBERT E. NOTARI (24), *College of Pharmacy, The Ohio State University, Columbus, Ohio 43210*

J. PALMER (4), *Netherlands Cancer Institute, 1066 CX Amsterdam, The Netherlands*

B. D. RATNER (36), *Center for Bioengineering, and Department of Chemical Engineering, University of Washington, Seattle, Washington 98195*

A. REMBAUM (12, 14), *Jet Propulsion Laboratory, California Institute of Technology, Pasadena, California 91109*

G. RICHARDS (12), *Jet Propulsion Laboratory, California Institute of Technology, Pasadena, California 91109*

J. E. DE ROO (4), *Department of Pharmacy, University of Amsterdam, 1018 TV Amsterdam, The Netherlands*

W. C. J. ROSS (16), *Biology Section, Institute of Cancer Research, Royal Cancer Hospital, Chester Beatty Laboratories, London SW3 6JB, England*

ANNE M. ROUSSELL (40), *Diabetes Unit, Massachusetts General Hospital, Boston, Massachusetts 02114*

ULF SCHRÖDER (9), *Department of Tumor Immunology, The Wallenberg Laboratory, University of Lund, S-220 07 Lund, Sweden*

A. SCHWARTZ (14), *Becton Dickinson Research Center, Research Triangle Park, North Carolina 27709*

ANDREW E. SENYEI (5), *Department of Obstetrics and Gynecology, University of California, Irvine Medical Center, Orange, California 92668*

JANE E. SHAW (33), *ALZA Corporation, 950 Page Mill Road, Palo Alto, California 94304*

W. THOMAS SHIER (19), *Department of Medicinal Chemistry and Pharmacognosy, College of Pharmacy, University of Minnesota, Minneapolis, Minnesota 55455*

K. L. SMITH (37, 38), *Bend Research, Inc., 64550 Research Road, Bend, Oregon 97701*

BARBRA H. STEWART (28), *College of Pharmacy, The University of Michigan, Ann Arbor, Michigan 48109*

P. E. THORPE (16), *Drug Targeting Laboratory, Imperial Cancer Research Fund, Lincoln's Inn Fields, London WC2A 3PX, England*

THOMAS R. TICE (8), *Applied Sciences Department, Southern Research Institute, Birmingham, Alabama 35255*

E. TOMLINSON (3, 4), *Ciba-Geigy Pharmaceuticals, Horsham, West Sussex RH 12 4AB, England*

IAN S. TROWBRIDGE (18), *Department of Cancer Biology, The Salk Institute for Biological Studies, San Diego, California 92138*

KATSUO UNNO (11), *Department of Pharmacy, Akita University Hospital, Akita 010, Japan*

J. M. VARGA (20), *Division of Cancer Biology and Diagnosis, National Cancer Institute, National Institutes of Health, Bethesda, Maryland 20205*

KENNETH J. WIDDER (5), *Molecular Biosystems, Inc., 11180 Roselle Street, San Diego, California 92121*

BRUCE D. WIGNESS (40), *Department of Surgery, University of Minnesota Hospitals, Minneapolis, Minnesota 55455*

DONALD L. WISE (32), *Dynatech Research and Development Company, 99 Erie Street, Cambridge, Massachusetts 02139*

ANTHONY F. YAPEL, JR. (1), *Biosciences Laboratory, 3M Company, St. Paul, Minnesota 55144*

Preface

Rapid growth and increasing interest in the science of targeting biologically active agents to specific and desired sites of action have resulted in a proliferation of scientific meetings and journals devoted to this topic. The concept of drug targeting technology offers a rational selective method for treating various diseases, including cancer, infectious diseases, and enzyme deficiencies in animals and man. Because drug and enzyme targeting is a relatively new field, the scientists who have made major contributions to its growth have come from diverse backgrounds. Drug delivery technology brings together contributors from a broad spectrum of scientific disciplines, including pharmacology, biochemistry, enzymology, organic chemistry, biophysics, lipid chemistry, physiology, anatomy, and pathology.

In Drug and Enzyme Targeting we have attempted a comprehensive coverage of the different methods that have been developed in the drug delivery field. Our aim is to provide scientists interested in this field with a complete compendium of detailed techniques. This volume (Part A) deals with microspheres, polymer systems, and drug conjugates. For the sake of completeness, we have included a contribution on infusion pumps, a significant advance in sustained medical therapeutics. Part B will deal with liposomes, receptor-mediated targeting, and cell carriers.

We wish to thank our editorial advisory board, Drs. Ernest Beutler, Roscoe Brady, Gregory Gregoriadis, Robert Langer, and Anthony Sinkula, for their assistance. We would also like to thank the very capable staff of Academic Press for their efficiency, professionalism, and cooperation in the production of these volumes.

Kenneth J. Widder
Ralph Green

METHODS IN ENZYMOLOGY

EDITED BY

Sidney P. Colowick and Nathan O. Kaplan

VANDERBILT UNIVERSITY
SCHOOL OF MEDICINE
NASHVILLE, TENNESSEE

DEPARTMENT OF CHEMISTRY
UNIVERSITY OF CALIFORNIA
AT SAN DIEGO
LA JOLLA, CALIFORNIA

I. Preparation and Assay of Enzymes
II. Preparation and Assay of Enzymes
III. Preparation and Assay of Substrates
IV. Special Techniques for the Enzymologist
V. Preparation and Assay of Enzymes
VI. Preparation and Assay of Enzymes (*Continued*)
Preparation and Assay of Substrates
Special Techniques
VII. Cumulative Subject Index

METHODS IN ENZYMOLOGY

EDITORS-IN-CHIEF

Sidney P. Colowick and Nathan O. Kaplan

VOLUME XVIII. Vitamins and Coenzymes (Parts A, B, and C)
Edited by DONALD B. McCORMICK AND LEMUEL D. WRIGHT

VOLUME XIX. Proteolytic Enzymes
Edited by GERTRUDE E. PERLMANN AND LASZLO LORAND

VOLUME XX. Nucleic Acids and Protein Synthesis (Part C)
Edited by KIVIE MOLDAVE AND LAWRENCE GROSSMAN

VOLUME XXI. Nucleic Acids (Part D)
Edited by LAWRENCE GROSSMAN AND KIVIE MOLDAVE

VOLUME XXII. Enzyme Purification and Related Techniques
Edited by WILLIAM B. JAKOBY

VOLUME XXIII. Photosynthesis (Part A)
Edited by ANTHONY SAN PIETRO

VOLUME XXIV. Photosynthesis and Nitrogen Fixation (Part B)
Edited by ANTHONY SAN PIETRO

VOLUME XXV. Enzyme Structure (Part B)
Edited by C. H. W. HIRS AND SERGE N. TIMASHEFF

VOLUME XXVI. Enzyme Structure (Part C)
Edited by C. H. W. HIRS AND SERGE N. TIMASHEFF

VOLUME XXVII. Enzyme Structure (Part D)
Edited by C. H. W. HIRS AND SERGE N. TIMASHEFF

VOLUME XXVIII. Complex Carbohydrates (Part B)
Edited by VICTOR GINSBURG

VOLUME XXIX. Nucleic Acids and Protein Synthesis (Part E)
Edited by LAWRENCE GROSSMAN AND KIVIE MOLDAVE

VOLUME XXX. Nucleic Acids and Protein Synthesis (Part F)
Edited by KIVIE MOLDAVE AND LAWRENCE GROSSMAN

VOLUME XXXI. Biomembranes (Part A)
Edited by SIDNEY FLEISCHER AND LESTER PACKER

VOLUME XXXII. Biomembranes (Part B)
Edited by SIDNEY FLEISCHER AND LESTER PACKER

VOLUME XXXIII. Cumulative Subject Index Volumes I–XXX
Edited by MARTHA G. DENNIS AND EDWARD A. DENNIS

VOLUME XXXIV. Affinity Techniques (Enzyme Purification: Part B)
Edited by WILLIAM B. JAKOBY AND MEIR WILCHEK

VOLUME XXXV. Lipids (Part B)
Edited by JOHN M. LOWENSTEIN

VOLUME XXXVI. Hormone Action (Part A: Steroid Hormones)
Edited by BERT W. O'MALLEY AND JOEL G. HARDMAN

VOLUME XXXVII. Hormone Action (Part B: Peptide Hormones)
Edited by BERT W. O'MALLEY AND JOEL G. HARDMAN

VOLUME XXXVIII. Hormone Action (Part C: Cyclic Nucleotides)
Edited by JOEL G. HARDMAN AND BERT W. O'MALLEY

VOLUME XXXIX. Hormone Action (Part D: Isolated Cells, Tissues, and Organ Systems)
Edited by JOEL G. HARDMAN AND BERT W. O'MALLEY

VOLUME XL. Hormone Action (Part E: Nuclear Structure and Function)
Edited by BERT W. O'MALLEY AND JOEL G. HARDMAN

VOLUME XLI. Carbohydrate Metabolism (Part B)
Edited by W. A. WOOD

VOLUME XLII. Carbohydrate Metabolism (Part C)
Edited by W. A. WOOD

VOLUME XLIII. Antibiotics
Edited by JOHN H. HASH

VOLUME XLIV. Immobilized Enzymes
Edited by KLAUS MOSBACH

VOLUME XLV. Proteolytic Enzymes (Part B)
Edited by LASZLO LORAND

VOLUME 85. Structural and Contractile Proteins (Part B: The Contractile Apparatus and the Cytoskeleton)
Edited by DIXIE W. FREDERIKSEN AND LEON W. CUNNINGHAM

VOLUME 86. Prostaglandins and Arachidonate Metabolites
Edited by WILLIAM E. M. LANDS AND WILLIAM L. SMITH

VOLUME 87. Enzyme Kinetics and Mechanism (Part C: Intermediates, Stereochemistry, and Rate Studies)
Edited by DANIEL L. PURICH

VOLUME 88. Biomembranes (Part I: Visual Pigments and Purple Membranes, II)
Edited by LESTER PACKER

VOLUME 89. Carbohydrate Metabolism (Part D)
Edited by WILLIS A. WOOD

VOLUME 90. Carbohydrate Metabolism (Part E)
Edited by Willis A. Wood

VOLUME 91. Enzyme Structure (Part I)
Edited by C. H. W. HIRS AND SERGE N. TIMASHEFF

VOLUME 92. Immunochemical Techniques (Part E: Monoclonal Antibodies and General Immunoassay Methods)
Edited by JOHN J. LANGONE AND HELEN VAN VUNAKIS

VOLUME 93. Immunochemical Techniques (Part F: Conventional Antibodies, Fc Receptors, and Cytotoxicity)
Edited by JOHN J. LANGONE AND HELEN VAN VUNAKIS

VOLUME 94. Polyamines
Edited by HERBERT TABOR AND CELIA WHITE TABOR

VOLUME 95. Cumulative Subject Index Volumes 61–74 and 76–80
Edited by EDWARD A. DENNIS AND MARTHA G. DENNIS

VOLUME 96. Biomembranes [Part J: Membrane Biogenesis: Assembly and Targeting (General Methods; Eukaryotes)]
Edited by SIDNEY FLEISCHER AND BECCA FLEISCHER

Section I

Microencapsulation Techniques

[1] Albumin Microspheres: Heat and Chemical Stabilization

By ANTHONY F. YAPEL, JR.

Introduction

Medical science has long recognized the need to control, regulate, and target the release of drugs in the body. In general, the aim has been to provide less frequent drug administration, constant and continuous therapeutic levels of drug in the systemic circulation or at a specific target organ site, and a reduction in undesirable drug side effects. During the past decade, considerable progress has been made in this regard. A wide variety of drug delivery systems have been designed and evaluated, including drug carriers based on proteins, polysaccharides, synthetic polymers, erythrocytes, DNA, and liposomes, to name a few.[1,2] These drug delivery systems have taken many forms, one of the more popular configurations being that of microspheres. Microspheres as used in drug delivery are discrete, micrometer-sized spherical particles containing an entrapped drug. They can be prepared from a variety of carrier materials; one of the most utilized is serum albumin from humans or other appropriate species. Drug release from serum albumin microspheres can be sustained and controlled by various stabilization procedures generally involving heat or chemical cross-linking of the protein carrier matrix. It is the purpose of this chapter to outline representative procedures and their influence on modifying the sustained release characteristics of drugs incorporated in the albumin microsphere matrix.

Albumin Microsphere Drug Delivery Systems: General Considerations

Intravenous injection can provide efficient targeting of albumin microspheres to either the lung or the liver. Their ultimate location will depend strictly on the size range of the microspheres. When injected intravenously, microspheres 15–30 μm or larger will pass through the heart and then into the capillary bed of the lungs where they will deposit with 99% efficiency. If microspheres about an order of magnitude smaller (e.g., 1–3

---

[1] J. R. Robinson, ed., "Sustained and Controlled Release Drug Delivery Systems." Dekker, New York, 1978.
[2] G. Gregoriadis, ed., "Drug Carriers in Biology and Medicine." Academic Press, New York, 1979.

METHODS IN ENZYMOLOGY, VOL. 112

μm) are injected intravenously, they will pass into the reticuloendothelial system where they will deposit with about 90% efficiency in the liver. Intravenous injection of microspheres less than 1 μm in diameter results in a mixed distribution of the tiny particles among the liver (80–90%), spleen (5–8%), and bone marrow (1–2%). Microsphere targeting to these organs can be accomplished on the basis of size alone.

The lung, liver, and to some extent, the spleen are the only organs that can be reached with relative ease upon simple intravenous injection of appropriately sized microspheres. However, specialized angiographic methods can be used to target drug-loaded spheres to other organs, to specific sites within organs, or to solid tumors.

The rate of drug release from injected albumin microspheres and the rate at which the microspheres themselves degrade in the body are controlled by their degree of cross-linking. Microsphere cross-linking can be accomplished by heat denaturation of the albumin or through use of chemical cross-linking agents such as formaldehyde or glutaraldehyde.

Heat Cross-Linking Methods

In 1972 Evans described both batch and continuous process methods for preparing heat cross-linked albumin microspheres.[3,4] The microspheres prepared using his procedures were free flowing and nonagglomerated. Their size could be varied in the range from 0.5 μm to 1 mm. The rate of biodegradation of the albumin microspheres in vivo could be adjusted by varying the time and temperature of heat cross-linking. A representative batch preparation and heat cross-linking procedure utilized by Evans is outlined below.

Method 1—Batch Process for Preparing Heat Cross-Linked Microspheres

Serum albumin of an appropriate species is dissolved in water to achieve a solution of final protein concentration of about 25% (w/v). Four milliliters of this solution are injected with a 25-gauge hypodermic needle into 1 liter of stirred cottonseed oil which has been preheated to 30–50° and which is stirred at 500 rpm with a 2.5-in. propeller-type stirrer. A water-in-oil emulsion forms and microspheres 10–20 μm in diameter are ultimately obtained. Slower stirring speeds result in larger microspheres being formed, and vice versa. Stirring is continued while raising the temperature of the oil bath to 110° or higher or until the microspheres are

[3] R. L. Evans, U.S. Patent 3,663,685, to 3M Company (1972).
[4] R. L. Evans, U.S. Patent 3,663,687, to 3M Company (1972).

TABLE I
Effect of Cross-Linking Temperature on
Albumin Microsphere Degradation[a]

Cross-linking temperature (°C)	50% Solubilization time[b]
135	24 hr
160	84 hr
170	4 days
190	30 days

[a] From Evans.[4]
[b] Microspheres are heat cross-linked at the temperatures specified for 40 min. They are 50% biodegraded and solubilized in body fluids in the times shown as judged by histological techniques.

essentially dehydrated (as can be determined by filtering a small number of the particles from the oil bath to determine whether they are still tacky). After drying by dehydration at an appropriately selected temperature above 100°, the microspheres are filtered from the oil bath on filter paper and washed with diethyl ether. The resulting serum albumin microspheres are 10–20 μm in diameter and are an unagglomerated, free-flowing tan powder. They are insoluble but swellable in water, their degree of swellability being inversely proportional to the extent of heat cross-linking.

Evans has found that the resulting *in vivo* biodegradability of heat cross-linked albumin microspheres is both time and temperature dependent.[4] Heat cross-linking at 70–80° for 2 hr, for example, is equivalent to heat cross-linking at 120° for 1 hr. Additional degradability data for heat cross-linked albumin microspheres are presented in Table I.

The general heat cross-linking procedures outlined above can be used to prepare microspheres which degrade *in vivo* over time periods ranging from minutes to many months. For example, heating at 105° for 40 min produces microspheres which began to biodegrade and dissolve within 15 min after intravenous injection. On the other hand, microspheres heated at 180° for 18 hr just start to degrade 6 months after injection. Intermediate temperatures and cross-linking times will produce microspheres with intermediate degradation rates. Hence, it is possible to tailor the degradation characteristics of heat cross-linked albumin microspheres to meet specific drug delivery or diagnostic needs.

For drug delivery applications, the serum albumin chosen is generally that of the species into which the microspheres will be injected. However, egg albumin, hemoglobin, and a number of other proteins and polysaccharides can be spheroidized and insolubilized using the same or slight modifications of the above procedure. A variety of hydrophobic liquids can be utilized in the spheroidizing oil bath. Vegetable oils such as cottonseed oil, corn oil, olive oil, etc. are preferred; however, low melting animal fats, mineral oils, inert hydrocarbons, and halogenated hydrocarbons can also be used.

Diagnostic Applications for Heat Cross-Linked Albumin Microspheres

Albumin microspheres prepared using techniques similar or identical to those described above can be labeled with various radioisotopes to produce diagnostic agents which upon intravenous injection are particularly useful for assessing the vascular integrity of (or detecting the presence of thrombi in) such organs as the lung, liver, and spleen or for detecting the presence of tumors in these organs. In some cases, the radiolabeled microspheres can be utilized for therapeutic purposes (e.g., tumor treatment).

A variety of methods have been described in the literature for radioactively labeling albumin microspheres for use in diagnostic applications. These include prelabeling the native albumin with a radionuclide (e.g., ^{125}I, ^{131}I, ^{59}Fe, ^{99m}Tc) prior to formation of the microspheres,[5] or alternatively, labeling the albumin microspheres after preparation by equilibrating them for a sufficient period of time in an aqueous solution of a radionuclide.[4] Radioactivity has also been imparted to albumin microspheres by incorporating relatively water-insoluble salts of radionuclides (e.g., ^{131}AgI, $Ba^{35}SO_4$) into the microspheres during the spheroidization process[3] or by absorbing radionuclides onto suitable water-insoluble carriers (e.g., hydroxides of iron, chromium, or aluminum) which are subsequently incorporated into the microspheres during their formation process in the spheroidizing oil bath.[3] Kits which provide the user with considerable convenience in labeling albumin microspheres with radioisotopes have likewise been described.[6] Detailed procedures for labeling albumin microspheres with radiochemicals are discussed elsewhere in this volume.[7]

[5] I. M. Grotenhuis and D. O. Kubiatowicz, U.S. Patent 3,663,686, to 3M Company (1972).
[6] D. O. Kubiatowicz, U.S. Patent 3,707,353, to 3M Company (1972).
[7] J. J. Burger, E. Tomlinson, J. E. de Roo, and J. Palmer, this volume [4].

Drug Entrapment in Heat Cross-Linked Albumin Microspheres

Yapel has described heat cross-linking methods for preparing and stabilizing albumin microspheres which exhibit a biphasic release of entrapped drugs.[8] Such biphasic release is thought to be desirable for injectable drug delivery systems, since a therapeutic "loading dose" of drug can be provided initially in a fast release phase followed by a subsequent slower sustained release of drug necessary to maintain therapeutic blood levels. The basic procedure utilized by Yapel to entrap a variety of drugs in albumin microspheres via heat cross-linking methods is outlined below using 5-fluorouracil, an antitumor agent, as a representative drug. Zolle[9] and Kramer[10] also have described related methods for preparing drug-loaded, heat cross-linked albumin microspheres.

Method 2—Procedure for Entrapping Drugs in Heat Cross-Linked Albumin Microspheres

One gram of human serum albumin is dissolved in 2.0 ml of deionized water by stirring with a magnetic stirrer. To the dissolved albumin is added 0.1 g of 5-fluorouracil (5-FU) and stirring is continued for an additional 15 min. Since 5-FU at this concentration does not completely dissolve in the albumin solution, the aqueous mixture is placed in a standard tissue grinder of 10-ml capacity and further dispersed to ensure homogeneous distribution of the remaining undissolved 5-FU particles throughout the albumin solution.

The mixture is then immediately injected with a tuberculin syringe equipped with a 20-gauge hypodermic needle into 500 ml of cottonseed oil contained in a 600-ml stainless steel beaker at room temperature while stirring with a 1.5-in. propeller-type stirrer at approximately 2300 rpm. Stirring rates are monitored with a tachometer since the rate of stirring determines to a significant extent the ultimate particle size distribution of the resulting microspheres. Stirring is continued while the oil bath temperature is raised with a 500-W immersion heater to a temperature of 140° over a 15-min time period. The bath is held at this temperature for 1 hr while stirring is maintained at 2300 rpm.

The oil bath and its contents are then cooled to room temperature and the resulting microspheres separated from the oil on Whatman No. 5 filter paper using vacuum filtration. Final traces of oil are removed from the

[8] A. F. Yapel, U.S. Patent 4,147,767, to 3M Company (1979).
[9] I. Zolle, U.S. Patent 3,937,668 (1976).
[10] P. A. Kramer, *J. Pharm. Sci.* **63,** 1646 (1974).

drug-containing microspheres by washing them several times with 30-ml aliquots of heptane. The microspheres produced by the above process contain 9.6% 5-FU (w/w) and exist as an unagglomerated, free-flowing, light tan powder with a particle size distribution of 10–60 μm.

Factors Regulating Drug Release from Microspheres

With regard to the above generalized procedure, several clarifying points may be helpful. For example, the concentration of the albumin solution used to prepare the microspheres will have some effect on the ultimate drug release characteristics. Typically, microspheres are prepared from 20–50% (w/v) solutions of albumin in water, although they have been prepared from solutions as low as 2–5% in concentration and as high as 70–80%. Particularly for heat cross-linked microspheres, those prepared from higher concentration albumin solutions are more dense and release incorporated drugs somewhat more slowly than those prepared from lower concentration solutions, all other things being equal.

Following the addition of drug to the albumin solution prior to spheroidization, the mixture is allowed to equilibrate for a period of time ranging from 15 to 60 min. During this equilibration step, a certain percentage of the drug molecules may be bound to sites on the native albumin molecules. The actual amount bound depends on such factors as the nature of the binding sites on the albumin, the amount and the polarity of the electrostatic charge, if any, on the drug and the albumin carrier, the concentration of both drug and carrier in the equilibrating solution, the equilibrium constant between carrier sites and drug molecules, the molecular weight and water solubility of the drug, the pH, the temperature, and other mass-law considerations.

During heat cross-linking of the microspheres, the oil bath is typically heated to a temperature of 110–180° over a period of 15–30 min and then held at this temperature for at least 20 additional min. Heat treatments in excess of 10 hr are rarely required. During this heating process, which removes water from and ultimately insolubilizes the spheres (via probable scrambling of disulfide bonds and formation of lysinoalanine cross-links), that portion of insoluble drug or excess soluble drug not chemically bound to actual sites on the protein carrier is entrapped within the hardened and cross-linked albumin sphere matrix.

The particle size range of the microspheres can be regulated by varying the rate of injection of the drug/albumin solution or dispersion into the oil bath and/or by varying the stirring speed of the bath as previously discussed. The addition of small amounts (1–2%) of a surfactant (e.g.,

Tween 80) to the starting albumin solutions will also influence ultimate sphere size through its effects on surface and interfacial tensions. The biodegradability and porosity of the microspheres are controlled by varying the time and temperature of the heating process in oil. Other things being equal, higher temperatures and longer heating times generally produce harder, less porous, and more slowly degradable spherules (see Table I).

In general, a given weight of very tiny microspheres (e.g., 1–5 μm) containing a specific amount of drug will release that drug faster than the same weight of larger microspheres (e.g., 50–100 μm) containing the same amount of drug because of the much larger surface area and shorter drug diffusion path length of the smaller spheres. Thus, variations in microsphere size can be used in conjunction with the other parameters discussed above to aid in regulating the rate of drug release from albumin spherules.

It should be pointed out that albumin microspheres prepared according to Method 1 above can be loaded with low levels of drug by simply equilibrating the microspheres with stirring in a solution (preferably aqueous to enhance microsphere swelling) of the drug of choice. Microsphere drug loadings achieved by this equilibration technique are generally low (<5% of sphere weight) and must be reserved for highly potent drugs.

Biphasic Drug Release from Heat Cross-Linked Albumin Microspheres

As previously discussed, Yapel has found that the release of drugs from heat cross-linked albumin microspheres is frequently biphasic in character.[8] This biphasic release behavior is dependent to some degree on the water solubility of the entrapped drug. Highly water-soluble drugs exhibit very noticeable biphasic release characteristics. Relatively water-insoluble drugs such as steroids, which diffuse much more slowly from the albumin microsphere matrix, exhibit less pronounced biphasic release properties. The concentration of drug incorporated in the microspheres also appears to play a major role in governing the biphasic character of that drug's release from the albumin carrier.

For example, studies have been carried out in which a series of batches of heat cross-linked albumin microspheres were loaded according to Method 2 with increasing quantities of the water-soluble drug L-epinephrine ranging from 9.1 to 87.5% of the total sphere weight.[8] In each case, the heat cross-linking temperature was 140° and the cross-linking time 4 hr. Drug release profiles from the microspheres were measured using an *in vitro* flow cell method (pumping rate = 30 ml/hr) to monitor L-

TABLE II

SUMMARY OF BIPHASIC RELEASE DATA FOR HEAT CROSS-LINKED ALBUMIN
MICROSPHERES LOADED WITH VARYING CONCENTRATIONS OF L-EPINEPHRINE[a]

HSA (%)	L-Epinephrine (%)	K_{fast} ($\times 10^2$ min^{-1})	K_{slow} ($\times 10^3$ min^{-1})	$t_{1/2}$(fast) (min)	$t_{1/2}$(slow) (min)
90.9	9.1	1.09	1.92	63.5	361.8
76.9	23.1	1.15	4.21	60.4	164.8
66.7	33.3	1.86	14.58	37.2	47.5
40.0	60.0	2.80	15.46	24.7	44.8
12.5	87.5	3.47	25.62	19.9	27.1

[a] All microspheres prepared according to Method 2 under cross-linking conditions of 140° for 4 hr.

epinephrine release from the microspheres into pH 7.6 phosphate-buffered saline solution (137 mM NaCl, 27 mM KCl, 8 mM NaHPO$_4$, 1 mM KH$_2$PO$_4$).

Analysis of the resulting drug release profiles demonstrated that L-epinephrine is released from the microspheres in two distinct steps, a fast phase, t(fast), followed by a much slower sustained release phase, t(slow). Each of these steps could in turn be adequately characterized by the following first-order equation, in which the rate of drug release is proportional to the concentration of drug remaining in the microspheres.

$$-dC/dt = KC \tag{1}$$

In these equations, C is the concentration of drug remaining in microspheres at time t; C_0, initial concentration of drug at time $t = 0$; K, rate constant for drug release; and t, time. Eq. (1) can be integrated and rearranged to give Eq. (2), where C/C_0 is the fraction of drug remaining in the microspheres at time t.

$$\ln(C/C_0) = -Kt \tag{2}$$

The half-time ($t_{1/2}$) of the first-order rate process represented by Eq. (2) is given by Eq. (3).

$$t_{1/2} = 0.693/K \tag{3}$$

The release data for each preparation of microspheres were analyzed according to Eq. (2) using a least-squares procedure to fit straight lines through the initial and final portions of each release curve.[8] Values for $t_{1/2}$ and K for the fast and slow drug release steps from each of the sphere preparations were determined, and the results are summarized in Table II and Fig. 1.

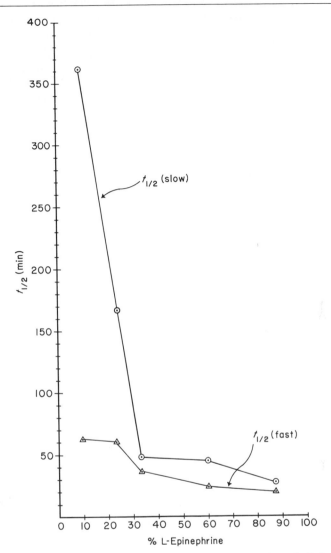

FIG. 1. Effect of L-epinephrine loading concentration on drug release half-times for heat cross-linked albumin microspheres. All microsphere samples prepared according to Method 2 using cross-linking conditions of 140° for 4 hr.

From Fig. 1 and Table II it is evident that drug release from albumin microspheres prepared according to Method 2 can be tailored to exhibit biphasic release properties, particularly when the drug loading is less than 30% of the total sphere weight. From Fig. 1, L-epinephrine release rates for both the fast and slow release processes increase with increasing drug loading of the microspheres. However, when drug loadings exceed approximately 30%, the biphasic release properties of the albumin microspheres are significantly diminished (i.e., the fast and slow release rate processes tend to approach each other in magnitude). Similar drug release behavior has been observed for a number of drugs belonging to a variety of drug classes.[8]

Chemical Cross-Linking Methods

The heat insolubilization techniques discussed above are particularly advantageous in that it is possible to incorporate both dispersions of water-insoluble drugs and solutions of water-soluble drugs by entrapment. The major drawback of the heat insolubilization method relates to the fact that during the microsphere formation and drug entrapment process, the spherules must often be heated to temperatures in excess of 110° to obtain sustained drug release characteristics. Although higher temperatures may be of little consequence for highly stable drugs, they can potentially lead to degradation and loss of drug efficacy for less stable drugs. To alleviate this problem, room temperature chemical cross-linking techniques have been developed. These methods typically involve the use of formaldehyde and glutaraldehyde, both of which are excellent cross-linking and hardening agents for albumin.

General Considerations—Chemical Cross-Linking

Lipophilic drying alcohols such as *n*-butanol, *sec*-butanol, and 2-ethylhexanol can be readily dissolved in vegetable oils such as cottonseed oil, a standard bath medium for preparing albumin microspheres. Water-soluble cross-linking agents such as formaldehyde and glutaraldehyde can in turn be dissolved in the drying alcohols. Specifically, up to 25% by volume glutaraldehyde (in the form of a 25% aqueous solution) or formaldehyde (in the form of a 37% aqueous solution) can be dissolved in a lipophilic drying alcohol such as *n*-butanol. If approximately 1–40 parts by volume of such a formaldehyde/butanol or glutaraldehyde/butanol solution are mixed with 70–500 parts of cottonseed oil, a homogeneous bath solution containing both dissolved cross-linking agent and drying alcohol is obtained. If an aqueous solution of 25–50% serum albumin is now injected into this bath while it is stirring at 1200–1500 rpm, microspheres in the 20- to 100-μm size range are produced. Smaller microspheres can be obtained by decreasing the concentration of albumin in the solution injected into the bath or by employing higher stirring speeds. During the sphere stabilization process, the solvent water within the microspheres partitions into the drying alcohol while the spheres themselves are cross-linked by the glutaraldehyde or formaldehyde. After at least 20 min of contact with the cross-linking medium, the spheres can be filtered from the bath. The degree of microsphere cross-linking can be adjusted by varying the length of time the spheres are kept in contact with the cross-linking agent and/or by varying the amount of cross-linking agent in the oil bath. The general biphasic *in vitro* drug release characteristics observed for these chemically cross-linked microspheres are similar to those observed for heat cross-linked spheres. An example of a typical procedure useful for preparing chemically stabilized albumin microspheres is given in Method 3, in which glutaraldehyde is employed as the cross-linking agent for spheres containing L-epinephrine.[8]

Method 3—Chemical Cross-Linking at Room Temperature

One gram of human serum albumin is dissolved in 2.0 ml of deionized water at room temperature by stirring with a magnetic stirrer. To the dissolved albumin is added 0.15 g of L-epinephrine (free base) and 0.0725 g of L(+)-ascorbic acid to stabilize the drug. Stirring is continued for an additional 15 min.

The mixture is then immediately injected through a 20-gauge hypodermic needle into a stirred homogeneous bath consisting of 500 ml cottonseed oil, 13 ml of *n*-butanol, and 2.0 ml of 25% glutaraldehyde. The bath is contained in a 600-ml stainless steel beaker and stirred at room temperature with a 1.5-in. propeller-type stirrer at approximately 1200 rpm. Stirring is continued for 4 hr during which time the resulting microspheres are dewatered by the *n*-butanol and cross-linked by the glutaraldehyde present in the bath. After this time period, the hardened microspheres are separated from the bath medium on Whatman No. 5 filter paper using vacuum filtration. Final traces of oil are removed from the microspheres by washing them several times with 100-ml aliquots of heptane.

The above procedure produces drug-containing microspheres composed of 81.8% albumin, 12.3% L-epinephrine, and 5.9% L(+)-ascorbic acid by weight. The microspheres exist as a free-flowing, nonagglomerated yellow-orange powder with individual microspheres in the size range 10–80 μm.

Method 3 is especially advantageous in that the albumin spheroidization, dewatering, and cross-linking can all be accomplished in a single, convenient step. Interestingly, if the cross-linking agent is omitted in the above procedure, chemically dewatered, free-flowing albumin microspheres can still be isolated from the oil bath by filtration. However, these microspheres are sensitive to water and will dissolve on contact with it.

Method 3 is particularly useful for entrapping heat-sensitive water-soluble or water-insoluble drugs in albumin microspheres, although it can also be used for drugs which are heat stable. A modification of Method 3 may be useful for entrapping drugs which can be safely heated to temperatures up to 105°. In this modification, the albumin/drug mixture is injected into a lower temperature oil bath (<105°) and the resulting microspheres partially insolubilized. Alternatively, the albumin/drug mixture can be spheroidized and chemically dewatered without cross-linking as discussed above. The resulting microspheres are isolated by filtration, washed, and then placed in a closed desiccator and exposed to vapors of glutaraldehyde or formaldehyde for a minimum of 20 min. The formaldehyde and/or glutaraldehyde can be placed in the bottom of the desiccator in the form of commercially available 37 and 25% aqueous solutions, respectively. Aldehyde vapor treatments of this type can be continued for up to several days if desired, with longer exposures leading to denser, less porous, more water-insoluble microspheres. After vapor cross-linking, excess formaldehyde or glutaraldehyde can be removed from the treated albumin microspheres by aspiration or vacuum desiccation.

Additional Chemical Cross-Linking Methods

In addition to formaldehyde and glutaraldehyde, other cross-linking agents may be utilized in preparing albumin microspheres according to Method 3. For example, many di-, tri-, and tetravalent metallic cations are often suitable for this purpose. More specifically, cations such as Fe^{3+} and Al^{3+} can be readily dissolved in *n*-butanol, *sec*-butanol, 2-ethylhexanol, and other lipophilic drying alcohols which in turn can be dissolved in cottonseed or other vegetable oils. The resulting solution serves as a one-step cross-linking bath for albumin microspheres.

A variety of additional protein cross-linking agents are known and can be utilized as suitable alternative cross-linking agents. Some of these agents are water soluble and some are water insoluble.[8,11,12] Water-insoluble cross-linkers can be dissolved directly in the hydrophobic oil bath and utilized in this form to stabilize microspheres formed following injection

[11] M. Rubinstein and S. Simon, U.S. Patent 4,101,380, to Research Products Rehovot Ltd. (1978).

[12] P. M. Abdella, P. K. Smith, and G. P. Rover, *Biochem. Biophys. Res. Commun.* **87**(3), 734 (1979).

of an aqueous solution of albumin into the bath. Water-soluble cross-linking agents can be incorporated directly into the albumin solution immediately prior to injection into the oil bath, or alternatively, dissolved in the alcohol which functions as a dewatering agent when present in the oil. It should be evident that various combinations of heat and chemical cross-linking techniques can also be employed to achieve the sustained release properties desired for drug-containing albumin microspheres.

Royer *et al.*[13–15] have recently described the preparation of drug-loaded albumin microspheres using mild chemical cross-linking techniques which are claimed not to lead to physical denaturation of the protein during the stabilization process. In their procedure, the spheroidization bath temperature does not exceed 37° (preferred temperature, 4°), and the pH of the albumin solution is maintained between 5.5 and 8.5. The recommended starting albumin solution concentration range is 10–30%. Glutaraldehyde and other bifunctional cross-linking agents[12] are used to harden the resulting microspheres and are added directly to the starting albumin solution immediately prior to its injection into the spheroidizing oil bath. Royer has used variations of this general technique to produce not only albumin microspheres but also albumin microcapsules and albumin implant materials.[13] An example of the chemical cross-linking procedure utilized by Royer to prepare steroid-containing albumin microspheres is given in Method 4 below.

Method 4—Chemical Cross-Linking at Low Temperatures

Rabbit or bovine serum albumin (600 mg) is dissolved in 2.4 ml of a solution containing 0.1% sodium dodecyl sulfate in 1 mM sodium phosphate buffer (pH 7.5). The solution is cooled to 4° and 39 mg of a finely divided steroid (progesterone or norgestrel) are dispersed in the solution. This is immediately followed by the addition of a 5% (v/v) aqueous solution of glutaraldehyde. The resulting dispersion is rapidly pipetted into 150 ml of a rapidly stirred 1:4 mixture of corn oil and petroleum ether contained in a 250-ml beaker. The stirring is continued for 15 min, at the end of which time solid microspheres settle to the bottom of the beaker. The spheroidizing bath mixture is decanted and the microspheres washed three times with petroleum ether followed by drying in a vacuum desiccator. The dried microspheres are incubated twice for 15-min time periods in 25 ml of a solution containing 0.1% (w/v) serum albumin dissolved in

[13] G. P. Royer, U.S. Patent 4,349,530, to Ohio State University (1982).
[14] T. K. Lee, T. D. Sokolski, and G. P. Royer, *Science* **213**, 233 (1981).
[15] G. P. Royer and T. K. Lee, *J. Parenter. Sci. Technol.* **37**(2), 34 (1983).

0.05 *M* Tris buffer at pH 8.6. The microspheres are then washed with 500 ml of 1 m*M* HCl in a sintered glass funnel, the acid is washed off with distilled water, and the microspheres are again dried in a vacuum desiccator. The resulting steroid-containing microspheres are dark brown in color and 100–200 μm in diameter. When examined under a microscope at 100-fold magnification, crystals of the drug can be seen entrapped within the albumin microsphere matrix. If desired, the density and porosity of the albumin matrix can be increased or decreased by correspondingly increasing or decreasing the concentration of glutaraldehyde or other bifunctional cross-linking agent added to the original albumin solution.

Progesterone-containing albumin microspheres prepared according to Method 4 were suspended in corn oil and then injected into rabbits both subcutaneously and intramuscularly by Royer *et al.*[13–15] Following an initial burst of drug, sustained levels of approximately 1 ng/ml progesterone were measurable in the blood of the test animals for at least 20 days following injection. Examination of the injection sites demonstrated that the microspheres administered intramuscularly had completely biodegraded by the end of 2 months with no adverse immunological response. Microspheres injected subcutaneously also disappeared completely but at a slower, unspecified rate.[13]

Goosen *et al.*[16] have also used the procedure of Method 4 with glutaraldehyde as the cross-linking agent to entrap insulin crystals in bovine serum albumin microspheres. The final microsphere preparation contained approximately 20% insulin by weight. A single subcutaneous injection of these microspheres produced elevated blood insulin levels in diabetic rats for more than 2 months. Subsequent histological studies indicated that the microsphere implant sites were surrounded with a fibrous capsule which may have further retarded insulin release from the microspheres. In contrast to the observations of Royer *et al.*[13–15] with progesterone-containing microspheres, complete *in vivo* degradation of the insulin-containing microspheres took more than 5 months. Goosen *et al.*[16] speculate that species differences in the test animals used in the two studies may account for the longer microsphere biodegradation times they observed. Alternatively, the altered body chemistry of the diabetic rats utilized in their study may have made degradation of the microspheres more difficult.

Royer[13] has also used the procedure of Method 4 to incorporate enzymes in albumin microspheres. Utilizing glutaraldehyde as the cross-

[16] M. F. A. Goosen, Y. F. Leung, S. Chou, and A. M. Sun, *Biomater., Med. Devices, Artif. Organs* **10**(3), 205 (1982).

linking agent, microspheres containing either asparaginase or yeast alcohol dehydrogenase have been prepared with retention of up to 15% of the original enzymatic activity.

Finally, Longo et al.[17,18] have recently described an interesting glutaraldehyde cross-linking technique for producing drug-loaded albumin microspheres which are highly dispersible in aqueous media because of their unique hydrophilic surface properties. These properties apparently result from the use of concentrated polymer solutions as the spheroidizing bath media. Examples of the bath media include 25–30% solutions of poly(methyl methacrylate) dissolved in a 1 : 1 mixture of chloroform/toluene or polyoxyethylene–polyoxypropylene copolymer dissolved in chloroform. Details of this method are described in the following chapter.[19]

Magnetically Responsive Microspheres

Magnetically responsive microspheres prepared from albumin and other proteins, polysaccharides, and polypeptides are attracting increasing interest for use as targetable drug delivery systems and as convenient carriers which can be utilized in a variety of in vitro diagnostic applications. In general, the heat and chemical cross-linking procedures outlined above in Methods 1–4 can be utilized to prepare "magnetic microspheres" by simply dispersing a sufficient amount of magnetically responsive material (e.g., Fe_3O_4, barium ferrite, etc.) in the initial albumin solution (which may or may not contain a drug) used in the microsphere preparation. The magnetic particles are entrapped with high efficiency within the microspheres formed during the subsequent heat and/or chemical cross-linking procedures previously described. The resulting microspheres are potentially useful in both drug delivery and diagnostics applications.

The preparation of magnetically responsive albumin microspheres containing additional binding agents (e.g., Staphylococcus protein A, ion-exchange resins, etc.) which make them particularly useful in cell sorting, bacterial or virus separations, and diagnostic tests involving antigen–antibody reactions has been described.[20-22] Preparation procedures and

[17] W. E. Longo, H. Iwata, T. A. Lindheimer, and E. P. Goldberg, J. Pharm. Sci. 71(12), 1323 (1982).

[18] W. E. Longo and E. P. Goldberg, Abstr. 10th Int. Symp. Controlled Release Soc., 1983, p. 245.

[19] W. E. Longo and E. P. Goldberg, this volume [2].

[20] A. F. Yapel, U.S. Patent 4,169,804, to 3M Company (1979).

[21] D. S. Ithakissios, U.S. Patent 4,115,534, to 3M Company (1978).

[22] A. E. Senyei and K. J. Widder, U.S. Patent 4,230,685, to Northwestern University (1980).

methods for using magnetic albumin microspheres in targeted drug delivery applications have likewise been described in considerable detail by Senyei and Widder.[23-25] Additional discussion of this subject material is presented elsewhere in this volume.[26]

[23] K. J. Widder and A. E. Senyei, U.S. Patent 4,247,406 (1981).
[24] A. E. Senyei and K. J. Widder, U.S. Patent 4,357,259, to Northwestern University (1982).
[25] K. J. Widder, A. E. Senyei, and D. G. Scarpelli, *Proc. Soc. Exp. Biol. Med.* **158,** 141 (1978).
[26] A. E. Senyei, C. F. Driscoll, and K. J. Widder, this volume [5].

[2] Hydrophilic Albumin Microspheres

By WILLIAM E. LONGO and EUGENE P. GOLDBERG

Introduction

A new method for preparing hydrophilic chemically cross-linked albumin microspheres is presented here. This method was developed to overcome problems associated with the relatively hydrophobic albumin microspheres produced by current vegetable oil emulsion methods.[1,2]

There are many reasons for regarding albumin microspheres with increased hydrophilicity as advantageous: (1) they may exhibit enhanced surface physical and chemical properties *in vivo*; (2) they do not require the surfactants currently used to prepare aqueous dispersions which may influence tissue interactions, drug release, and activity[3-5]; (3) hydrophilicity facilitates aqueous chemical modification; and (4) high concentrations of water-soluble drugs may be incorporated into the microspheres after synthesis which is not readily possible with state-of-the-art hydrophobic albumin microspheres.

The hydrophobicity of albumin microspheres produced by current oil emulsion techniques may be due to preferential organization of the albumin molecules at the vegetable oil interface during synthesis. Thermal albumin denaturation may enhance this effect.[6] There is evidence that hydrophobic interactions between vegetable oil and albumin result in

[1] B. Rhodes, I. Zolle, J. Buchanan, and H. Wagner, *Radiology* **92,** 1453 (1969).
[2] P. Kramer, *J. Pharm. Sci.* **63,** 1646 (1976).
[3] T. Lee, T. Sokoloski, and G. Royer, *Science* **213,** 4503 (1982).
[4] A. Senyei, S. Reich, C. Gonczy, and K. Widder, *J. Pharm. Sci.* **70,** 328 (1981).
[5] K. Widder, A. Senyei, and D. Ranney, *Cancer Res.* **40,** 3512 (1980).
[6] H. Kimelberg, *Mol. Cell. Biochem.* **10,** 171 (1976).

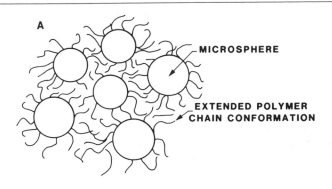

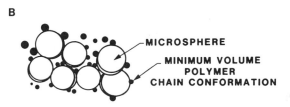

FIG. 1. Schematic representation of steric stabilization.

adsorption of vegetable oil molecules during denaturation or cross-linking of the albumin microspheres.[7] These hydrocarbon compounds may adhere to the albumin even after washing, thus affecting the hydrophobicity of the microspheres. In contrast, the method described here, involving steric stabilization of aqueous albumin dispersions in organic polymer solutions, produces a versatile new class of relatively hydrophilic albumin microspheres.

Synthesis of Hydrophilic Albumin Microspheres

The technique for the synthesis of hydrophilic albumin microspheres is fundamentally different from other current procedures in that (1) a concentrated high molecular weight polymer solution is employed to stabilize the aqueous albumin dispersion and (2) the cross-linking reagent is added via the organic phase of the dispersion.

The concentrated polymer solution affords a medium for producing stable aqueous albumin dispersions which are then readily cross-linked. Polymer molecules act to "sterically stabilize" the albumin microspheres as shown schematically in Fig. 1. This is consistent with studies on steric

[7] K. Widder, G. Flouret, and A. Senyei, J. Pharm. Sci. 68, 79 (1979).

stabilization of colloids such as the work of Heller[8] on the stabilization of aqueous gold sols in poly(ethylene glycol) solutions. Heller demonstrated that microparticle stabilization was due to adsorption of macromolecules at the surface of the microparticle, making it impossible for the particles to agglomerate because they could not closely approach one another. Inelastic collisions between the microparticles, which are responsible for aggregation or destabilization of microparticles, are thereby minimized. It has also been shown that steric stabilization is favored by increasing polymer molecular weight and polymer concentration. This has been confirmed in our studies on microsphere preparation.

In other albumin microsphere procedures involving glutaraldehyde cross-linking, the dialdehyde is added to the aqueous albumin phase and requires cooling to avoid immediate cross-linking before formation of the dispersion.[4,5,9] In contrast, the technique described here involves glutaraldehyde addition via the organic phase. The glutaraldehyde is thereby presented to the aqueous human serum albumin (HSA) dispersion from the organic phase with a resulting high concentration of glutaraldehyde at the surface of the microspheres and lower concentrations in the interior. This tends to produce a case hardening effect, i.e., a higher cross-link density near the surface and a higher concentration of surface aldehyde groups. A higher concentration of mono-reacted dialdehyde at the surface tends also to increase the hydrophilicity of the microspheres due to polar aldehyde and carboxyl groups (the latter formed by oxidation). Moreover, the available aldehyde groups can be used as chemical handles for further surface and bulk modification. The reactive CHO groups are readily quenched or capped with compounds containing primary amino groups such as aminoalcohols or aminoacids (e.g., glycine). Glycine quenching, for example, increases the carboxyl group concentration thereby increasing anionicity and hydrophilicity. The high concentration of reactive aldehyde groups remaining after cross-linking also allows many types of chemical modification including coupling with enzymes, antibodies, proteins, and amino-functional drugs.

Experimental Methods

Preparation of Hydrophilic Human Albumin Microspheres (HSA/MS)

The general procedure is described and discussed in detail elsewhere.[10,11] In a typical synthesis, HSA (0.150 g; Sigma) was dissolved in

[8] W. Heller, *Pure Appl. Chem.* **12**, 249 (1966).
[9] G. Royer, U.S. Patent 4,349,530 (1982).
[10] W. Longo, H. Iwata, T. Lindheimer, and E. P. Goldberg, *J. Pharm. Sci.* **71**, 1323 (1982).
[11] E. P. Goldberg, W. Longo, and H. Iwata, *In* "Microspheres and Drug Therapy" (E. Tomlinson and S. Davies, eds.). Elsevier, Amsterdam (in press).

0.5 ml distilled water in a 16 × 125-mm test tube. This solution was added dropwise to a 25 wt% solution of poly(methyl methacrylate) (PMMA; Polyscience; intrinsic viscosity 1.4) in a mixture of 1.5 ml chloroform and 1.5 ml toluene in a screw-cap test tube. The mixture was dispersed with a vortex mixer (Vortex Genie, Scientific Industries) for 2 min at a power setting of 9. Glutaraldehyde for cross-linking, 1.0 ml (25 wt%, Polyscience), and 1.0 ml of toluene were combined in a 13 × 100-mm test tube. The two phases were dispersed by ultrasonication (Heat Systems-Ultrasonics, model W-374) with a microtip power head attachment for 20 sec at 50 W. The resulting saturated solution of glutaraldehyde in toluene was added to the albumin dispersion and mixed with a rotary mixer (Labquake Labindustries) at room temperature for 8 hr. The cross-linked HSA/MS were washed to remove all PMMA by the addition of 10.0 ml of acetone, briefly agitated, and then centrifuged (2000 rpm, 2 min). The supernate was discarded and the HSA/MS pellet resuspended with an additional 10.0 ml of acetone. This wash procedure was repeated eight times. After the last wash, the HSA/MS were allowed to air dry. The product was obtained in 75% yield as a brown, free-flowing powder. The HSA/MS had an average diameter of 30 μm.

Chemical Modification of Hydrophilic HSA/MS

The HSA/MS produced by the above procedure are chemically reactive; i.e., reactive aldehyde groups on the surface and in the interior of the microsphere are available for further chemical modification. They can be readily capped using amino alcohols or amino acids. A typical procedure is as follows. The HSA/MS were resuspended in 5.0 ml of 1.0 M glycine-HCl. The HSA/MS glycine solution was mixed at room temperature for 22 hr with a rotary mixer. Microspheres were removed from the glycine solution by centrifugation. After decanting the supernate, the MS pellet was resuspended in a 50-ml polypropylene centrifuge tube (Corning) with 45.0 ml of water (pH 3.0), agitated, and centrifuged (2000 rpm, 2 min). This procedure was repeated three times. The wash process was repeated again with water at pH 7.0. After the last water wash, MS were dehydrated with acetone (washed four times in 10.0 ml volumes) and allowed to air dry. The glycine-quenched MS product was a free-flowing, yellowish brown powder obtained in 70% yield.

Adriamycin Bound Hydrophilic HSA/MS

Adriamycin-HCl (AD) (Farmitalia) was dissolved in water and 5.0 ml of the dark red solution containing 10.46 mg AD was combined with 9.99 mg of unquenched HSA/MS in a screw-cap test tube. The pH of the mixture was adjusted from 4.0 to 5.7 by the addition of 0.1 N NaOH and mixed on a rotary mixer at 4° in the dark for 11 hr. The mixture was then centrifuged (2000 rpm, 2 min) and the light red supernate was carefully removed with a Pasteur pipet and saved for analysis. The dark red pellet was resuspended in 10.0 ml of water, briefly stirred, and centrifuged, and the supernate saved for analysis. This was repeated five times. After the last wash, the AD-HSA/MS were dehydrated with acetone and allowed to air dry. The product was a dark red, free-flowing powder. The concentration of AD in the wash was determined spectrophotometrically indicating the concentration of AD bound to the MS to be 18 wt%. This procedure was repeated for HSA/MS that had been chemically modified with glycine (quenched). The concentration of AD bound to the quenched HSA/MS was also 18 wt%.

In Vitro Drug Release: Dynamic Column Elution Method

A column elution method was used to measure *in vitro* drug release. Water, 2.0 ml, was added to 11.6 mg of dry AD-HSA/MS. The resulting red slurry was pipetted into a 140 × 7-mm glass column. The ends of the column were fitted with chromotograph caps packed with glass wool. These were attached to threaded zero-volume collectors and connected to 10-mm I.D. Teflon tubing. Care was taken to ensure that all the AD-HSA/MS were transferred into the column. The column was placed in a circulating water bath at 37°. Physiological saline was pumped through the column at 0.4 ml/min with an HPLC pump (Altex 110A). Fractions were collected as a function of time at 4°, and analyzed by UV/VIS spectroscopy.[12]

Results and Discussion

The synthesis described here typically produces HSA/MS that have a round, smooth, spherical geometry as shown in Figs. 2 and 3. In addition to poly(methyl methacrylate), other polymers such as Poloxamer 188 [poly(ethylene oxide)–poly(propylene oxide) block copolymer], cellulose acetate butyrate, and polycarbonate were shown to stabilize albumin dispersions, indicating the versatility of the procedure. X-Ray photoelectron (ESCA) characterization of HSA/MS surfaces indicated that no polymer remained on the HSA/MS surface after the washing procedure.

Washed, lyophilized, or air-dried MS were readily resuspended in a variety of aqueous media such as distilled water, physiological saline, phosphate buffer, or acetate buffer. They were easily wetted and dispersed without the need for surfactants. Hydrophilicity was confirmed

[12] F. L. DiMarco and F. Arcamore, *Arzneim.-Forsch.* **25,** 368 (1975).

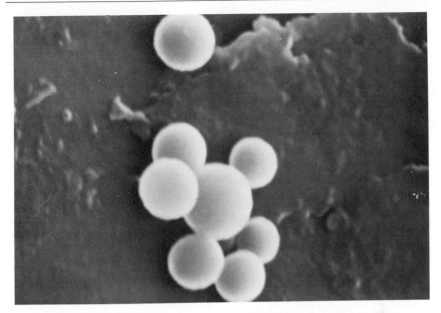

FIG. 2. Scanning electron micrograph of HSA/MS, 0.8 μm average diameter.

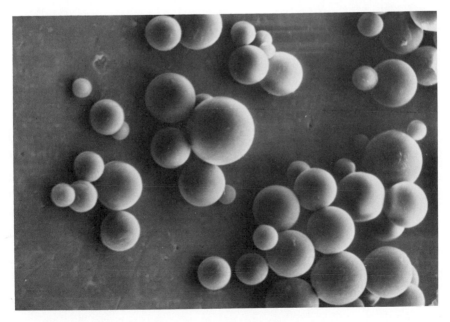

FIG. 3. Scanning electron micrograph of HSA/MS, 15 μm average diameter.

HYDRATION OF BSA/MS AS A FUNCTION OF CROSS-LINK DENSITY
USING GLUTARALDEHYDE

Glutaraldehyde concentration (mmol)	Mean diameter, dehydrated (μm)	Mean diameter, hydrated (μm)	Water uptake (mg)/ BSA/HSA (mg)	Hydration (%)
2.50	14	17	4.6	82
0.50	13	21	6.3	86
0.25	10	23	11.5	92
0.13	17	37	15.4	94

using a quantitative capillary rise technique for measuring relative hydrophilicity by water wetting in a MS-packed column. The cross-link density of HSA/MS is also easily controlled by varying the concentration of cross-linking agent (e.g., glutaraldehyde). As shown in the table,

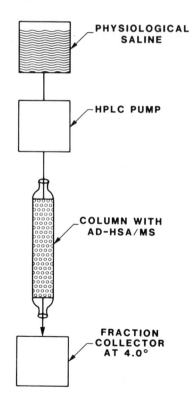

FIG. 4. Schematic of *in vitro* dynamic flow system for drug release.

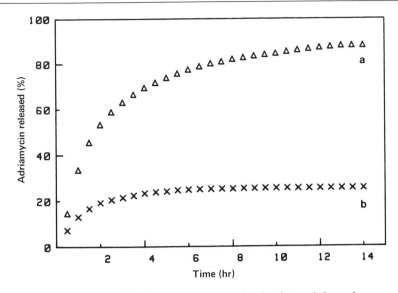

FIG. 5. *In vitro* release of AD from albumin/MS, showing fast and slow release, average diameter 27 μm. (a) Physically associated drug, quenched; (b) covalently bound drug, unquenched.

changes in cross-linking are accompanied by changes in hydration. Reducing cross-link density increases MS porosity, hydration, and swelling of dried preparations.

The uptake of water-soluble drugs by the hydrophilic HSA/MS was demonstrated by the preparation of adriamycin (AD)-loaded compositions. Binding of 18–20 wt% AD by simple addition of MS to aqueous AD solutions occurred by two mechanisms: (1) covalent attachment of AD by reaction of amino groups with reactive aldehyde groups on unquenched HSA/MS, and (2) physical association of AD with glycine-quenched HSA/MS. To determine if AD penetrated the interior of the MS during binding or was primarily bound to the surface, AD-HSA/MS were mounted in epoxy and serial sections cut with an ultramicrotome. Sections mounted on TEM grids were examined by optical microscopy. The red AD was present throughout the sections, indicating that drug had diffused uniformly throughout the HSA/MS.

AD release rates from AD-HSA/MS were evaluated using the dynamic column flow system as shown in Fig. 4. Release rates were controllable depending upon the two binding mechanisms: (1) slow release by hydrolysis of covalently attached AD and/or (2) fast release of physically associated drug. Shown in Fig. 5 is a comparison of AD release data from

18% AD-HSA/MS with covalent drug attachment (curve b) and from 18% AD-HSA/MS with physically associated drug (curve a).

Summary

A method has been developed for preparing unique hydrophilic HSA/MS. Important aspects of this synthesis include addition of the cross-linking agent (glutaraldehyde) in the organic phase and use of concentrated polymer solutions as dispersion media. The polymer solutions afford excellent steric stabilization of aqueous albumin microdispersions for microsphere synthesis. Steric stabilization of dispersions by polymer solutions was shown to be a function of polymer concentration and molecular weight.

The HSA/MS prepared by this method are hydrophilic and easily dispersed in a variety of aqueous media without surfactants. Chemical modifications are easily accomplished using available reactive aldehyde groups remaining after cross-linking. Although hydrophilicity of the microspheres is advantageous for many drug delivery applications, in some instances (such as the use of MS in adjuvant immunotherapy or vaccine preparations) some hydrophobicity may be desirable. For this purpose, surface modifications to produce controlled hydrophobicity is easily achieved by covalent coupling with appropriate reagents (e.g., fatty amines).

Adriamycin was bound to HSA/MS by both physical association (to 18 wt%) and covalent binding (also to 18 wt%). *In vitro* release of drug was measured for the MS using a dynamic flow method. Two distinct release mechanisms could be achieved depending on the type of drug bonding used: (1) slow by hydrolytic degradation of covalent bonds and (2) fast by release of physically adsorbed drug.

This new and versatile synthesis of hydrophilic HSA/MS opens up many new opportunities for producing chemically modified MS containing high concentrations of therapeutic agents. Use for immunodiagnostic and adjuvant compositions is also suggested. Because HSA/MS preparations are easily dispersed and injected, sustained and controlled drug release applications are of special interest, especially for localized chemotherapy.

Acknowledgment

The work reported here was supported in part by the State of Florida Biomedical Engineering Center. We wish to thank Dr. Arcamone of Farmitalia for generous samples of adriamycin and M. Smith for her help with this chapter.

[3] Incorporation of Water-Soluble Drugs in Albumin Microspheres

By E. TOMLINSON and J. J. BURGER

Introduction

Previous chapters in this volume have described methods for the chemical and heat stabilization of albumin microspheres intended for drug targeting and controlled release. As discussed recently by Tomlinson,[1] more than 17 therapeutic roles can be identified for microsphere carriers, ranging from cancer chemotherapy through diseases of the reticuloendothelial system to inhalation therapy, etc. Over 90 drug types have been reported as being incorporated into microspheres, with over 25 different carrier matrix materials being studied. The matrix material most used is human serum albumin (HSA), which has been chosen because it should meet many of the requirements of the ideal drug carrier given by Widder et al.[2] (Table I).

For the past 3 years we have been examining (in particular) the pharmaceutical and biopharmaceutical aspects of albumin microspheres intended both for intraarterial targeting to tumors (i.e., second-order targeting[2]) and for controlled release at locally accessed sites, such as lungs and joints (i.e., first-order targeting[2])[3–5] These aspects are outlined in Table II. In this present contribution we describe some of our findings on the nature and extent of incorporation of water-soluble compounds into albumin microspheres.

Microsphere Manufacture

There appear to be two basic methods for the production of albumin microspheres. First, either a thermal denaturation at elevated tempera-

---

[1] E. Tomlinson, *Int. J. Pharm. Technol. Prod. Manuf.* **4**, 49 (1983).

[2] K. J. Widder, A. E. Senyei, and D. F. Ranney, *Adv. Pharmacol. Chemother.* **16**, 213 (1979).

[3] E. Tomlinson, J. J. Burger, E. M. A. Schoonderwoerd, J. Kuik, F. C. Schlötz, J. G. McVie, and S. N. Mills, *J. Pharm. Pharmacol.* **34**, 88P (1982).

[4] E. Tomlinson, J. J. Burger, J. G. McVie, and K. Hoefnagel, *in* "Recent Advances in Drug Delivery Systems" (J. M. Anderson and S. W. Kim, eds.). Plenum, New York, 1984, pp. 199–208.

[5] E. Tomlinson, J. J. Burger, E. M. A. Schoonderwoerd, and J. G. McVie, *Int. J.-Pharm.* (in press).

METHODS IN ENZYMOLOGY, VOL. 112

TABLE I
IDEAL TARGETABLE DRUG SYSTEM
CHARACTERISTICS[a]

Restricted drug distribution to target
Prolonged control
Ready access to tissue parenchyma
Uniform carrier target tissue distribution

Controllable and predictable rate of drug release
High capacity for drugs and drug types
Drug release unaffecting drug action
Therapeutic amounts of drug released
Minimal drug leakage during carrier transit to target
Drug protected

Biocompatible surface properties
Host protected from agent's allergic properties
Biodegradable carrier
No carrier-induced modulation of disease state

Easy to prepare

[a] From ref. 2.

tures (95–170°) or chemical cross-linking in vegetable oil or isooctane (etc.) emulsions. The latter method—which is claimed to produce "hydrophilic" microspheres[6]—depends on chemical cross-linking in a water-in-oil emulsion using concentrated polymer solutions as the dispersing phase. Second, non-drug-bearing microspheres may be prepared using either a simple one-step preparative method involving thermal denaturation of protein aerosol in a gas medium,[7] or an aerosol step followed by denaturation in oil.[8] Although this latter method appears to be promising for large-scale manufacture of drug-bearing microspheres, the discussion which now follows concerns itself only with drug incorporation in albumin microspheres manufactured using a water-in-vegetable oil emulsion, and should not *a priori* be extrapolated to other manufacturing techniques.

Preparation

Human serum albumin microspheres may be prepared in the range 0.2–100 μm using the manufacturing procedures now outlined. Highly

---

[6] W. E. Longo, H. Iwata, T. A. Lindheimer, and E. P. Goldberg, *J. Pharm. Sci.* **71,** 1323 (1982).
[7] M. Przyborowski, E. Lachnik, J. Wiza, and I. Licińska, *Eur. J. Nucl. Med.* **7,** 71 (1982).
[8] A. M. Millar, L. McMillan, W. J. Hannan, P. C. Emmett, and R. J. Aitken, *Int. J. Appl. Radiat. Isot.* **33,** 1423 (1982).

TABLE II
Pharmaceutical and Biopharmaceutical
Considerations in the Development
of a Microsphere Product[a]

Core material
Route of preparation with respect to points 3–7
Size
Drug incorporation
Type and amount of drug
Drug release (*in vitro* and *in vivo*)
Drug stability during preparation and storage
Microsphere stability (*in vitro* and *in vivo*)
Storage
Surface properties
Presentation (e.g., free-flowing, freeze-dried powder)

[a] From ref. 5.

purified olive oil (125 ml) is added to a flat-bottomed glass beaker (diameter 60 mm, height 110 mm) equipped with four baffles (4 mm depth) positioned against the wall of the beaker. A motor-driven four-bladed axial-flow impeller is placed in the center of the beaker, two-thirds into the oil, such that there is a distance of 3 mm between the baffles and the impeller blade. After prestirring the oil for 30 min at the desired speed, an aqueous solution of HSA and drug in isotonic buffer (pH 7) is added dropwise from a syringe to the olive oil. Generally, this aqueous phase (including cross-linkers, see later) has a final volume of 0.5 ml. The resulting water-in-oil emulsion is stirred for an additional period of time depending upon the method of stabilization. It has been found that the design of the mix-cell is critical for producing a reproducibly uniform product. Figure 1[5] shows that manufacture using an nonbaffled cell leads to microspheres having a broad size range with no detectable maxima, whereas with a baffled cell, well-defined microspheres are produced. The most critical factor affecting size in a baffled cell is the rate of stirring, with, for the cell described above, the following relationship being found:

$$\bar{d} = -67 \log \text{rpm} + 218 \qquad (n = 8; \; r = 0.99) \qquad (1)$$

This refers to microspheres of mean diameter ($\bar{d}$, in micrometers) in the range 1–50 μm. Calculations for the above system at an rpm of 1800 show that the Reynolds number (Re) for the system is approximately 500, which gives streaming in the cell in the transitional area between the laminar flow (Re less than 10) and fully developed turbulent flow (Re greater than 10^4). Table III gives the influence of other manufacturing variables on microsphere size.[5]

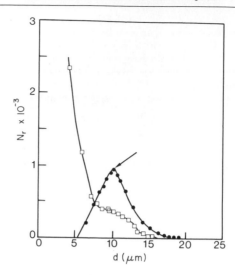

Fig. 1. Relative number of butadione-stabilized albumin microspheres per milligram ($N_r \times 10^{-3}$) versus sphere diameter (d), determined using a model z_B Coulter counter equipped with a Channelyser. Points refer to spheres produced in a nonbaffled (□) and baffled (●) cell. (Manufacture was at 1120 rpm.)[5] Arrow shows mean, $\bar{d} = 10.3$; $\sigma = 2.5$.

TABLE III

EFFECT OF MANUFACTURING VARIABLE ON THE MEAN
DIAMETER OF HUMAN SERUM ALBUMIN MICROSPHERES[a]

Factor	Change	Mean diameter[b]
Oil viscosity[c]	Decrease	Increase
Oil amount	Increase	Increase
Protein amount	Increase	Increase
Aqueous phase[d]	Increase	Increase
Stirring speed	Increase	Decrease
Surfactant[e]	Addition	No effect

[a] From ref. 5.
[b] Standard factors were 125 ml olive oil and 0.5 ml of a 25% w/v HSA solution.
[c] The viscosity of the oil was varied by addition of petroleum ether 100-140 to give an end volume of 125 ml.
[d] Addition of 1 ml of a 12.5% w/v HSA solution.
[e] Addition of 0.1% sodium dodecyl sulfate to the aqueous HSA solution.

Nonstabilized Microspheres. These are prepared at room temperature (20 ± 3°) by addition of 0.5 ml of a (generally) 25% (w/v) HSA solution containing dissolved drug to 125 ml of oil. After stirring for 90 min, 60 ml of diethyl ether is added to the water-in-oil emulsion. Stirring is continued for an additional 10 min, after which the formed microspheres are isolated by centrifugation for 5 min at 2000 rpm followed by decanting of the supernate. After resuspending the microspheres in diethyl ether and collecting them on a 0.8-μm polycarbonate filter, residual oil is removed by further washing with diethyl ether.

Stabilization with 2,3-Butadione. After stirring the oil-in-water emulsion produced upon mixing 0.5 ml of (generally) 25% HSA solution containing drug with 125 ml of olive oil for 90 min at the desired stirring speed, the emulsion is transferred to a sealed beaker (to prevent subsequent loss of cross-linker through evaporation). Stirring is continued at 40 rpm, and then butadione is added in either a 3 or 15 ml volume. After reaction for a desired period, the microspheres are centrifuged, washed, and collected as above. Butadione-stabilized microspheres may also be produced by suspending 40 mg of nonstabilized spheres in 100 ml of diethyl ether containing various concentrations of cross-linker.

Stabilization with Glutaraldehyde. To 125 ml of olive oil is added 0.4 ml of a (generally) 25% aqueous HSA solution containing drug. After stirring for 15 min at the desired speed, 0.1 ml of an aqueous glutaraldehyde solution is added, so as to produce a final (aqueous) glutaraldehyde concentration of between 1 and 5%. Stirring is continued at the desired rate for the desired time, after which the microspheres are collected and washed as described above.

Heat Denaturation. This is a similar method to that described for nonstabilized microspheres, except that the temperature of the oil is between 90 and 170°, with denaturation times of between 5 and 60 min.

Lyophilization and Storage

After collection, microspheres are lyophilized at 10–20 mtorr and −60° for 18 hr. Upon freeze-drying the sample, vials are sealed with an air tight seal, and stored at 4° in the dark.

Fractionation

Freeze-dried microspheres are fractioned by sieving using microsieves obtained from Veco Inc. (Eerbeek, The Netherlands). Approximately 80 mg of freeze-dried microspheres are suspended in 50 ml of chloroform, ultrasonicated for 5 min, and then sieved between the appropriate-sized

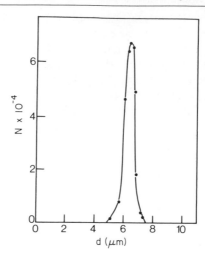

FIG. 2. Size distribution of stabilized albumin microspheres sieved between 5- and 10-μm microsieves. Plot is of number of microspheres per milligram ($N \times 10^{-4}$) versus sphere diameter (d). Mean, $\bar{d} = 6.5$; $\sigma = 0.3$.

microsieves. After sieving, the microspheres are again freeze-dried and stored as before. Using standard available sieves, microspheres can be collected in 1-μm fractions between 1 and 20 μm, and in 5-μm fractions above 20 μm. Figure 2 gives the size distribution of microsieves fractionated in this way, and Fig. 3a is a scanning electron micrograph showing the approximate monodispersity of sieved microspheres.

Production of Minimicrospheres

Albumin microspheres can be readily produced at sizes between 0.1 and 1 μm by incorporation of a homogenization step. A method which we have developed is to add 0.25 ml of a 25% HSA aqueous solution to 100 ml of high viscosity oil (850 cSt, centistokes), homogenize the emulsion with a minihomogenizer (e.g., an Ultraturrax) for 5 min, add 60 ml diethyl ether, stir for 5 min using a magnetic stirrer, and then collect and wash in the normal manner. Figure 4 is a scanning electron microscopy photograph of such minimicrospheres.

Assessment of Drug Incorporation

A number of methods have been proposed for the estimation of the amount of drug associated with albumin microspheres. First is an indirect method which determines the amount of drug to be found *remaining* in the

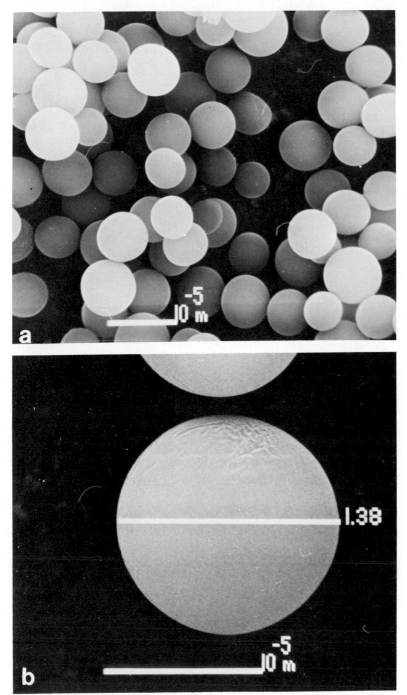

FIG. 3. Sieved butadione-stabilized microspheres. (a) Spheres sieved between 5 and 10 μm; (b) sodium cromoglycate-bearing sphere of 13.8 μm diameter.

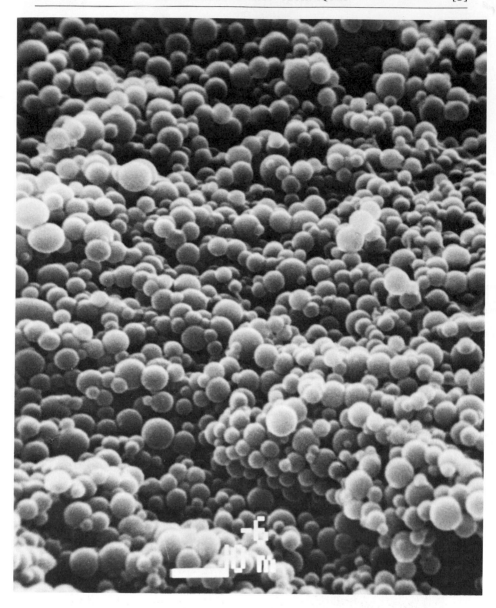

FIG. 4. Albumin minimicrospheres.

system after collection of the microspheres (including in the emulsion and the washings). Although perhaps of use in some cases, mass balance methods are prone to error. A second method is to use radiolabeled drug. This enables the amount of drug to be directly estimated, and depends only upon (1) the availability of radiolabeled compound, and (2) the acceptance of contamination of apparatus.

A third method is to digest the manufactured microspheres with a mixture of 5% HCl in ethanol.[9] Although this often leads to erroneous results due to breakdown of the drug and/or resulting microsphere products interfering with the drug assay, this can be of some use. A fourth method, and one which we favor, is to measure the incorporation of drug in *nonstabilized* microspheres. Cross-linking and heat stabilization have an insignificant effect on incorporation, and hence this method can be readily used to estimate the amount of drug associated with the microsphere. The method involves dissolving the (freeze-dried) nonstabilized microsphere in aqueous buffer (a step which is almost instantaneous), then assaying the amount of drug in solution. Controls have to be carried out to determine the influence of dissolved protein on the assay of the drug, but this is relatively simple to perform.

Physical Characteristics with Incorporated Drug

Appearance

Drug incorporation to a level of 25% by weight has little effect on the visible surface characteristics of albumin microspheres, though, as seen from Fig. 3b, surface crennelations can be observed at drug levels above this.

Swelling

When placed into a 0.1% Tween 80 saline solution at 20°, freeze-dried albumin microspheres prepared by chemical cross-linking swell considerably. With the systems studied by us swelling is complete within 2 min, is uninfluenced by the extent of cross-linking, and, for microspheres in the range of 1–60 μm diameter, leads to an increase in mean microsphere diameter of between 20 and 80%. The larger the microsphere, the greater is the extent of swelling. The incorporation of highly water-soluble drugs into the microspheres can lead to a greater extent of swelling. For example, we have found[5] that 30-μm lyophilized microspheres, containing ap-

[9] K. J. Widder, A. E. Senyei, and D. G. Scarpelli, *Proc. Soc. Exp. Biol. Med.* **158,** 141 (1978).

TABLE IV
ZETA POTENTIALS OF HUMAN SERUM ALBUMIN MICROSPHERES[a]

Microsphere	Zeta potential at 25° (mV) Suspension medium			Mean diameter (μm)
	0.1% Tween 80 in physiological salt solution	0.1% Tween 80 in plasma	Plasma	
0.1 mol dm^{-3} SCG, ultrasonicated[b]	−18.5	−9.5	−8.9	21.7
0.1 mol dm^{-3} SCG	−8.7	−7.2	−6.9	23.0
Non-drug containing, ultrasonicated	−18.9	−7.1	−8.2	21.7
Non-drug containing	−13.0	−5.3	−7.5	21.7

[a] From ref. 5. Microspheres stabilized by cross-linking for 4 hr with 5% glutaraldehyde.
[b] Approximately 40 mg freeze-dried microspheres suspended in 15 ml 0.1% Tween 80 physiological salt solution, followed by ultrasonication for 5 min, filtration, washing with water, ethanol, and ether, and then freeze drying.

proximately 25% by weight of the dianionic drug sodium cromoglycate (SCG) and prepared using cross-linking with glutaraldehyde for 18 hr, increased in diameter on average by 75% when placed in saline, whereas similar non-drug-bearing microspheres increased in diameter by only approximately 40%. This is obviously of therapeutic relevance for those cases where microsphere size is critical.

Surface Charge

Since the colloid and surface characteristics of microspheres are of importance to their presentation and biological fate, it is of interest to examine the influence of incorporated drug on the surface charge of albumin microspheres. Table IV gives the zeta potentials for a number of different microspheres measured in various media. It is seen that SCG-bearing and non-drug-bearing albumin microspheres are negatively charged, and that the magnitude of the zeta potential is dependent upon both the media and the presence of (ionized) drug. In plasma, all microspheres have a fairly similar zeta potential ranging from −5.3 to −9.5 mV. However, in physiological salt solution the potentials are much higher. Of particular interest is that although microspheres containing sodium cromoglycate have a higher zeta potential after ultrasonication, drug-containing microspheres not subject to ultrasonication have a much lower zeta potential. These data suggest that upon ultrasonication a surface layer of

TABLE V
EFFECT OF COMPOUND INCORPORATION ON MICROSPHERE SIZE[a]

Compound	Aqueous phase concentration (mol dm^{-3})	Mean diameter (μm)
Sodium cromoglycate	—	8.0
	0.02	7.5
	0.05	8.8
	0.15	12.7
	0.19	14.0
Sodium iodohippurate	—	12.7
	0.12	11.4
	0.32	11.4
	0.52	10.1
	0.71	14.0
	0.91	23.0

[a] From ref. 5. Sodium cromoglycate-bearing microspheres were nonstabilized, and were manufactured at 1215 rpm. Sodium iodohippurate microspheres were stabilized with 5% glutaraldehyde for 1 hr, and were manufactured at 825 rpm.

cromoglycate is removed. This, of course, has implications for the release of drug, and indicates that for water-soluble compounds, microspheres should first be ultrasonicated, recollected, and then injected.

Microsphere Size

Incorporation of water-soluble drugs leads to alterations in the size of the formed microspheres. Table V shows that under similar manufacturing conditions, this effect depends on whether the concentration of drug in the aqueous disperse phase is approaching the maximal aqueous solubility (in the presence of albumin). Thus at levels well below this maxima, an increase in drug concentration in the manufacturing vessel leads to a reduction in microsphere diameter, whereas as the limiting solubility is approached, the size of the microsphere begins to increase. This latter effect is presumed to be due to the formation of microsuspensions of drug which begin to limit the minimum achievable size of formed sphere.

Physicochemical Considerations

Use of a water-in-oil technique for producing drug-bearing albumin microspheres leads to constrictions on the maximal amount of drug which may be incorporated. This amount is largely dependent upon the aqueous

Oil phase	Aqueous disperse phase

$$D \rightleftharpoons D + P \rightleftharpoons D \cdot P$$
$$\downarrow \qquad\qquad \downarrow$$
$$[(P)_D \quad + \quad (D \cdot P)]^{MOS}$$

SCHEME I. Physical (P) and physicochemical (D·P) incorporation of drug, D, into protein, P, microspheres [MOS].

solubility of the drug in the presence of dissolved protein. However, this only applies to microspheres intended for second- or third-order targeting (i.e., of diameter 0.1–10 μm, ranging in use from sieve plate passage through reticuloendothelial system uptake to capillary blockage). As may be gathered from Table V, for larger spheres, a microsuspension/emulsion technique can be used,[10] leading to direct sphere incorportion of (micronized) drug powder.

Considering thus a nonsuspension technique, in any two-phase system the distribution of drug between water and oil is given by its liquid/liquid distribution coefficient (K_d), such that

$$K_d = [D]^o/[D]^w \tag{2}$$

where [D] is the concentration of drug, and superscripts o and w refer to oil and water phases, respectively. For liquid solutes there is a general relationship between drug aqueous solubility (S) and its octanol/water distribution coefficient, i.e.,[11]

$$\log S = -1.07 \log K_d + 0.67 \tag{3}$$

Although for solids this relationship is perturbed due to the fact that crystal interaction energies have no effect on drug distribution but do on drug solubility, Eq. (3) is a reasonable approximation for many drug types. Figure 5 and Scheme 1 illustrate those physicochemical factors coming into play during drug incorporation into albumin microspheres.

It is clear that as the distribution coefficient increases, the aqueous solubility of the drug decreases, leading to a reduction in the maximum amount of drug which may be dissolved in the disperse phase of the emulsion. Since the phase–volume ratio of water to oil is 250 (see above), then, for the case where there is *no protein* present, the maximum amount of drug in the aqueous phase at equilibrium can be estimated. Figure 5

---

[10] T. K. Lee, T. D. Sokoloski, and G. P. Royer, *Science* **213**, 233 (1981).
[11] S. H. Yalkowsky and W. Morozowich, *in* "Drug Design" (E. J. Ariëns, ed.), Vol. 9, p. 198. Academic Press, New York, 1980.

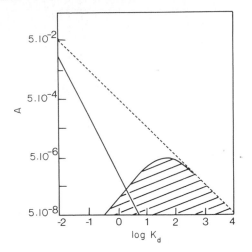

FIG. 5. Relationship between drug solubility, liquid/liquid distribution, and nonspecific protein binding. Dashed line is the maximum amount (A, in moles) of drug possible in 0.5 ml of the aqueous disperse phase of the microsphere manufacturing emulsion [calculated using Eq. (3)]. The solid linear line gives the amount of drug remaining in the aqueous phase after distribution to the oil phase [estimated using Eq. (2), and assuming a phase volume ratio of 250]. The hatched area gives the maximum amount of drug able to be protein bound (in the absence of an oil phase), estimated from Eq. (4).

shows that under this assumption, compounds having a K_d greater than 0.1 will be found practically entirely in the oil phase. However, Scheme I shows that the specific and nonspecific protein binding of the drugs needs also to be considered. For the latter case, the following relationship between drug octanol/water K_d values and protein binding has been found.[12]

$$\log 1/C = 0.7 \log K_d + 2.6 \tag{4}$$

where C represents the molar concentration of solute necessary to produce a 1:1 complex with pure bovine serum albumin (2.5×10^{-5} mol dm^{-3}). By making a number of assumptions concerning the nature of the binding, protein binding capacity, and solute molecular weight, one can show (Fig. 5) the relationship between drug distribution coefficient, maximal aqueous solubility, and amount of drug able to be bound to the protein during microsphere manufacture. Thus, in the *absence of oil,* on the basis of nonspecific protein binding alone, only drugs with distribution coefficients above approximately 75 will be significantly associated with the protein. Figure 5 indicates that considering only solute distribution,

[12] J. M. Vandenbelt, C. Hansch, and C. Church, *J. Med. Chem.* **15,** 787 (1972).

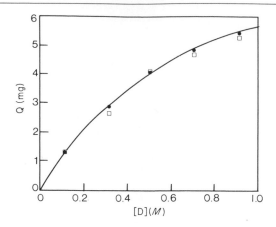

FIG. 6. Incorporation of sodium iodohippurate into 10- to 25-μm albumin microspheres. Q is amount of drug incorporated into 10 mg of freeze-dried microsphere, [D] is the molar concentration of sodium iodohippurate initially in the aqueous disperse phase of the emulsion, and the line drawn is for a theoretical 100% incorporation of all present drug. Data points are mean experimental results (coefficient of variation 0.5%) using the 5% HCl/ethanol (●) and nonstabilized (□) microsphere methods, respectively.[5] See text for further discussion.

phase–volume ratio, and amounts of drug, compounds having log K_d values in the range −1.5 to 1.5 should not be incorporated into albumin microspheres. However, our results show this not to be the case. For example, approximately 50% of 5-fluorouracil added to an aqueous disperse phase during microsphere manufacture becomes incorporated into the formed spheres, even though this compound has a log K_d of −0.85. Indeed at maximal aqueous solubility of 5-fluorouracil, 2.5% by weight of albumin microspheres formed are drug.

These observations mean that in many cases drug associated with albumin microspheres must be entrapped within the matrix, probably in regions of bound water.

During the washing, collection, and sieving of microspheres, attention has to be given to the solubility of the drug in, for example, diethyl ether and chloroform. Certainly for highly water-soluble drugs this shall not present a problem, but for compounds with higher distribution coefficients other washing and sieving solvents should be sought.

Drug Incorporation

Figure 6 gives the incorporation of sodium iodohippurate (SIH) into albumin microspheres. The points shown are the experimentally found

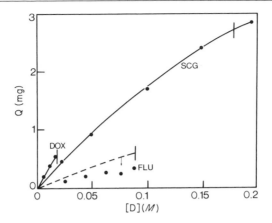

FIG. 7. Incorporation of some water-soluble compounds into 10- to 20-μm albumin microspheres, showing relationship between the amount (Q) of compound incorporated per 10 mg of formed sphere vs the original disperse phase compound concentration [D]. SCG, DOX, and FLU refer to sodium cromoglycate, doxorubicin, and 5-fluorouracil, respectively. Lines represent theoretical 100% incorporation, with the broken line being that for FLU. Vertical bars give the maximum aqueous solubility of the compound in the presence of 25% (w/v) HSA.[5]

incorporations using both the nonstabilized microsphere method and the HCl/ethanol digestion method. The drawn line is the theoretical line for 100% incorporation of the original amount of drug in the disperse phase of the emulsion. This theoretical line is not linear since at higher amounts of drug incorporation the weight of the drug contributes significantly to the overall weight of the microsphere. Figure 6 shows that for SIH, more than 54% by weight of the formed microsphere can be drug, and for other compounds 87% by weight incorporations have been reported.[13]

Yapel, in his extensive patent[13] on drug-bearing albumin microspheres, showed that the normal biphasic release of epinephrine from microspheres becomes monoexponential as the amount of drug rises above 35%, suggesting that at high drug concentrations release is largely dependent upon solution of the drug in the external medium. Our own results[5] with SIH are that even at incorporation levels greater than 50% (by weight) the release is still biphasic, indicating that the trend toward monoexponential drug release at high levels of drug incorporation is compound dependent.

Other extents of drug incorporation are given in Fig. 7. Here, the relationships between amount of drug present in the microsphere against

[13] A. F. Yapel, U.S Patent 4,147,767 (1979).

the original concentration of drug in the aqueous phase of the emulsion are shown. For highly water-soluble compounds such as SCG and SIH, large amounts of drug may be incorporated. For doxorubicin, incorporations of 16% by weight have been reported,[9] although in our studies use of a doxorubicin/lactose mixture permitted only maximal incorporations of 6% (by weight) to be achieved. For SCG and SIH, incorporation is practically 100% of the original amount. (The theoretical lines in Figs. 6 and 7 are not superimposed since drug molecular weights are not fully considered.) For doxorubicin, incorporation is only 95% of the original amount, and for 5-fluorouracil this falls to 50%. These variations are due to the various physicochemical effects discussed above.

Effect of Protein Amount on Drug Incorporation

Although Figs. 6 and 7 show that the extent of drug incorporation is limited by their aqueous solubilities, it is possible to dramatically alter the percentage level of incorporation. This is achieved by a *reduction* in the amount of protein used. For example, we have found that with an initial aqueous phase concentration of SCG of 0.1 M, a reduction in the concentration of HSA in the aqueous phase of the emulsion from 25% (w/v) to 1% (w/v) still results in 100% of the original drug amount being incorporated—which means that the amount of formed microsphere which is drug rises from 16 to more than 83% (by weight). This is attractive with regard to drug dosing, although it results in (1) production of smaller microspheres (Table III), and (2) different release rates of drug.

Concluding Remarks

This chapter has examined the relationship between drug physicochemical structure and its incorporation into albumin microspheres. It is shown that incorporation of water-soluble compounds can have an effect on the size of microspheres formed, their eventual swelling in water, their physical appearance, and their microelectrophoretic mobility. It is also shown that drug associated with the microspheres is probably entrapped within the matrix of the sphere, and that nonspecific protein binding has little effect on the incorporation of water-soluble compounds. Only briefly discussed has been the release of water-soluble drugs from albumin microspheres. Notwithstanding the possibility for manufacture of "hydrophilic" microspheres,[6] it seems evident to us[5] that if albumin microspheres are to be used for drug targeting and sustained release of water soluble compounds, that the very large "burst" effect—which leads to between 60 and 93% of incorporated drug being lost within 5 min of contact with water—mitigates for a covalent link between drug and microsphere matrix. The extent of drug incorporation may then be much different to that discussed here.

Acknowledgment

E. M. A. Schoonderwoerd, F. Juanita, R. Wong, and J. Bakker are warmly thanked for their technical assistance. This project supported in part by the Queen Wilhelmina Cancer Funds (Project GU 82-8).

[4] Technetium-99m Labeling of Albumin Microspheres Intended for Drug Targeting

By J. J. Burger, E. Tomlinson, J. E. de Roo, and J. Palmer

Introduction

Albumin microspheres are perceived as potential carriers of drugs for targeting and controlled release. Examination of the biological fate of these carriers is an essential part of their study. Apart from tedious histological and autoradiology examination, the use of gamma scintigraphy affords a most suitable means of following the kinetics of the carrier *in vivo* and without recourse to invasion.

Various γ-emitting labels are available for labeling albumin microspheres, including ^{97}Ru,[1] ^{111}In and ^{113}In,[2] ^{67}Ga,[3] ^{68}Ga,[4] ^{46}Sc, ^{85}Sr, ^{113}Sn and ^{57}Co,[5] and ^{51}Cr.[6] However, technetium-99m does have some advantages (for diagnostic imaging purposes) over these, including a short half-life, a detection level of 140 keV with the gamma camera, and its safety and availability. Accordingly, this label is used extensively in diagnostic imaging with albumin microspheres, including studies on the phagocytic properties of the Kupffer cells,[7] the rate and pattern of gastric emptying,[8]

[1] G. Subramanian and J. G. McAfee, *J. Nucl. Med.* **11,** 365 (1970).

[2] P. L. Hagan, G. E. Krejcarek, A. Taylor, and N. Alazraki, *J. Nucl. Med.* **19,** 1055 (1978).

[3] D. J. Hnatowich and P. Schlegel, *J. Nucl. Med.* **22,** 623 (1981).

[4] F. Brady, *Annu. Rep. MRC Cyclotron Unit* p. 44 (1981).

[5] M. H. Laughlin, R. B. Armstrong, J. White, and K. Rouk, *J. Appl. Physiol.: Respir., Environ. Exercise Physiol.* **52,** 1629 (1982).

[6] R. P. Hof, *Triangle* **21,** 29 (1982).

[7] S. N. Reske, K. Vyska, and L. E. Feinendegen, *J. Nucl. Med.* **22,** 405 (1981).

[8] M. C. Theodorakis, G. A. Digenis, R. M. Beihn, M. B. Shambhu, and F. H. Deland, *J. Pharm. Sci.* **69,** 568 (1980).

METHODS IN ENZYMOLOGY, VOL. 112

regional blood flow,[6] peritoneovenous shunt potency,[9] lung ventilation,[10] cerebral perfusion imaging,[11] brain scans,[12] and reticuloendothelial system structure and function.[13] These extensive studies and the concomitant growth of computer software for ^{99m}Tc use have prompted us to examine whether ^{99m}Tc labeling of HSA microspheres intended for drug targeting can be satisfactorily used to follow their biological fate.

This contribution discusses ^{99m}Tc labeling procedures described in the literature, the mechanism of labeling, and the labeling and biological fate of chemically stabilized HSA microspheres.

Labeling of Heat-Denatured HSA Microspheres with ^{99m}Tc

To label HSA microspheres with technetium, it is necessary to first convert heptavalent Tc(VII) to one of its lower oxidation states.[14] Tc(VII) is readily available in soluble form from a ^{99}Mo generator as the sodium or ammonium salt in isotonic saline. Various methods have been developed to cause its subsequent reduction.

One of the first approaches for labeling of 12- to 44-μm HSA spheres incorporated ferric hydroxide into the microspheres which was then reduced by ascorbic acid in an acidic environment.[15,16] This method suffered from the fact that the reduction mixture of spheres, TcO_4^-, and reducing agent had to be boiled for several minutes. The method was later modified to improve the labeling efficiency, by using sodium thiosulfate as reducing agent (although boiling was still necessary).[17] For smaller spheres (0.5–0.7 μm) the ferric chloride/ascorbic acid method can be used without the need for initial incorporation of ferric chloride into the spheres.[13,15,18] It appears that the ferric hydroxide and chloride are themselves reduced to the ferrous forms, which can itself reduce pertechnetate, so leading to a higher labeling efficiency.

[9] G. Madeddu, N. G. D'Ovidio, A. R. Casu, R. Mura, C. Costanza, N. Lai, and H. H. LeVeen, *J. Nucl. Med.* **24**, 302 (1983).

[10] A. M. Millar, L. McMillan, W. J. Hannan, P. C. Emmett, and R. J. Aitken, *Int. J. Appl. Radiat. Isot.* **33**, 1423 (1982).

[11] H. Etani, K. Kimura, S. Yoneda, Y. Tsuda, M. Nakamura, K. Kataoka, Y. Iwata, and H. Abe, *J. Nucl. Med.* **24**, 136 (1983).

[12] J. A. Burdine, R. E. Sonnemaker, L. A. Ryder, and H. J. Spjut, *Radiology* **95**, 101 (1970).

[13] U. Scheffel, B. A. Rhodes, T. K. Natarajan, and H. N. Wagner, *J. Nucl. Med.* **13**, 498 (1972).

[14] C. L. DeLigny, W. J. Gelsema, and M. H. Beunis, *Int. J. Appl. Radiat. Isot.* **27**, 351 (1976).

[15] West German Patent 1,916,704 (1969).

[16] I. Zolle, B. A. Rhodes, and H. N. Wagner, *Int. J. Appl. Radiat. Isot.* **21**, 155 (1970).

[17] M. A. P. C. van de Poll and M. G. Woldring, *Pharm. Weekbl.* **108**, 741 (1973).

[18] U.S. Patent 3,663,687 (1972).

Iron-free HSA spheres of diameter 16–30 μm may be labeled by reduction of pertechnetate by sodium thiosulfate in the presence of HCl (pH 1.0),[12,19] although this again requires incubation at 100°—presumably to convert the sulfate into the sulfite, which acts as the reducing agent.[17]

Use of Sn(II) salts as reducing agents obviates the use of high temperatures. For example, Raban[20] gave a method for labeling of albumin microspheres with ^{99m}Tc by reduction of pertechnetate with stannous chloride in HCl. The use of HCl was later circumvented[11] by first soaking the spheres with stannous chloride, which gave a labeling efficiency of up to 95%.

Albumin microspheres prepared by thermal denaturation of 2-μm spheres produced from a protein aerosol have been labeled by Millar *et al.*[10] according to the method of Yeates,[21] which involves addition of a mixture of stannous chloride in acid and an aqueous sodium acetate solution to microspheres, followed by addition of sodium pertechnetate. The observation of a 99% labeling efficiency in the absence of microspheres (at pH 4.7) has led Millar *et al.* to examine the influence of pH, concentration of stannous chloride, and amount of microspheres on the labeling. These workers found that a kit containing 0.4 mg of 2-μm diameter HSA microspheres, 1 μg of $SnCl_2 \cdot 2H_2O$ and 1 ml of 25 mM HCl gave optimal labeling (of 88.3%). This kit has been later modified[22] to include 10 μg of the tin compound. Finally, a further general method for the labeling of HSA microspheres produced by thermal denaturation has been recently published,[23] in which either stannous tartrate is incorporated into the microspheres during a one-step preparative technique involving protein aerosol formation in a gas medium, or stannous chloride is applied as a coating on the surface of the formed spheres.

Comparative Advantages and Disadvantages

Labeling methods using either sodium thiosulphate, incorporated ferric hydroxide/and ascorbic acid, or free ferric chloride/and ascorbic acid as reducing agents followed by incubation at elevated temperatures have disadvantages for imaging of HSA microspheres intended for drug targeting. First, incubation at 100° to obtain maximum labeling leads to a further denaturation of the protein, which can influence the release rate of incor-

[19] West German Patent 2,115,066 (1971).

[20] P. Raban, V. Gregora, J. Šindelář, and J. Alvarez-Cervera, *J. Nucl. Med.* **14,** 344 (1973).

[21] D. B. Yeates, A. Warbick, and N. Aspin, *Int. J. Appl. Radiat. Isot.* **25,** 578 (1974).

[22] W. J. Hannan, P. C. Emmett, R. J. Aitken, R. G. Love, A. M. Millar, and A. L. Muir, *J. Nucl. Med.* **23,** 872 (1982).

[23] M. Przyborowski, E. Lachnik, J. Wiza, and I. Licińska, *Eur. J. Nucl. Med.* **7,** 71 (1982).

porated drugs.[24] Second, because these agents are not able to reduce all of the pertechnetate, labeled microspheres have to be separated from the free label by, for example, centrifugation. Indeed, sometimes the labeled microspheres need to be washed several times with an aqueous solution to remove remaining free pertechnetate.[12]

As stated previously, the use of stannous chloride simplifies the labeling method because both the incubation step is not necessary and the reduction of pertechnetate is complete. However, its use does have some disadvantages. For almost all of the methods published, an oxygen-free hydrochloric acid environment is required in order to prevent a rapid oxidation of stannous chloride to water-insoluble tin oxides. Furthermore, the labeling of the microspheres has to be carried out at pH 1.0–2.0, since at higher hydrogen ion concentrations labeled colloids are formed, probably from water-insoluble Sn(IV) chloride, ^{99m}Tc(IV) oxide, and stannous oxide.[1] These colloids will interfere with the correct interpretation of the biological distribution of correctly labeled HSA microspheres, especially because they themselves have a size range of 0.5–2.0 μm. Fortunately, formation of such colloids may be avoided by carrying out the labeling at pH 1.6,[10] although the resulting acidic suspensions cannot be used directly systemically. In order to do so the pH of the suspension needs to be raised. The question which arises is, does this also produce unwanted colloid? Oppenheim et al. have found that for gelatin nanoparticles (0.4 μm), upon raising the pH of the labeling mixture to pH 6.0–6.5 with sodium hydroxide solution, colloids do not form.[25] However, our experience with 20- to 30-μm HSA microspheres is that raising of the pH results in the formation of a labeled mixture of microspheres and colloid, which was only evident upon intravenous injection of the mixture into rabbits.

The easiest and most used method for removing hydrochloric acid containing the oxidized form of stannous chloride is centrifugation followed by collection and washing of the spheres. This reduces the yield of labeled spheres. In addition, with regard to the present use of HSA microspheres, the possible influence of hydrochloric acid on protein structure (and hence drug release) is unknown.

Mechanism of Labeling with Technetium-99m

The mechanism of labeling is not completely known. Reduction of the water-soluble [^{99m}Tc]pertechnetate can result in production of the lower

[24] E. Tomlinson, J. J. Burger, E. M. A. Schoonderwoerd, and J. G. McVie, Int. J. Pharm. (in press).
[25] R. C. Oppenheim, J. J. Marty, and N. F. Stewart, Aust. J. Pharm. Sci. 7, 113 (1978).

oxidation states (III), (IV), and (V). For the isotope ^{99}Tc, complexes Tc(III)–DPTA (i.e., diethylenetriaminepentaacetic acid), Tc(IV)–Sn diphosphonate, and Tc(V)–citrate form upon reduction of Tc(VII) with Sn(II) in the presence of the respective ligands. Ascorbic acid and ferric ion reduce the heptavalent technetium to the (V) state. Certainly, the most stable oxidation state is the Tc(IV) ion, which is due probably to its d^3 configuration. In aqueous solutions at pH 1.1, Tc(IV) is present as the ions TcO^{2+} and TcO_2H^+. At a pH of 2.6, Tc(IV) is present as both TcO_2H^+ and an insoluble hydrated oxide. At higher pH (4.9), Tc(IV) is only present as the insoluble hydrated oxide. Despite a great deal of study, the valence state of ^{99m}Tc in HSA complexes is unknown, although the (V) valence state has been suggested as the most likely.[14] It has been demonstrated[26] that reduction of [^{99m}Tc]pertechnetate with stannous chloride in the presence of DTPA results in complexes in which the ^{99m}Tc ion is in the (IV) state, and that to prevent hydrolysis of ^{99m}Tc(IV), the complex must be kept in an acidic environment.

Some additional light on the nature of the labeling of HSA microspheres can be obtained from studies of the ^{99m}Tc labeling of albumin macroaggregates using stannous chloride.[27] Again, this work suggests that, depending upon the pH, complexes of macroaggregated albumin with TcO^{2+} and TcO_2H^+ are formed, with labeling at high pH resulting in a ^{99m}TcO$_2$ colloid. The amount of stannous chloride needed to reduce the [^{99m}Tc]pertechnetate has been found to be critical for efficient labeling, such that if the amount of reducing agent exceeds a determined level, the binding between the macroaggregate and the reduced technetium-99m was lowered, despite the fact that no free pertechnetate was present. This was explained as being due to a competition between the positive tin ions of valency (II) and (IV), and the reduced technetium ion of valency (IV), for binding sites on the protein.

Labeling of Chemically Cross-Linked HSA Microspheres with Technetium-99m

Although the labeling of HSA microspheres produced by various thermal denaturation techniques has been frequently studied, little information has been published concerning the labeling of HSA microspheres stabilized by chemical cross-linking agents. Elsewhere in this volume [3] we describe our findings on the incorporation of highly water-soluble drugs into HSA microspheres in the range 0.1–100 μm. The release of drug from such microspheres may be manipulated by the extent of chemi-

26 W. C. Eckelman, G. Meinken, and P. Richards, *J. Nucl. Med.* **13**, 577 (1972).
27 L. O. M. J. Smithuis and J. B. F. Spijkers, *Pharm. Weekbl.* **108**, 869 (1973).

cal cross-linking of the spheres,[24] affording an attractive means of producing a targetable drug carrier with controlled drug release. In this present section we describe some of our own findings on the labeling of HSA microspheres stabilized using either 1,5-glutaraldehyde or 2,3-butadione as cross-linking agents.

Depending upon the degree of cross-linking, such microspheres contain either none or a reduced number of terminal amino groups (arising from the lysine residues of the protein), compared to thermal denatured HSA microspheres. This can have an influence on the complexation of any reduced technetium(IV) ions with the surface of the spheres.

Preparation

The full details of microsphere preparation are given in [3], this volume. Typically, microsphere stabilization is carried out using, for example, 0.27 M 2,3-butadione for 4 hr, or 5% glutaraldehyde for 1 hr. After manufacture, the spheres are freeze-dried, and sieved through microsieves using chloroform as the suspending medium. The exact mean diameter of these sieved spheres has been determined using a Coulter counter (model z_B) equipped with a Channelyser, although it will be seen elsewhere that suspending non-drug-containing freeze-dried spheres in isotonic saline increases their diameter by approximately 30%. After labeling of the microspheres in the ways outlined below, the efficiency of labeling has been examined in two ways: first, by paper chromatography, using an acetone or a saline solution as eluent, followed by analysis with a Squibb QC Analyser (Model QC 10), and second, by analyzing the activity of the microspheres suspended in a 0.1% Tween 80 isotonic saline solution using a Pitman Isotope Assay Calibrator (Type 238).

In order to examine the *in vivo* distribution of HSA microspheres, labeled spheres were suspended in 0.1% Tween 80 isotonic saline, and administered (in this case) intravenously to New Zealand White rabbits through their ear vein. Distribution of the spheres was followed using a Siemens Rota II gamma camera.

Sodium Thiosulfate as Reducing Agent

To examine the labeling of chemically cross-linked HSA microspheres, spheres stabilized with 2,3-butadione have been produced, such that their freeze-dried size range was 20–25 μm, with a mean diameter of 22.9 μm. Intravenous administration of these spheres via the ear vein can be expected to result exclusively in capillary blockage of the lungs.[28,29] Any colloid of less than 3 μm (formed either during or after labeling) can pass through the lung capillaries and accumulate in the liver and spleen. In this way, the presence of labeled species other than labeled HSA microsphere can be observed.

Labeling of chemically cross-linked spheres has been carried out using a simplification of a published technique.[12] Thus, using solutions filtered through a 0.2-μm polycarbonate filter, and performing the labeling under argon, the following scheme was followed:

1. Add 1 mg of freeze-dried microspheres to a glass vial. Close the vial.
2. Add 2 ml of ^{99}TcO$_4^-$ (approximately 8 mCi) in normal saline containing 0.1% Tween 80.
3. Add 0.1 ml of aqueous sodium thiosulfate solution containing (per liter) 0.1 g Na$_2$CO$_3$, 1 ml Tween 80, and 1.94 g Na$_2$S$_2$O$_3$·5H$_2$O. Suspend the contents of the vial by ultrasonication for 1 min.
4. Add 0.1 ml of 2 M HCl, and incubate for 10 min at 100°. Cool.
5. Centrifuge the vial at 800 rpm for 3 min, and remove the supernate.
6. Add 2 ml of saline containing 0.1% Tween 80, and resuspend the microspheres using further ultrasonication (2 min). Repeat the centrifugation step and discard the supernate.
7. Add 2 ml of saline containing 0.1% Tween 80 to the vial.

Using this procedure we have labeled HSA microsphere fractions with mean diameters of 6.6, 10.9, 17.2, and 22.9 μm, although only approximately 30% of the radioactivity initially added could be recovered in the form of microspheres. However, the labeling percentage of all microspheres studied was greater than 99.9%. In addition, a similar labeling procedure carried out in the absence of microspheres gave no detectable radioactivity in the final solution (step 7), proving the absence of radiolabeled colloids.

Step 5 is most interesting. Here, the activity of the supernate amounts to approximately 66% of the initial activity. Also, examination of the supernate (pH 1.0) using paper chromatography shows a labeling percentage of more than 99%. This means that although reduction of pertechnetate by sodium thiosulfate is complete, only a certain amount of the reduced Tc(IV) ions have complexed with the microspheres.

[28] B. A. Rhodes and B. Y. Croft, in "Basics of Radiopharmacy," p. 56. Mosby, St. Louis, Missouri, 1978.
[29] M. Kanke, G. H. Simmons, D. L. Weiss, B. A. Bivins, and P. P. DeLuca, J. Pharm. Sci. 69, 755 (1980).

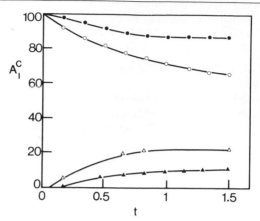

FIG. 1. Total ^{99m}Tc activity corrected for decay and background, expressed as a percentage of the amount of activity initially injected A_i^c versus time t (hours), in the lungs (circles) and bladder (triangles). Closed datum points are for 20- to 30-μm butadione-stabilized microspheres, using sodium thiosulfate as reducing agent, and open datum points are results found using a commercially available 3M HSA microsphere kit.

In vivo studies have been carried out with 0.25 mg of labeled microspheres suspended in 0.5 ml isotonic saline containing 0.1% Tween 80. Figure 1 gives the γ activity in the lungs and bladder after intravenous injection. No accumulation of label in the liver and spleen could be de-

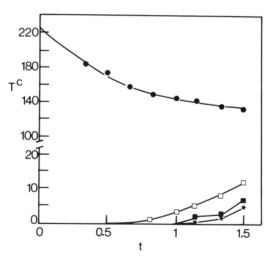

FIG. 2. ^{99m}Tc distribution after intravenous injection of labeled HSA microspheres of size 5–10 μm. Plot gives total counts (decay corrected) T^c versus time t (hours), and shows activity in the lungs (●), bladder (□), liver (■), and spleen (★).

TABLE I

RELATIVE DISTRIBUTION OF RADIOACTIVITY AFTER INTRAVENOUS
INJECTION OF CHEMICALLY STABILIZED HSA MICROSPHERES LABELED
WITH ^{99m}TC USING SODIUM THIOSULFATE AS REDUCING AGENT

Microsphere fraction (μm)	Time (min)	Relative distribution (%)[a]			
		Lungs	Liver	Spleen	Bladder
5–10	10	97.3 (9.8)	1.8 (1.3)	0.9 (0.9)	—
5–10	90	77 (8.8)	5.5 (2.3)	4.5 (2.1)	13 (3.6)
10–15	10	96.9 (9.8)	2.2 (1.5)	0.9 (0.9)	—
10–15	90	71.4 (8.4)	9.2 (3.0)	8.2 (2.9)	11.2 (3.3)
15–20	10	100 (10)	—	—	—
15–20	90	88.9 (9.4)	—	—	11.1 (3.3)
20–25	10	100 (10)	—	—	—
20–25	90	88.0 (9.4)	—	—	11.1 (3.3)

[a] Values in parentheses are coefficients of variation.

tected, which indicates that a formation of labeled colloid during or after labeling can be discounted. Since these microspheres were extensively cross-linked, it is likely that the activity found in the bladder after 90 min is due to a loss of label from the spheres, and not to breakdown of the spheres. Figure 3 also shows the time course of γ activity in the lungs and bladder for HSA microspheres of similar size supplied in kit form by the 3M company. It is seen that label is much more quickly lost from the lungs with these latter microspheres. In addition, use of a 3M microsphere kit leads to significant activity being measured in the liver, spleen, kidneys, and thyroid.

Identical results were found with the fraction of size 15–20 μm, but with smaller sized fractions activity in both the spleen and the liver could be found (Fig. 2). The appearance of activity in these two organs increased with time. Table I summarizes the distribution of activity over various organs with different size fractions of spheres. It is interesting to note that, irrespective of size, the activity in the bladder after 90 min is the same.

Stannous Chloride as Reducing Agent

The tedious procedure required for labeling using sodium thiosulfate has led us to examine whether stannous chloride may be used with the labeling of chemically stabilized HSA microspheres. The procedure is simple. Stannous chloride is added to HSA microspheres under argon in a closed vial. After addition of a pertechnetate saline solution containing

TABLE II
APPARENT LABELING WITH TECHNETIUM-99m[a]

| Component presence | | Labeling percentage |
Microspheres	Stannous chloride	
+	+[b]	99.9
+	−	None[c]
−	+[b]	99.9[d]
−	−	None[c]

[a] +, Present; −, absent.

[b] The amount of $SnCl_2$ has no influence.

[c] 2,3-Butadione and glutaraldehyde cannot reduce pertechnetate.

[d] Argon, Tween 80, and ascorbic acid had no influence.

0.1% Tween 80, the contents of the vial are ultrasonicated for 5 min. However, although a labeling percentage of 99.9% could be found using paper chromatography, intravenous injection of 0.25 mg HSA microspheres (2 mCi) resulted in accumulation of activity in lungs, liver, spleen, kidneys, bladder, and spinal column irrespective of the diameter of the microspheres. Obviously this is due to the formation of a radiocolloid during labeling of the microspheres. This phenomenon has been studied further by examination of the labeling under different circumstances. Thus, Table II shows that in all cases in which stannous chloride is present, even in the absence of microspheres, radiocolloids are formed upon the addition of pertechnetate. Using hydrochloric acid (pH 0.5) in place of isotonic saline (pH 5.5) results in a labeling percentage of 99.9% (as determined using paper chromatography). Because formation of colloid is impossible at such a low pH, this result means that the colloid is formed during the chromatography, and thus the paper chromatography technique used is insufficient for distinguishing between labeled microspheres and soluble, reduced Tc(IV) ions.

Interestingly, excellent results for the labeling of nanoparticles using a similar technique have been claimed.[25] However, using the suggested method[24] of labeling at pH 2.2 followed by raising of the pH to 6.0–6.5, with HSA microspheres 20–30 μm in diameter, 5 min after intravenous injection we find all activity entrapped in the liver and spleen, a finding which shows that radiocolloid has been formed. (Again, in the absence of microspheres the labeling percentage was 99%.)

In an attempt to avoid colloid formation either during or after labeling, stannous chloride has been incorporated into HSA microspheres. Since stannous chloride forms insoluble complexes with phosphate buffer, and a mixture of stannous chloride and albumin is insoluble in water (of pH 7), conventional ways[24] of incorporation of solute cannot be used. This may be accomplished by realizing that (1) a mixture of stannous chloride and albumin is soluble in water at pH 2, and (2) complexation of stannous chloride with either DTPA or ethylenediaminetetraacetic acid in water, followed by addition of HSA, results in clear solutions. These clear solutions can then be used as the disperse phase of the manufacturing emulsion (see this volume [3]). Using the DTPA method, and levels of incorporation of $SnCl_2$ of between 10 and 60 μg per milligram of formed microsphere, we have found that it is not possible to label HSA microspheres. Most of the pertechnetate was to be found as the water-soluble TcO_4^- ion [instead of the microsphere-complexed reduced Tc(IV) ion]. We assume that this is due to an inaccessibility of the incorporated reducing agent.

Because both ⁹⁹ᵐTc labeling with free stannous chloride results in the formation of radiocolloids, and labeling does not take place with incorporated $SnCl_2$, we examined whether coating of HSA microspheres with reducing agent can lead to successful labeling.

Coating was studied by soaking HSA microspheres in solutions of stannous chloride in HCl (pH 1.6) under argon for 30 min. After filtration, the spheres were washed twice with HCl (pH 1.6) and then with ethanol. The ratios of stannous chloride to microspheres (by weight) were 0.06, 0.12, and 0.24. ⁹⁹ᵐTc labeling was carried out by the addition of 4 ml of [⁹⁹ᵐTc]pertechnetate (8 mCi) in saline containing 0.1% Tween 80 to 1 mg of soaked, washed (and then freeze-dried) spheres. After ultrasonication for 5 minutes, the labeling percentage was found using paper chromatography to be approximately 80%. Intravenous injection of 0.25 mg (2 mCi) of labeled spheres gave visualization of the lungs, kidneys, and bladder, with no activity being observed in either the liver or the spleen. Figure 3 gives the gamma camera results for HSA microspheres soaked with differing quantities of stannous chloride.

Typically, the disappearance of activity from the lungs is biphasic, with the amount being lost during the first phase being greater than the amount of free pertechnetate (determined by paper chromatography) present in the injected suspension of HSA microspheres. It is highly probable that this is due to release of complexed label from the surface of the microsphere, and that this is mediated through enzyme attack. Figure 4 shows that the disappearance of activity from the lungs approximately parallels its accumulation in the bladder. The method does not produce radiocolloids.

From Fig. 3 it is clear that the relative concentration of stannous

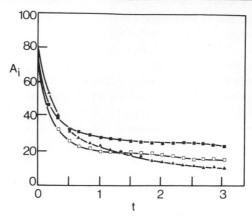

Fig. 3. Effect of SnCl$_2$ concentration on ^{99m}Tc activity in the lungs after intravenous injection of 15- to 25-μm HSA microspheres, showing the relationship between the activity (decay but not background corrected) expressed as a percentage of the initially injected amount of activity A_i versus time t (hours). Spheres (100 mg) soaked with 6 mg (■), 12 mg (□), and 24 mg (▲) SnCl$_2$, respectively.

chloride is important with regard to the efficiency of the labeling. With a ratio of 0.06 (see previous), approximately 70% of the radioactivity originally in the lungs has disappeared after 3 hr, whereas with ratios of 0.12 and 0.24, 80 and 88%, respectively, of the activity have disappeared after

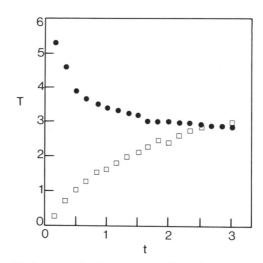

Fig. 4. Relationship between the disappearance of activity from the lungs (●) and that accumulated in the bladder (□), showing the total counts (decay but not background corrected) $T \times 10^{-5}$ versus time t (hours), using microspheres described in Fig. 3 with 6 mg SnCl$_2$.

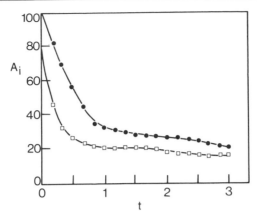

FIG. 5. Effect of washing on activity residence in the lungs, showing the relationship between A_i (see Fig. 3) and time t (hours). Open points are as for Fig. 3, whereas closed points are for similarly prepared spheres but washed and ultracentrifuged twice (see text).

this time. After 40 min, the rate of loss of activity from the lungs is comparable for all ratios, with a half-life of 3.5 hr being found. Whether this is due to a release of label or is a result of microsphere breakdown is not certain, although considering the extent of stabilization of the HSA microspheres used, then the former reason is more likely. We consider that the differences between the profiles given in Fig. 3 may be explained by two types of interaction between Tc(IV) and the microsphere, such that with higher amounts levels of $SnCl_2$ coating, the availability of binding sites for the more strongly complexed Tc(IV) becomes less.

To remove free, unreduced pertechnetate, soaked microspheres were centrifuged after labeling, and the supernate discarded. After resuspending the spheres in saline solution containing 0.1% Tween 80, a labeling percentage of 99.98% was found by paper chromatography. Figure 5 gives the rate of disappearance of activity from the lungs for washed and unwashed HSA microspheres. Unlike Fig. 4, these data have not been corrected for background activity. It is clear that (1) further washing of the microspheres results in all of the original injected activity becoming entrapped (associated with microspheres) in the lungs, and (2) this results in a constant displacement of the loss/time profile upwards.

Concluding Remarks

This study has shown that human serum albumin microspheres, cross-linked by 2,3-butadione and glutaraldehyde, can be efficiently labeled with the radioisotope technetium-99m. It has been found that reduction with sodium thiosulfate needs to be carried out at 100° in an acidic environment for several minutes, and although the method leads to long retention of label on the microspheres (Fig. 1), the cleaning procedures are a problem and lead to only a 30% labeling efficiency.

The use of stannous chloride as reducing agent has many advantages, although, to avoid the formation of radiocolloid either during or after labeling, it needs to be coated on the surface of the microspheres. Such coated HSA microspheres can be labeled with technetium-99m at neutral pH in saline. Figure 5 shows that the amount of nonreduced pertechnetate may be removed by a washing step.

Further studies are in progress both to increase the labeling efficiency and to reduce the initial large release of label from the spheres. Studies are also under way to label the microspheres with [125]I.

Acknowledgments

We gratefully acknowledge the receipt of a postdoctoral fellowship for J. J. B. from the Queen Wilhelmina Cancer Funds (Project GU 82-8). The assistance of K. Hoefnagel, J. H. Smit and E. M. A. Mulder is warmly acknowledged.

[5] Biophysical Drug Targeting: Magnetically Responsive Albumin Microspheres

By Andrew E. Senyei, Charles F. Driscoll, and Kenneth J. Widder

Introduction

Biophysical approaches to specific drug targeting encompass a wide range of modalities, including local implantation of drug reservoirs,[1] hyperthermia,[2] arterial perfusion,[3] arterial chemoembolization,[4] intracavity injection,[5] and the use of extracorporeal magnetic fields.[6] Magnetic target-

---

[1] G. N. Pandya and A. Scommegna, *Int. Congr. Ser.—Excerpta Med.* **246,** 103 (1972).

[2] G. Hahn, *Cancer Res* **39,** 2264 (1979).

[3] B. Cady, *Ann. Surg.* **178,** 156 (1973).

[4] T. Kato, R. Nemoto, H. Mori, and I. Kumagai, *Cancer* **46,** 14 (1980).

[5] R. F. Ozols, R. C. Young, J. L. Speyer, P. H. Sugarbaker, R. Greene, J. Jenkins, and C. E. Myers, *Cancer Res.* **42,** 4265 (1982).

[6] K. J. Widder, A. E. Senyei, and D. G. Scarpelli, *Proc. Soc. Exp. Biol. Med.* **158,** 141 (1978).

ing has several unique advantages[7] that allow for effecting very high local concentration of therapeutic agents. A variety of magnetically responsive drug carriers have been proposed as vehicles for chemotherapeutic agents. Widder *et al.*[6] used albumin microspheres containing magnetite as vehicles to target adriamycin to a specific body region in a rat model system. Other matrix materials have been examined, including starch,[8] ethyl cellulose,[9] ethyl oleate based emulsion,[10] and natural cells such as erythrocyte ghosts.[11]

However, of all these matrix materials, albumin seems to best fulfill many of the requirements for an ideal drug carrier. These requirements include restriction of pharmacologic activity to the target area[12]; appropriate size to allow uniform vascular distribution at the target site[13]; controllable drug release rates[14]; accommodation of a wide variety of therapeutic agents[7]; biocompatability and biodegradability with minimal toxicity[6]; demonstration of therapeutic efficacy *in vivo*[15]; and practical properties, such as long shelf life, suspendability, and ease of large-scale production.

The physical laws underlying the intravascular magnetic guidance of magnetically responsive drug carriers were initially described by Senyei *et al.*[16] However, much of the complex *in vivo* behavior of the microspheres remains to be elucidated, especially the mechanisms involved in microsphere retention at the target site after the magnetic field is removed. Recently investigators[17] utilized intravital microvideo techniques to study capture of microspheres in the hamster mesentary circulation under various magnetic field conditions. On a larger scale, Richardson *et al.*[18] have demonstrated uniform perfusion and magnetic capture of ^{99}Tc-labeled albumin microspheres in a spontaneous osteosarcoma tumor test system.

[7] K. J. Widder, A. E. Senyei, and D. F. Ramey, *Adv. Pharmacol. Chemother.* **6,** 213 (1979).

[8] K. Mosbach and U. Schorder, *FEBS Lett.* **102,** 112 (1979).

[9] T. Kato, R. Nemoto, H. Mori, S. Sato, K. Unno, M. Honma, M. Okada, and T. Minowa, *ICRS Med. Sci.* **7,** 621 (1979).

[10] Y. Morimoto, K. Sugibayashi, and M. Akimoto, *Chem. Pharm. Bull.* **31,** 279 (1983).

[11] U. Zimmerman, G. Pilwat, and B. Esser, *J. Clin. Chem. Clin. Biochem.* **16,** 135 (1978).

[12] A. E. Senyei, S. D. Reich, C. Gonczy, and K. J. Widder, *J. Pharm. Sci.* **70,** 389 (1981).

[13] K. J. Widder, R. M. Morris, G. Poore, D. P. Howard, and A. E. Senyei, *Eur. J. Cancer Clin. Oncol.* **19,** 135 (1983).

[14] K. J. Widder, C. Flouret, and A. E. Senyei, *J. Pharm. Sci.* **68,** 79 (1979).

[15] K. J. Widder, R. M. Morris, G. A. Poore, D. P. Howard, and A. E. Senyei, *Proc. Natl. Acad. Sci. U.S.A.* **78,** 579 (1981).

[16] A. E. Senyei, K. Widder, and G. Czerlinski, *J. Appl. Phys.* **49,** 3578 (1978).

[17] A. E. Senyei, K. Kwang, C. F. Driscoll, K. J. Widder, and M. Intaglietta, *in* "Microspheres in Medicine " (unpublished). University of Amsterdam, 1983.

[18] R. Richardson and J. Bartlett, *in* "Microspheres in Medicine " (unpublished). University of Amsterdam, 1983.

The magnetic forces operant in holding at a given site are still incompletely understood. A section of this chapter will be devoted to further elucidating these concepts.

Other therapeutic and diagnostic applications of magnetic microspheres are beyond the scope of this chapter but deserve brief mention for the sake of completeness. Extracorporeal magnetic fields have been employed by Hsieh et al.[19] to modulate the release of albumin from an ethylene–vinyl acetate copolymer matrix in which large magnetic microspheres were embedded. This technique holds promise for feedback regulation of release rates for biologically active polymers such as insulin. Stauffer et al.[20] have reported similar techniques for production of local hyperthermia using a magnetic device.

Magnetic microspheres offer unique advantages as solid supports for a variety of applications, including cell sorting,[21,22] solid phase immunoassays,[23] supports for immobilized enzymes, and bioaffinity adsorbents.[24] The unique concentrating effect of magnetic fields on magnetically responsive particles has a broad spectrum of applicability.

Synthesis of Magnetically Responsive Albumin Microspheres

The synthesis of magnetically responsive albumin microspheres can be readily accomplished by modification of standard emulsion phase separation polymerization techniques. The microspheres form after the solid albumin matrix is stabilized either by appropriate heat denaturation or by chemical cross-linking with various carbonyl reagents. The degree of matrix stabilization can be varied by altering the parameters of temperature and time in the case of heat denaturation, or the time of exposure to chemical cross-linkers. Widder et al.[14] reported in detail a methodology which may be applied to a variety of therapeutic agents. The predominant considerations for entrapment include the aqueous solubility of the drug, minimal partition to nonpolar solvents, heat stability, molecular weight (<1000), affinity and binding to albumin, and (in the case of heat-labile

[19] D. S. T. Hsieh, R. Langer, and J. Folkman, Proc. Natl. Acad. Sci. U.S.A. 78, 1863 (1981).
[20] P. R. Stauffer, T. C. Cetas, and R. C. Jones, in "Cancer Therapy by Hyperthermia, Drugs and Radiation" (L. A. Dethlefsen and W. C. Dewey, eds.), p. 483. U.S. Dept. HHS, Public Health Service, Bethesda, Maryland, 1982.
[21] K. J. Widder, A. E. Senyei, H. Ovadia, and P. Y. Paterson, Clin. Immunol. Immunopathol. 14, 395 (1979).
[22] P. C. Kronick, in "Methods of Cell Separation" (N. Catsimpoolas, ed.), Vol. 3, p. 115. Plenum, New York, 1980.
[23] M. Pourfarzaneh, R. S. Kamel, J. London, and C. C. Dawes, Methods Biochem. Anal. 28, 267 (1982).
[24] P. J. Halling and P. Dunhill, Enzyme Microb. Technol. 2, 196 (1980).

drugs) the degree of reactivity with the carbonyl-stabilizing agents. The following section describes the synthesis of adriamycin-containing magnetically responsive albumin microspheres.

Heat-Stabilized Adriamycin-Containing Microspheres

An aqueous solution is prepared containing (per milliliter) 250 mg of human serum albumin (Sigma Chemical), 0.02 ml of ^{125}I bovine serum albumin (specific activity of 1.51 mCi/mg, 1.42 mCi/ml New England Nuclear if radioactive labeling is desired), 32 mg of bulk purified doxorubicin hydrochloride (adriamycin, Adria Laboratories), and 72 mg of magnetite suspension (10–20 nm diameter particles of Fe_3O_4 in aqueous suspension, Ferrofluidics Corp.). A 5-ml aliquot of this suspension is added to 30 ml of cottonseed oil (Sargent Welch) at 4°, and the resultant emulsion is homogenized by sonication (Branson sonifier model 185) at 100 W for 1 min at 4°. The homogenate is then added dropwise into 100 ml of stirred (1600 rpm) cottonseed oil preheated to a desired temperature (110–165°).

The duration of heating usually can be varied depending on temperature selected. After 10 min the oil is allowed to cool to 25° with constant stirring. The microspheres are washed free of oil by adding 60 ml of anhydrous ether (Mallinckrodt Chemicals), centrifuging (Sorval model RC2-B) for 15 min at 2000 g, and decanting the supernate. After the fourth wash, the pellet of microspheres is allowed to air dry in the dark for 24 hr at 4°. The resultant microspheres can be lyophilized to remove any remaining water and stored at 4° until they are to be used. Shelf life in excess of 6 months can be anticipated depending on the therapeutic agent encapsulated.

Carbonyl-Stabilized Microspheres

An oil homogenate is prepared as described above. The homogenate is then added dropwise into 100 ml of constantly stirred (1600 rpm) cottonseed oil at 25°. After 10 min of stirring, the microspheres are washed free of oil with ether. The microspheres are then resuspended in ether (40 mg of microspheres/100 ml of ether) containing either 0.2 M 2,3-butanedione (Aldrich Chemical Co.) or 0.1 M formaldehyde (Scientific Products). The 2,3-butanedione is soluble in the ether phase; however, formaldehyde must be extracted into the ether phase. This is accomplished by the vigorous shaking for 10 min of a 1:5 solution of 37% aqueous formaldehyde–ether with the addition of a saturating amount of ammonium sulfate.

The ether suspension is stirred rapidly for 15–60 min (the time selected to give the desired release rate). Excess carbonyl reagent is removed by immediately adding 100 ml of ether, pelleting the microspheres by centrifuging at 2000 g for 10 min, and decanting the supernate. The microspheres are washed four times and subsequently lyophilized and stored at 4°.

Suspension of Microspheres

A standard technique has been developed to allow ease of suspension of microspheres from the dry state to aqueous suspension. The microspheres are suspended initially in 95% ethanol as a wetting agent (0.1 ml/ 100 mg), this is sonicated for 1 min, and then the ethanol is removed. Microspheres may then be suspended in 0.1% phosphate buffer immediately prior to injection (usually 2–30 mg/ml) or other appropriate suspending media.

Characterization of Microspheres

The physical properties including size, drug content, drug release rate, bioactivity (parent drug released), and magnetic responsiveness are crucial parameters that must be critically controlled for valid *in vivo* testing of the carrier. Recommended methodologies for quality control are listed below.

Sizing. Scanning or transmission electron microscopy can give a qualitative assessment of overall morphology (see Fig. 1), but cannot easily quantify the distribution of sizes. Sizing by analysis of forward angle light scattering (Microtrac, Leeds Northrop) gives quantitative information on the distribution of sphere diameters. Typical distributions used for targeting to capillary beds include spheres from 0.3 to 2.5 μm, with a mean of 0.9 μm. Larger microspheres are trapped at inappropriately large-caliber vessels in the circulatory tree. Smaller spheres lack sufficient volume for magnetic responsiveness.

Magnetic Responsiveness. A simple *in vitro* flow apparatus[16] can assess magnetic responsiveness under flow conditions approximating *in vivo* flow velocities. Alternatively, close control of magnetite content to 18% (w/w) (as determined by atomic absorption spectroscopy) ensures uniform responsiveness if size is well controlled.

Parent Drug Assay. Appropriate methodology is dictated by the specific drug entrapped. Adriamycin is best evaluated by high-pressure liquid chromatography. Bioassays on released drug may also be employed, but give little information as to the presence of either nonactive or active break-down products that may have been formed during the polymerization process. However, for antineoplastic agents such as adriamycin in an

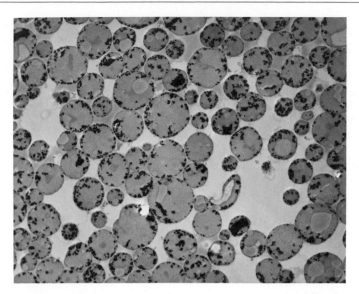

FIG. 1. A transmission electron photomicrograph of the magnetic microspheres reveals approximately 1-μm spheres with Fe_3O_4 particles embedded within the matrix material. Note the peripheral orientation of the particles.

in vitro cell culture, assay can be very informative with regard to correlation of released drug and bioactivity. Widder *et al.*[25] described a system using a Fisher 344 rat fibrosarcoma line *in vitro* cell culture assay system. The tumor had been initially induced with methylnitrosourea and was adapted for growth in cell culture in RPMI 1640 (Grand Island Biological Co.) containing 15% fetal calf serum (Reheis Chemical Co., Phoenix). The effect of inhibition of cellular RNA synthesis was estimated from the decrease of radiolabel incorporation ([5-^{3}H]uridine) over a 0- to 240-min time interval. Essentially, parent nonencapsulated drug and that released from the microspheres had similar inhibitory activity.

Toxicity. Minimal toxicity has been noted with albumin microspheres used for lung scans in humans for over 15 years. Magnetically responsive microspheres are also expected to show similar safety. The dosing must be limited in rate if aggregation at the injection site is to be avoided. Optimal dosing parameters need to be experimentally determined for the specific target region. Short-term applications of magnetic fields less than 10,000 gauss (Gs) have not shown any measurable adverse effects on biological tissue. In general, encapsulation of parent drug is expected to

[25] K. J. Widder, A. E. Senyei, and D. F. Ranney, *Cancer Res.* **40**, 3512 (1980).

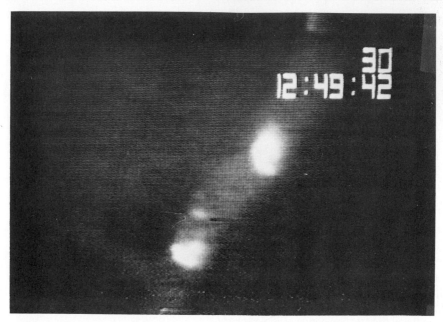

Fɪɢ. 2. Intravital videomicroscopic demonstration of microspheres forming "patches" on the wall of a 20-μm diameter venule in a hamster mesentery preparation. The microspheres were fluorescently labeled and exposed to an external magnetic field applied at the bottom of the venule (magnification ×350).

greatly minimize systemic toxicity, especially if the drug is restricted to one specific area of the body.

Forces on a Microsphere in Blood Flow

In this section, we consider the forces exerted on a microsphere when it is in a cylindrical tube of flowing viscous liquid, such as a capillary or venule. The magnetic force is well defined, and is reviewed below. The viscous drags and torques are proportional to the blood viscosity, which depends on hematocrit fraction H. Theory and experiment indicate a 3-fold viscosity increase for $H = 40\%$, as compared to plasma alone. The viscous shear force and rolling torque for a microsphere attached to a wall can be estimated from theory and experiment, as shown below. An additional force due to hydrostatic pressure drop in the tube is also considered. These forces may preclude magnetic holding in capillaries, but are much smaller in venules, where holding is observed experimentally (see Fig. 2).

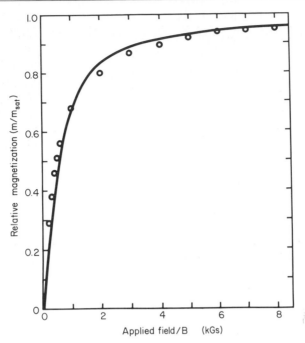

FIG. 3. Magnetization of microspheres as a function of the applied magnetic field.

The magnetic force F_m on a microsphere can be written[26,27]

$$F_m = V\rho_\mu f M_{sat} L(\varepsilon B) \nabla B$$
$$= 7.1 \times 10^{-8} d^3 f L(\varepsilon B) \nabla B$$

This force represents the induced magnetic moment of the microsphere interacting with the field gradient ∇B (in kGs/cm). The mass of magnetite is given by the volume of the microsphere V, times the density ρ_μ (typically 1.5 g/cm³), times the fraction f of magnetite by weight (typically 0.2). Magnetite can have a maximum magnetic moment $M_{sat} = 90$ emu/g, and $L(\varepsilon B)$ gives the fraction of this "saturation" moment which is induced by an applied field B.

As shown in Fig. 3, fields of 1 kGs are sufficient to induce a strong magnetic moment in our typical microspheres.[28] (The details of this curve depend on the sizes of the magnetite grains in the microspheres.) If there

[26] J. D. Jackson, "Classical Electrodynamics," §5.7. Wiley, New York, 1967.

[27] D. B. Montgomery, *J. Appl. Phys.* **40,** 1039 (1969).

[28] C. F. Driscoll, R. M. Morris, A. E. Senyei, K. J. Widder, and G. S. Heller, *Microvasc. Res.* **27,** 353 (1984).

is also a spatial gradient to the magnetic field, a force will act on the microspheres. For convenience, we consider a scaled force defined as

$$\mathscr{F}_m = d^3 f L(\varepsilon B)\, \nabla B$$

where d is the sphere diameter in micrometers. A scaled force $\mathscr{F}_m = 1$ would act on a 1-μm 20% magnetite sphere in a strong field (such that $L \approx 1$), if the field gradient is 5 kGs/cm. This corresponds to $F_m = 7.1 \times 10^8$ dyn acting to accelerate or keep the microsphere moving.

When a microsphere is moving relative to a uniform viscous fluid, it experiences a drag force. This force is given by Stokes' law,[29] as

$$F_v = 6\pi\eta a v$$
$$= 7.5 \times 10^{-6}\,(\eta/\eta_0)dv \text{ (dyn)}$$

Here η is the viscosity of the fluid [$\eta_0 = 0.8 \times 10^{-2}$ P (poise) for water], a is the sphere radius (cm), d is the sphere diameter (μm), and v is the relative velocity (cm/sec). This can be scaled to the same units as the scaled magnetic force, as

$$\mathscr{F}_v = 1.06(\eta/\eta_0)\, dV_{100\,\mu m}$$

where $V_{100\,\mu m}$ is the velocity in units of 100 μm/sec. A sphere with diameter 1 μm would be pulled through water at a velocity of 100 μm/sec by a scaled magnetic force of 1.06. Thus, this is the velocity range of interest for our magnetic forces.

In applying formulae such as the above, one must be sure that the blood does indeed act like a viscous fluid. Experimental measurements have determined that blood is *not* simply viscous in velocity profiles with low shear.[30] However, in capillaries and venules, the shear rates are very high ($dV/dr \approx 500$ sec^{-1}). For shear rates greater than 200 sec^{-1}, experiments suggest that the only effect of the hematocrit is to increase the apparent viscosity of whole blood over that of the plasma alone. We take

$$\eta_P = 0.011 \text{ P}$$
$$\eta_B = \eta_P(1 + 5\,H)$$
$$= 0.03 \text{ P} \qquad \text{(at } H = 40\%)$$

We will use this value $\eta_B = 3.75\eta_0$ for all further drag considerations.

When a microsphere is attached to the wall of a cylindrical tube, Stokes' law does not correctly predict the axial drag (or "shear force")

[29] J. Happel and H. Brenner, "Low Reynolds Number Hydrodynamics," §7-2, §7-3. Nordhoff Int., Leyden, 1973.

[30] R. L. Whitemore, "Rheology of the Circulation." §4-17, Chapter 6. Pergamon, Oxford, 1968.

that it experiences due to viscosity. The force in this case can be written as[29]

$$F_{vw} = 6\pi\eta a \left[V_c \frac{d}{R} \right] \left(\frac{25}{16} \right)$$

$$= 11.7 \times 10^{-6} \left[\frac{\eta_B}{\eta_0} \right] \left(\frac{d}{R} \right) dV_c \text{ (dyn)}$$

Here, R is the tube radius in microns. This equation is the limiting form of a mathematical expression as the sphere approaches the wall. Although the expression is not strictly valid when the sphere touches the wall, the expression is continuous and there appears to be no singular viscous effects associated with actually touching the wall. The term in square brackets is merely the flow velocity at one sphere radius out from the wall, since the velocity profile in the tube is $V(r) = V_c(1 - r^2/R^2)$. Except for the geometric factor of 25/16, this is obviously the same drag effect as in Stokes' law. This shear force on a sphere attached to a wall has been verified experimentally to within 50%.[31] It can be written in scaled force units as

$$\mathscr{F}_{vw} = 1.65 \left(\frac{\eta_B}{\eta_0} \right) \left(\frac{d}{R} \right) dV_{C\ 100\ \mu m}$$

Consider blood flowing in a capillary ($R = 4\ \mu m$) with central velocity $V_{C\ 100\ \mu m} = 10$, giving average velocity $\frac{1}{2}V_c = 0.05$ cm/sec. A 1-μm sphere should feel a scaled shear force of $\mathscr{F}_{vw} = (1.65)(3.75)(0.25)(10) = 15$. That is, a scaled magnetic force of 15 (which is exceedingly large) directed against the flow would be required to hold the particle. Such a force might arise from a standard microsphere in a field gradient of 75 kGs/cm. The shear force is large because the blood viscosity is 3.75 times greater than water, and because the microsphere is a large fraction of the radius of the tube.

The shear forces at the wall will be substantially lower in venules than in capillaries, which would favor holding in venules. Venules have about the same flow velocities as capillaries, but the venule diameter is typically four times greater. This gives shear forces at the wall four times smaller. Furthermore, the large magnitude of the viscous forces suggests that irregularities or crevices in the vessel walls (as would be expected in tumors) could radically enhance the probability of a microsphere sticking. Indeed, if the microsphere can "bury" itself in a gap in the vessel wall, there will be essentially no shear force acting on it. This is exactly the situation encountered in most tumor vasculatures.

[31] G. W. Schmid-Schoenbein, Y.-C. Fung, and B. J. Zweifach, *Circ. Res.* **36**, 173 (1975).

A microsphere near a wall also experiences a rolling torque due to viscosity. This torque is given analytically as[29]

$$\tau_{vw} = 6\pi\eta a \left[V_c \frac{d}{R} \right] \left(\frac{2}{3} a \right)$$

$$= 2.5 \times 10^{-10} \left(\frac{\eta_B}{\eta_0} \right) \left(\frac{d}{R} \right) d^2 V_c \text{ (dyne-cm)}$$

It can be thought of as a force of magnitude F_{vw} acting with an effective moment arm of $(32/75)a$. The torque will tend to roll the microsphere along the wall in the direction of the blood flow.

A different type of axial force arises due to the hydrostatic pressure gradient (∇P) of the flow. The flow generates a pressure drop along the tube, making the pressure on the upstream side of the microsphere greater than on the downstream side; this results in a net force downstream. This ∇P force can be written as

$$F_{\nabla P} = \pi a^2 (4/3) a \nabla P$$

$$= 6\pi\eta a \left[V_c \frac{d}{R} \right] \frac{2}{9} \frac{d}{R}$$

Thus, it is a factor of 0.14 (d/R) smaller than the shear force F_{vw}, and so will generally be negligible.

When a microsphere is stationary against the wall of a blood vessel, the forces and torques on the particle necessarily sum to zero. This means that binding forces between the sphere and the endothelial cells, in combination with magnetic forces (when present), exactly equal the hydrodynamic forces. However, very little is known about this chemical/biological binding especially when cytotoxic drugs such as adriamycin are present.

We have considered the forces on a single microsphere, but magnetic holding is observed to occur in patches (see Fig. 2). This is to be expected, for the following reason. Consider a microsphere stuck to the wall. A second microsphere sticking immediately behind the first will experience much less drag than it would in isolation, because the first has "broken the flow" for it. There would also be advantage to being immediately adjacent to other microspheres. Unless there is some sticking force between microsphere surfaces, as there is between microsphere and endothelial cell, one would expect the patch to be only one microsphere thick. In general, one would expect a patch of microspheres which is N wide by N long to experience a total drag force about N (rather than N^2) times that on a single sphere. For patches of the order of 10 × 10 as observed in venules, this drag reduction (perhaps a factor of 10) can be highly significant. Thus, patch formation allows for optimum magnetic holding and at least partially explains the postmagnetic retention of microspheres in venules.

Conclusions

Magnetically responsive microspheres may find utility as carriers for a wide variety of therapeutic agents that are used in the treatment of localized disease processes. The ability to restrict drug activity to a specific body region allows for administration of small doses and therefore decreased side effects. Particular applications including treatment of localized cancers (whether curative or palliative) or ablation of aberrant physiological functions (hyperthyroidism, for example) are a few of the many areas in which the concentrating ability of this targeting system could be of value. Clinical application, however, is the ultimate test for any drug targeting system. Results to date in both small and large animals are very encouraging. Hopefully further investigations will reveal a variety of novel applications in areas not discussed in this review.

[6] Attachment of Monoclonal Antibodies to Microspheres

By LISBETH ILLUM and PHILIP D. E. JONES

Introduction

An improvement in therapeutic efficacy of drugs by selective targeting to specific sites would be of clinical importance. One method by which this may be achieved is by antibody-mediated targeting of the drug.[1] Although drugs may be coupled directly to antibodies, only a small number of functional groups are available per antibody molecule which can be successfully used without significant loss of antibody activity.[2] This effectively limits the molar drug-to-antibody ratio to 10:1. To increase this

[1] G. F. Rowland, *in* "Targeted Drugs" (E. P. Goldberg, ed.), p. 57. Wiley, New York, 1983.
[2] G. F. Rowland, R. G. Simmonds, J. R. F. Corvalan, R. W. Baldwin, J. P. Brown, M. J. Embleton, C. H. J. Ford, K. E. Hellström, I. Hellström, J. T. Kemshead, C. E. Newman, and C. S. Woodhouse, *Protides Biol. Fluids* **30,** 315 (1983).

ratio, drugs have been conjugated to carrier molecules such as dextran and human serum albumin.[3,4] However, the covalent linking of drugs to such complexes may also impair the functional activity of the drug. The use of drug-loaded polymeric microspheres coated with antibody (immunomicrospheres) may circumvent these problems.

No studies have yet been published using immunomicrospheres loaded with drugs. The established methods for preparing protein-coated microspheres which have been used in radioimmunoassays, immunofluorescence assays, microscopy, and cell separation techniques[5] are generally applicable for the preparation of microspheres coated with monoclonal antibody.

Monoclonal Antibodies

Immunoglobulins are bifunctional molecules which not only bind to antigens, but also initiate a number of other biological phenomena such as complement activation and histamine release by mast cells, activities in which the antibody acts as a directing agent. These two kinds of functional activities are localized to different portions of the molecule: the antigen-binding activity to the Fab and the biological activities to the Fc portion of the molecule. Structurally they have a tetrameric arrangement of pairs of identical light and heavy polypeptide chains held together by noncovalent forces and usually by interchain disulfide bridges. Each chain is made up of a number of loops or domains of more or less constant size. The N-terminal domain of each chain has greater variation in amino acid sequence than the other regions, and it is this which imparts the specificity to the molecule. There are five types of heavy chains, which distinguish the class of immunoglobulins IgM, IgG, IgD, IgA, and IgE, and two types of light chains. In many species there is often more than one version of some of the heavy chain classes, which impart distinct physicochemical and biological characteristics to the molecule.

The introduction by Köhler and Millstein[6] of methods for producing hybrid myelomas (hybridomas) that synthesize monoclonal antibodies against single antigenic determinants has allowed the production of apparently inexhaustible supplies of pure, specific, standardized antibody. The technique typically involves the fusion of spleen cells from a suitably

[3] Y. Tsukada, V. K. D. Bischof, N. Hibi, M. Hirai, E. Hurwitz, and M. Sela, *Proc. Natl. Acad. Sci. U.S.A.* **79**, 621 (1982).

[4] M. C. Garnet, M. J. Embleton, E. Jacobs, and R. W. Baldwin, *Int. J. Cancer* **31**, 661 (1983).

[5] A. Rembaum and W. J. Dreyer, *Science* **208**, 364 (1980).

[6] G. Köhler and C. Milstein, *Eur. J. Immunol.* **6**, 292 (1975).

immunized donor with myeloma cells of an *in vitro* adapted enzyme-deficient cell line. Hybridoma cells are selected under *in vitro* culture conditions which allow the growth of only the cell hybrids. Hybridoma clones producing the desired antibody are then identified and isolated (reviewed by Goding[7]). Because of the nature of the immunization/fusion techniques, monoclonal antibodies belonging to the IgG or IgM class are most commonly produced.

The extreme specificity of monoclonal antibodies is obviously one of their most attractive properties. However, the very fact that hybridoma antibodies are monoclonal means that they may not behave like conventional antisera.[2]

Microspheres

Monoclonal antibodies can be attached to microspheres by means of direct coupling if functional groups capable of covalently bonding with proteins, e.g., aldehyde groups, are available on the surface of the microspheres. However, microspheres often carry a variety of groups on the surface such as carboxyl, hydroxyl, and/or amino groups, which either can be linked to antibodies employing a coupling reagent or can be modified to give reactive aldehyde groups. Microspheres can be made from many different materials and in many different ways in order to obtain functional groups (carboxyl, ester, hydroxyl, sulfonate, amino, and amide).

The so-called latex polymer systems are made by an aqueous polymerization process. For example, polystyrene latex particles prepared without emulsifier using persulfate initiator and bicarbonate buffer had, after cleaning, only strong acid sulfate groups on the surface, after hydrolysis, only weak acid carboxyl groups, and after oxidation, only uncharged hydroxyl groups.[8] Other functional groups can be obtained by using different initiators. In this way, microspheres with different functional groups can be prepared which can be reacted with, for example, amino groups on proteins.

Reactive groups can also be incorporated into microspheres by an emulsion copolymerization process in which, for example, a mixture of water-soluble and water-insoluble acrylic monomers containing hydroxyl and carboxyl functional groups (methyl methacrylate, methacrylic acid, and 2-hydroxyethyl methacrylate) are polymerized in the presence of an

---

[7] J. W. Goding, *J. Immunol. Methods* **39**, 285 (1980).
[8] A. A. Kamel, M. S. El-Aasser, and J. W. Vanderhoff, *J. Dispersion Sci. Technol.* **2**, 183 (1981).

emulsifier and a cross-linking agent.[9] In the same way, microspheres can be polymerized from acrylamide and $N'N'$-methylenebisacrylamide in buffer (pH 7.4) using a catalyst system consisting of N,N,N',N'-tetramethylethylenediamine and ammonium peroxodisulfate.[10] Microspheres with native aldehyde functional groups are particularly desirable since the chemical derivatization procedure is simplified to a one-step reaction between the aldehyde group and a primary amino group on the protein molecule. Furthermore, it is considered advantageous for immunomicrospheres to be hydrophilic, in order to avoid nonspecific interactions which are generally hydrophobic.[11] Hydrophilic microspheres with functional aldehyde groups can be produced by polymerization of glutaraldehyde[12] and of acrolein[13,14] at high pH. Microspheres polymerized from natural materials such as albumin and gelatin contain, due to their nature, functional groups on the surface, e.g., amino groups and carboxyl groups.

In a recent paper, Longo *et al.*[15] described a method to prepare hydrophilic human serum albumin microspheres. The cross-linking of the albumin phase with glutaraldehyde produced free aldehyde groups on the surface of the microspheres. Besides the ability to couple directly to antibodies, the aldehyde groups also facilitated surface chemical coupling of 2-aminoethanol or aminoacetic acid for increased hydrophilicity.

Methods of Attachment of Monoclonal Antibodies to Microspheres

This section describes the different approaches that may be applied in attaching monoclonal antibodies to microspheres. The table lists commonly used methods and the reactive groups and reactions involved.

Nonspecific Adsorption

Hydrophilic immunomicrospheres are considered to be more suitable for cell targeting; however, hydrophobic particles may be easily coated with immunoglobulins by physical adsorption, and as such are likely to be rendered more hydrophilic. At their isoelectric point, immunoglobulins bind firmly to the surface of such particles by van der Waals–London

[9] R. S. Molday, W. J. Dreyer, A. Rembaum, and S. P. S. Yen, *J. Cell Biol.* **64**, 75 (1975).
[10] P. Arturson, T. Laakso, and P. Edman, *J. Pharm. Sci.* **72**, 1415 (1983).
[11] A. Rembaum, S. P. S. Yen, E. Cheong, S. Wallace, R. S. Molday, I. L. Gordon, and W. J. Dreyer, *Macromolecules* **9**, 328 (1976).
[12] S. Margel, *Ind. Eng. Chem. Prod. Res. Dev.* **21**, 343 (1982).
[13] S. Margel, V. Beitler, and M. Ofarim, *J. Cell Sci.* **56**, 157 (1982).
[14] M. Kumakura, M. Suzuki, S. Adachi, and I. Kaetsu, *J. Immunol. Methods* **63**, 115 (1983).
[15] W. E. Longo, H. Iwata, T. A. Lindheimer, and E. P. Goldberg, *J. Pharm. Sci.* **71**, 1323 (1982).

forces.[16,17] Furthermore, they tend to bind to hydrophobic surfaces via the Fc portion, leaving the Fab immunoreactive site free to interact with antigen.[18] Total saturation of the particle surface seems to be important to prevent denaturation of the antibody.[19,20]

Monoclonal antibodies have been nonspecifically adsorbed onto the surface of poly(hexyl cyanoacrylate) nanoparticles.[21] The nanoparticles [~10^{13} particles in phosphate-buffered saline with 0.02% (w/v) sodium azide] have been coated by overnight incubation at 4° with an excess of anti-osteogenic sarcoma monoclonal antibody (1 mg/ml). Also present in the absorption mixture was fluorescein-conjugated bovine serum albumin (0.5 mg/ml) to act as a label for the nanoparticles. Using gel filtration on Sepharose CL-4B, the coated particles were separated from the unadsorbed proteins as the exclusion peak. The capacity of the antibody-coated nanoparticles to bind specifically to appropriate target cells *in vitro* was demonstrated using flow cytofluorimetric techniques.

Although physical washing of the nanoparticles would appear to indicate that the antibody is strongly bound to the surface of the particle, competitive displacement of adsorbed proteins has been shown to occur in other systems,[22] which may impose limitations on the usefulness of this technique for *in vivo* targeting. Indeed we have found that in the presence of serum, the antibody-coated nanoparticles fail to bind to their target cells *in vitro* or to localize *in vivo*.[21,23]

Specific Adsorption

Antibody can be adsorbed noncovalently onto the surface of microspheres by means of a ligand which interacts specifically with the intact or modified antibody.

Protein A. Staphylococcus aureus protein A (MW 42,000) has the capacity to bind with high affinity, specifically to the Fc portion of the

[16] C. F. Van Oss and J. H. Singer, *RES, J. Reticuloendothel. Soc.* **3**, 29 (1966).

[17] P. Bagchi and S. M. Birnbaum, *J. Colloid Interface Sci.* **83**, 460 (1981).

[18] C. F. Van Oss, C. F. Gillman, and A. W. Neumann, "Phagocytic Engulfment and Cell Adhesiveness," p. 1. Dekker, New York, 1975.

[19] S. Kochwa, M. Brownell, R. E. Rosenfield, and L. R. Wasserman, *J. Immunol.* **5**, 981 (1967).

[20] B. W. Morrissey and C. C. Han, *J. Colloid Interface Sci.* **65**, 423 (1977).

[21] L. Illum, P. D. E. Jones, J. Kreuter, R. W. Baldwin, and S. S. Davis, *Int. J. Pharm.* **17**, 65 (1983).

[22] J. L. Brash and V. J. Davidson, *Thromb. Res.* **9**, 249 (1976).

[23] L. Illum, P. D. E. Jones, and S. S. Davis, *in* "Microspheres and Drug Therapy" (S. S. Davis, L. Illum, J. G. McVie, and E. Tomlinson, eds.), p. 353. Elsevier North-Holland Biomedical Press, Amsterdam, 1984.

METHODS OF ATTACHMENT OF MONOCLONAL ANTIBODIES TO MICROSPHERES, REACTIVE GROUPS, AND REACTIONS INVOLVED

Functional group required on the surface of the microspheres	Principle of the attachment of the monoclonal antibody	Antibody–microsphere systems	Examples references	Comments
"None" (M)	Direct adsorption	Poly(alkyl cyanoacrylate) nanoparticles–antibody	21, 27	Possible competitive displacement by blood proteins. Simple procedure, no chemical agents. Hydrophobic surface required
	Indirect adsorption via protein A	Poly(alkyl cyanoacrylate) nanoparticles–antibody	27	Simple procedure. No chemical agents. Spacer effect. Efficient method
		Human serum albumin microspheres–antibody	25, 26	
	Indirect adsorption via avidin–biotin	Methacrylate microspheres–antibody	28	Requires covalent conjugation of avidin to microspheres and biotin to antibody
Aldehyde (M)–CHO	Direct coupling	Polyacrolein microspheres–antibody	13, 14	Simple procedure. Efficient method
		Polyaldehyde microspheres–antibody	12, 30	
		Human serum albumin microspheres–antibody	15	

Functional group	Coupling method	Product	Ref.	Comments
Carboxylic $\text{(M)}-COOH$	Coupling via carbodiimide	Methacrylate microspheres–antibody	9, 11	Simple procedure. Inter- and intra-molecular cross-linking of antibody can occur. Quite efficient
	Coupling via EEDQ	CM-Sephadex polymer–enzyme CM-Cellulose polymer–enzyme	37 37	
	Coupling via Woodward's reagent K	Polyacrylyl polymer–enzyme	38	Degree of binding easy to control
Hydroxyl $\text{(M)}-OH$	Coupling via cyanogen bromide	Methacrylate microspheres–antibody	35	Very toxic material. The coupling reaction is highly dependent on pH. Not very efficient
Amino $\text{(M)}-NH_2$	Coupling via glutaraldehyde	Ethylenediamine-derivatized methacrylate microspheres–antibody	11, 28	Efficient. Derivatization of microsphere often required
	Coupling via SPDP	Liposomes–antibody	39	Complicated reactions
Dextran $\text{(M)}-O-CH_2$	Coupling with a dextran bridge	Dextran–antibody Poly(alkyl cyanoacrylate)–dextran–nanoparticles–antibody	40 23	

majority of subclasses of IgG.[24] This has been exploited by several investigators[25-27] who have incorporated protein A into the polymerization mixture during microsphere manufacture on the assumption that protein A would coat the particle surface. Due to the high binding affinity of protein A to the Fc portion of IgGs, antibodies can be coupled to the microspheres after very brief incubation at 37° without the use of chemical coupling agents which might reduce antigen binding activity.

Protein A has been used for linking monoclonal antibody against HLA-BW6 to human serum albumin microspheres.[26] Protein A was included during the polymerization process. A suspension of 50 mg HSA, 18 mg Fe_3O_4, and 12 mg protein A was prepared in 1 ml distilled water, added to 40 ml cottonseed oil, and cooled to 4°. The cold homogenate was immediately added dropwise to 100 ml cottonseed oil at 125° and the albumin microspheres condensed and hardened in a 10-min period. Microspheres (0.5 mg) were coated with 1 mg anti-HLA-BW6 antibody by incubation at 37° for 40 min. The immunomicrospheres were then washed once with saline–Tween 80 and three times with RPMI + 5% FCS.

An important consideration in the use of this system is that the complex of protein A with certain subclasses of IgG can activate the complement system. This may limit its usefulness *in vivo*.

Biotin–Avidin. Avidin is a 68,000 MW glycoprotein with four high-affinity binding sites for biotin. When biotin is conjugated to proteins, avidin will form essentially irreversible complexes with these biotinylated proteins.

This feature has been exploited by Kaplan *et al.*[28] to link biotinylated monoclonal antibody to polymethacrylate microspheres bearing acetylated avidin. Acetylated avidin, being less basic than avidin, showed lower nonspecific binding to cells while retaining its capacity to bind biotin. The acetylated avidin was coupled to the microspheres by a two-step glutaraldehyde procedure (see below). The monoclonal antibody was biotinylated following precipitation with 40% saturated ammonium sulfate. The antibody was redissolved in phosphate buffered saline. $NaIO_4$ was added to 0.2 mg (0.2 ml) of the immunoglobulin to a final concentration of 1 mM. After 30 min at 0° the process was terminated by passing

[24] J. J. Langone, *Adv. Immunol.* **32,** 158 (1982).

[25] K. J. Widder, A. E. Senyei, H. Ovadia, and P. Y. Paterson, *J. Pharm. Sci.* **70,** 387 (1981).

[26] J. Kandzia, M. J. D. Anderson, and W. Müller-Ruchholtz, *J. Cancer Res. Clin. Oncol.* **101,** 165 (1981).

[27] P. Couvreur and J. Aubry, *in* "Topics in Pharmaceutical Sciences" (D. D. Breimer, ed.), p. 305. Vol. II. Elsevier Biomedical Press, Amsterdam, 1984.

[28] M. R. Kaplan, E. Calef, T. Bercovici, and C. Gitler, *Biochim. Biophys. Acta* **728,** 112 (1983).

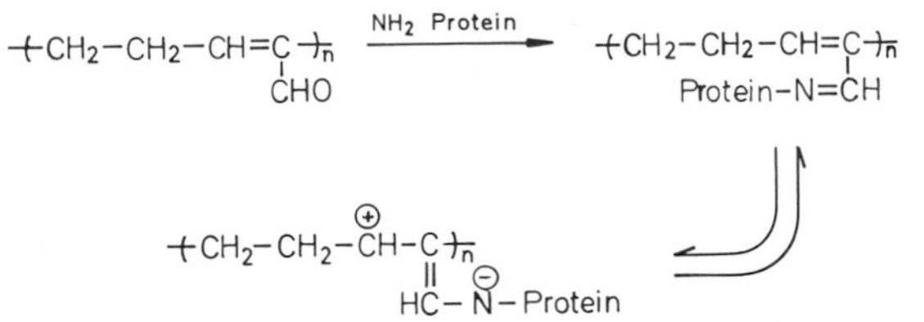

$$+CH_2-CH_2-CH=C\frac{}{}\Big)_n \quad \xrightarrow{\text{NH}_2 \text{ Protein}} \quad +CH_2-CH_2-CH=C\frac{}{}\Big)_n$$

with CHO below the first structure and Protein—N=CH below the second.

$$+CH_2-CH_2-\overset{\oplus}{CH}-C\frac{}{}\Big)_n$$

with HC— $\overset{\ominus}{N}$ —Protein below.

SCHEME I. Reaction mechanism for the coupling of protein to functional aldehyde groups via polyglutaraldehyde.

down a Sepharose G-25 column. To the pooled protein peak, 1.5 mg biotin hydrazide was added and reacted for 30 min at room temperature. Then 1 mM NaCNBH$_3$ was added and the reduction carried out for 16 hr at 4° until the reaction was terminated by passing down a Sephadex G-25 column.

Direct Coupling of Monoclonal Antibodies to Microspheres

Direct linking between antibodies and microspheres can, for example, take place if functional aldehyde groups are available on the surface of the microspheres (e.g., polyacrolein, polyglutaraldehyde).

In an early study by Rembaum et al.[29] of polyglutaraldehyde-coated microspheres, the polymer was described as having relatively long chains extending from the microspheres into the surrounding aqueous medium, thereby facilitating the heterogeneous reaction with protein. It was shown that a single polyglutaraldehyde molecule (chain) was able to react with several protein molecules.

The polymer contains repeating units of conjugated aldehyde groups which form stable bonds with primary amino groups in the protein, e.g., lysine residues (Schiff base reaction) (Scheme I).

The normal procedure for the coupling process between polyglutaraldehyde and protein has been described by Margel.[12] Polyglutaraldehyde microspheres (1 mg in 0.2 ml of phosphate-buffered saline, pH 7.4) were shaken for 3 hr at 4° with a suitable concentration (~100 μg) of human IgG. Unbound antibody (IgG) was then separated from the microspheres of size 0.6 μm by centrifugation four times at 750 g for 20 min, as described below in the section on purification. The remaining free aldehyde groups of the microsphere–antibody conjugate were quenched with 1% glycine solution for 3 hr at room temperature.

To determine the extent of reaction between the microspheres and the antibody, Margel et al.[30] and Rembaum et al.[29] used fluorescent human

[29] A. Rembaum, S. Margel, and J. Levy, J. Immunol. Methods 24, 239 (1978).
[30] S. Margel, S. Zisblatt, and A. Rembaum, J. Immunol. Methods 28, 341 (1979).

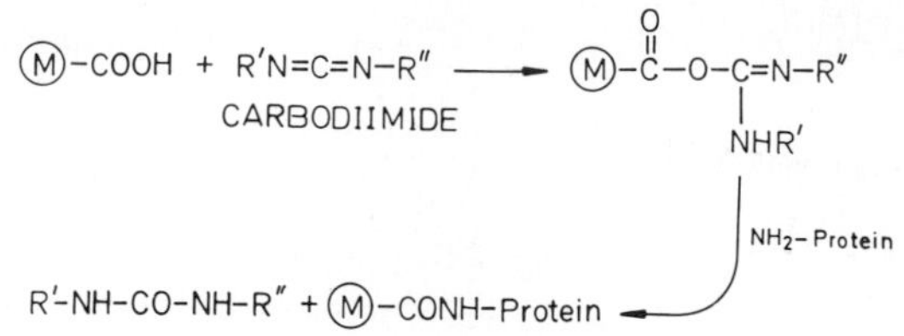

SCHEME II. Reaction mechanism for the coupling of protein to functional carboxyl groups via carbodiimide.

IgG as a model. Following the reaction, the suspension containing free and microsphere-bound fluorescent IgG was centrifuged. The decrease in fluorescence intensity in the supernatant was taken as a measure of degree of covalent binding of IgG to the microspheres.

The rate of the reaction between the aldehyde group and the primary amino group is influenced by pH. At a lower pH the reactivity of the aldehyde group toward nucleophiles (here NH_2) will be enhanced but at the same time the NH_2 will be converted to the unreactive species NH_3^+. Hence the rate of addition will show a maximum at a moderately acid pH, falling off sharply at each side.[31]

Coupling of Monoclonal Antibodies to Microspheres via a Reagent

When the microspheres of choice do not carry groups which can link directly to monoclonal antibodies it is possible to obtain antibody attachment by means of different methods such as the carbodiimide method, the cyanogen bromide method, and the glutaraldehyde method.

Carbodiimide Method. A water-soluble carbodiimide derivative can be used to activate carboxyl groups present on the surface of the microspheres and thereby couple the amino group on the antibody molecule to the carboxyl group through an amide linkage[32] (Scheme II). Alternatively, an amino group on the microsphere can be coupled to a carboxylic acid group on the antibody via this method.

This coupling procedure has been used for latex spheres carrying carboxyl groups and goat anti-rabbit IgG antibodies.[9] Ten milligrams of 1-ethyl-3-(3-dimethylaminopropyl)carbodiimide (EDC) were added to 50 mg of microspheres and 1 mg of antibody in 2 ml of 0.1 M NaCl at pH 7.0 and at 4°. After 2 hr the coupling reaction was stopped by the addition of excess (0.2 ml) of 0.1 M glycine ($H_2N–CH_2–COOH$) solution, pH 8.0. When EDC is used, the resulting biproduct is a derivative of "urea" and

[31] A. Rembaum, U.S. Patent 4,267,234 (1981).
[32] T. L. Goodfriend, L. Levine, and G. D. Fasman, *Science* **144,** 1344 (1964).

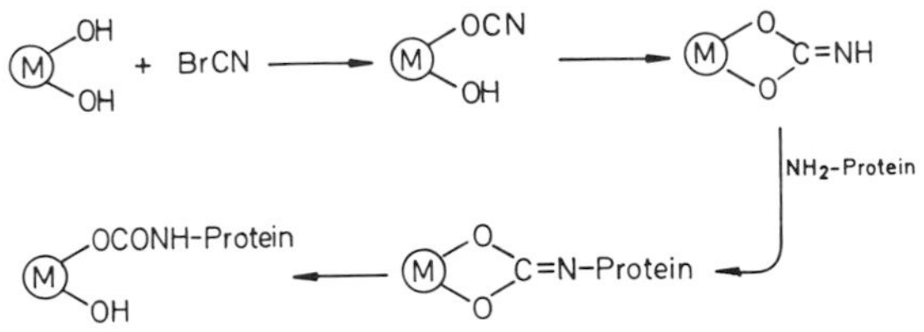

SCHEME III. Reaction mechanism for the coupling of protein to functional hydroxyl groups via cyanogen bromide.

is water soluble. The degree of coupling is dependent upon the density of the reactive groups, i.e., carboxyl groups on the polymer. Sufficient amounts of reactive groups should be present to provide adequate coupling of antibody.

When the reaction was carried out at pH 7 for 1 hr at 4°, a linear dependence was found between the concentrations of IgG and the degree of binding to the microspheres. Six antibody molecules were found per latex sphere and at a IgG concentration of 1.4 mg/ml and a carbodiimide concentration of 2 mg/ml.[9]

It is important that the pH of the reaction medium is kept at neutral pH. At a too low pH the amine is protonated and does not react readily, whereas at a higher pH the carbodiimide may decompose. The coupling procedure is simple to perform and has the advantage of occurring under very mild conditions and of conjugating the antibody directly to the microsphere without interposing additional groups between the two. However, a drawback of the method is that as proteins contain both amino groups and carboxyl groups it will be possible that both intra- and intermolecular cross-linking in the antibody molecule can occur.[33]

Cyanogen Bromide Method. In this reaction method hydroxyl groups present on the surface of the microspheres are first activated with cyanogen bromide at alkaline pH. They are then coupled to amino groups on the antibody at pH values between 7 and 10 using the established procedure for the similar conjugation of cross-linked dextrans with proteins[34] (Scheme III).

The coupling procedure has been described for latex spheres carrying hydroxyl groups and antibody.[35] Latex spheres (20–50 mg/ml) adjusted to pH 10.5 with NaOH were activated at 25° with 10 mg of CNBr per millili-

[33] T. Ghose, A. H. Blair, K. Vaughan, and P. Kulkarni, *in* "Targeted Drugs" (E. P. Goldberg, ed.), p. 1. Wiley, New York, 1983.
[34] R. Axén, J. Poráth, and S. Ernback, *Nature (London)* **214**, 1302 (1967).
[35] S. P. S. Yen, A. Rembaum, R. W. Molday, and W. J. Dreyer, "Emulsion Polymerization," ACS Symp. Ser., Vol. 24, p. 236. Am. Chem. Soc., Washington, D.C., 1979.

ter of suspension. The reaction mixture was maintained between pH 10 and 11 by the slow addition of 2 N NaOH. After 15 min, the suspension was diluted with an equal volume of cold 0.1 M borate buffer (pH 8.5) and equilibrated at 4°. A solution of purified antibody in 0.1 M borate buffer (pH 8.5) was added to the latex suspension to give a final protein concentration of 0.5 mg/ml. The reaction was allowed to proceed at 4° for at least 4 hr after which it was terminated by the addition of an equal volume of 0.1 M glycine buffer at pH 8.5.

As with the carbodiimide method, a linear relationship was found for the cyanogen bromide method between the concentration of IgG and the amount of IgG per latex sphere for a reaction at pH 7.0. Due in part to the low efficiency of the reaction at neutral pH, the amount of antibody linked to the spheres were lower than for the carbodiimide and the glutaraldehyde methods.[9]

A drawback of the method is that the cyanogen bromide is very toxic and that the coupling reaction is highly dependent on the pH.

Glutaraldehyde Method. The glutaraldehyde reaction method can be used to link amino groups present on the microsphere to amino groups on the antibody molecule. Functional amino groups on the surface of microspheres are not as common as, for example, carboxyl or hydroxyl groups unless the microsphere is polymerized in the presence of an amine that would provide the surface grouping. Amino groups in the form of diamino compounds can also be derivatized onto the surface of the microsphere by means of the carbodiimide method, before the reaction with glutaraldehyde takes place.[35] The linking of the diamino compound (e.g., diaminoheptane) to the microsphere surface with the amino acid group of the compound protruding into the surrounding medium also works as a spacer, facilitating the approach of, and subsequent reaction with, the glutaraldehyde molecule.

The mechanism of the glutaraldehyde reaction is as yet not fully understood. Richards and Knowles[36] proposed a mechanism in which the reaction with amino groups is believed to occur through the reactive groups on the α,β-unsaturated aldehyde polymers that are found to be present in the aqueous solutions of glutaraldehyde, to give stable Michael-type adducts (Scheme IV).

The glutaraldehyde coupling reaction has been used for linking protein to methyl methacrylate microspheres (50 nm in diameter) derivatized with ethylenediamine in a two-step linking procedure.[11,28] Twenty microliters of 2.5% glutaraldehyde solution was added to 0.5 ml of a 7.5 mg/ml microsphere suspension. The suspension was stirred for 30 min and then

[36] F. M. Richards and J. R. Knowles, *J. Mol. Biol.* **37,** 231 (1968).

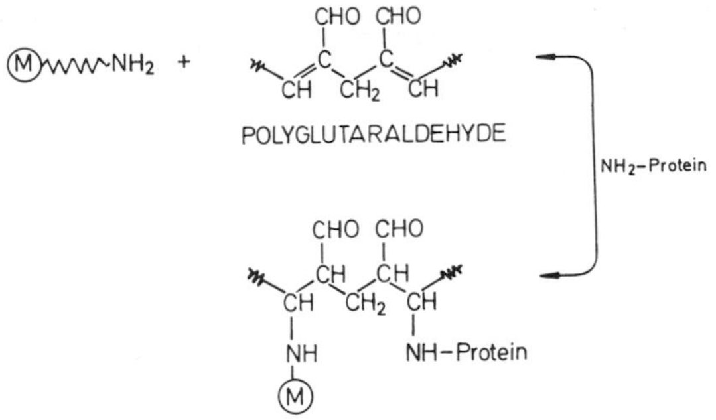

SCHEME IV. Reaction mechanism for the coupling of protein to functional amino groups via glutaraldehyde.

passed on to a Sephadex G-25 column equilibrated with a 10 mM phosphate buffer, pH 7.4. The eluted microspheres were mixed with 0.5–1 mg protein and left for 150 min. Glycine and cyanoborohydride were then added to give final concentrations of 0.2 M and 1 mM, respectively, and the reaction continued for 16 hr at 4°. Free protein was separated from microsphere bound protein by passing the mixture through a Sepharose CL-4B column equilibrated with 0.1% gelatin in PBS.

The advantage of the two-step reaction procedure over the one-step method, in which the coupling of antibody to microspheres takes place in the presence of excess glutaraldehyde, is that it is possible to avoid the intra- and intermolecular cross-linking of the antibody. In the described two-step reaction method the antibodies are first added to activated microspheres after excess glutaraldehyde has been removed. Compared to the cyanogen bromide and carbodiimide methods, the glutaraldehyde reaction method was found by Molday et al.[9] to be the most effective in binding antibody to copolymer acrylic latex spheres.

EEDQ Method. Another method of interest for coupling antibodies to microspheres is the *N*-ethoxycarbonyl-2-ethoxy-1,2-dihydroquinoline (EEDQ) method described by Sundaram.[37] In this two-step reaction method the carboxylic groups on the microspheres are first activated with EEDQ to form a mixed carbonic anhydride which can then react with amino groups on a protein (Scheme V).

The method has been used for coupling enzymes (i.e., urease V, trypsin) to CM-cellulose and Sephadex CM-50.[37] EEDQ (0.6–1.0 g) was dissolved in 10 ml of 10% acetone and added slowly to 1 g of polymer at pH 4 and 0° within 20–30 min. Excess of EEDQ was then destroyed by

[37] P. V. Sundaram, *Biochem. Biophys. Res. Commun.* **61,** 717 (1974).

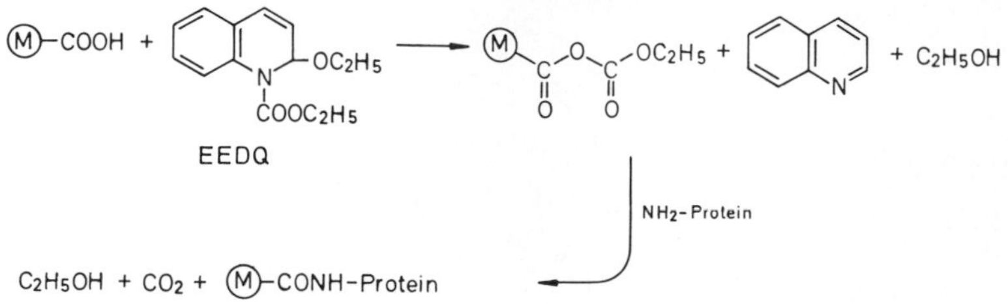

SCHEME V. Reaction mechanism for the coupling of protein to functional carboxyl groups via EEDQ.

quickly lowering the pH to 1.0 and maintaining it there for 90 sec using concentrated HCl. After washing with ice-cold 0.1 M HCl the polymer was added to the protein solution (pH 8.0) and left at 0° for half an hour to react. Finally the mixture was washed with phosphate buffer, pH 8.0. The noncovalently adsorbed protein was removed by washing with 1 M NaCl, followed by washing with ice-cold deionized water. The amount of protein bound to the Sephadex-CM-50 polymer was about 20% (w/w) and the specific activity retained by the protein after coupling was found to be in the order of 49%.

Woodward's Reagent K (WRK) Method. In this method, reported by Patel et al.[38] for coupling proteins to polymers, carboxyl groups on the polymer are reacted with N-ethyl-5-phenylisoxazolium 3′-sulfonate (Woodward's reagent K) to form active ester derivatives that can react with protein amines to form amide bonds (Scheme VI).

Identical procedures were used[38] to couple α-chymotrypsin to each of both poly(acrylic acid), carboxymethylcellulose, and poly(L-glutamic acid) and the method should be applicable to other polymers with carboxyl groups. One gram of 15% poly(acrylic acid) solution was diluted to 10 ml with water and cooled to 5°; 1.7 ml of 1 N NaOH and a suitable amount (10 mg) of Woodward's reagent K dissolved in water were then added and stirred for a few minutes. Then 300 mg of α-chymotrypsin in 20 ml of cold water was added and stirring continued overnight in the cold.

The activity per unit protein decreased as the quantity of the WRK in the reaction was increased, probably due to disruption of the tertiary structure of the protein through cross-linking with the polymer.[38] The advantage of this method is that the number of reactive groups on the polymer is strictly controlled by the amount of WRK used in the reaction process. Furthermore, the by-product and unreacted starting material can easily be removed by centrifugation and washing with cold water (5°).

[38] R. P. Patel, D. V. Lopiekes, S. P. Brown, and S. Price, *Biopolymers* **5,** 577 (1967).

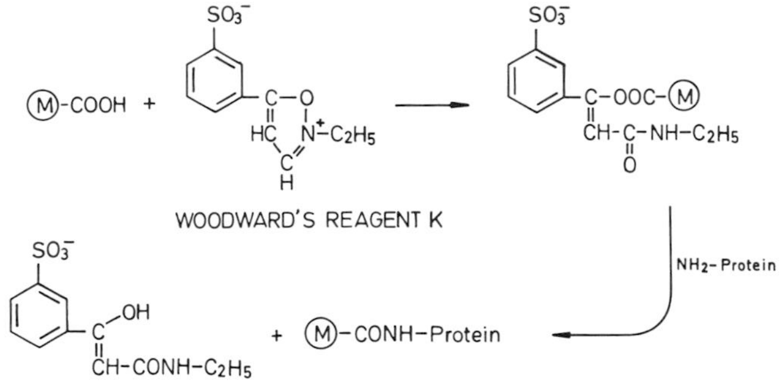

SCHEME VI. Reaction mechanism for the coupling of protein to functional carboxyl groups via Woodward's reagent K.

Application of Other Coupling Methods to Microspheres

SPDP Method. N-Hydroxysuccinimidyl 3-(2-pyridyldithio)propionate (SPDP) has been used to couple monoclonal antibody to liposomes.[39] SPDP can react with molecules containing primary amino groups, e.g., antibodies, and introduce dithiopyridyl groups through the formation of amide bonds. Protein-bound dithiopyridine (DTP) can easily be activated by reduction with dithiothreitol under mild conditions (acidic pH). Thiolated proteins can then react with other DTP-substituted molecules and produce covalently coupled products (Scheme VII).

The coupling of antibody to SPDP was achieved as follows.[39] The antibody in phosphate buffer (0.1 M, pH 7.5, 0.1 M NaCl) was incubated with SPDP (5–20 mol/mol protein) for 30 min at room temperature and transferred to acetate buffer (0.1 M, pH 4.5, 0.145 M NaCl) by gel filtration (Sephadex PD-10). The DTP–protein was then incubated with dithiothreitol (50 mM final concentration) for 20 min at room temperature and passed through a Sephadex PD-10 column in buffer (10 mM HEPES, 145 mM NaCl, pH 7.4). It should be possible in the same way to react amino groups on a microsphere with SPDP to produce DTP-substituted molecules and then to couple this covalently with the derivatized antibody.

Coupling with a Dextran Bridge. It is possible to link monoclonal antibodies to anticancer agents using dextran as a bridge. In the binding procedure dextran is oxidized to give functional aldehyde groups that can react with amino groups on the antibody molecule and also on the anticancer agent through Schiff base linkages.[40] This method is now being evalu-

[39] J. Barbet, P. Machy, and L. D. Leserman, *J. Supramol. Struct. Cell. Biochem.* **16**, 243 (1981).
[40] R. Arnon and M. Sela, *Immunol. Rev.* **62**, 5 (1982).

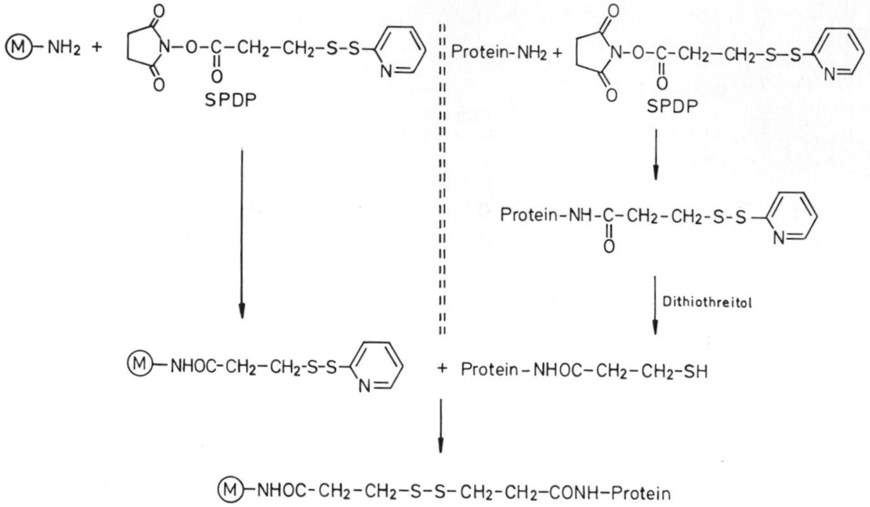

SCHEME VII. Reaction mechanism for the coupling of protein to functional amino groups via SPDP.

ated in our laboratory to link monoclonal antibodies to poly(alkyl cyano-acrylate) nanoparticles polymerized with dextran as a stabilizer (Scheme VIII).

The glucosidic residues in the dextran sticking out from the surface of the particles are oxidized by excess amounts of sodium periodate (2 hr, room temperature, in the dark) to give reactive aldehyde groups. The excess of periodate is removed by either dialysis or centrifugation and washing. The product can then be incubated with antibody at a suitable

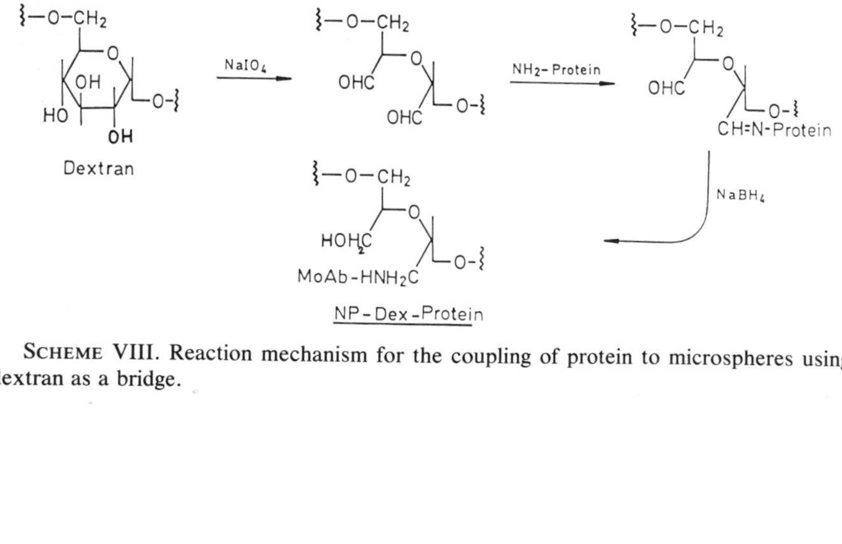

SCHEME VIII. Reaction mechanism for the coupling of protein to microspheres using dextran as a bridge.

pH (5–8) in the dark overnight at 4°, forming Schiff base linkages. The linkages are reduced with sodium borohydride to form more stable bonds.

Purification and Characterization of Monoclonal
Antibody–Microsphere Conjugates

Once the monoclonal antibody has been attached onto the microsphere it is important to separate the antibody–microsphere conjugates from free antibody and other reaction reagents. Also, the efficiency of the binding process and the immunochemical activity of the immunomicrospheres should be determined.

Separation Techniques

Free antibodies are most conveniently separated from microsphere-bound antibodies by means of physical techniques such as differential sedimentation, centrifugation, centrifugation using density gradients, and gel filtration.

The differential sedimentation technique can be used for larger microspheres. The microspheres are centrifuged under appropriate conditions of gravitational force and time. The centrifugation can be repeated with intermediate washings. For example, 0.6-μm polyglutaraldehyde microspheres were centrifuged four times at 750 g for 20 min.[12] For smaller microspheres, centrifugation on a discontinuous gradient can be used as a way of separating free and bound antibody. In this method, the microsphere–antibody mixture is layered on a discontinuous gradient, consisting of a 10–20% (w/w) sucrose solution on top of a high density 58% (w/w) sucrose solution. The microsphere–antibody conjugates will form a narrow band at the interface between the two sucrose solutions after high-speed centrifugation. The size of the microsphere and the concentration of the sucrose solution in the upper layer, i.e., viscosity, will determine the time required for sedimentation. As an example, methacrylate microspheres of 110 nm sedimented through an upper layer of 10% sucrose at 10^5 g in less than 1 hr. The unbound antibody stayed near the top of the tube. It is normally useful to repeat the centrifugation two to three times in order to remove all the free antibody.[35]

For small microspheres the free antibody can be separated by filtering the mixture through a Sepharose gel and monitoring the separation spectrophotometrically. Antibody-coated poly(hexyl cyanoacrylate) nanoparticles (170 nm) were separated from the adsorption mixture by gel filtration on Sepharose CL-4B columns (17 × 1.5 cm) equilibrated with PBS containing 0.02% (w/v) NaN$_3$. The elution of nanoparticles and antibody from the column was monitored continuously by UV absorption. Two distinct peaks were obtained, the first of these corresponding to the nanoparticles and the second to the free antibody.

The three techniques can also be used for assaying the efficiency of the binding of antibody to the microspheres by using radiolabeled or fluorescent labeled antibody.

Immunoreactivity of Antibody–Microsphere Conjugates

The immunoreactivity of antibody–microsphere conjugates to target cells can be evaluated *in vitro* by using fluorescent labeled microspheres and flow cytofluorimetry[21] or by using radioisotopic techniques in which the isotope is incorporated into the microsphere.[27]

Acknowledgments

The authors would like to thank S. Douglas and S. S. Davis for helpful assistance.

[7] Nylon-Encapsulated Pharmaceuticals

By JAMES W. MCGINITY and GEORGE W. CUFF

Introduction

The commercialization of microencapsulated drug products during the past few years suggests a growing interest in this field of technology as a viable method of drug delivery. The properties of polyamides for use as membrane materials in drug or enzyme containing microcapsules have been the subject of several reports.[1-16] Polyamides are prepared by an

[1] T. M. S. Chang, *Science* 146, 524 (1964).
[2] T. M. S. Chang, F. C. MacIntosh, and S. G. Mason, *Can. J. Physiol. Pharmacol.* 44, 115 (1966).
[3] T. M. S. Chang, *Science* 3, 62 (1967).
[4] T. M. S. Chang and M. J. Poznansky, *J. Biomed. Mater. Res.* 2, 187 (1968).
[5] T. M. S. Chang, *Nature (London)* 229, 117 (1971).
[6] L. A. Luzzi, M. A. Zoglio, and H. V. Maulding, *J. Pharm. Sci.* 59, 338 (1970).
[7] A. W. Jenkins and A. T. Florence, *J. Pharm. Pharmacol.* 25, 57P (1973).
[8] J. W. McGinity, A. B. Combs, and A. N. Martin, *J. Pharm. Sci.* 64, 889 (1975).
[9] A. F. Florence and A. W. Jenkins, *in* "Microencapsulation" (J. R. Nixon, ed.), p. 39. Dekker, New York, 1976.
[10] M. D. DeGennaro, B. B. Thompson, and L. A. Luzzi, *in* "Controlled Release Polymeric Formulations" (D. R. Paul and F. F. Harris, eds.), p. 195. Am. Chem. Soc., Washington, D.C., 1976.

interfacial polymerization reaction which may be defined as a reaction that takes place at the interface between two or more immiscible liquids. Polymerization should take place wherever the reactive precursors encounter each other. Theoretically, the site of the reaction may be located exactly at the liquid interface, or in either phase. Evidence shows that most water-insensitive polymers form and grow on the organic solvent side of the interface.[17] Although some reactions do occur primarily at the interface, the polymer accumulation occurs in the organic phase. No examples of polymerization in the aqueous phase are known.

The Schotten–Baumann reaction of an acid halide and a compound containing an active hydrogen ($-NH_2$, $-OH$, $-SH$, etc.) provides the basis for numerous interfacial polycondensation reactions. Several reports[17–22] have described the development of numerous interfacial polycondensation reactions at immiscible interfaces as well as the effects of important variables on both stirred and unstirred systems.

Production and Properties of Nylon Microcapsules

Nylon 610, the most popular and perhaps the most extensively studied polyamide, is prepared by the interfacial polymerization reaction between 1,6-hexamethylenediamine (HMD) and sebacoyl chloride (SC) as shown in the following equation[18]:

$$H_2N(CH_2)_6NH_2 + Cl\overset{O}{\overset{\|}{C}}(CH_2)_8\overset{O}{\overset{\|}{C}}Cl \longrightarrow \left[HN(CH_2)_6NH\overset{O}{\overset{\|}{C}}(CH_2)_8\overset{O}{\overset{\|}{C}} \right]_n + HCl$$

HMD SC Nylon 610

[11] L. A. Luzzi, in "Microencapsulation" (J. R. Nixon, ed.), p. 193. Dekker, New York, 1976.
[12] J. W. McGinity, A. B. Combs, and A. Martin, in "Microencapsulation—New Techniques and Applications" (T. Kondo, ed.), p. 57. Techno Inc., Tokyo, 1979.
[13] T. M. S. Chang, in "Biomedical Applications of Immobilized Enzymes and Proteins" (T. M. S. Chang, ed.), Vol. 1, p. 93. Plenum, New York, 1977.
[14] T. M. S. Chang, in "Biomedical Applications of Immobilized Enzymes and Proteins" (T. M. S. Chang, ed.), Vol. 1, p. 281. Plenum, New York, 1977.
[15] J. W. McGinity, A. Martin, G. W. Cuff, and A. B. Combs, J. Pharm. Sci. 70, 372 (1981).
[16] G. W. Cuff and J. W. McGinity, U.S. Patent pending.
[17] P. W. Morgan and S. L. Kwolek, J. Polym. Sci. 40, 299 (1959).
[18] E. Wittbecker and P. Morgan, J. Polym. Sci. 40, 289 (1959).
[19] R. Beaman, P. Morgan, C. Koller, and E. Wittbecker, J. Polym. Sci. 40, 329 (1959).
[20] M. Katz, J. Polym. Sci. 40, 337 (1959).
[21] V. Shashoua and W. Eareckson, J. Polym. Sci. 40, 343 (1959).
[22] P. Morgan and S. Kwolek, J. Polym. Sci. 62, 33 (1963).

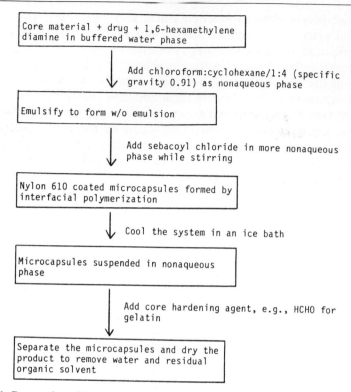

FIG. 1. Preparation of drug-containing nylon 610 microcapsules by the interfacial polymerization process.

To prepare microcapsules by interfacial polymerization, the aqueous phase to be encapsulated should contain the water-soluble monomer, HMD. This phase is then emulsified in an organic phase. Additional organic phase containing a second monomer (SC) is then added. The reaction occurs at or near the interface forming a polymeric membrane. A general scheme outlining this process is seen in Fig. 1. Important variables of these systems include the choice and concentration of reactants, composition and volume of the immisicible phases, stirring rate, and the nature of material to be encapsulated. For example, to prepare a polyamide, an acid halide is chosen that is nearly insoluble in water to prevent its hydrolysis. The diamine chosen must be water soluble but exhibit an appreciable partitioning into the organic phase. Therefore, the choice of diamine and organic phase must complement each other. The density of the organic solvent determines the microcapsule flotation properties after formation. One should take advantage of this in choosing a recovery process for the microcapsules. The chloroform/cyclohexane (1 : 4, v/v) system has a density close to one. The choice of organic phase affects the reactivity of the reactants and hence the reaction rate. Unhindered polycondensation reactions occur very rapidly, having rate constants from 10^2 to 10^6 liter/mol/sec. The organic phase also affects the solubility and swelling of the polymer and thereby the extent of polymerization and molecular weight of the polymer. By choosing an organic solvent that interacts favorably with the polymer, one can extend the polymerization period before precipitation, thereby enhancing the degree of polymerization.[18] The organic solvent may also affect the partitioning of material to be encapsulated.[15]

The rate at which the initial emulsion is stirred determines the droplet size, hence the microcapsule size. Optimal concentrations and ratios of reactants exist for different systems and depend on the nature of the system and its intended application. The organic phase volume must be sufficient to permit polymer swelling to occur during stirring. A general rule is to use a total volume of organic phase 10 times that of the aqueous phase.[18] The material to be encapsulated should not react with either monomer or the reaction by-product and should not diffuse into the organic solvent. Liquids, dispersions of solids, or oil–water emulsions may be encapsulated in the aqueous phase. The aqueous phase also serves to accept the polycondensation by-product.

An attractive feature of the nylon polyamides is that slight monomer structural changes such as carbon chain length, addition of side chains, or cross-linking agents may alter the properties of the microcapsule wall. Nylon was the name given by the DuPont Corporation to the linear polyamides first developed in their laboratories by Carothers.[23] The first nylons were aliphatic and differed only in the ratio and arrangement of the methylene and amide groups. Thus, a nomenclature was developed to readily distinguish these polymers.[24,25] The name Nylon was followed by either one or two numbers. The first number, n, indicates the number of carbon atoms in the diamine. The second number, m, indicates the number of carbon atoms in the dibasic acid. A single number indicates the number of carbon atoms in the amino acid or lactam which is polymerized to give the nylon. However, with the development of various copolymers and more advanced and sophisticated types of polyamides[26] the recom-

[23] W. H. Carothers, in "Collected Papers of W. H. Carothers" (H. Mark and G. Whitley, eds.). Wiley (Interscience), New York, 1940.
[24] M. Kohan, in "Nylon Plastics" (M. Kohan, ed.), p. 5. Wiley (Interscience), New York, 1973.
[25] H. Allcock and F. Lampe, "Contemporary Polymer Chemistry," p. 563. Prentice-Hall, Englewood Cliffs, New Jersey, 1981.
[26] R. Lenk, J. Polym. Sci. 13, 355 (1978).

mended names of the International Union of Pure and Applied Chemistry[27] must be used, even though these are more burdensome. Not all polyamides are nylons, but the distinction is not easily made.[28] The recent development of new polyamides having special properties has created even more possibilities.

In 1964, Chang[29] reported the microencapsulation by interfacial polymerization of an aqueous erythrocyte hemolysate using a nylon 610 membrane. The aqueous phase was a mixture of 50% hemolysate and 50% of 0.4 M HMD in a 0.45 M sodium carbonate–sodium bicarbonate buffer pH 9.8. This was emulsified in five times its volume of a chloroform/cyclohexane (1 : 4, v/v) mixture with 15% Span 85. The same volume of this organic solvent containing 0.018 M SC was then added and stirred for 3 min at 0°. The microcapsules were transferred to an aqueous phase by centrifuging and resuspending in a series of media including a 50% Tween 20 aqueous solution. Chang has published numerous reports[1–5,13,14,29] detailing the preparation, properties, and applications of these semipermeable microcapsules with emphasis on their biomedical application as artificial cells. Chang used a chloroform/cyclohexane (1 : 4) organic solvent because it did not readily denature proteins, because it had a specific gravity close to one, and because of the desirable nylon membrane properties attributed to the partitioning of HMD into this system.[2] The nylon 610 membrane was thin but strong enough for the desired purposes. Chang used an emulsification period of 1 min before adding the SC, then continued the stirring for another 3 min. He stated that the reaction time was important in determining the physical properties of the membrane prepared, and cited the work of Morgan and Kwolek.[17] The reaction was stopped by adding organic solvent to the mixture in order to dilute the reactant concentration, then centrifuged immediately.

Chang found that the microcapsule size was never uniform, and in any batch, varied over a fairly wide range. However, the mean diameter for a batch was successfully related to the stirring speed. At higher speeds, smaller diameters and less variation was observed. A "Virtis 45" homogenizer (Virtis Co.) produced microcapsules having mean diameters of approximately 2 μm. The edges and tip of the high shear homogenizer blade were blunted to prevent microcapsule destruction. At very high speeds, the microcapsules were irregularly shaped. Large microcapsules of 5 mm or more in diameter were prepared individually from diamine solution dropped from a needle or pipet into the unstirred SC solution.

[27] International Union of Pure and Applied Chemists, *ibid,* **8,** 257 (1952).
[28] J. Sprauer and J. Harrison, *in* "Nylon Plastics" (M. Kohen, ed.), p. 536. Wiley (Interscience), New York, 1973.
[29] T. M. S. Chang, *Science* **146,** 524 (1964).

Chang[3] noted that the membrane composition could be varied by doubling the concentration of SC and omitting HMD from the aqueous phase. Thus, even nylon membranes must have some cross-linked protein in their structures when the hemolysate is microencapsulated. A permanent negative surface charge was developed on the microcapsule by using a sulfonated diamine such as 4,4'-diamino-2,2'-diphenyldisulfonic acid.

Chang et al.[2] emphasized that a high concentration of protein in the aqueous phase maintained the integrity of the microcapsules. Aqueous solutions of protein prevented the collapse of the microcapsules, due to the osmotic pressure of the protein solution. The protein also decreased the surface tension, which facilitated preparation of the microcapsules. The authors reported that mammalian erythrocytes were lysed by the alkaline diamine phase and the chloroform/cyclohexane mixture. However, by first emulsifying the erythrocytes in a silicone fluid, they could be successfully microencapsulated.

The preliminary work of Chang and co-workers[2] with nylon microcapsules with an average diameter 80 μm suggested that the membrane thickness was about 0.2 μm as determined from electron microscope measurements. The membrane flexibility was microscopically observed by passing microcapsules through a thin capillary. They easily deformed to pass through the constriction, then rapidly regained their previous shape. Preliminary work on the permeability of the membranes demonstrated rapid crenation and folding of the membrane in hypertonic solution. This was reversed in a nonhypertonic environment. Later work by Chang and Poznansky[4] closely examined the permeability characteristics of the microcapsules. The diffusion of urea, creatinine, creatine, uric acid, glucose, sucrose, and salicylic acid into nylon microcapsules of 207 diameter was studied. Rapid diffusion resulted in a $t_{0.5}$ of 3.4 sec for urea and the studies enabled calculation of the permeability constants for various materials. The equivalent pore radius of the membrane was estimated to be 18 Å for the nylon 610 microcapsules. These early findings by Chang and co-workers[2-4] provided the basis for drug encapsulation studies involving nylon membranes.

Vandegaer and Meier[30] patented a method for the production of nylon 610 microcapsules (and other polymer types) by an interfacial polycondensation process. The aqueous phase containing the diamine was injected into an organic solvent which contained the SC. The unique aspect of this method was that the aqueous diamine was injected at the lower end of a column of organic solvent. As the microcapsules surfaced, the polycondensation reaction progressed. Completely coated microcapsules

[30] J. Vandegaer and F. Meier, U.S. Patent 3,464,926 (1964).

were recovered from the surface of the column or from a special outlet on the column. The same workers in 1971 patented methods to recover the microcapsules as nonaggregated particles.[31] This was accomplished by adding finely divided, solid colloidal agents such as talc, clays, activated attapulgite, magnesium aluminum silicate, alumina, calcium carbonate, and magnesia. The best results were obtained when the agents were dispersed in both phases.

Jay and Edwards[32] studied the mechanical properties of nylon 610 membranes from microcapsules prepared by the method of Chang et al.[2] They found that membrane stiffness was independent of the microcapsule diameter and that the yield point of nylon was constant. These results were interpreted to indicate a uniform thickness of nylon around the entire microcapsule. Jay and Sivertz[33] further supported this idea by studies of the membrane resistance. These workers suggested that because the reaction was not stopped by depletion of the reactants, but by dilution of the sebacoyl chloride, the membranes should be of uniform thickness. The investigators cleverly designed microelectrodes from Kimax glass capillaries of 1 mm o.d. These were filled with 3 M KCl and connected to the input of an amplifier by a silver–silver chloride wire in a closed 3 M KCl-filled electrode holder. The reference electrode was a low resistance ground in the Ringer's solution that microcapsules were stored in, prior to testing. Using a micromanipulator, the workers microscopically observed the tip of a suction pipet touching the surface of a microcapsule which had been measured using a calibrated eyepiece. The microcapsule was then impaled by the microelectrode. An oscilloscope was used for calibration and a recorder gave a hard copy of the data. The specific resistivity of the microcapsule membrane was estimated to be $5.6 \times 10^3 \ \Omega \ cm^{-2}$.

Shiba et al. in 1970[34] prepared nylon 610 microcapsules from an aqueous solution containing bovine serum albumin (BSA) by a method similar to that of Chang et al.[2] Piperazine or HMD was the diamine and p-phthaloyl dichloride or SC was the diacid chloride used in the preparation of polyamide combinations. In general, an increase in concentration of BSA produced smaller microcapsules. The partition coefficient of HMD was determined under conditions simulating the microencapsulation process, and found to be greatest at 0.1% BSA. There was no acid chloride added during the determination of the diamine partition coefficient. The diamine in the organic solvent was determined by titration with 0.01 N alcoholic HCl using a visual indicator. Shiba et al.[34] also reported that as

[31] J. Vandegaer and F. Meier, U.S. Patent 3,575,882 (1971).
[32] A. Jay and M. Edwards, Can. J. Physiol. Pharmacol. **46,** 731 (1968).
[33] A. Jay and K. Sivertz, J. Biomed. Mater. Res. **3,** 577 (1969).
[34] M. Shiba, S. Tomioka, M. Koishi, and T. Kondo, Chem. Pharm. Bull. **18,** 803 (1970).

the stirring speed increased, the transfer of diamine to the organic phase was found to increase in the absence of Span 85. These results reflect the fact that equilibrium for the partitioning was not attained during the stirring time employed. Shigera et al.[35] studied the effects of variables on microcapsule size for nylon 610 and other systems. They concluded that the reactant type affected the microcapsule size and that nylon 610 microcapsules were intermediate in size between those formed for piperazine combined either with 2,2'-dichlorodiethyl ether (forming a polyether) or with phthaloyl dichloride. Reactant concentration and temperature also affected the microcapsule size. The authors concluded that large microcapsules were formed as a result of lowered membrane strength.

Luzzi and co-workers[6] in 1970 discussed the preparation of nylon 610 microcapsules and their potential application as controlled release systems for drugs. In preparing the microcapsules, a Waring blender was operated at high speed for 10 sec after the addition of the SC, then stirring was continued for 1 min at low speed. The possible adverse effect of the blender at high speed on the integrity of the microcapsules was not addressed by the authors. The microcapsules were separated and spray-dried at 125°, and some were tableted. Claims of controlled release were made for the products; however, the in vitro dissolution data obtained were difficult to assess since the workers used a stirring rate of 6 rpm. The authors reported that vacuum-dried microcapsules had to be scraped from the collection jar and spatulated in order to obtain a powder. Their fears that some of the microcapsules may have been ruptured were probably well founded.

In 1972, Mori et al.[36] optimized conditions for microencapsulation of asparaginase in nylon 610 microcapsules. Presence of proteins stabilized the asparaginase after it was encapsulated. The optimal conditions for microencapsulation of this enzyme were reported as 450 μmol SC and 800 μmol HMD in cyclohexane/chloroform (5 : 1), an aqueous core with pH of 8.4 and containing casein and L-aspartic acid and a total reaction time of 3 min.

McGinity et al.[8] reported the combination of Chang's method with the method of Tanaka et al.[37] to prepare microcapsules having a core of formaldehyde-cross-linked gelatin surrounded by a nylon 610 membrane. The microcapsules were air dried at room temperature and displayed ideal physical characteristics for formulation purposes. They were gritty and dense and, because of the nylon coating, they did not adhere together.

[35] Y. Shigera, M. Koishi, and T. Kondo, Can. J. Chem. 48, 2047 (1970).
[36] T. Mori, T. Sato, Y. Matuo, T. Tosa, and I. Chibata, Biotechnol. Bioeng. 14, 663 (1972).
[37] N. Tanaka, S. Takino, and I. Utsimi, J. Pharm. Sci. 52, 664 (1963).

The capsules had excellent flow properties and could be made of very small diameter by controlling the stirring speed during nylon formation.

Nylon microcapsules of sulfathiazole sodium containing unhardened gelatin, various cellulose gums, proteins, alginates, and other carrier materials were generally difficult to separate. In addition, they did not possess the superior physical characteristics of the formalin-treated nylon gelatin capsules. Gelatin micropellets containing sulfathiozole sodium were prepared by the method of Tanaka et al.[37] These pellets showed poor flow properties even after several rinses in benzene. They tended to adhere together and were difficult to wet.

Dissolution studies with nylon microcapsules containing sulfathiazole sodium embedded in the formalized-gelatin core suggested that nylon membranes prepared by interfacial polymerization had limited utility as a controlled-release barrier for small soluble drug molecules.[8]

In 1976, DeGennaro and co-workers[10] reported on the effects of different cross-linking agents on the release of sodium pentobarbital from nylon microcapsules. Glycerin was added to the aqueous phase in order to plasticize the nylon. The Schotten–Baumann reaction would predict that a competition between the glycerin and diamine for the reaction with the acid chloride might occur. The reaction should occur primarily with the HMD and SC because of the solubility of the HMD in the organic solvent where the reaction is known to occur. However, the coating material may contain small amounts of glycerin copolymerized in the nylon and this amount may be expected to increase as the stirring time prior to the addition of the SC increased. DeGennaro and co-workers could attribute no discernable pattern of release to the various cross-linking agents employed.

Florence and Jenkins[9] dissolved phenothiazine derivatives in polyethylene glycol 400 (PEG) containing HMD as the internal phase of nylon 610 microcapsules. The application of scanning electron microscopy suggested the nylon 610 microcapsules of trifluoperazine embonate in PEG 400 were solid, with very deep folds. Following *in vitro* dissolution tests, these were not ruptured but exhibited a more open appearance. The solid structure of the microcapsules was attributed by the authors to the PEG 400 in the aqueous phase.

Watanabe and Hayashi[38] discussed the microencapsulation of emulsions and oils by an interfacial polymerization method. In 1978, Madan[39] discussed interfacial polymerization in a series of articles on microencapsulation in which various types of polymers were discussed from a prepar-

---

[38] A. Watanabe and T. Hayashi, *in* "Microencapsulation" (J. Nixon, ed.), p. 13. Dekker, New York, 1976.
[39] P. Madan, *Pharm. Technol.* **2,** No. 9, 68 (1978).

ative view point. Madan stated that the presence of impurities such as drugs may be important if one needs to obtain optimal results of the polymerization.

McGinity et al.[12] described the application of interfacial polymerization to pharmaceuticals and discussed the influence of formulation variables and dissolution factors on release of sodium sulfathiazole from the microcapsules. The variables examined included the particle size, amount of surfactant in the dissolution media, rate of agitation during dissolution, added electrolyte to the dissolution media, and pH of the media as well as the amount of formalin added. Drug release from the microcapsules followed first-order kinetics. The release rate was not significantly affected by volumes from 5 to 20 ml formalin, but addition of 40 ml formalin decreased the rate of drug release. The rate of release also was found to increase with decreasing microcapsule size fractions and with increasing concentrations of surfactant added to the dissolution medium. A moderate increase in drug release rate was found as the agitation rate increased. Further, the reproducibility between different batches was excellent. Interesting dissolution patterns were observed when increasing amounts of electrolyte (NaCl) were added to an acidic dissolution medium (0.1 N HCl). The effect of adding electrolyte up to 0.1 M reached a maximum at 0.01 M NaCl. All other concentrations of sodium chloride gave slower dissolution rates. These results suggested a dual effect upon drug release. From 0.001 to 0.01 M NaCl, the release of sulfathiazole sodium from the microcapsules increased. From 0.01 to 0.1 M, there was a decrease in dissolution rates. This influence of electrolyte on release rates also complicated the interpretation of media pH effects since it was found that phosphate and acetate buffer media at the same pH did not yield comparable dissolution rates.

In 1979, Ishizaka and co-workers[40] reported on the permeability of polyamide microcapsules toward different ions, with interesting results. The authors found that for nylon 610 and poly(1,4-terephthaloylpiperazine), the order of rate of permeability is potassium, sodium, and rubidium in decreasing order. The authors also studied the thermal behavior of aqueous microcapsule suspensions by differential scanning calorimetry. They concluded that water "was highly structured on and around the hydrophobic polymers forming the microcapsule membrane." The reported value for membrane thickness was about 10 times that reported by other workers.

Lim and Moss[41] in 1979 reported on a process for the development of semipermeable microcapsules using a variety of polyamides and nylons.

[40] T. Ishizaka, M. Kioshi, and T. Kondo, *J. Membr. Sci.* **5**, 283 (1979).
[41] F. Lim and R. Moss, *in* "Microencapsulation—New Techniques and Applications" (T. Kondo, ed.), p. 167. Techno Inc., Tokyo, 1979.

The process was a two-step polymerization that was performed at room temperature. The authors cited problems with other techniques such as poor recovery of specific activity in biologically active materials, frequent aggregation of microcapsules, and the need to maintain a low temperature for the reaction to proceed. Not all of their criticisms are completely justified, however. Lim and Moss employed human red blood cell hemolysate solutions as "carrier polymers." The process is the same interfacial polymerization process used by other workers except that after the microcapsules were formed, they were separated from the bulk organic solvent and instead of drying were resuspended in fresh organic solvent. At this point, more of the organic media containing acid chloride was added. The wall was subjected to treatment by excess acid chloride. The authors reported that very useful microcapsules were recovered and the benefits of the method included ease of recovery without clumping and control of reaction temperature was not required. Further, excellent enzyme stability was reported for the microencapsulated enzymes. The authors believed that the second treatment with acid dichloride corrected macroporous imperfections in the surface of the microcapsules.

Rosenthal and Chang[42] in 1980 reported on a method of complexing nylon 610 membranes with lipid or Na^+,K^+-ATPase. The permeability coefficients of the lipid-incorporated membranes for Na^+ and Rb^+ were dramatically lowered. The addition of valinomycin did not increase the sodium permeability of the treated membranes, but did increase the Rb^+ permeability. Further, after incorporation into the microcapsule membrane, the Na^+,K^+-ATPase retained 50–100% of its original sensitivity to oubaine.

In 1981, the influence of various matrix materials on the properties of nylon 610 microcapsules containing a variety of pharmaceuticals incorporated into the core was reported.[15] Anionic, cationic, nonionic, quaternary, and amphoteric compounds were used in the formulation studies. Matrix materials of formalized gelatin, calcium sulfate, and calcium alginate were all reported to yield microcapsules with superior handling characteristics. The preparation of microcapsules containing calcium alginate employed freeze-drying procedures. Lyophilization was not necessary with the formalized gelatin and calcium sulfate systems. The spherical nature of the microcapsules was verified by scanning electron microscopy. The plate-like surface of the nylon microcapsules containing formalized gelatin and morphine sulfate is shown in Fig. 2. After lyophilization for 2 days, microcapsules containing calcium alginate were obtained as shown in Fig. 3. Lyophilization for periods longer than 4 days resulted

[42] A. Rosenthal and T. M. S. Chang, *J. Membr. Sci.* 6 (1980).

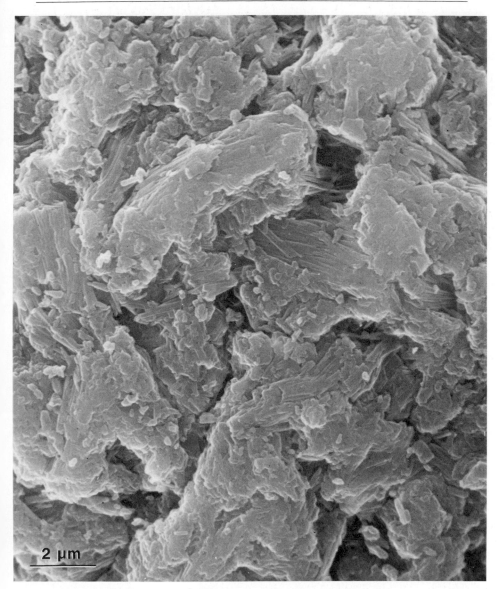

2 μm

FIG. 2. Scanning electron micrograph of the surface of a nylon microcapsule containing a formalized gelatin matrix. (From ref. 15; reproduced with the permission of the copyright owner.)

Fig. 3. Scanning electron micrograph of nylon microcapsules containing morphine sulfate in a calcium alginate matrix (lyophilized for 2 days). (From ref. 15; reproduced with the permission of the copyright owner.)

15 µm

FIG. 4. Scanning electron micrograph of a nylon microcapsule containing morphine sulfate in a calcium alginate matrix (lyophilized for 5 days). (From ref. 15; reproduced with the permission of the copyright owner.)

in disruption of the nylon membrane and separation from the matrix, as seen in Fig. 4. Very porous membranes were obtained for microcapsules containing a calcium sulfate matrix.[15]

One of the major problems associated with the methods of Chang and co-workers[1,2] to microencapsulate drugs is that the aqueous phase containing the diamine must be alkaline in order for the nylon membrane to form. For some drugs the alkalinity of the aqueous phase may create a favorable environment for partitioning of drug from the microcapsule into the organic phase, as was shown by McGinity et al.[15] For several drugs, the pH inside the microcapsule influenced the ionic character of the encapsulated drug. To determine the extent of partitioning into the organic phase, the microcapsules and organic phase were assayed for drug content. Equilibrium conditions were established rapidly with all of the drugs studied. As expected with anionic drugs, little sulfathiazole sodium and sodium salicylate appeared in the organic phase.

Great difficulty was experienced in encapsulating the quaternary drugs, methantheline bromide and benzalkonium chloride, in nylon; in each case, microcapsules were not formed. Preliminary studies showed that another quaternary compound, pralidoxime iodide, could be encapsulated in nylon with a formalized gelatin matrix, although the capsules were sticky and difficult to separate. Quaternary compounds apparently interfere with the reaction to form nylon.

Both phenytoin and diazepam formed suspensions in the aqueous phase containing hexamethylenediamine and the matrix components. Phenytoin can form a sodium salt at very high pH, so a small amount of a diamine salt may have formed in the hexamethylenediamine solution. Negligible quantities of drug were recovered in the organic phase, probably due to the very low solubility of phenytoin in both cyclohexane and chloroform. This was not the case with diazepam. Its yellow color was quite noticeable in the organic phase, and analysis confirmed the partitioning of the drug from the microcapsules.

Interesting results were seen with both the xanthine derivatives and the cationic drugs. The three xanthine derivatives included caffeine, theophylline, and theobromine. The caffeine and theobromine data can be explained on the basis of the relative drug solubilities in the organic phase. Caffeine has a solubility of 1 g in 5.5 ml of chloroform whereas theobromine is practically insoluble in both chloroform and cyclohexane. Theophylline microcapsules did not form when the matrix was calcium sulfate. The addition of theophylline to the calcium sulfate suspension and amine solution resulted in a very thick suspension, which was diluted with a small quantity of water. However, on addition to the organic phase, the aqueous suspension of calcium sulfate and theophylline did not emulsify in the organic phase and adhered strongly to the sides of the glass vessel and the stainless-steel stirring apparatus, thus preventing the formation of microcapsules. The addition of formalin solution to the gelatin–theophylline microcapsules significantly decreased the amount of drug remaining in the capsules.

The two basic drugs studied were diphenhydramine hydrochloride and morphine sulfate. Small quantities of morphine were recovered in the organic phase. However, significant partitioning of the drug from the microcapsules was found only with the antihistamine due to its high solubility in chloroform.[15]

By encapsulating pH indicators, Cuff and co-workers[43] examined the change in core pH during the preparation of microcapsules containing formalized gelatin, calcium sulfate, and calcium alginate. The pH of the final product could be adjusted to the acidic region by the addition of dilute hydrochloric acid to the reaction vessel. Several other variables affecting the nylon wall formation were examined to determine optimum processing conditions. Increased agitation produced a decrease in microcapsule size. This caused an increase in nylon weight recovered. The increased recovery of nylon was due in part to nylon formation over a larger surface area. Varying the amounts of hexamethylenediamine (HMD) and sebacoyl chloride (SC) used also affected the total weight of nylon formed. In general, more nylon is recovered as the level of each reactant increases. However, the molar ratio of HMD : SC determined the total amount of nylon formed when SC was present in excess as shown in Fig. 5.

The partitioning of drug into the organic phase was also experienced by Beal and co-workers,[44] who studied the preparation and properties of polyamide microcapsules for delivery of pilocarpine nitrate to the eye. The low yield of nylon 610 microcapsules may have resulted from the drug interfering with nylon production. The polyphthalamide microcapsules prepared by these authors contained an aqueous phase buffered to pH 9.8. At this pH level pilocarpine, being very soluble in chloroform, readily partitioned into the organic phase causing a 50% reduction in payload. To overcome such problems, an improved method of nylon encapsulation was recently developed by Cuff and McGinity.[16] The aqueous phase may be buffered to an optimal level for the drug to be encapsulated, thus minimizing drug partitioning into the organic phase.

In the improved method of preparing nylon microcapsules, the aqueous phase to be encapsulated does not contain either monomer required

[43] C. W. Cuff, A. B. Combs, and J. W. McGinity, *J. Microencap.* (in press).
[44] M. Beal,   N. E. Richardson,   and J. N. Stanforth,   *Int. Conf. Pharm. Technol.,   4th,* Vol. 3, p. 171 (1983).

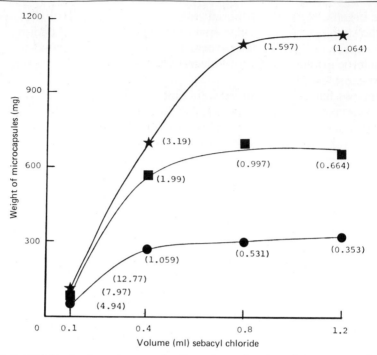

Fig. 5. Weight of nylon microcapsules formed as a function of the concentrations of hexamethylediamine and sebacoyl chloride with no matrix added. The mole ratio for HMD : SC is shown in parentheses. (★) 0.6 M HMD; (■) 0.25 M HMD; (●) 0.133 M HMD. (From ref. 43; reproduced with the permission of the copyright owner.)

for the production of nylon. Therefore, no pH constraint is seen. Further, the potential interaction of hexamethylenediamine with any active agents in the aqueous phase is prevented. This has the effect of ensuring drug stability as well as avoiding the possible interference of nylon formation which results if the water-soluble monomer is not available for polymerization. This improved method of microencapsulation should also prove useful as a means of encapsulating enzymes or cell components with minimal adverse effect on the stability of these substances.

Future Studies with Pharmaceuticals

The wide range of drug diffusion rates from nylon microcapsules as reported in the literature is a function of the drug itself, the core matrial, and the properties of the nylon membrane. By encapsulating drugs in various matrix materials, drug release may be carefully controlled from the dense, free-flowing microcapsules. In addition to controlled-release, these microparticulates offer several other pharmaceutical advantages, including taste masking, separation of incompatible drugs, and improved drug stability. It is our opinion that for nylon microcapsules to become viable drug delivery systems, further research into the physicochemical properties of the nylon membrane is necessary. Optimization studies for nylon formation are also needed. Other problems that have been alluded to in the literature include the solubility of the polyamide in the organic solvent, establishing optimal times for completion of the polymerization reaction, and interaction of the monomers with the encapsulated moiety. For nylon microcapsules to be useful as an oral delivery system, the stability of the membrane in gastrointestinal fluids must be characterized. In a parenteral dosage form, the nylon must biodegrade, and further studies on the toxicity of the monomers are needed to verify safety as well as efficacy.

[8] Poly(lactide-*co*-glycolide) Microcapsules for Controlled Release of Steroids

By Donald R. Cowsar, Thomas R. Tice,
Richard M. Gilley, and James P. English

The development of controlled-release parenteral doses for contraceptive steroids and other potent drugs has now progressed to the stage of human trials.[1-4] Clinics around the world are presently testing injectable doses of microcapsules that continuously release progestins at precise rates for durations of either 3 or 6 months.[5,6] The technology of microencapsulation had its beginnings in the 1950s, and since that time, numerous microcapsule-based products have emerged. Only recently, however,

[1] L. R. Beck, D. R. Cowsar, D. H. Lewis, J. W. Gibson, and C. E. Flowers, Jr., *Am. J. Obstet. Gynecol.* **135,** 419 (1979).

[2] L. R. Beck, D. R. Cowsar, and D. H. Lewis, *Prog. Contracept. Delivery Syst.* **1,** 63 (1980).

[3] L. R. Beck, V. Z. Pope, D. R. Cowsar, D. H. Lewis, and T. R. Tice, *Prog. Contracept. Delivery Syst.* **1,** 79 (1980).

[4] L. R. Beck, V. Z. Pope, C. E. Flowers, Jr., T. R. Tice, D. H. Lewis, A. B. Moore, and R. M. Gilley, *Biol. Reprod.* **28,** 186 (1983).

[5] L. R. Beck, R. Aznar, C. E. Flowers, G. Z. Lopez, D. H. Lewis, and D. R. Cowsar, *Am. J. Obstet. Gynecol.* **140,** 798 (1981).

[6] L. R. Beck, C. E. Flowers, Jr., V. Z. Pope, W. H. Wilborn, and T. R. Tice, *Am. J. Obstet. Gynecol.* **147,** 815 (1983).

have microencapsulation processes been refined to the degree that they can be used to produce reliable, preprogrammed, drug-delivery systems.

This chapter addresses long-acting, drug-delivery systems comprising steroidal drugs microencapsulated in aliphatic polyester resins. Because the aliphatic polyesters slowly degrade, owing to hydrolysis of ester linkages, when they are exposed to water they disappear from the tissues soon after (or during) drug release. Many drug and biologic substances, other than the steroids, can be microencapsulated in the biodegradable polyesters to produce injectable controlled-release doses.[7] The specific copolymer resins and the appropriate microencapsulation processes must be chosen carefully, however, to achieve the rates and durations of drug release desired.

When steroid hormones are microencapsulated in the biodegradable copolyesters, in which the drug is homogeneously dispersed within the resin matrix to form a monolithic microcapsule, and the microcapsules are injected either intramuscularly or subcutaneously, the drugs are slowly released into the tissues principally by a diffusion mechanism. That is, the crystalline drugs slowly dissolve in the copolyester resin and then diffuse through the resin into the tissues. Because drug diffusion in the copolymer resin is much slower than drug dissolution in the copolymer, the diffusion rate is the release-rate-controlling factor. If the copolymer did not degrade during drug release, the release rate would slowly decline with time because the diffusion path length slowly increases as drug is slowly depleted from the microcapsules. The declining release rates can be avoided by selecting copolyesters that degrade at the appropriate rate during drug release. As the microcapsules break into fragments, short diffusion path lengths are restored, and a nearly constant or zero-order release is obtained.

The microcapsules described in this chapter were designed to release norethisterone at a nearly constant rate for 3 months (Fig. 1). The 85 : 15 poly(DL-lactide-co-glycolide) resin was selected as the biodegradable excipient because the onset of resin fragmentation occurs at about 50 days postinjection. Total resorption of the resin by the tissues is complete in about 180 days.

Preparation of Lactide–Glycolide Copolymers

The preparation of high molecular weight polymers and copolymers of lactic acid and glycolic acid is extremely difficult using conventional condensation techniques due to the stepwise nature of the condensation reac-

---

[7] T. R. Tice and D. H. Lewis, inventors, U.S. Patent 4,389,330 (1983).

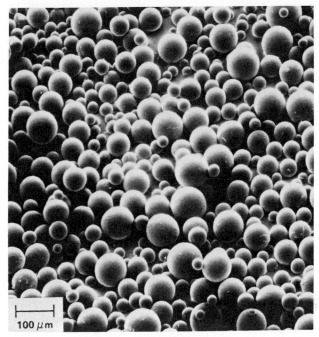

FIG. 1. Photomicrographs obtained by SEM of 3-month NET microcapsules.

tion. High molecular weight polymers and copolymers of these materials are therefore typically prepared by ionic, ring-opening, addition polymerization of cyclic dimers of these materials.

The cyclic dimer of glycolic acid is glycolide, and the cyclic dimer of lactic acid is lactide. Both dimers are prepared by a two-step process as shown in Fig. 2. The acid is first condensed to form low molecular weight polymers, and the polymers are subsequently depolymerized at elevated temperatures with the cyclic dimer being the product formed preferentially. The cyclic dimers are then purified and finally polymerized by ring-opening, addition polymerization to high molecular weight polymers and copolymers. The synthesis of an 85 : 15 random copolymer of DL-lactide and glycolide is shown in Fig. 3, and the experimental details are given below.

Glycolide

The glycolide monomer was prepared and purified by the following method. Excess water was first distilled from 450 g of 67% aqueous glycolic acid in a 500-ml, three-necked, round-bottom flask, equipped with a

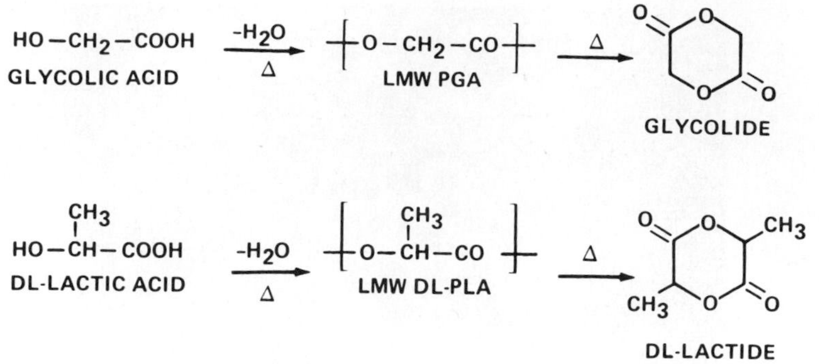

FIG. 2. Synthesis of glycolide and DL-lactide monomers.

heating mantle, a distillation head, a thermometer, and a condenser. A water aspirator was used to reduce the pressure as the solution was boiled. After the excess water had distilled (~150 g), the flask was heated to about 200° to remove additional water by dehydration of the glycolic acid. After no further water evolved, the flask was allowed to cool to room temperature under vacuum. At this point, 3 g of antimony trioxide catalyst was added to the flask. The distillation head and condenser were removed, and the flask was connected to two receiving flasks and a trap arranged in series. The receiving flasks and trap were cooled by dry ice: isopropanol baths. The second receiving flask is actually a trap. The pressure was then reduced to about 2 mm Hg with a vacuum pump, and the reaction flask was heated to 260–280° to distill the crude glycolide. The material that distilled between 110 and 130° was collected in the first receiving flask (~195 g).

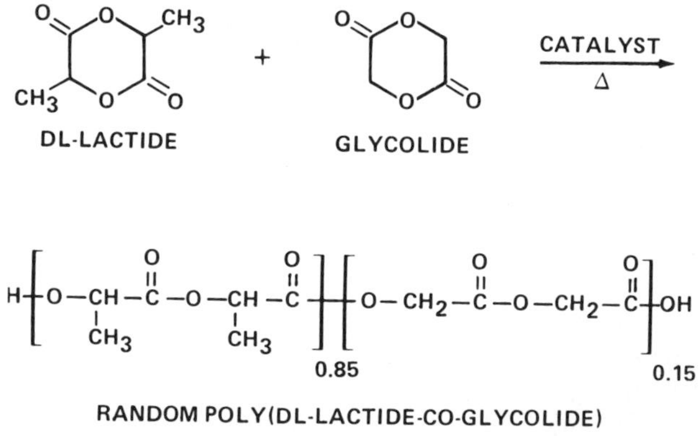

FIG. 3. Ring-opening copolymerization of DL-lactide and glycolide monomers.

The crude glycolide was purified as follows: The crystalline mass was first pulverized and then washed by slurrying it at room temperature in 400 ml of isopropanol. After the washed glycolide was collected on a filter and vacuum dried, the following precautions were observed to protect the glycolide from atmospheric moisture during subsequent recrystallizations. All glassware was oven dried (overnight at >150°), dry ethyl acetate (stored over molecular sieves) was used, and a glove box filled with nitrogen was used to filter the recrystallized monomer. The crude glycolide was combined with a volume of ethyl acetate approximately equal to three-fourths its weight. The mixture was then heated to reflux to dissolve the gycolide, cooled slowly to room temperature, and finally cooled in an ice-water bath to allow crystallization. The monomer was recrystallized three times in this manner. After each recrystallization the glycolide crystals were collected by vacuum filtration in a glove box, and, after the final recrystallization, the product was dried at room temperature under a vacuum of <2 mm Hg in a vacuum desiccator. The final yield was approximately 120 g. The melting point was 82–84°. The Van't Hoff purity was determined by differential scanning calorimetry and was found to be >99.5% glycolide.

DL-*Lactide*

DL-Lactide was purchased from Gallard-Schlesinger Chemical Manufacturing Corp., Carle Place, NY. The crude DL-lactide was purified by the following recrystallization procedure; precautions were taken to protect the product from moisture during and after the purification. Oven-dried glassware (overnight at >150°) and dry ethyl acetate (stored over molecular sieves) were used. Final filtering procedures were conducted in a glove box filled with dry nitrogen.

The crude DL-lactide (200 g) was first combined with 200 ml of ethyl acetate in a 1-liter beaker. The mixture was gently heated on a stirring hot plate to dissolve the lactide, and the hot mixture was quickly filtered through a sintered-glass (extra coarse) element to remove insoluble materials. The filtered solution was then distilled under vacuum to reduce the solvent volume to about half the weight of the lactide. The filtered material was allowed to cool slowly to room temperature, and then it was cooled for 2 hr in an ice bath. The product was collected by vacuum filtration in a glove box. The recrystallization of the product from ethyl acetate was repeated three more times in a 500-ml boiling flask fitted with a reflux condenser and a drying tube. A volume of ethyl acetate equal to about half the weight of the lactide was used for each recrystallization. Filtration of the hot ethyl acetate solution of the lactide, however, was not repeated.

After the final filtration to collect the purified lactide, the product was dried at room temperature in a desiccator under vacuum (2 mm Hg). The dry purified lactide was then characterized and stored in an oven-dried glass jar in a desiccator until needed. The final yield was ~125 g. The melting point was 124–126°. The Van't Hoff purity was determined by differential scanning calorimetry and was found to be >99.5% DL-lactide.

Copolymerization of DL-Lactide and Glycolide

The polymerization apparatus was a 300-ml, three-neck, round-bottom flask equipped with a mechanical stirrer and a gas inlet tube. All glassware was dried in an oven at 150° overnight and cooled in a dry-nitrogen atmosphere. Both the loading and assembly of the apparatus were also conducted in a dry-nitrogen atmosphere (glove box). Pure DL-lactide (90.1 g) and pure glycolide (9.9 g) were combined to form a mixture comprising 88 mol% of lactide and 12 mol% of glycolide. The flask was heated in an oil bath at 140–145° until the monomers had partially melted. The flask was kept under positive nitrogen pressure, and stirring was initiated. After the monomers had completely melted, catalyst (0.05% by weight of stannous octoate) was added to the mixture with a microsyringe. Mechanical stirring was continued and maintained until the polymerizate became too viscous to stir (~30 min). The stirrer was raised from the melt, and heating was continued for a total of 16–18 hr. The resulting copolymer was cooled to room temperature, and the flask was broken away. Excess glass was removed by submersion into liquid nitrogen. While still cold, the block of copolymer was broken into several smaller pieces. The polymer was then dissolved in 800 ml of dichloromethane, filtered through a sintered-glass filter, and finally precipitated into excess methanol.

After precipitation from dichloromethane into methanol, the purified copolymer was then dried in vacuo at 30–40°, ground with a Wiley Mill, characterized, and stored in a desiccator over Drierite until needed. The yield was about 90 g.

The polymer was characterized by first determining its inherent viscosity in chloroform at 30° at a concentration of 0.5 g/dl. Typical values range from 0.6 to 0.8 dl/g. However, the molecular weight can be adjusted downward if desired by briefly autoclaving the ground polymer at 121° and 15 psig steam pressure.

The final ratio of monomers incorporated into the copolymer was determined by proton nuclear magnetic resonance spectroscopy. For this analysis the polymer was dissolved in a 50:50 weight ratio of hexafluoroacetone and trifluoroacetic acid. The values obtained are typically

85–86% lactide and 14–15% glycolide. This difference from the theoretical is due to the more rapid incorporation of the glycolide into the copolymer.

Preparation of Microcapsules

We have routinely prepared microcapsules containing the steroid norethisterone (NET) using two different microencapsulation processes: solvent evaporation and solvent evaporation/solvent extraction. Both of these processes have produced high-quality NET microcapsules. The exact procedure that we have used to prepare 10-g batches of NET microcapsules by these processes follows.

Solvent Evaporation Process

First, 500 g of 5 wt% aqueous poly(vinyl alcohol) (PVA) is prepared by slowly pouring 25 g of PVA (Vinol 205C, Air Products and Chemicals, Allentown, PA) into a 1-liter beaker containing 475 g of distilled, deionized water. During this addition, the water is being stirred at about 1000 rpm with a 2-in. stainless-steel impeller to ensure that the PVA is well wetted and does not clump. After stirring the solution for 15 min at room temperature, it is then heated (with continued stirring) to 90° by means of a hot plate to facilitate the dissolution of the PVA. After the PVA solution has cooled, it is placed into a 500-ml resin kettle (Pyrex, Fisher Scientific Company, Norcross, GA) and chilled to 1° with an ice-water bath surrounding the resin kettle.

In a separate 4-oz. cream jar, 7.5 g of 85 : 15 poly(DL-lactide-*co*-glycolide) (DL-PLG) is dissolved in a mixed solvent comprising 16.3 g of acetone and 29.9 g of chloroform. After the copolymer has dissolved, this solution is cooled to 1° with an ice-water bath to minimize the solubility of NET in the mixed solvents. Next, 2.5 g of micronized NET (Ortho Pharmaceutical, Raritan, NJ, or Schering AG, Berlin, West Germany) is added to the copolymer solution, and the mixture is stirred vigorously for 30 min (at 1°) with a magnetic stir bar to obtain a homogeneous drug dispersion. It is important to note that micronized NET must be used with this process so that microcapsules of the desired size for injection (<250 μm) can be obtained.

This drug dispersion (organic phase) is then slowly poured through a 5-mm-bore funnel into the resin kettle containing the PVA solution to form an oil-in-water emulsion. The discontinuous oil phase is microdroplets consisting of polymer, drug, and the two volatile solvents. During this addition, the PVA solution is stirred at 800 rpm with a truebore stirrer fitted with a 2.5-in. Teflon turbine impeller (Ace Glass Inc., Vineland, NJ) that is powered by a Fisher Stedi-speed motor. When the organic phase is added to the resin kettle, the acetone quickly diffuses from the organic microdroplets into the PVA solution. This causes a thin polymer film to form at the interface between the organic droplets and the aqueous dispersing media and increases the viscosity of the oil microdroplets, thereby trapping the NET crystals inside the embryonic microcapsules.

After stirring for 10 min, the resin kettle is closed, and the stir rate is reduced to 600 rpm. The pressure inside the resin kettle is then reduced by 50 torr every 5 min until a pressure of 200 torr is reached. The conditions of 200 torr, 600 rpm, and 1° are maintained for 24 hr to allow the volatile solvents to evaporate slowly. When the microcapsules have hardened, the contents of the resin kettle are centrifuged at 25 g for 15 min. The resulting pellet of microcapsules is washed thoroughly with deionized water. The microcapsules are then wet sieved through stainless-steel sieves and dried in a vacuum desiccator maintained at room temperature.

A series of photomicrographs taken through a light microscope (Fig. 4) show the formation of NET microcapsules during various phases of the solvent evaporation microencapsulation process. Figure 4A shows the microcapsules immediately after the polymer–NET dispersion is added to the resin kettle. At this stage, the embryonic microcapsules are very soft and consist of a thin copolymer film around NET crystals that are dispersed in a viscous polymer/chloroform mixture. After evaporation of the volatile solvents has been carried out for about 1 hr, the microcapsules become more opaque and smaller due to the loss of the solvents (Fig. 4B). After about 5 hr of evaporation, the microcapsules are completely opaque (Fig. 4C). At this point the microcapsules are fairly hard, but they still contain a considerable amount of chloroform. When all of the chloroform has been removed, the microcapsules appear as opaque spheres with rough irregular surfaces (Fig. 4D).

Closer examination of a hard microcapsule from a photomicrograph taken by SEM shows that the microcapsule walls are continuous polymer films with no evidence of unencapsulated drug on their surfaces (Fig. 5). The slight surface irregularity seen in the microcapsules is due to the NET crystals just beneath the surface. The cross-sectional view in Fig. 5 shows how the micronized NET is distributed throughout the polymer excipient.

Solvent Evaporation/Solvent Extraction Process

We have also prepared NET microcapsules by a solvent evaporation/ solvent extraction microencapsulation process. Basically, this process

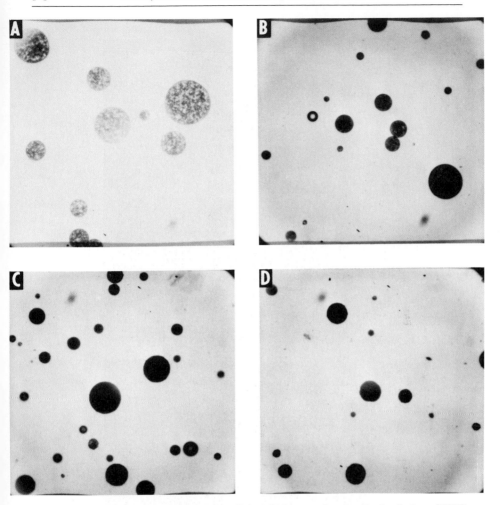

FIG. 4. Photomicrographs obtained by light microscopy showing the hardening of NET microcapsules as the volatile solvents are removed by solvent evaporation. See text for discussion.

involves totally dissolving the NET in a dichloromethane/polymer solution instead of dispersing micronized crystals of drug in an acetone/chloroform mixture. This polymer–NET solution is added to the resin kettle as before, and an oil-in-water emulsion is formed. The volatile organic solvent is then evaporated until the microcapsule walls have become elastic. At this time, the contents of the resin kettle are poured into a large volume of deionized water to quickly extract the remaining organic sol-

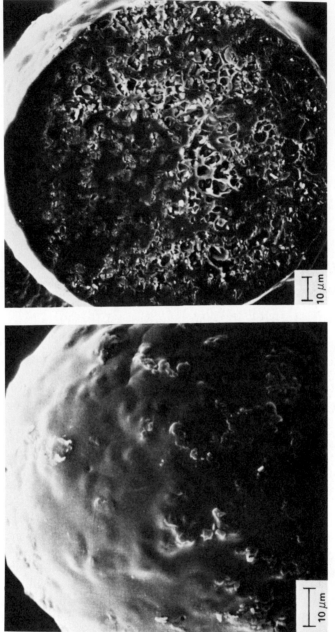

FIG. 5. Photomicrographs obtained by SEM of 25%-loaded NET microcapsules prepared by the solvent evaporation microencapsulation process.

vent from the microcapsules. The purpose of this extraction step is to quickly harden the microcapsule walls. If the extraction step is performed too soon, i.e., before the microcapsule walls have become sufficiently elastic, the microcapsules will agglomerate when they are poured into the extraction water. On the other hand, if the extraction step is performed too late, the NET will form large crystals that protrude through the micro-capsule walls.

NET microcapsules prepared by this process are shown in Fig. 6. These microcapsules are smooth, spherical particles. Moreover, these microcapsules show no signs of drug crystals underneath the surface, indicating that the NET is better encapsulated in these microcapsules than in microcapsules prepared by the solvent evaporation process. From the cross-sectioned view, no NET crystals are seen inside of the micro-capsules.

Characterization of Microcapsules

Core Loading

We determine the drug content (i.e., core loading) of steroid microcap-sules by dissolving the microcapsules in dichloromethane and spectro-photometrically assaying for drug. The general procedure that we use to determine the core loading of NET microcapsules follows: Triplicate 7-mg samples of microcapsules (± 0.1 mg) are dissolved in 100 ml of di-chloromethane. The absorbance of each solution is then measured at 240 nm. Applying Beer's Law and assuming additive absorbances from copolymer and drug, the amount of NET in each solution is determined from Eq. (1)

$$C_1 = \frac{A_3 - E_2C_3 - (I_1 + I_2)}{(E_1 - E_2)} \qquad (1)$$

where C is the concentration (g/dl); E, extinction coefficient (dl/g $\cdot$ cm); A, absorbance at 240 nm; and I, the sum of the Beer's Law plot intercepts for DL-PLG and NET. The subscripts refer to (1) NET, (2) DL-PLG, and (3) microcapsules. This equation mathematically subtracts the small con-tribution of the DL-PLG to the absorbance. The core loading of each sample is then calculated from Eq. (2).

$$\text{Core loading, wt\% NET} = \frac{(\text{NET concentration, g/dl})(1 \text{ dl})}{\text{Microcapsule sample size, g}} \times 100 \qquad (2)$$

It is important to accurately determine the core loading of the micro-capsules because the drug content is one of the factors that affect the rate

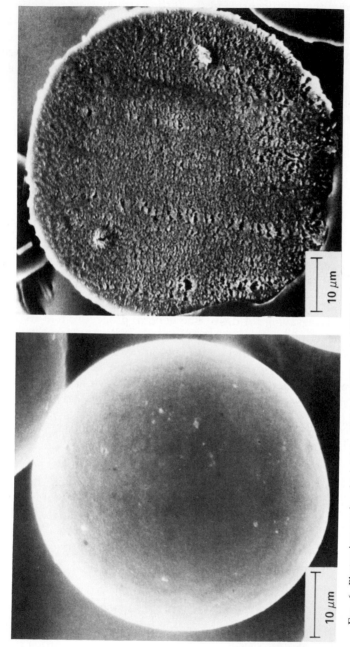

FIG. 6. Photomicrographs obtained by SEM of 25%-loaded NET microcapsules prepared by the solvent evaporation/solvent extraction microencapsulation process.

of release of NET from the microcapsules. In monolithic systems, such as NET microcapsules prepared by the above microencapsulation processes, higher loadings of drug will increase the porosity of the microcapsules, resulting in faster rates of release of the drug. More specifically, the drug near the outer walls of highly loaded microcapsules diffuses out of the microcapsules, leaving voids. These resulting voids make it possible for drug centrally located in the microcapsules to diffuse out of the microcapsules faster. Because monolithic microcapsules release drug at a slower rate with time (theoretically), the formation of interconnected voids effects a more constant rate of release of drug.

In Vitro Release Kinetics

We routinely determine the *in vitro* release kinetics of individual batches of NET microcapsules. The procedure that we use is based on diffusion of drug through solvent swollen walls and does not account for biodegradation of the microcapsule excipient. The *in vitro* release profiles, therefore, do not demonstrate a one-to-one correlation with the *in vivo* rate of release. They are useful, however, for determining the quality of individual batches of microcapsules and demonstrating batch-to-batch reproducibility.

When designing *in vitro* tests, several factors must be considered. First, the drug being measured must be sufficiently soluble in the receiving fluid; that is, a solvent should be used such that "infinite sink" conditions are maintained, i.e., the concentration of drug in the receiving fluid is at most only 10% of the drug's saturation solubility. Second, the release of drug from the microcapsules should be accelerated so that a meaningful release profile can be obtained in 1–2 weeks. (It is important that individual batches of microcapsules be quickly evaluated so that microcapsule optimization can be achieved in a timely manner.) One way to accelerate the *in vitro* release rate for microcapsules is to choose a receiving fluid that will plasticize the copolymer excipient, causing the microcapsules to swell and allowing the drug to diffuse through the walls at a rapid rate. Another method is to vary the pH of the receiving fluid so that the polymer excipient is hydrolyzed at a rate faster than is seen *in vivo*, resulting in an accelerated release of drug. Other factors that should also be addressed when designing *in vitro* release systems include temperature, analytical techniques, and adequate agitation of the receiving fluid, to prevent "boundary layer" effects.

The *in vitro* release system we have designed for NET involves agitating the microcapsules at 37° in a receiving fluid comprising 27.5 wt% ethanol and 72.5 wt% water. Aliquots of this receiving fluid are re-

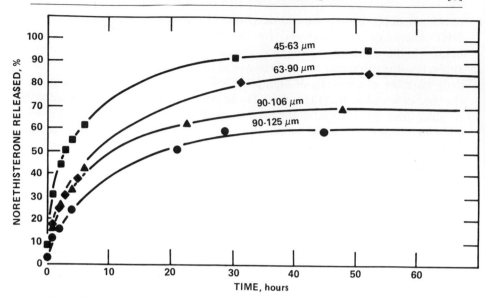

FIG. 7. *In vitro* release of norethisterone from 25%-loaded microcapsules into a receiving fluid consisting of 27.5 wt% aqueous ethanol maintained at 37°.

moved periodically and assayed spectrophotometrically for NET. The experimental procedure that we use to measure the *in vitro* release kinetics of NET microcapsules is as follows: Triplicate 5-mg samples of NET microcapsules (±0.1 mg) are weighed into 8-oz. bottles containing 200 ml of 27.5 wt% aqueous ethanol that has been prewarmed to 37°. The bottles are then sealed with a Teflon-lined screw cap and placed in a shaker bath (Eberbach Corp., Midland, MI) maintained at 37° and oscillated at 120 cycles/min. Samples of the receiving fluid are removed periodically, and their absorbances are measured at 247 nm with a Perkin-Elmer Model 575 spectrophotometer and 1-cm cuvettes. The amount of NET released (%) at each time point is calculated from Eq. (3),

Amt. NET released, %

$$= \frac{[(A - I)/E](\text{Receiving fluid, dl})}{(\text{Amt. microcapsules, g})(\text{Core loading}/100)} \times 100 \quad (3)$$

where A is the absorbance of the receiving fluid; E, the extinction coefficient for NET in 27.5 wt% aqueous ethanol (dl/g · cm); and I, the Beer's Law plot intercept for NET in 27.5 wt% aqueous ethanol.

Figure 7 shows the *in vitro* release kinetics for various-sized fractions of a typical batch of 25%-loaded NET microcapsules prepared by the

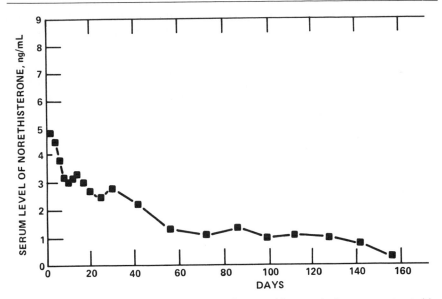

FIG. 8. Mean serum levels of immunoreactive norethisterone in five women treated by intramuscular injection of sterilized microcapsules (mean dose = 1.502 mg of norethisterone per kilogram).

solvent evaporation process described above. These microcapsules were prepared with the same polymer excipient and have diameters ranging from 45 to 125 μm. As expected from theory, microcapsules with smaller diameters, i.e., higher surface area per unit mass, release NET at a faster rate.

In Vivo Release Kinetics

Figure 8 shows the serum levels of NET in women following intramuscular injection of about 100 mg of microencapsulated NET. The microcapsules comprised 21.0 wt% NET and 79.0 wt% 85:15 DL-PLG. The microcapsules were sterilized by exposing them to a minimum dose of 2 Mrad of γ radiation. The *in vivo* release profile shows a slightly higher rate of release of NET during the first month, followed by a more constant rate of release thereafter. The target serum level of NET to inhibit ovulation is 1 ng/ml. If the same amount of NET but in unencapsulated form was administered, the initial serum levels of NET would be very high (>11 ng/ml), and the level of NET in the serum would be below detectable limits within a month's time.

We believe that the initial release of NET from the microcapsules (up to about day 50) is due to diffusion of NET through the DL-PLG excipient. Thereafter, the release of NET is due to biodegradation of the DL-PLG excipient and diffusion of NET through smaller fragments of the microcapsules as a result of this biodegradation. If the DL-PLG did not degrade during NET release, the release rate would slowly decline with time because the diffusion path length slowly increases as NET is depleted from the microcapsules. Another advantage with a microcapsule formulation that degrades during drug release is that the excipient is gone shortly after drug release is completed.

[9] Crystallized Carbohydrate Spheres for Slow Release and Targeting

By ULF SCHRÖDER

Introduction

Biologically active substances, such as drugs, are often eliminated or inactivated, or gain low access to target areas, and repeated administration becomes necessary. In order to avoid these problems there is an increasing interest in employing carrier systems for slow release and targeting, thus avoiding frequent administrations and negative side effects and promoting higher concentrations in the target area.

When working with carrier systems for slow release and targeting, the following criteria should preferably be fulfilled before possible use in humans.[1] The matrix should be chemically inert in biological systems. The matrix should be biologically well characterized. The matrix should be nontoxic and nonimmunogenic. The matrix should have the ability of being eliminated from the body by normal routes. The matrix preparation should be administered easily (i.e., by injection). The matrix preparation should be capable of entrapment and release of biologically active substance of various molecular weights.

Traditionally, synthetic or semisynthetic polymers have been designed to be used as biodegradable matrix systems in agricultural as well as in veterinary and human medical areas. This kind of design involves covalent cross-linking of a polymer and often also a covalent attachment

[1] R. S. Langer and N. A. Pappas, *Biomaterials* **2**, 201 (1981).

METHODS IN ENZYMOLOGY, VOL. 112

of the active substance to the matrix. However, covalently cross-linked biodegradable matrices are potentially harmful to organisms, due to the possible production of toxic by-products from the degrading matrix. Furthermore, because the covalent attachment itself might introduce changes in the chemical structure of the substance, cumbersome and often expensive toxicological testing is required. In the area of slow-release preparations for low molecular weight substances there are a number of methods available.[2] However, less progress has been made in order to find a suitable system for proteins, peptides, and other water-soluble substances. It has even been more difficult to find a combination of targeting and slow release of biologically active substances.

In one type of slow release systems used for proteins, the proteins are coupled to a matrix by a covalent linkage, achieved after the creation of reactive ligands on the matrix with subsequent attachment of the protein. Because of steric hindrance, all the reactive ligands will not react with the protein and are subsequently deactivated, normally by allowing ethanolamine or merceptoethanol to be coupled to the reactive ligands. However, these substituted polymers may enhance the immunological system, creating antibodies against the coupled protein.

On the other hand, carrying out the activation procedure with the protein creates hapten groups on the protein, also with the possibility of forming antibodies against the protein. Furthermore, it has been shown that low molecular weight dextran, covalently attached to a protein, is strongly immunogenic.[3] All of these factors are equally unwanted if a slow release of proteins is desired.

One way of avoiding the problems connected with covalent attachment of the proteins is to use entrapment of proteins into a nonbiodegradable polymer matrix, from which the proteins are released by diffusion through pores in the matrix.[2] However, the need of surgical insertion and removal of the matrix makes this approach for slow release of proteins less favorable, particularly in long-term usage such as in the delivery of insulin or human growth hormone.

However, the factors obtained in the use of covalent attachment, as discussed above, may be useful and desirable if a slow release system for antigens is needed for enhancement of the immunological response against a protein. There has been extensive efforts during the last decade to find safer adjuvants with good effectiveness for use in human vaccination. The adjuvants that have been tested range from inorganic compounds such as alum and silica, through organic material such as sub-

[2] R. S. Langer, *Chem. Eng. Commun.* **6,** 1 (1980).
[3] W. Richter and L. Kagedahl, *Int. Arch. Allergy Appl. Immunol.* **42,** 887 (1972).

stances of plant origin (e.g., vegetable oils, alginate, saponins, or carotenoids) to bacterial toxins and cell wall components. In some of the above systems, the adjuvant effect is partly due to the sustained release of antigens from the matrix, i.e., mineral oil, alum and silica.[4] The above-mentioned entrapment procedure using a nonbiodegradable polymer matrix has also been utilized for antigens with a good effect in the production of antibodies against the entrapped antigen.[5]

The most popular adjuvant used in animals, Freund's, has not been considered safe in humans because of the long-term persistence of mineral oil in animals, formation of disseminated focal granulomas, induction of autoimmune reactions, and potentiation of plasma cell tumours in mice. Further, using aluminum hydroxide in vaccination in humans, Slater[6] has reported granulomas which might restrict future use of adjuvants in humans to some extent.

Working with matrix systems and slow release it would also be desirable if the immunological response could be altered and/or enhanced with the help of different kinds of immunomodulators. Examples of such immunomodulators are DEAE-dextrans or sulfate-substituted dextrans. It has been shown that DEAE-dextran enhances humoral antibody production, thereby protecting animals against infection.[7] Dextran substituted with sulfate groups has, furthermore, an adjuvant effect on the T-cell-mediated immunological response.[8]

In attempting to design "magic bullets" for drug delivery, soluble and particulate carrier systems have been used. An example of soluble carriers would be antibodies with attached drugs, and particulate systems include all types of microspheres. The difficulties in avoiding uptake of carriers by the RES is, however, a problem if the goal is to target microspheres, with their drug contents, to other areas in the body. By co-entrapping magnetite in microspheres and using arterial injection together with external magnetic fields, this problem can be substantially reduced, thus obtaining targeting of the magnetic microspheres in a desired area of the body.

Described in this chapter is a simple technique for the preparation of a slow release biodegradable matrix system, in which both drugs, proteins, and magnetite particles are entrapped in carbohydrate spheres stabilized

[4] F. Borek, in "The Antigens" (M. Sela, ed.), Vol. 4, p. 369. Academic Press, New York, 1979.

[5] I. Preis and R. S. Langer, J. Immunol. Methods 28, 193 (1979).

[6] D. N. Slater, J. C. E. Underwood, T. E. Durrant, T. Hopper, and I. P. Hopper, Br. J. Dermatol. 107, 103 (1982).

[7] G. Wittman, B. Dietzschold, and K. Bauer, Arch. Virol. 47, 225 (1975).

[8] R. E. McCarthy, L. W. Arnold, and G. F. Babcock, Immunology 32, 963 (1977).

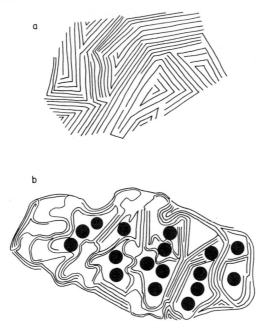

FIG. 1. (a) Schematic view of a perfect crystalline polymer matrix, in which the polymers are attached to each other via noncovalent bonds, such as hydrogen bonds, van der Waals forces, or in some cases ionic bonds. (b) A schematic view of a partially crystalline structure, in which drugs, proteins, or magnetite particles are entrapped into the matrix.

by crystallization (Fig 1). The term *crystallization* is used in a general sense, implying that all chemical bonds except covalent bonds are used in the stabilization process. However, the most important of these bonds are hydrogen bonds, but the other weak bonds such as van der Waals forces and, in some circumstances, ionic bonds are also involved.[9]

Material

Unsubstituted as well as substituted (DEAE and sulfate) dextrans with various molecular weights from 10,000 (T10) and up to 500,000 (T500) were obtained from Pharmacia AB, Uppsala, Sweden; starch and dextrins from STADEX AB, Malmö, Sweden; the Gafac detergents from AB Svenska Gaf, Stockholm, Sweden; the Pluronic detergents F-68 and L-35 from BASF Wyandotte Corp., MI; and the Tween detergents from ICI United States Inc. Crystalline monocomponent insulin (26.8 IU/mg) and

[9] U. Schröder, *Biomaterials* **5,** 100 (1984).

plasmin (3.2 Novo units/mg) were kindly supplied by Novo IAS, Copenhagen, Denmark. Leukocyte-interferon (4 Munits/ml) came from Danish Interferon Foundation, Roskilde, Denmark. Vincristine was supplied by Eli Lilly and Company, Stockholm. ^{3}H-Labeled vincristine, ^{125}I-labeled thyroxine, and ^{125}I-labeled insulin were obtained by the Radiochemical Centre, Amersham, England. Biosynthetic human growth hormone (HGH 2.5 IU/mg) as well as the ^{125}I-labeled hormone (specific activity 120 μCi/μg) were kindly supplied by KabiVitrum, Stockholm. Monoclonal antibodies against PHA were kindly supplied by Dr. C. A. K. Borrebaeck, Department of Pure and Applied Biochemistry, University of Lund. The bee venom as well as the purified antigens from schistosomiasis adult worm were kindly supplied by Dr. H. Schröder, Pharmacia. Magnetite particles (suspended in water as a stable colloidal suspension; Ferrofluid EMG 805 HGMS) were purchased from Ferrofluidics Corp., Nashua, NH. Tresylklorid (2,2,2-trifluoroethanesulfonyl chloride) was from Fluka and the vegetable oils (corn, cottonseed, or rapeseed) were purchased in the local grocery store.

Crystallized Carbohydrate Spheres (CCS) for Slow Release

Entrapment of substances in carbohydrate spheres for slow release was performed according to the following procedure. One gram of an aqueous carbohydrate solution was thoroughly mixed with 200 μl of an aqueous solution of the substance to be entrapped. The substances were dissolved in their normal solvents such as buffers and water. Thirty milliliters of an emulsifying medium (corn, rapeseed, or cottonseed oil) was added at a temperature of 4° and a coarse preemulsion was formed by magnetic stirring. The preemulsion was further homogenized by probe (0.15 mm) sonication (Ultrasonic Ltd., model A 350 G) for 40 sec while cooling in an ice bath or by high-pressure homogenization. The microdroplet suspension was then slowly poured into 200 ml of acetone, containing 0.1% (w/v) Tween 80, while stirring at 1000 rpm. The carbohydrate precipitated in the acetone, forming crystallized carbohydrate spheres, which were washed by centrifugation at 3000 g with 4 × 50 ml of 0.1% Tween 80–acetone. Finally the CCS was suspended in 2 ml 1% Tween–acetone and allowed to dry at room temperature. If sterile CCS were wanted, the last wash and the drying was performed under sterile conditions.

In order to prevent spontaneous crystallization caused by the high concentrations of the carbohydrate, these solutions were freshly prepared before use. The dextrans, the mono- and disaccharides, and the dextrins were dissolved in water by gentle heating. Starch can be dissolved in dimethyl sulfoxide (DMSO), formamide, or any other solvent that has a high dielectric constant. The highest concentration of the carbohydrate that can be used is determined by the increasing viscosity of the solutions, making them difficult to suspend in the emulsifying medium. This concentration corresponds to approximately 0.5 g/ml for dextran T500, and 1 g/ml for the dextran T10 and the dextrin Awedex W90. Glucose, sucrose, maltose, and other low molecular weight carbohydrates can be used up to a concentration of about 2.5 g/ml. At the lowest concentration, which is about one-tenth of those mentioned above, the CCS are rapidly redissolved when suspended in water.

Two different emulsifying media can be used, vegetable oil or xylene–chloroform in a ratio of 4:1, containing 2.5% (w/v) of the detergents Pluronic F-68 and L-35, respectively. The organic solvent system gives slightly smaller spheres and the carbohydrate solution is more readily suspended. However, due to the toxicity of the organic solvents, the vegetable oils are preferred.

The carbohydrates preferentially used are the dextrans, because of their well-known and well-documented use of plasma volume expanders in the clinic for almost 25 years, and the dextrins (acid-hydrolyzed starch) because of their biocompatability and biodegradability.

Homogenization can be achieved by either probe sonication or high-pressure homogenization. Because of its simplicity and high reproducibility, the high-pressure homogenization is the method of choice, especially when larger amounts of entrapped material are required. Other advantages of the method are the possibilities of making CCS of various sizes and of creating by a recycling procedure very narrow size distributions.

Assay of the Biological Activity of the Entrapped Proteins

In making CCS from a 20% dextran T10, they are rapidly dissolved when suspended in water, thus making determinations of the biological activity of the proteins after entrapment and subsequent dissolution possible. Insulin was allowed to stimulate adipocytes to incorporate [³H]glucose and the production of ³H-labeled fatty acids was determined by liquid scintillation.[10] The anti-viral effect of interferon was assayed in a plaque assay.[11] Plasmin was detected using the substrate H-D-Val-Leu-Lys-pNA[12] (S-2251, Kabi Diagnostika, Stockholm, Sweden). β-Galactosidase was determined with the substrate ONPG (Sigma).[13] The mono-

[10] A. J. Moody, M. A. Stan, M. Stan, and J. Gliemann, *Horm. Metab. Res.* **6**, 12 (1974).
[11] J. B. Campbell, M. A. Kochman, and S. L. White, *Can. J. Microbiol.* **21**, 1247 (1975).
[12] P. Friberger, M. Knös, S. Gustavsson, L. Aurell, and G. Claesson, *Haemostasis* **7**, 138 (1978).
[13] G. R. Craven, E. Steers, Jr., and C. B. Anfinsen, *J. Biol. Chem.* **240**, 2468 (1965).

clonal antibody against PHA was determined in an ELISA technique, using peroxidase-labeled anti-mouse IgG (Dakopatt, catalog no. P161) as second antibody and ABTS (Sigma A-1888) as chromogene and H_2O_2 as substrate to the enzyme.[14]

When the biological activity of the protein insulin, the enzymatic activity of the enzymes plasmin and β-galactosidase, or the binding capacity of the monoclonal antibody were assayed, all retained approximately 70% of their activity compared to the activity added for entrapment. Furthermore, when radioactively labeled insulin or HSA were entrapped, 70% of the added radioactivity was found to be entrapped within the spheres, implying that all of the proteins entrapped retained their biological activity (see table).

Hydrophobic proteins were normally entrapped with a lower yield when dextran as the matrix and the vegetable oils as emulsifying systems were used. This was seen with interferon, human growth hormone, and myoglobin, for which only about 20–40% of the protein was found to be entrapped. However, using the dextrin Awedex W90 as a matrix and a 0.1% of the emulgator Gafac PE-410 in the oil, the entrapment efficiency could be increased and enabled an entrapment of 100% of human growth hormone.

Furthermore, no significant difficulties were seen in the preparation of the spheres whether the final protein content was 0.2% (dry weight) or 35%, tested with a 3 g/ml solution of dextran T500 and ovalbumin.

Crystallized dextran spheres, made from a 0.5 g/ml solution of dextran T40 and suspended in water, withstand complete erosion over a period of more than 8 months at room temperature.

Release Studies

For *in vitro* release studies, 100 mg of dried CCS with entrapped radioactively labeled proteins or drugs was suspended in 10 ml of PBS, containing 1% (w/v) HSA, at room temperature. Samples (100–300 μl) of released substances were collected from the suspension at various intervals by filtration through a membrane with a molecular weight cut-off of 100,000 (Millipore PTHK 07610).

Testing of *in vitro* release of radioactively labeled proteins showed that 50% release was achieved in 15 days for human growth hormone, in 8 days for HSA and myoglobin, and in about 4 days for insulin (Fig. 2). With the use of the dextrins as matrix material for the CCS, one would anticipate that the amount of α-amylase would influence the release rate. However, no such influence could be detected even at enzyme concentra-

[14] C. A. K. Borrebaeck and M. E. Etzler, *J. Biol. Chem.* **256,** 4723 (1981).

TABLE (Protein Entrapped in CCS and Used in Recovery and Slow Release Studies)

Protein Entrapped in CCS and Used in Recovery and Slow Release Studies[a]

| | Recovery studies | | | | Slow release studies | |
Substance	MW	Conc. (mg/ml)	Radiolabeled recovery (%)	Biological activity (%)	Carbohydrates	Concentration (g/ml)
Insulin	6,000	0.92	70	70	Dextran T500	0.35
Myoglobin	16,000	83	35	—	Dextran T500	0.35
Interferon	18,000	N. D.	N. D.	20	N. D.	
Human growth hormone	21,000	2	100	N. D.	Dextrin (Awedex W90)	0.80
Human growth hormone	21,000	2	35	N. D.	Dextran T500	0.35
Human serum albumin	66,000	83	70	N. D.	Dextran T500	0.35
Plasmin	90,000	3	N. D.	75	N. D.	
Monoclonal antibody	15,000	N. D.	N. D.	65	N. D.	
β-Galactosidase	480,000	20	N. D.	70	N. D.	
Vincristine sulfate	850	5	75	N. D.	Glucose	2.5
Thyroxine	777	0.1	10	N. D.	Dextran T10	0.80

[a] In the recovery studies the proteins were entrapped in CCS made from a 20% (w/v) solution of dextran T10. N. D., not determined.

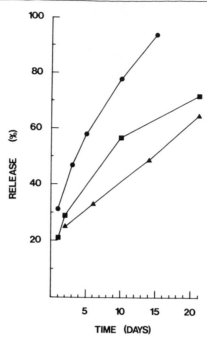

FIG. 2. Release of entrapped insulin (●), HSA and myoglobin (■), and biosynthetic human growth hormone (▲) in PBS containing 1% HSA at room temperature. CCS (100 mg) contained 1 mg of insulin, 40 mg of HSA, 4 mg of myoglobin, and 2 mg of growth hormone, respectively. The insulin, HSA, and the myoglobin were entrapped in dextran T500 and the growth hormone was entrapped in dextrin.

tions 10,000 times higher than in normal human blood serum. The initial burst effect that was seen during the first 24 hr was probably due to the spheroid shape of the matrix, giving a high surface area.[1] In these studies, no significant variations in the release rate were seen at the different entrapment levels.

Assay of the Immunological Response to Entrapped Antigens

For immunization, groups of five mice were used and each mouse was injected at a single site, subcutaneously above the left hind leg with 0.4 ml of the antigen preparations. Blood samples were taken from the tail with 100 μl microcaps and centrifuged before analysis. Analysis of the antibody response was performed with serial duplicate dilutions of the samples with an ELISA technique, using the same method as above when analyzing PHA in the slow release experiments. Using a pooled serum control, corrections for day-to-day variations in the assays were made as

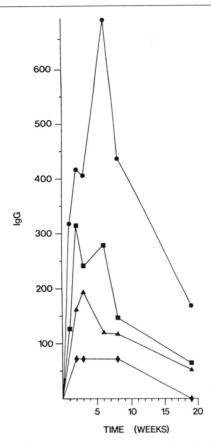

FIG. 3. IgG response after subcutaneously injection in mice of 80 μg ovalbumin, entrapped in dextran spheres: 65% of dextran was sulfate dextran (●), 35% of dextran was DEAE-dextran (■), unsubstituted dextran (▲). Control, ovalbumin dissolved in water (♦).

well as a subtraction of background values before the absorbance values from the serial dilutions were plotted on semilogarithmic paper; the obtained dilution factors for the antigen preparations were determined from the parallel lines.

As a model system of a well-known antigen, ovalbumin was chosen and entrapped into dextran spheres.[15] The mice given a single subcutaneous injection of 80 μg of ovalbumin entrapped in dextran spheres had, after about 1 month, a peak in antibody concentration, which slowly declined to a level comparable to the highest level obtained in the mice that received 80 μg of the antigen dissolved to water (Fig. 3). When

[15] U. Schröder and A. Ståhl, **70,** 127 (1984).

injecting the ovalbumin dissolved in water together with 5% solution of the dextran, no enhancement of the IgG antibody level as compared to water could be detected. As anticipated, co-entrapment of DEAE– or sulfate–dextran gave an even better enhancement of the IgG level compared to CCS made of non-substituted dextrans (Fig. 3). When injecting the antigen and the substituted dextran in soluble form, the antibody response was again at a considerably lower level than the CCS formulations. The literature presents a broad spectrum of various immunomodulators, most of which can be entrapped or associated to the CCS, thus permitting the most favorable enhancement of the immune response against an antigen.

When bee venom was entrapped in the CCS and compared with mice that received the antigen dissolved in water, the CCS preparation of antigen was shown to enhance the IgG level approximately to the same O.D. as with ovalbumin, suggesting approximately the same enhancement as with ovalbumin. When injecting 80 μg of the antigen dissolved in water and with subsequent reinjection 2 weeks later with the same dose, the IgG level, another 2 weeks later, reached the same level as that achieved with a single injection of 8 μg entrapped in the CCS, assayed 4 weeks after the injection.

Injecting 400 μg of bee venom entrapped in CCS (which is four times the LD_{50} in mice), the IgG response increased rapidly. For comparison, mice injected with 80 μg bee venom suspended in FCA is shown in Fig. 4. No mouse in this high dose group, receiving the CCS–bee venom preparation, showed any visible side reactions, i.e., inflammatory response or necrosis at the injection site, compared to the mice receiving the bee venom in FCA, which had to be terminated after 8 weeks. This implies that by using CCS preparations when injecting high doses of toxic antigens an escalation in the toxic side reactions may be avoided and at the same time a high antibody response can be achieved.

When the antigens from schistosomiasis were tested in CCS, the antibody response was of the same magnitude as that for ovalbumin.

Magnetic Microspheres for Drug Targeting

When preparing magnetic dextran spheres,[16] 600 μl Ferrofluid EMG 805 HGMS was mixed with 400 μl of water and 200 mg of dextran T500 (or 500 mg dextran T40), and heated to 70° for 1 hr. Instead of dextran, it is also possible to use 1 g of glucose or maltose, 250 mg of dextrin Awedex W-25, or any other carbohydrate that might be of special interest. The Ferrofluid carbohydrate suspension was emulsified in an emulsifying medium consisting of 25 ml of vegetable oil, 1.25 g Gafac PE-510, and 5 ml of

---

[16] U. Schröder, A. Ståhl, and L. G. Salford, submitted for publication.

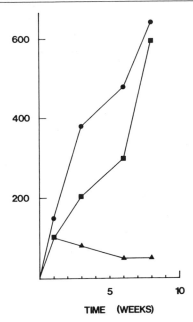

F$_{IG}$. 4. IgG response after subcutaneous injection of 80 μg of bee venom dissolved in water (▲), 80 μg suspended in Freund's complete adjuvant (■), or 400 μg entrapped in dextran spheres (●).

chloroform with the probe sonicator for 35 sec. The carbohydrates were stabilized by crystallization by pouring the emulsion into acetone containing 0.1% Tween 80 while stirring the acetone at 1000 rpm. The magnetic carbohydrate spheres were washed three times with 50 ml of the acetone–Tween solution and then suspended in 2 ml of a 5% solution of Tween 80 in acetone and allowed to dry while being gently agitated at room temperature.

Magnetic microspheres with entrapped ^{125}I-labeled HSA were used in animal studies to evaluate the possibility of using these magnetic carbohydrate spheres for drug targeting. In a series of experiments on rats, a suspension of the microspheres was injected intraarterially or intravenously while a selected part of the body was exposed to a magnetic field of 0.8–1.4 T (tesla) produced by a bipolar magnet. Thus, it was possible to arrest the microspheres in the temporal muscle of the rat after intravenous injection. By the use of intraarterial injection, the microspheres could be arrested in the rat tail, with the highest concentration of microspheres in the vicinity of the pole edges of the magnet. The major efforts have, however, been directed toward the use of magnetically responsive microspheres in the brain (for use against brain tumors). By intraarterial injection into the internal carotid artery, the microspheres could be arrested in a part of one hemisphere of the brain. Using radioactively labeled magnetic carbohydrate microspheres, it was proved that the microspheres could be concentrated 100-fold compared to the negligible amounts of spheres in the brain when the magnetic field was not used.

With the goal of arresting magnetically responsive carbohydrate microspheres in a selected area of the body, it is of utmost importance to utilize the maximum field gradient. The fact that these carbohydrate microspheres can be arrested in such a highly vascularized tissue as that of the brain gives promise for the future use of this method in both tumors and other tissues in any part of the body.

Summary

Crystallized carbohydrate spheres have been prepared for use as a matrix for entrapment of biologically active substances, such as insulin, interferon, and growth hormone, which, when released, retain their biological activities, and where the carbohydrates are well known with documented low toxicity in human, as well as in the veterinary medical fields. The crystallized carbohydrate spheres containing entrapped antigens with co-entrapped immunomodulators may also be considered as a possible matrix of choice in immunization with weak antigens in humans, where the adjuvant effect presumably is a combination of slow release and macrophage stimulation, which is due to the small size of the spheres. By co-entrapment of magnetite particles it is possible to target magnetically responsive carbohydrate microspheres to the brain, for example, or any other area in the body, thus promoting higher concentration of drug in the target area. The method of preparation, using the high-pressure homogenization technique, is easily scaled up, and with the use of commercially available devices, one could easily fabricate tons of the CCS per day. Release of the entrapped substances probably occurs via erosion of the matrix, rendering degradation products that are nonmodified as compared to the starting material. A further advantage of this concept is that there is no enzymatic degradation of the matrix influencing the release, since the parameters capable of breaking the hydrogen bonds stabilizing the crystallized structure are mainly pH and ionic strength, both of which are stable in a human body.

Acknowledgments

This work is partially supported by the Swedish Board for Technical Development (83-4403) and the Swedish Cancer Society (83-211).

[10] Poly(alkyl acrylate) Nanoparticles

By Jörg Kreuter

Nanoparticles are solid colloidal particles ranging in size from 10 to 1000 nm (1 μm). They consist of macromolecular materials in which the active principle (drug or biologically active materials) is dissolved, entrapped, and encapsulated, and/or to which the active substance is adsorbed or attached.[1] This definition includes most particles or microspheres dealt with in this volume, although the term *nanoparticles* was first used only for polyacrylic nanoparticles produced by emulsion polymerization with both a continuous aqueous or organic phase[2–5] or for gelatin or albumin nanoparticles produced by a desolvation process.[6,7]

This chapter deals with the preparation of poly(alkyl acrylic) and poly(alkyl cyanocrylic) nanoparticles. Poly(alkyl acrylic) nanoparticles are much more slowly biodegradable than poly(alkyl cyanoacrylate) nanoparticles. Therefore, the latter are much more suitable for drug delivery purposes. In addition, one of these materials, poly(butyl cyanoacrylate), has been used extensively in surgery as tissue glue for over 15 years. This material seems to be nontoxic and well tolerated. The biodegradability and also the toxicity increase with decreasing side-chain length. After intravenous injection into mice, over 75% of the poly(butyl cyanoacrylate) nanoparticle radioactivity has left the body within 24 hr,[8] while with poly(hexyl cyanoacrylate) nanoparticles, 80% of the radioactivity still remains in the body after 3 days (unpublished observation). The biodegradation time, however, can be monitored by mixing different cyanoacrylates.[9]

Poly(methyl methacrylate), on the other hand, is the material of choice for the use of nanoparticles as an adjuvant for vaccines.[10,11] The efficacy

[1] J. Kreuter, *Pharm. Acta Helv.* **58**, 196 (1983).
[2] H. Kopf, R. K. Joshi, M. Soliva, and P. Speiser, *Pharm. Ind.* **38**, 281 (1976).
[3] H. Kopf, R. K. Joshi, M. Soliva, and P. Speiser, *Pharm. Ind.* **39**, 993 (1977).
[4] J. Kreuter, *Pharm. Acta Helv.* **53**, 33 (1978).
[5] P. Couvreur, B. Kante, M. Roland, P. Guiot, P. Baudhuin, and P. Speiser, *J. Pharm. Pharmacol.* **31**, 331 (1979).
[6] J. J. Marty and R. C. Oppenheim, *Aust. J. Pharm. Sci.* **6**, 65 (1977).
[7] J. J. Marty, R. C. Oppenheim, and P. Speiser, *Pharm. Acta Helv.* **53**, 17 (1978).
[8] L. Grislain, P. Couvreur, V. Lenaerts, M. Roland, D. Deprez-Decampeneere, and P. Speiser, *Int. J. Pharm.* **15**, 335 (1983).
[9] P. Couvreur, B. Kante, M. Roland, and P. Speiser, *J. Pharm. Sci.* **68**, 1521 (1979).
[10] J. Kreuter and P. P. Speiser, *Infect. Immun.* **13**, 204 (1976).
[11] J. Kreuter and E. Liehl, *J. Pharm. Sci.* **70**, 367 (1981).

METHODS IN ENZYMOLOGY, VOL. 112

of polyacrylic adjuvants increases with increasing hydrophobicity and decreasing particle size.[12] Poly(methyl methacrylate) nanoparticles are also biodegradable although very slowly.[13]

Poly(methyl methacrylate) Nanoparticles

Poly(methyl methacrylate) nanoparticles are preferentially produced by emulsifier-free polymerization in aqueous media, because detergents may interact with and damage certain biological materials or change the immunological response.[1,14,15] The mechanism for this type of polymerization is concisely reviewed in *Pharmaceutica Acta Helvetiae*.[1] The polymerization can be initiated with γ rays[10,16] or with potassium peroxodisulfate.[17]

Materials and Methods

Purification of the Monomers. One hundred milliliters methyl methacrylate was first extracted three times with 20 ml of a solution of 5 g NaOH and 20 g NaCl in 100 ml twice distilled water and then extracted another three times with 20 ml twice distilled water.

Polymerization with γ Rays. A certain amount (between 0.1 and 1.5%) of purified methyl methacrylate is dissolved in twice-distilled water and irradiated with 500 krad γ rays in a ^{60}Co source.[16] The resulting nanoparticle suspension can be used or stored as such or in lyophilized form. Instead of water, buffers can be used.

Biologically active materials, as for instance antigens, can be entrapped within the particles by polymerization of the monomer in the presence of the active material.[10,11] In this case, the solution or suspension of the biologically active material in water or buffer is used in the above formulation instead of the twice distilled water.

Alternatively, the biologically active material may be adsorbed to the nanoparticles after polymerization,[10,11] especially if the material is labile to γ-irradiation or heating (see below). In this case, lyophilized or centrifuged (680 g or 1000 rpm in a laboratory centrifuge for 15 min) nanoparticles are resuspended in the solution or suspension of the biologically active material.

[12] U. Berg, "Immunstimulation durch hochdisperse Polymersuspensionen," Diss. Eidgenössische Technische Hochschule No. 6481, p. 84. Zürich, 1979.
[13] J. Kreuter, M. Nefzger, E. Liehl, R. Czok, and R. Voges, *J. Pharm. Sci.* **72**, 1146 (1983).
[14] D. Gall, *Symp. Ser. Immunobiol. Stand.* **6**, 49 (1967).
[15] H. Bachmayer, *Intervirology* **5**, 191 (1969).
[16] J. Kreuter and H. J. Zehnder, *Radiat. Eff.* **35**, 161 (1978).
[17] U. Berg, "Immunstimulation durch hochdisperse Polymersuspensionen," Diss. Eidgenössische Technische Hochschule No. 6481, p. 34. Zürich, 1979.

TABLE I

INFLUENCE OF MONOMER CONCENTRATION, INITIATOR CONCENTRATION,
AND TEMPERATURE ON THE PARTICLE SIZE
OF POLY(METHYL METHACRYLATE) NANOPARTICLES[a]

Potassium peroxodisulfate concentration (mmol)	Particle size (nm)							
	Methyl methacrylate concentration (mmol) at two temperatures							
	10		33.75		80		156.25	
	65°	85°	65°	85°	65°	85°	65°	85°
0.3	85	72	129	128	181	170	256	262
1.65	98	88	151	169	212	193	248	248
3.0	92	72	135	149	223	177	250	258

[a] From ref. 17.

The presence of another colloid (either material to be entrapped or protection colloids) during polymerization (for instance 2% *Streptococcus mutans* or 1% casein) protects the polymer particles from flocculation[1] and may lead to an increased particle size and a very homogeneous particle size distribution.[1]

Polymerization with Potassium Peroxodisulfate. The second possibility for the production of poly(methyl methacrylate) nanoparticles is the polymerization by initiation with potassium peroxodisulfate at elevated temperatures (60 or 80°).[17] High initiator concentrations (>5 mmol) lead to precipitation and flocculation comparable to that observed with γ-ray-initiated polymerization. Lower initiator concentrations lead to the formation of a stable latex. The particle size increases significantly with increasing amounts of monomers (Table I). The molecular weights of the nanoparticles increase with increasing monomer concentration, decreasing initiator concentration, and decreasing temperature.[17,18]

Degassed twice-distilled water is added into a jacketed round- or flat-bottom flask that can be closed partially with a funnel in order to minimize evaporation. Nitrogen is then bubbled through the water for 15 min and then a certain amount of monomer (see Table I) is added and dissolved in the water by stirring with a magnetic stirrer. The temperature is then raised gradually (about 2°/min) to about 45°. Then the potassium peroxodisulfate dissolved in a small amount of water (this amount should not exceed 4% of the volume of the aqueous monomer solution) is added to the monomer solution and the temperature is further increased with

[18] V. Bentele, U. E. Berg, and J. Kreuter, *Int. J. Pharm.* **13**, 109 (1983).

stirring to 65 or 85°. The temperature is then kept at this level for 2 hr. The resulting nanoparticles suspension can then be treated as described under polymerization with γ rays.

Poly(alkyl cyanoacrylate) Nanoparticles

Poly(alkyl cyanoacrylate) nanoparticles are generated by an anionic polymerization mechanism by base induction. The OH^- ions of water are sufficient to induce the polymerization of these cyanoacrylates. At neutral pH and hence high hydroxyl ion concentrations, the polymerization rate is too rapid to allow discrete particle formation. At very low pH (<1), the polymerization time is greatly extended and the polymerized particles become swollen with monomer resulting in a greater particle size and a higher polydispersity.[19] Douglas et al.[19] observed a minimum in particle size at a pH of 2 and a minimal polydispersity for pH >2.5. The influence of the monomer concentrations on the particle size was more complex: the size increased from about 100 nm at very low concentrations to about 130 nm at 1% butyl cyanoacrylate. Then it decreased again to a minimum of 110 nm at 2% and further increased to about 160 nm at 7%.[19] Suspensions with concentrations above 7% were not free flowing. The stirring rate, electrolytes, the acid used with the exception of phosphoric acid, and the polymerization temperature had no significant influence on the particle size. The molecular weight decreased with decreasing pH and increased with increasing surfactant concentrations, but was in general very low (~3000).[20]

Materials and Methods

Examples of polymerization media are given in Table II. Empty poly(cyanoacrylate) nanoparticles can be produced by omitting the drug. After dissolution of all components, a desired amount of monomer (methyl, ethyl, butyl, or hexyl cyanoacrylate; 0.1–5%) is added and the polymerization mixture stirred with a paddle or a magnetic stirrer at room temperature for at least 3 hr for butyl cyanoacrylate and lower side-chain cyanoacrylates and for at least 4 days for hexyl cyanoacrylate. After this time, the suspension is filtered through a G 1 glass frit or a 1-μm membrane filter in order to remove lumps of poly(cyanoacrylate) that can occasionally form. The polymerization mixture is then brought to the desired pH with 1 N NaOH and to isotonicity with NaCl or phosphate buffer and stirred for at least another 2 hr. Certain drugs, such as strong

[19] S. J. Douglas, L. Illum, S. S. Davis, and J. Kreuter, J. Colloid. Interface Sci. **101,** 149 (1984).
[20] M. A. El-Egakey, V. Bentele, and J. Kreuter, Int. J. Pharm. **13,** 349 (1983).

TABLE II
POLYMERIZATION MEDIA FOR THE PREPARATION
OF POLY(ALKYL CYANOACRYLATE) NANOPARTICLES

	Polymerization system				
Acid	0.1 N HCl 10.0%	0.1 N HCl 10.0%	Citric acid 0.5%	1 N HCl 10.0%	1 N H$_3$PO$_4$ 1.0%
Surfactant	Polysorbate 20 0.5%	Pluronic L 63 0.8%	—	Polysorbate 20 0.5%	Pluronic F 68 0.1%
Other excipients	—	Dextran 70 0.2%	Dextran 70 1.0% Calcium chloride 0.01%	—	Dextran 70 1.0%
Drugs	Dactinomycin Methotrexate Vinblastine	Dactinomycin	Doxorubicin 0.1%	5-Fluorouracil 1.0%	Vincamine 0.2%
References[a]	5, 9, a	b, c	d	e	f

[a] Key to references: numbers refer to references given in text; (a) P. Couvreur, B. Kante, V. Lenaerts, V. Scailteur, M. Roland, and P. Speiser, *J. Pharm. Sci.* **69,** 199 (1980); (b) B. Kante, P. Couvreur, V. Lenaerts, P. Guiot, M. Roland, P. Baudhuin, and P. Speiser, *Int. J. Pharm.* **7,** 45 (1980); (c) B. Kante, P. Couvreur, G. Dubois-Krack, C. De Meester, P. Guiot, M. Roland, M. Mercier, and P. Speiser, *J. Pharm. Sci.* **71,** 786 (1982); (d) P. Couvreur, B. Kante, L. Grislain, M. Roland, and P. Speiser, *J. Pharm. Sci.* **71,** 790 (1982); (e) J. Kreuter and H. R. Hartmann, *Oncology* **40,** 363 (1983); (f) P. Maincent, Thesis, Université Paris-Sud, Paris (1982).

bases (for instance vindesine) that would be incorporated into the polymer chain by covalent binding, or drugs that are too insoluble at low pH, can be added shortly before or after neutralization.

The particle size of the cyanoacrylate nanoparticles is close to the pore size of many filters used for sterile filtration and the particles are very elastic. This can lead to the clogging of the filters. Therefore, it is better to work under aseptic conditions if sterile products are desired. It has to be mentioned that the monomers have a strong bactericidal action causing sterility of the polymerization mixture after addition of the monomers. The polymer has no bactericidal action at all.

If the particles have to be stored at a pH above 5, the suspension should be lyophilized in order to ensure their stability.[21]

Magnetic Poly(cyanoacrylate) Nanoparticles. Magnetic poly(cyanoacrylate) nanoparticles can be produced as follows[22]: After the dissolution of 1 g of glucose and 1 g of citric acid in 100 ml of distilled water, 0.7 g of magnetite particles of a size between 10 and 50 nm is dispersed by

[21] J. Kreuter, *Pharm. Acta Helv.* **58,** 242 (1983).
[22] A. Ibrahim, P. Couvreur, M. Roland, and P. Speiser, *J. Pharm. Pharmacol.* **35,** 59 (1983).

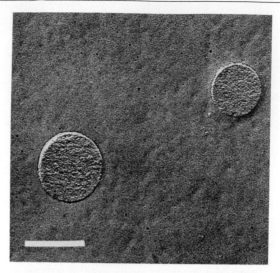

FIG. 1. Transmission electron micrograph of poly(methyl cyanoacrylate) nanoparticles after freeze-fracturing. Bar = 200 nm. (From ref. 23. Reproduced with permission of the copyright holder.)

ultrasonication for 15 min. Then the suspension is passed through a glass filter (pore size 9–15 μm) to avoid magnetite agglomerates. Dactinomycin (27.4 mmol) and then 1.5 ml isobutyl cyanoacrylate monomer are added and stirred with a magnetic stirrer under simultaneous ultrasonication for 3 hr, and filtered through a glass filter (pore size 9–15 μm).

To separate magnetized nanoparticles, this suspension is then pumped through a magnetic field at a rate of 0.3 ml/min with a peristaltic pump. After removal of the magnetic field, the nanoparticles attached to the internal surface of the tubes are removed and washed with an aqueous solution containing 0.7% NaCl and 0.2% $CaCl_2 \cdot 2H_2O$. The resulting suspension can be homogenized by ultrasonication for 15 min and is again filtered through the above glass filter.

Results

Poly(alkyl acrylate) nanoparticles normally have a particle size from between 50 and 300 nm. They possess a solid, rather porous polymer matrix of low density (Fig. 1). Their density is considerably lower than that of normal polymer beads or rods made with the same monomers.[23]

[23] J. Kreuter, *Int. J. Pharm.* **14,** 43 (1983).

TABLE III
ELECTROPHORETIC MOBILITY OF NANOPARTICLES[a]

Polymer	Electrophoretic mobility (μm cm sec^{-1} v^{-1})		
	Water	Phosphate-buffered saline, pH 7.4	Human serum
Poly(methyl methacrylate)	-2.76 ± 0.57	-1.30 ± 0.07	-0.25 ± 0.05
Poly(acrylamide)	<-0.35	—	—
Poly(methyl cyanoacrylate)	-2.33 ± 0.10	-1.64 ± 0.12	-0.23 ± 0.02
Poly(ethyl cyanoacrylate)	-2.18 ± 0.16	-1.32 ± 0.16	-0.23 ± 0.02
Poly(butyl cyanoacrylate)	-2.01 ± 0.16	-0.87 ± 0.07	-0.19 ± 0.03

[a] From ref. 23. Data given represent mean $\pm$ SD.

This low density yields a rather high sorption capacity for these nanoparticles.[21,24]

Poly(alkyl acrylate) nanoparticles are X-ray amorphous. The hydrophobicity increases in the order poly(methyl cyanoacrylate) < poly(ethyl cyanoacrylate) < poly(butyl cyanoacrylate) < poly(methyl methacrylate).[23] The increase in hydrophobicity is coupled by an increase in negative charge. The negative charge almost disappears after suspension in serum (Table III). This is caused by the adsorption of blood components on the surface of the nanoparticles. This adsorption of blood components also leads to a reduction of the water contact angles from 73° for the pure material to 53° for the product treated with serum in the case of poly-(methyl methacrylate) and from 69 to 45° in the case of poly(butyl cyanoacrylate.[23] Among the substances absorbed from the blood most probably are also opsonins which trigger phagocytosis by macrophages and are therefore responsible for the rapid uptake of colloids by the reticuloendothelial system.

As a result, the body distribution of poly(alkyl acrylic) nanoparticles is the same as with other colloids or microspheres of a particle size below 1 μm. The particles are taken up by the reticuloendothelial system after intravenous injection and distributed mainly into the liver (60–90%), spleen (2–10%), and to a low degree, the bone marrow.[25,26] Agglomerates of nanoparticles can also distribute temporarily into the lungs.

[24] H. S. Yalabik-Kas, J. Kreuter, A. A. Hincal, and P. Speiser, *3rd Int. Congr. Pharm. Technol., Paris 1983* Proc. 1, p. 246 (1983).
[25] J. Kreuter, *Pharm. Acta Helv.* **58**, 217 (1983).
[26] J. Kreuter, U. Täuber, and V. Illi, *J. Pharm. Sci.* **65**, 838 (1979).

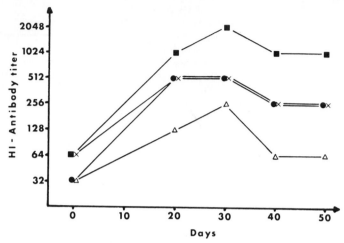

FIG. 2. Influence of different adjuvants on the antibody response of mice. Adjuvants: (■) 0.5% poly(methyl methacrylate) copolymerized with 0.5% poly(acrylamide), polymerization in presence of the antigen; (●) 0.5% poly(methyl methacrylate) copolymerized with 0.5% polyacrylamide, adsorption of antigen to empty polymer particles; (×) 0.2% Al(OH)$_3$; (△) fluid vaccine without adjuvant. Antigen: 62.5 CCA of A$_2$/Hong Kong X-31, whole virions. (From ref. 10. Reproduced with permission of the copyright holder.)

After subcutaneous administration of labeled poly(methyl methacrylate) nanoparticles, almost all of the radioactivity stayed at the injection site, while less than 1% of the administered dose was found in the residual body.[25] After 200 days, an increased distribution into the rest of the body with an increased excretion via the feces was observed. After 287 days, the radioactivity levels in all organs were more than 100 times higher than after 187 days. After oral administration of radiolabeled poly(methyl methacrylate) nanoparticles, 10–15% of the radioactivity was absorbed.[25]

Poly(methyl methacrylate) nanoparticles are, as mentioned in the introduction, very promising adjuvants. After polymerization in the presence of the antigen, the adjuvant effect was dependent on the poly(methyl methacrylate) concentration used, reaching an optimal concentration at 0.5% poly(methyl methacrylate).[10] The optimum, however, is dependent on the antigen used. The preparation produced by incorporation of the antigen into the optimal polymer concentration was considerably more effective than simple addition of the antigen to comparable poly(methyl methacrylate) preparations or to aluminum hydroxide (Fig. 2). The latter two adjuvants were approximately equivalent in effectiveness, although the reproducibility of the adjuvant effect was greatly enhanced with

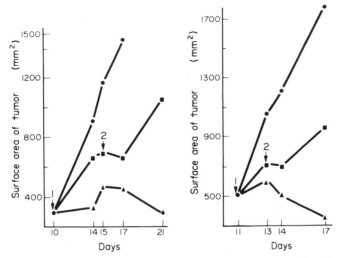

FIG. 3. Results from two different replicate experiments on tumor size of soft tissue sarcoma S 250 treated with free actinomycin D (■); or with actinomycin D adsorbed on nanoparticles (▲). Dosage: two injections (arrows) of 222 μg of actinomycin D per kilogram of rat. Control (●). (From ref. 31. Reproduced with permission of the copyright holder.)

poly(methyl methacrylate) nanoparticles over aluminum hydroxide.[27] Long-term experiments showed a prolonged antibody response of polymer adjuvants with incorporated or adsorbed influenza virus. Adsorption also yielded an optimal adjuvant effect with 0.5% poly(methyl methacrylate).[11] The antibody response was accompanied by a protection of the mice against infection with mice-adapted influenza virus. In addition, the polymer vaccines were much more stable against temperature inactivation than were vaccines with aluminum hydroxide or without adjuvants. Histological examinations of the tissue at the injection site 1 year after injection of poly(methyl methacrylate) nanoparticles into seven guinea pigs showed no abnormalities. Histological reactions were the same as in the fluid vaccine control.[28]

The most promising application of poly(alkyl cyanoacrylate) nanoparticles seems to be their use as carriers for antitumoral agents. A number of tumors exhibit an enhanced endocytotic activity. The enhanced endocytotic activity is expressed by the very pronounced increase in lipid clearance in the Lipofundin Clearance test in cancer patients.[29] The possible

[27] J. Kreuter and I. Haenzel, *Infect. Immun.* **19,** 667 (1978).
[28] J. Kreuter, R. Mauler, H. Gruschkau, and P. P. Speiser, *Exp. Cell Biol.* **44,** 12 (1976).
[29] G. Lemperle and M. Reichelt, *Med. Klin. (Munich)* **68,** 48 (1973).

TABLE IV

ANTITUMOR EFFECTS AND SUBACUTE TOXICITY OF 5-FLUOROURACIL
IN POLY(BUTYL CYANOACRYLATE) NANOPARTICLES[a]

Sample	Dose, ip (mg/kg)	Tumor weight (mg ± SD)	Body weight change (g)	White blood cell counts	Survivors
Controls		1790 ± 150	+5.2	>6000	10/10
5-FU/NP[b] (1 : 1)	6.25	1830 ± 80	+5.0	>6000	10.10
	12.5	1430 ± 150	+3.9	>6000	9/10
	25.0	800 ± 120	0	2850	6/10
5-FU/NP[b] (1 : 10)	6.25	1280 ± 110	+0.6	>6000	10/10
	12.5	940 ± 80	−0.1	>6000	9/10
	25.0	650 ± 80	−2.8	2340	5/10
5-FU in saline	12.5	1640 ± 140	—	—	10/10
	25.0	930 ± 60	+4.0	4900	10/10
	37.5	610 ± 90	−6.1	2070	4/10

[a] From ref. 32.
[b] 5-Fluorouracil bound to nanoparticles.

use of nanoparticles as endocytotic and lysosomotropic carriers was demonstrated by Couvreur *et al.*[30]

First results with nanoparticles as carriers for cytotoxics were very promising. Brasseur *et al.*[31] observed a considerable decrease in tumor growth of a soft tissue carcinoma of rats (S 250) with actinomycin D adsorbed to poly(methyl cyanoacrylate) nanoparticles in comparison to the free drug (Fig. 3). Nanoparticles without drug had no influence on the tumor growth.

Kreuter and Hartmann[32] showed an enhanced efficacy of 5-fluorouracil against Crocker sarcoma S 180 by binding to nanoparticles (Table IV). The increased efficacy, however, was coupled with a higher toxicity of the drug measured by induced leukopenia, body weight loss, and premature death.

In contrast, the binding of doxorubicin to nanoparticles resulted in a considerable decrease in the heart accumulation and yielded a significant decrease in acute toxicity of this drug.[33]

These results demonstrate that poly(alkyl cyanoacrylate) nanoparticles hold great promise as carriers for antitumor agents.

[30] P. Couvreur, P. Tulkens, M. Roland, A. Trouet, and P. Speiser, *FEBS Lett.* **84,** 323 (1977).

[31] F. Brasseur, P. Couvreur, B. Kante, L. Deckers-Passau, M. Roland, C. Deckers, and P. Speiser, *Eur. J. Cancer* **16,** 1441 (1980).

[32] J. Kreuter and H. R. Hartmann, *Oncology* **40,** 363 (1983).

[33] P. Couvreur, B. Kante, L. Grislain, M. Roland, and P. Speiser, *J. Pharm. Sci.* **71,** 790 (1982).

[11] Ethylcellulose Microcapsules for Selective Drug Delivery

By TETSURO KATO, KATSUO UNNO, and AKIO GOTO

Microencapsulation is designed to protect, separate, or change the diverse function of substances encased within a small particle of a diameter less than approximately 500 μm. Depending on the materials and structures of the shell, microcapsules can alter the physical properties of the encased substances so that a desired availability is achieved while at the same time protecting the activity of the encased substances from the environment. The release of the internal substances may be achieved by erosion, dissociation, or semipermeability of the shell materials used. Thus, encasing of medicinals within microcapsules is usually employed for the purpose of masking taste, protecting against the environment, and/ or influencing the release rate of the encapsulated substances. In most cases, microencapsulation of therapeutic drugs is simply designed for oral administration as a long-acting dosage form.[1,2]

However, when the potential of microencapsulation is considered, this technique must be best applied to selective delivery of drugs with a low therapeutic index such as anticancer drugs.[3] If a microencapsulated anticancer drug, for example, were selectively distributed in a cancerous lesion, then a controlled release of the cytotoxic agent would lead to total killing of cancer cells in the target area without systemic drug toxicity. In this respect, intravascular administration of microcapsules may be the most acceptable and effective way of site specificity.

It has been demonstrated that more than 90% of microparticles smaller than 1.4 μm are removed by the reticuloendothelial system, mainly in the liver and spleen, when they are intravenously injected, and almost 100% of microparticles larger than 10 μm are entrapped in the lungs.[4-6] The reticuloendothelial clearance and the lung embolization are the major

[1] J. R. Nixon, "Microencapsulation." Dekker, New York, 1976.

[2] L. A. Luzzi, *J. Pharm. Sci.* **59**, 1367 (1970).

[3] T. Kato, *in* "Controlled Drug Delivery" (S. D. Bruck, ed.), Vol. II, p. 189. CRC Press, Boca Raton, Florida, 1983.

[4] G. C. Ring, A. S. Blum, T. Kurbatov, and G. L. Nicolson, *Am. J. Physiol.* **200**, 1191 (1961).

[5] G. V. Taplin, D. E. Johnson, E. K. Dore, and H. S. Kaplan, *J. Nucl. Med.* **5**, 259 (1964).

[6] M. Kanke, G. H. Simmons, D. L. Weiss, B. A. Bivins, and P. P. deLuca, *J. Pharm. Sci.* **69**, 755 (1980).

METHODS IN ENZYMOLOGY, VOL. 112

problems in the intravascular targeting of colloidal drug carriers. To overcome these problems and achieve targeting at the same time, the authors and associates developed ethylcellulose microencapsulation of anticancer drugs for selective intraarterial use, proposing a concept of "chemoembolization."[3,7-9] The microcapsules with a considerably larger size when infused into tumor-supplying arteries embolize the arterioles, with gradual release of the entrapped drugs.

Approach to intravascular targeting by arterial chemoembolization is limited to cancer therapy at present, and a variety of polymeric materials other than ethylcellulose can be used as the shell of the microcapsules. This chapter briefly describes the method of ethylcellulose microencapsulation and the experimental as well as clinical results of intraarterially administered microcapsules.

Preparation of Ethylcellulose Microcapsules

The rationale for selecting ethylcellulose as the shell material is that this substance forms a stable, semipermeable membrane and is commonly used as an additive to foods and drugs because of its inert nature. Under certain conditions, ethylcellulose dissolved in an organic solvent accumulates over particulates of water-soluble substances to make a capsular membrane, the phenomenon being described as coacervation or phase separation.[10]

Original Method

Mitomycin C (MMC; Kyowa Hakko Kogyo Co. Ltd., Tokyo) was chosen as a prototype anticancer drug in our early stage of investigations. MMC is a typical cytotoxic antibiotic which does not need hepatic microsomal activation, has a broad spectrum against human solid tumors, and also exhibits good stability against heat and chemicals.

Two grams of MMC powder with a mean particle diameter of approximately 1 μm was dispersed in a solution containing 0.5 g ethylcellulose, 0.5 g polyethylene, and 500 ml cyclohexane at 80°, and the mixture was gradually cooled to room temperature with gentle stirring at 380 rpm. In this process, polyethylene promotes the phase separation of ethylcellulose, and MMC particles are encapsulated with ethylcellulose. The micro-

[7] T. Kato and R. Nemoto, *Proc. Jpn. Acad., Ser. B* **54,** 413 (1978).
[8] T. Kato, R. Nemoto, H. Mori, and I. Kumagai, *Cancer* **46,** 14 (1980).
[9] T. Kato, R. Nemoto, H. Mori, M. Takahashi, Y. Tamakawa, and M. Harada, *J. Am. Med. Assoc.* **245,** 1123 (1981).
[10] H. G. Bungenberg de Jong, *in* "Colloid Science" (H. R. Kruyt, ed.), Vol. II, p. 339. Elsevier, Amsterdam, 1949.

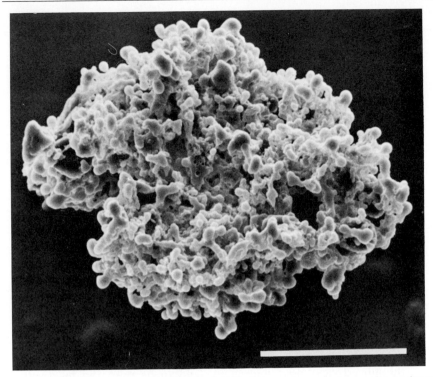

Fɪɢ. 1. Scanning electron micrograph of ethylcellulose microcapsule encasing peplomycin. Drug content was 80% (w/w). Bar indicates 100 μm.

capsules thus prepared were rinsed with n-hexane several times, and air-dried at 45° for 6 hr so that the polyethylene, cyclohexane, and n-hexane were completely removed.[7]

MMC microcapsules were proved to consist, on average, of 80% (w/w) of biologically active MMC as the core and 20% (w/w) of ethylcellulose as the shell, with a mean particle size of 224 μm ranging from 106 to 441 μm. The release rate of MMC from the microcapsules dispersed in unstirred physiological saline at 37° was 31% of the total encased MMC at 6-hr incubation, showing a sustained-release property of this preparation.

Microencapsulation by this method, in general, forms an irregular particle with a rough, invaginated surface and a particle size greater than approximately 100 μm (Fig. 1). This indicates that ethylcellulose produces a thin coating over the considerably large particles of core substances aggregated during the process of phase separation. Nevertheless this preparation has proved to be acceptable for the purpose of arterial chemoembolization.

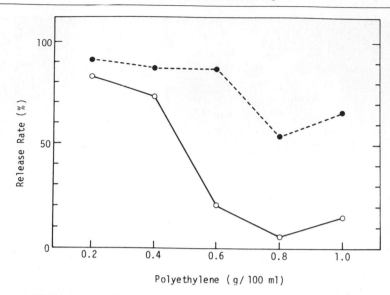

FIG. 2. Effect of polyethylene on release rate of methylene blue from microcapsules. Core : shell ratios were 2 : 0.5 (●) and 2 : 1 (○), respectively.

Modifications in Microencapsulation

Although ethylcellulose microcapsules can be prepared by a rather simple technique as described above, further improvements in respect to particle size, yield of microcapsules, and control of drug release are needed. Unfortunately, however, much of the available information relating to microencapsulation appears in the patent literature and is inadequately described.

This section describes data derived from our recent investigations. The results have revealed that the type of chemical and its concentration in the microencapsulation system definitely influence the release rate, particle size, and the resultant yield of the microcapsules. In these studies, methylene blue was used instead of MMC as the entrapped drug to facilitate determining release rates. Two grams of lactose was mixed with 1% methylene blue and was dispersed in 100 ml cyclohexane. The release rate was expressed as the percentage of the total encased drug after 10 min incubation in unstirred physiological saline at room temperature. Only microcapsules with a particle size less than 350 μm collected through a 42-mesh screen were used.[11]

It is generally appreciated that an increase of shell materials may

[11] A. Goto, H. Murota, Y. Katsurada, M. Kondo, K. Unno, and T. Kato, to be published.

decrease the release rate of encased substances. In fact, when the content of ethylcellulose in the system was increased from 0.5 to 1 g, the release rate was markedly inhibited. But, at the same time, our study demonstrated that the concentration of polyethylene definitely influenced the release rate, the maximum inhibition of release being achieved at the concentration of 0.8% (Fig. 2).

It was also found that addition of cholesterol in the solvent raised the yield of the microcapsules smaller than 350 μm in proportion to the cholesterol concentration, of which maximum value was limited to approximately 1% at room temperature. In an experimental system in which 2 g lactose was dispersed in 100 ml cyclohexane dissolving 0.5 g ethylcellulose, 0.2 g polyethylene, and 1 g cholesterol, the yield of microcapsules was increased up to 78%. The value could be well contrasted with the low yield of 21% in the microencapsulation without cholesterol. On the other hand, it seemed that cholesterol did not significantly affect the lactose content of the microcapsules (Fig. 3).

Further investigations confirmed these findings. Referring to the aforementioned experimental results, 2 g lactose was added in 100 ml cyclohexane dissolving 1 g ethylcellulose and 1 g cholesterol with varying concentrations of polyethylene. For the control, cholesterol was excluded. In the presence of cholesterol, addition of polyethylene gradually increased the yield of microcapsules. The yield was increased from 41% without poly-

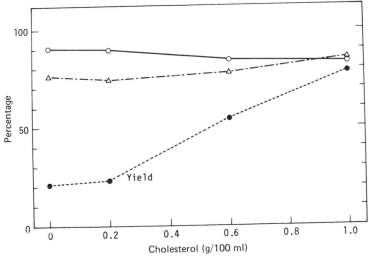

FIG. 3. Effect of cholesterol on release rate (○), encasement of core substance (△), and yield (●) in microencapsulation. Yield <350 μm. Lactose : ethylcellulose ratio was 2 : 0.5, and polyethylene concentration was 2% in the solvent.

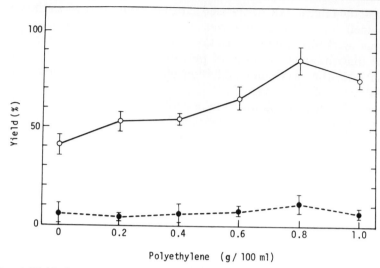

FIG. 4. Yield of microcapsules ($<350~\mu m$) as a function of cholesterol concentration [(○) 1%; (●) 0%] and polyethylene in the solvent. Core : shell ratio was 2 : 1.

ethylene up to 86% at the optimal polyethylene concentration of 0.8%, while no significant increase in the yield was observed without cholesterol (Fig. 4). Even with the presence of cholesterol, the release rate was shown to be clearly inhibited by adding polyethylene. The release rate was decreased from 68% without polyethylene to the value as low as 2% with 0.8% polyethylene. The experiment also revealed that cholesterol in the solvent significantly inhibited the release rate (Fig. 5).

The results may indicate that polyethylene promotes the formation of a stable ethylcellulose membrane, and that cholesterol facilitates making a monodispersion of the core particulates in the solvent, thus leading to an increase in uniform aggregation of the particulates. The fine aggregation of the core particulates may, at the same time, enhance the effect of polyethylene on formation of a stable membrane with uniform thickness. These functions of both polyethylene and cholesterol during the process of phase separation will contribute to increase of the yield and decrease of the release rate.

Ferromagnetic Microcapsules

An innovative drug delivery system, magnetic control of drug carrier complexes, has been initiated. The rationale for this approach is to guide the intravascular or intraluminal microcapsules into desired sites and/or

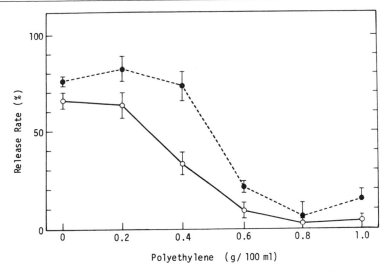

FIG. 5. Release rate of methylene blue from microcapsules as a function of cholesterol concentration [(○) 1%; (●) 0%] and polyethylene in the solvent. Core : shell ratio was 2 : 1.

to retain them at target lesions by means of external magnetic force. For this purpose, ferromagnetic ethylcellulose microcapsules containing both MMC and zinc ferrite were prepared based on the principle of coacervation.

Ethylcellulose (1 g) and polyethylene (0.5 g) were dissolved in cyclohexane (100 ml), and MMC powder (2 g) was dispersed in the solution. By cooling to room temperature with gentle stirring, MMC particles were encapsulated with ethylcellulose. MMC microcapsules thus prepared were then mixed with 100 ml of n-hexane containing 0.5 g zinc ferrite ($Zn_{20}Fe_{80} \cdot Fe_2O_3$) with mean particle size of 1.6 μm, and heated again to 45°C. The mixture was cooled with gentle stirring, whereby the ferrite particles were attached to the capsular surface (outer-type microcapsules).[12] The mean particle size was 307.9 ± 34.5 (SD) μm and ferrite particles solidly fixed to the capsular surface. The microcapsules consisted, on average, of 50% (w/w) of biologically active MMC as the core, and 34% of ethylcellulose and 16% of ferrite as the shell. With this method, the ferrite content was limited to within 16%. On the other hand, when ferrite particles were added to the initial solvent, the ferrite particles were shown to be entrapped within the ethylcellulose membrane, thus

[12] T. Kato, R. Nemoto, H. Mori, K. Unno, A. Goto, M. Harada, and M. Homma, *Proc. Jpn. Acad., Ser. B* **55**, 470 (1979).

generating an "inner-type" microcapsule. This method could increase the ferrite content up to 50%, thus enhancing the magnetic responsiveness of the microcapsules. The prototype microcapsules, consisting of 30% (w/w) MMC and 50% ferrite as the core and 20% ethylcellulose as the shell, had a mean particle size of 250 ± 43 μm.[13]

In vitro examinations revealed that both types of ferromagnetic MMC microcapsules had a sustained-release property and a sensitive responsiveness to conventional magnetic fields. Animal studies demonstrated that VX2 carcinoma transplanted in the rabbit hind limb was successfully treated with arterial infusion of ferromagnetic MMC microcapsules controlled with an extracorporeal magnet, and that VX2 carcinoma in the rabbit bladder wall responded to intravesically infused microcapsules under a magnetic field.[3,14] Independently of our research, Widder and associates also developed magnetic albumin microspheres with a mean particle size of 1 μm for the purpose of intracapillary targeting of an anticancer drug.[15,16]

Intraarterial Targeting of Microcapsules

Experimental results of intraarterial targeting of MMC microcapsules demonstrated the effectiveness of this approach.[3] Additional anticancer agents can be encapsulated other than MMC. Peplomycin (PEP; Nihon Kayaku Co. Ltd., Tokyo) microcapsules were prepared and tested in animals.[17]

Characteristics of PEP Microcapsules

PEP, a derivative of bleomycin, was encapsulated into ethylcellulose microcapsules by a phase separation method as mentioned above. The PEP content was 80% and the mean particle size was 235 ± 52 μm with a rough, invaginated surface (Fig. 1). Since PEP is highly water soluble, the release of the drug from the microcapsules was considerably more rapid as compared with MMC. The release rate of PEP in 200 ml of physiological saline (37%) stirring at 25 rpm was 50% at 30 min incubation and 90% at 7 hr incubation, respectively (Fig. 6).

[13] T. Kato, R. Nemoto, H. Mori, R. Abe, K. Unno, A. Goto, H. Murota, M. Harada, K. Kawamura, and M. Homma, *J. Jpn. Soc. Cancer Ther.* **16,** 1351 (1981).

[14] T. Kato, *Jpn. J. Cancer Chemother.* **8,** 698 (1981).

[15] K. J. Widder, A. E. Senyei, and D. F. Ranney, *Adv. Pharmacol. Chemother.* **16,** 213 (1979).

[16] K. J. Widder, R. M. Morris, G. Poore, D. P. Howard, and A. E. Senyei, *Proc. Natl. Acad. Sci. U.S.A.* **78,** 679 (1981).

[17] T. Kato, H. Mori, R. Abe, K. Etori, K. Unno, A. Goto, H. Murota, M. Harada, M. Shindo, and R. Chiba, *Artif. Organs* **11,** 213 (1982).

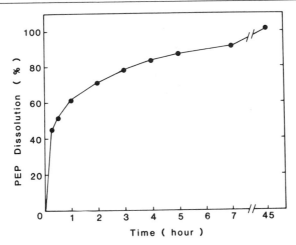

FIG. 6. Release rate of peplomycin from ethylcellulose microcapsules in 37° saline stirring at 25 rpm.

Pharmacokinetics and Cytotoxic Effect

PEP microcapsules containing 5 mg PEP suspended in physiological saline were infused into the left renal artery of Japanese white rabbits via a catheter threaded from the right femoral artery. For the control, 5 mg PEP in the usual dosage form was infused either alone or in combination with 5 mg placebo microcapsules.

PEP levels in the circulating blood of the animals infused with PEP microcapsules were significantly lower than those of the control animals infused with nonencapsulated PEP during the first 60 min after the infusion (Fig. 7). The peak PEP level, observed 1 min after the infusion, was 26.5 ± 3.18 µg/ml (mean ± SE) in the control group, while that in PEP microcapsule group was 12.25 ± 0.33 µg/ml. In terms of bioavailability, calculated from the area under the blood level curve, the amount of PEP released from the kidneys into the venous circulation in the PEP microcapsule group was estimated as 70% of that of the control. Since a considerable amount of PEP might be released from the microcapsules into the saline during the period of both the preparation of suspension and the infusion thereafter, further improvement of the microcapsules to control the initial release rate is needed in order to decrease the blood PEP levels.

However, bioassay of the tissue homogenate of the kidneys 3 hr after the infusion revealed that PEP concentration in the microcapsule group was 8.85 ± 1.38 µg/g, which is in contrast with the low concentration of 0.58 ± 0.07 µg/g in the control group. It may be that even though a large

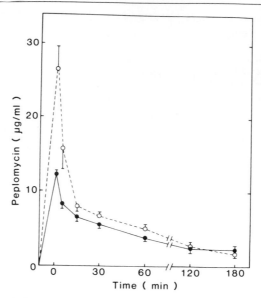

FIG. 7. Peplomycin levels in circulating blood after arterial infusion of peplomycin micro-capsules (●) and nonencapsulated peplomycin (○) into rabbit kidney.

amount of PEP was lost from the target area during the initial period after the microcapsule infusion, the capsular membrane as well as the blood stasis due to microembolization were responsible for retaining the activity of residual drug in the kidney.

Morphological examination proved that intraarterial PEP microcapsules, which were entrapped by the intrarenal arterioles, produced the most extensive coagulation necrosis involving the whole target organ when compared with other kinds of treatments. Nonencapsulated PEP caused only minimal changes, and while the placebo microcapsules even in combination with nonencapsulated PEP induced small foci of infarction, the majority of the kidney mass remained unaffected.

These results are in agreement with those obtained from intraarterial infusion of MMC microcapsules. The remarkably enhanced cytotoxic effects of the microencapsulated anticancer drugs must result from both infarction and prolonged drug action. Destruction of the surrounding endothelial lining and stromal tissues in the earlier stage of infarction may increase the extent of intraparenchymal migration of the drugs released from the microcapsules. With regard to the mode of action, this kind of treatment using microencapsulated cytotoxic agents was defined as "arterial chemoembolization." Animal study demonstrated that the decrease in drug level in circulating blood was responsible for preventing the systemic drug toxicity.[3]

Clinical Application of Microcapsules

Recent advances in angiography have permitted arteries in various sites to be readily catheterized under X-ray monitoring. Consequently, the microcapsules are able to be selectively infused into tumor-supplying arteries in a variety of organs. During the period from March 1978 to March 1982, 285 patients with advanced carcinoma in the kidney, liver, urinary bladder, prostate, bone, and lung were subjected to transcatheter arterial chemoembolization with microcapsules. The number of treatments was single in 67% of the patients, 2 in 21%, and 3 or more in 12%. Ninety percent of the patients received MMC microcapsules and the others received PEP microcapsules with a mean total dose of 21 mg expressed as the drug content.[9,18]

Of the 211 measurable tumors, 74 (35%) showed a marked tumor reduction greater than 50% in area, 77 (37%) had a tumor reduction less than 50%, and 60 (28%) did not respond. Side effects and complications which required medical care were experienced in approximately 10% of the patients. These included bone marrow depression, local pain, fever, anorexia, or local skin ulceration. Since the majority of the tumors were large and highly invasive, the results should be appraised in a positive light.

A preliminary controlled study in the treatment of renal cell carcinoma showed that intraarterial infusion of MMC microcapsules enhances the antineoplastic effect and reduces the systemic toxicity of the drug. These findings are consistent with the previous animal studies.[19]

Comment

Ethylcellulose microencapsulation by coacervation needs no special apparatus or expensive chemicals except for the drug to be entrapped. If necessary, an expensive core substance such as an anticancer drug can be readily recovered by dissolving the ethylcellulose membrane in organic solvents. Besides the chemicals, other technical factors such as temperature or speed of the stirrer may influence the microencapsulation. Optimal conditions of these factors should be examined for each substance to be encapsulated. Furthermore, various polymeric materials other than

[18] T. Kato, *Jpn. J. Cancer Chemother.* **10,** 333 (1983).
[19] T. Kato, R. Nemoto, H. Mori, M. Takahashi, and Y. Tamakawa, *J. Urol.* **125,** 19 (1981).

ethylcellulose could be used as the shell of the microcapsules with certain modifications in this method.

The particle size and the release rate should be determined depending on the application. For selective arterial administration, the microcapsules must be small enough to be infused through the catheter but large enough to avoid the undesirable migration into fine nontarget arteries such as those supplying the nerve fibers. In this respect, the size of the ethylcellulose microcapsules is acceptable for the clinical practice. On the other hand, in selective drug delivery, extremely slow drug release does not always satisfy a designed therapeutic effect. It should be realized that the drug release from microcapsules surrounded by compact tissue structures is much more inhibited than that in an aqueous environment, and also that the extent of drug diffusion in parenchymal tissues is proportionally influenced by the concentration gradient of drugs released from microcapsules. In this respect, it is preferable to design the release of the encased drugs at a rapid rate after the microcapsules reach the target tissue.

Selective drug delivery can be identified by three stages of targeting: first-, second-, and third-order targeting.[3,15] It is most likely that unless first-order targeting which involves the restricted drug distribution to vascular beds of target lesions is achieved, second- and third-order targeting will fail to provide any fruitful result. Selective arterial infusion of ethylcellulose microcapsules has been developed as a prototype of first-order targeting of anticancer drugs, and has proved to have practical utility in the treatment of cancer. Problems in second- and third-order targeting still remain to be settled.

[12] Polyacrolein Microspheres: Preparation and Characteristics

By M. CHANG, G. RICHARDS, and A. REMBAUM

Introduction

Antibody-coated microspheres or immunomicrospheres react in a highly specific way with target cells, viruses, or other antigenic agents. Microspheres incorporate compounds that are radioactive, fluorescent, magnetic, colored, or pharmacologically active. These various types of microspheres may be sensitized with pure, specific monoclonal antibodies obtained by the new hybridoma cell cloning techniques or with conventional antibody preparation.[1-4] Basically, microspheres that have been investigated so far can be classified into two types. (1) Nonactivated hydrophobic microspheres: antibodies are immobilized by physical adsorption; polystyrene and acrylic particles are in this category. The problem with hydrophobic binding is the "nonspecific stickiness" of proteins which leads to inaccurate results. (2) Hydrophilic microspheres with an additional activation step: agarose and poly(hydroxyethyl methacrylate) spheres functionalized with cyanogen bromide or toluene sulfonyl chloride or carboxylated polystyrene activated with carbodiimide are examples of this type.[5-7] Recently, in the course of preparation of novel immunomicrospheres, we found that polyacrolein spheres are superior to the above-mentioned two types of beads for the following reasons: (1) No activation step is required. The amine groups of the antibody react readily with aldehyde groups of polyacrolein spheres to form Schiff's bases. The polyacrolein microspheres remain active for protein binding when they are stored at 4° for at least 1 month. (2) Acrolein can be copolymerized with monomers such as hydroxyethyl methacrylate, styrene, or methacrylic acid to make the particle hydrophilic.[8] (3) Polyacrolein microspheres can be grafted onto preformed particles or polymeric sheets.[9]

Preparation of Polyacrolein Microspheres
  by Means of Ionizing Radiation

Acrolein polymerized quite readily under cobalt-60 radiation. The irradiation dose of 0.5 Mrad was found to be sufficient to obtain over 90% conversion. All polymerizations were carried out in 20-ml screw-capped scintillation vials. Typical ingredients were distilled water (15 ml), freshly distilled acrolein (0.5 ml), polyethylene oxide of molecular weight of 100,000 (0.1 g), and sodium lauryl sulfate (0.05 g). Distilled water and surfactants were introduced first; then a stream of nitrogen was passed for 5 min through the liquid to remove the dissolved oxygen. Acrolein was

[1] A. Rembaum and S. P. S. Yen, *J. Macromol. Sci., Chem.* **A13**(5), 603 (1979).

[2] A. Rembaum, *Br. Polym. J.* **10,** 275 (1978).

[3] A. Rembaum, R. C. K. Yen, D. H. Kempner, and J. Ugelstad, *J. Immunol. Methods.* **52,** 341 (1982).

[4] A. Rembaum and W. J. Dreyer, *Science* **208,** 364 (1980).

[5] T. Goodfriend, L. Levine, and G. Fassman, *Science* **144,** 1344 (1964).

[6] P. Cuatrecasas, *J. Biol. Chem.* **245,** 3059 (1970).

[7] E. Bergman, W. T. Tsatsos, and R. F. Fischer, *J. Polym. Sci.* **3,** 3685 (1965).

[8] A. Rembaum, U.S. Patent 4,413,070 (1983).

[9] A. Rembaum, R. C. K. Yen, U.S. Patent pending.

then added and the vial was placed in a cobalt-60 source for 4 hr (0.5 Mrad). The resulting polyacrolein was centrifuged at various speeds, washed with distilled water, dispersed, and centrifuged again. This procedure was repeated three times to ensure the removal of the unreacted monomer and water-soluble species. The size of the resulting particles was determined by scanning or transmission electron microscope and dynamic laser light-scattering methods.

The rate of this free radical polymerization was found to be proportional to the first power of the acrolein concentration, suggesting that the kinetics of polymerization is different from the conventional emulsion polymerization. This was not surprising because acrolein is soluble in water up to ~20% at room temperature which enables polymerization to follow solution reaction mechanism.

The polymer is obtained in the form of a stable dispersion of cross-linked microspheres. The size of the spheres can be controlled by the type of surfactant, surfactant concentration, monomer concentration, and radiation dosage rate. Relatively narrow distribution of particles from 100 Å to 3 μm can be achieved by varying these reaction parameters. The polyacrolein microspheres prepared are not soluble in water, acetone, chloroform, methanol, or acetic acid, but swelled in pyridine, dimethyl sulfoxide, and formamide.

Molecular Structures and Chemical Analyses of PA Microspheres

Acrolein being a difunctional monomer can undergo reaction at both its vinyl group as well as its aldehyde functionality.[10] Different initiation mechanisms yield polymers with different molecular structures. We employed a Fourier transform infrared spectrophotometer and solid-phase NMR equipped with cross-polarization magic angle to elucidate the structure difference. For easy comparison, two FTIR spectra of polyacrolein (PA I, radiation polymerized) and polyacrolein (PA II, base catalysis, see this volume [13]) and two NMR spectra of the same PA samples are shown in Figs. 1 and 2, respectively. Both PA I and PA II in FTIR spectra exhibit absorptions at 1725 cm^{-1}, which represents a carbonyl group. PA II initiated by alkali showed several additional peaks. For instance, the peaks at 3080 cm^{-1} are attributed to a hydrogen attached to a vinyl group. The carbon–carbon double bond appears in the vicinity of 1680 cm^{-1}. These two peaks, which represent vinyl unsaturation, decreased significantly upon radiation of PA II with 2.0 Mrad.

---

[10] R. C. Schultz, in "Encyclopedia of Polymer Science and Technology" (N. M. Bikales, ed.), Vol. 1, p. 60. Wiley, New York, 1964.

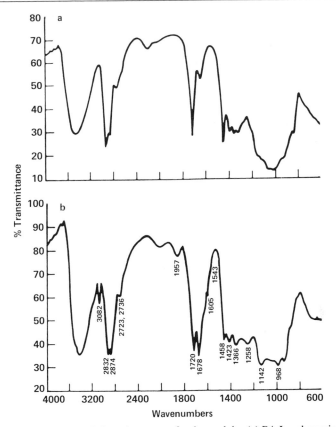

FIG. 1. Fourier transform infrared spectra of polyacrolein. (a) PA I, polymerization with
^{60}Co γ radiation; (b) PA II, polymerization with NaOH at pH 11.

The analysis of NMR and FTIR spectra of these two types of polyacrolein leads to similar conclusions; for instance, the infrared spectrum of
PA II exhibits carbon–carbon unsaturation at 3082 cm^{-1}. The unsaturation is also observable in the NMR spectrum, where such unsaturated
carbons are normally centered around 125 ppm. There are a number of
peaks in this region clearly indicating the complicated nature of the base-
catalyzed polymer. The spectrum of the radiation-initiated polymer is
simpler, with low-intensity peaks in the unsaturated region. When polyacrolein II was irradiated with 2.0 Mrad of γ radiation, the intensity of
these unsaturated carbon peaks was noticeably diminished, in particular
the peak at 165 ppm, relative to the peaks of saturated carbons in the
polymer. A similar change is noted in the FTIR spectrum with the reduction of the peak at 3082 cm^{-1}.

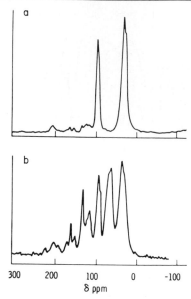

FIG. 2. Carbon magnetic resonance spectra of polyacrolein. (a) PA I; (b) PA II.

On the basis of these analyses, we proposed that by γ radiation, acrolein molecules polymerized primarily through their vinyl groups. The resultant polymer then exists in equilibrium with the acetal isomer as shown in the following figure:

Under alkaline conditions, Michael addition and carbonyl addition probably occurred simultaneously and competitively to give a complicated structure.

(The cross-links are omitted in the above figures.)

Preparation of Polyacrolein-Activated Hybrid Beads
  by Radiation Grafting

Polyacrolein can be grafted onto preformed microspheres to activate the surface of the preformed microsphere while retaining their original shape. In this study, preformed compact polystyrene microspheres used were 11.1 μm in diameter, the compact magnetic polystyrene microspheres were 1.5 μm in diameter, and magnetic porous polystyrene microspheres were 3 μm in diameter. Acrolein (Eastman Kodak) was distilled at atmospheric pressure at 53° before use. Poly(ethylene oxide) (PEO) of molecular weight 100,000 (Polysciences) and sodium lauryl sulfate (SLS, Fisher Scientific) were both used as received.

Cobalt γ radiation source of 0.12 Mrad/hr dosage rate was used for initiation of grafting and polymerization reaction.

A typical composition for this study was porous magnetic polystyrene, 0.1 g; 0.4% PEO/H_2O solution, 4.0 ml; 20% wt. of acrolein/H_2O, 2.0 ml; SLS, 0.01 g.

All reactions were carried out in 20-ml screw-capped scintillation vials. The mixtures were deaerated as previously described. Total irradiation time was about 7 hr. During the first 3 hr, samples were removed from the Co source every 20 min for 5 min shaking. For the remaining hours, this was repeated every 40 min.

After irradiation was completed, the crude product, in the case of nonmagnetic microspheres, was centrifuged at a speed of 1000 g for no more than 5 min so that homopolyacrolein particles (500–1000 Å) could be separated from the polyacrolein-grafted microspheres. The grafted product was washed with distilled water and resuspended in PEO solution. This process was repeated for at least four times to ensure the complete removal of homopolymer particles. When the grafting reaction was implemented on magnetic microspheres, the product was isolated by means of a magnetic field.

Scanning and transmission electron microscopes were used to study the morphology of products. Quantitative determination of the amount of polyacrolein in the microspheres was carried out by IR analysis using a Fourier transform infrared spectrophotometer (M091-0121 DIGILAB).

Radiation grafting of acrolein onto polymeric microspheres was evidenced by the infrared spectra of products which showed absorption peaks at 1725 cm^{-1} ascribed to the aldehyde carbonyl group. Furthermore, treatment of the radiation-grafted particles with m-aminophenol yielded fluorescent particles for all three systems. Fluorescence confirmed the presence of polyacrolein on the surface of the grafted spheres. A method of quantitative determination of the amount of polyacrolein in the grafted copolymer was developed, using the "peak height ratio" of Fourier transform infrared-selected wavelengths. For instance, to analyze the polyacrolein/porous magnetic polystyrene system, we selected the peaks at 1725 and 1540 cm^{-1}. A calibration curve was constructed by plotting the peak height ratio of 1725/1540 cm^{-1} of various polymer blends of known compositions. These polymer blends were made simply by mixing homopolyacrolein with untreated magnetic polystyrene. A linear relation between peak height ratio and concentration of polyacrolein was obtained. Using this technique, we found that aldehyde contents can be effectively increased by increasing (1) surfactant concentration, (2) frequency of mixing during radiation, (3) monomer concentration, and (4) surface area of the preformed polymeric microspheres.

We also examined the morphology of these products under the electron microscope. Theoretically, acrolein-treated polymeric microspheres may be in either of the following two forms:

1. A core–shell form where polymerization occurs at particle–water interface. Polyacrolein forms a thin film and wraps the preformed microsphere. Eventually, substrate particles "core" will be coated with a "shell" of polyacrolein. In this case the grafted spheres appear smooth in the scanning electromicrographs.

2. Hybrid beads morphology. This involves a two-step reaction. Polyacrolein was generated from aqueous-phase polymerization and formed small microspheres, which collided with substrate particles due to Brownian motion and which were subsequently grafted onto substrate particles by γ radiation.

Our electron microscope results showed that the actual morphology is in between these two forms. Electron micrographs can only display the hybrid beads structure. Since the grafted polyacrolein beads do not account for all the polyacrolein covalently bound to substrate polymer, we concluded that the remaining amount of polyacrolein not revealed by electron microscope is probably introduced into the core–shell form.

Chromatographic Characteristics of Polyacrolein Nanospheres

Nanospheres may be obtained by using high concentrations of SLS. Surfactant and residual monomers were removed from ^{60}Co-synthesized PA nanospheres by passing them twice through a Sephadex G25-150 column, equilibrated with 0.05% Tween 20 in water. The nanospheres had an average diameter of 100 Å (transmission electron microscope).

The eluate (0.1 ml) was reacted overnight with 0.5 ml of 1 mg/ml of

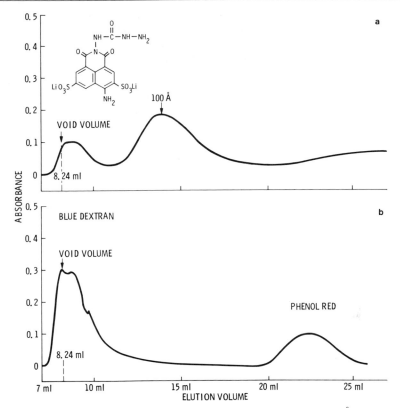

FIG. 3. Chromatographic separation of polyacrolein nanospheres (100 Å in diameter). (a) Absorbance at 427 nm; (b) absorbance at 600 nm, showing blue dextran (polydisperse, maximum MW $\geq 2 \times 10^6$) and phenol red (MW 354) peaks.

Lucifer yellow in 0.1 M sodium acetate, pH 5.3, containing 0.2% Tween 20 and 0.05% NaN_3. After the reaction, unbound Lucifer yellow was removed from the nanosphere suspension by again passing it twice through a Sephadex G25-150 column. The Sephadex had been equilibrated with .025 M potassium phosphate, pH 7.5, containing 0.05% Tween 20 and 0.1% NaN_3 (KPO$_4$–Tween).

The Lucifer yellow derivatized nanospheres were then chromatographed on a 1 × 26-cm bed volume AcA-34 (Ultragel) column. The column buffer was KPO$_4$–Tween. Phenol red was used as a marker to determine an approximate gel internal volume (V_i), and blue dextran was used to determine the void volume (V_0) of the column.

The results graphed in Fig. 3 show the successful separation of 100-Å nanospheres from the population of somewhat larger nanospheres. When mouse IgG (MW 150,000) and the F(ab')$_2$ fragment of goat anti-mouse IgG (MW 150,000) were chromatographed (separately) on this column (data not shown), their eluted peaks (measured at 280 nm) bracketed the 100-Å nanosphere peak, confirming the size of the nanospheres to be about the size of an antibody (IgG) molecule.

The ability to produce easily derivatized synthetic polymeric spheres that are of a similar size and shape as soluble globular protein molecules is of considerable significance. These spheres offer a possibility to be coated by large amounts of antibodies and other targeting substances such as hormones. Lucifer yellow is one example; it was chosen in this experiment not only because it has aldehyde binding groups but also because it is a fluorescent, useful biological tool. The nanospheres improve this stain by amplifying the fluorescent signal because several fluorescent molecules may bind to one nanosphere.

Binding of Antibodies to PA Microspheres and PA-Activated Hybrid Beads

The standard protocol for binding antibodies to microspheres involves incubation of microspheres and antibodies for 2 hr at room temperature with constant agitation either on a reciprocal shaker or by periodic mixing in a vortex mixer (every 5 min). Reactions were carried out in 0.1 M sodium acetate buffer, pH 5.3, containing 1% Tween 20. The presence of the "Tween" detergent minimizes hydrophobic binding of antibody to the microspheres, ensuring a maximum percentage of covalently bound antibody. The reactions were initiated by adding an excess of ^{125}I-labeled antibody (5–100 μg) to the appropriate amount of microspheres or particles (less than or equal to 1 mg by weight or about 10^9–10^{10} particles/ml). The reaction is quenched by the addition of bovine serum albumin (BSA).

[125]I-Labeled antibodies were prepared by the chloramine-T labeling method.[11] The immunological activity of radiolabeled specific antibodies was determined by reaction with the appropriate antigen bound to microtiter plates (solid-phase RIA).

Lyophilized mouse IgG and rabbit IgG were obtained from Sigma Chemical Co., St. Louis, MO. Affinity-purified goat anti-rabbit IgG was obtained from TAGO, Inc., Burlingame, CA. The mouse IG was resuspended and dialyzed (to remove chloride ion) against 0.05 M potassium phosphate buffer, pH 7.5. The rabbit IgG, being salt free, was resuspended in the same phosphate buffer and used "as is."

[11] F. C. Greenwood, W. M. Hunter, and J. S. Glouer, *Biochemistry* **89,** 114 (1963).

The goat anti-rabbit IgG was desalted in a Sephadex G25-150 (Sigma Chem. Co.) column equilibrated with 0.05 M potassium phosphate, pH 7.5, according to the method of Neal and Florini.[12]

The protein concentrations of the above reagents were determined by the Lowry[13] method, using bovine serum albumin as the standard (Sigma 905-10).

Na[125]I was obtained from New England Nuclear, Boston, MA. All immunoglobulin reagents were radiolabeled with ^{125}I using chloramine-T. Unreacted iodine was removed by desalting.

Buffers used for coupling protein to the microspheres were 0.1 M sodium phosphate, pH 5.3; 0.1 M potassium phosphate, pH 7.5; 0.1 M sodium bicarbonate, pH 9.4. These buffers were used with or without Tween 20 (ICI American, Inc., Wilmington, DE) as stated in the procedures below.

Buffers used for washing were either 0.5 M potassium phosphate, pH 7.5, with 0.05% Tween 20 (KPO$_4$–Tween), or KPO$_4$–Tween containing 0.15 M NaCl and 0.1% sodium azide (PBS–Tween).

Grafted and nongrafted samples of 11.1-μm compact polystyrene, or 3-μm magnetic porous polystyrene microspheres, were reacted with excess radiolabeled mouse IgG at pH 7.5 with 1.0, 0.1, and 0% Tween 20, for 2 hr at room temperature with continuous shaking, and then the samples were incubated overnight at 4°.

Results

Grafting of Acrolein. Acrolein grafting completely changed the binding characteristics of the 11.1-μm compact polystyrene (PS) microspheres (Table I). This is readily seen by comparing the amount of mouse IgG (MIG) bound in the presence and absence of Tween 20. Nonactivated PS binds protein hydrophobically. In RIA and ELISA systems, the mild detergent Tween 20 is extensively used to minimize hydrophobic binding of proteins to polystyrene (as well as to other plastics). In the present case, Tween 20 reduced the binding of MIG to the ungrafted compact PS microspheres to 5–8% of its "non-Tween" value. A look at the binding data for acrolein-grafted compact microspheres (Table I) shows that these microspheres no longer "act" like pure PS. Not only do they bind twice as much MIG in the absence of Tween, but there is also no sharp drop in binding in the presence of Tween 20: Tween only reduced the binding to

[12] M. W. Neal and J. R. Florini, *Anal. Biochem.* **55**, 328 (1973).
[13] D. H. Lowry, N. T. Rosebrough, A. L. Farr, and R. J. Randall, *J. Biol. Chem.* **193**, 265 (1951).

TABLE I
PROTEIN BINDING TO MICROSPHERES: EFFECT OF ACROLEIN GRAFTING

Type of microsphere	Polyacrolein (wt%)	Tween 20 buffer (%)	MIG[a] bound per 1×10^6 beads (μg)	MIG bound (% by weight of beads)
11.1-μm compact polystyrene	~3	0	15	1.8
		0.1	10	1.3
		1	9.8	1.2
	0	0	6.7	0.83
		0.1	0.34	0.03
		1	0.51	0.06
3-μm porous magnetic	15	0	1.9	9.5
		0.1	1.3	6.6
		1	1.2	5.9
	0	0	3.2	16.1
		0.1	1.4	7.0
		1	1.2	6.2

[a] MIG, Mouse IgG.

66% of its "non-Tween" value, suggesting that most of the binding of MIG to the grafted compact PS microspheres is covalent (via aldehyde) rather than hydrophobic in nature.

The best indicator of the effect of PA grafting on these compact PS microspheres is a comparison of the amount of MIG bound in the presence and absence of Tween 20. For the compact PS microspheres, this difference in grafted to nongrafted binding is a factor of 20 (0.51 vs 9.8 μg, and 0.06 vs 1.2; see Table I).

The inability of Tween 20 to cause a major reduction in the binding of the nongrafted magnetic porous PS microspheres is indicative that these microspheres are binding protein in a different manner (qualitatively or quantitatively) than pure PS. The grafting of even 15% (by weight) of acrolein on these microspheres had no effect on the MIG binding capacity in the presence of Tween, and actually reduced the MIG binding capacity in the absence of Tween. Though the quantity of MIG protein bound to the magnetic porous microspheres was not improved with PA grafting, it remains to be determined if there is a qualitative change in the protein bound to the microsphere, i.e., a change in the amount of denatured protein. It should be noted that all the acrolein-grafted microspheres tested, compact PS, magnetic porous PS, and the magnetic compact PS,

TABLE II
SPECIFIC ACTIVITY OF ANTIBODY AFTER COVALENT COUPLING TO
ACROLEIN-GRAFTED POLYSTYRENE BEADS[a]

| | First step | | Second step (pH 7.5) | | |
| | | | RAB. IgG bound | | |
pH	GαR[a] bound (μg)	No. of GαR molecules ($\times 10^{12}$ cm^{-2})	cpm	μg	Molar ratio, rabbit IgG (μg)/GαR (μg)
5.3	0.40	1.6	2064	0.16	0.4
7.5	0.18	0.7	1435	0.11	0.6
9.4	0.06	0.24	746	0.06	1.0

[a] Antibody, goat anti-rabbit IgG (GαR) reacted with 2.6×10^5 beads, 11 μm in diameter.

showed superior dispersibility after centrifugation in buffer without detergent.

Specific Activity of an Antibody after Covalent Coupling to Acrolein-Grafted Polystyrene (PS) Beads. Ten grams of goat anti-rabbit IgG (GαR) or radiolabeled GαR was reacted with 2.6×10^5 11.1-μm grafted compact PS microspheres for 50 min at room temperature at pH 5.3, 7.5, or 9.4 in the presence of 0.2% Tween 20. Then, excess of bovine serum albumin (BSA) was added to quench any further binding of GαR to the microspheres by competitively binding the remaining available aldehyde groups. The samples were then washed three times with PBS–Tween, and the amount of GαR bound determined by radioactive counting. Radiolabeled rabbit IgG (20 μg) in PBS–Tween containing 5% normal goat serum was then added to the microspheres, which had been reacted with the unlabeled-GαR. These were then incubated at room temperature for 5 hr with periodic vortexing. The samples were then washed and counted as above. Results are given in Table II.

Since the results of the previous experiments showed that the 11.1-μm compact polystyrene microspheres exhibited the most dramatic positive change in protein binding, and the presence of polyacrolein on the spheres could be confirmed by FTIR, these microspheres were chosen for testing in a "second-step" experiment. The ability of a microsphere to bind large amounts of protein is of little significance if the bound protein is denatured. This experiment was designed to give some indication of the ability of an active affinity-purified antibody to retain its affinity for antigen when bound to a grafted microsphere.

According to the manufacturer (TAGO, Inc.), this antibody reagent (goat anti-rabbit) contained 74% active antibody. Table II shows that the GαR remained very active upon being bound to the grafted compact microspheres. The specific binding of the radiolabeled rabbit IgG was determined by subtracting the counts bound to control microspheres (grafted compact PS coupled with BSA) from the counts bound to the grafted compact PS microspheres coupled with GαR IgG. The control counts were 178, 149, and 186 cpm at pH 5.3, 7.5, and 9.4, respectively. The maximum molar ratio of antigen bound in the second step to the antibody covalently bound in first step is 1 : 1. This indicates the bound goat anti-rabbit IgG is at least 50% active. (The maximum amount of binding could be expected to be two molecules of antigen per antibody molecule.)

Cell Separation by Means of Polyacrolein-Activated Magnetic Microspheres (Hybrid Microspheres)

It is believed that the grafting of polyacrolein on the surface of magnetic, antibody-coated microspheres would eliminate the antibody leaching in the recently described procedure[14] for removal of malignant cells from human bone marrow. An example of successful cell separation by means of polyacrolein-activated microspheres is described below.

Sheep red blood cells (SRBC) were washed three times in Dulbecco's phosphate-buffered saline (PBS) and subsequently sensitized by incubation with rabbit antiserum against SRBC (Cappel Lab, Cochranville, PA) for 30 min at 37°. After two more washes with PBS to remove excess rabbit antiserum, the SRBC were used for microsphere labeling. A similar procedure was applied to SRBC and chicken RBC (CRBC) for experiments involving magnetic separation of SRBC from CRBC. To obtain quantitative data of the efficiency of magnetic cell separation, the following procedure was used: 5 concentrations of hybrid magnetic microspheres (10^5 to 2×10^6) which had been coated with goat anti-rabbit IgG were added to tubes containing 5×10^6 sensitized SRBC and 5×10^8 CRBC in a total volume of 1.2 ml of PBS. Control experiments were set up using similar microspheres without goat anti-rabbit IgG bound to their surfaces. After incubation at 37° for 30 min, the RBC were gently agitated by tapping the tubes and a bar magnet was placed against the side of the tubes. The magnetic particles migrated toward the magnet immediately. Subsequently, nonattached cells in the suspension were removed by aspiration and 2 ml of PBS was added immediately to the tubes. The magnet

---

[14] J. G. Treleaven, F. M. Gibson, J. Ugelstad, A. Rembaum, T. Philip, G. D. Caine, and J. T. Kemshead, *Lancet* **1,** No. 8368 (1984).

was then removed and the magnetic particles again gently agitated by tapping the tubes. Care was taken to avoid aspirating any magnetically attracted particles. After at least eight repeated washings, the microspheres with attached SRBC were resuspended in 0.5 ml of PBS and an aliquot removed for microscopic examination in a hemacytometer. The washings were repeated until less than 10^3 CRBC/ml were observed. Vortexing for 30 sec was found to be sufficient to detach most of the SRBC from microspheres.

To reinforce the evidence presented above of cell separation with hybrid magnetic PS microspheres which had IgG bound on their surfaces, the following additional experiments were performed. (1) Hybrid magnetic PS microspheres which had not previously reacted with any fluorescent goat anti-rabbit IgG were incubated with SRBC sensitized with rabbit anti-SRBC. The microspheres were observed not to bind to any SRBC. (2) When hybrid magnetic PS microspheres with goat anti-rabbit IgG bound on their surfaces were mixed with SRBC which were not sensitized with rabbit anti-SRBC, a result similar to (1) was obtained. (3) Hybrid magnetic PS microspheres coated with goat anti-rabbit IgG did not bind to CRBC. Preincubation of CRBC with anti-SRBC did not result in any detectable amount of CRBC binding to microspheres, indicating that the antiserum used was specific for SRBC. (4) Hybrid magnetic PS microspheres with goat anti-rabbit IgG attached to their surface were observed to bind only to sensitized SRBC in a mixture containing 1% of the latter and 99% of CRBC. (5) Most sensitized SRBC remained attached to magnetic PS spheres after repeated magnetic separation from CRBC and washings in PBS.

The specific labeling of SRBC with a variety of polymeric spheres of various size and composition after PA was radiation grafted on their surface indicates the scope of this system as a cell marker. However, the specificity of these reagents in relation to different cell lines remains to be established. In the case of PS magnetic and nonmagnetic PS spheres, the hydrophobic properties of the spheres can result in physical adsorption of proteins without having to resort to covalent binding, in particular if the adsorption is performed at a specific pH of 7.8. Thus, nonspecific interactions may occur between such treated PS spheres and cell surfaces. Radiation grafting of PA onto the surfaces of PS spheres, however, allows covalent bonds to be formed with immunoglobulins and will permit four times as much protein to be bound per sphere. Subsequent quenching of excessive reactive aldehyde groups reduces any nonspecific interactions with cell surfaces. Poly(methyl methacrylate) (PMMA) spheres, in contrast, do not absorb detectable amounts of proteins. Therefore, radiation grafting of PA microspheres onto PMMA surfaces appears to be an ideal method to generate a reactive surface on these spheres for immunological applications.

Acknowledgment

This report represents one phase of research carried out at the Jet Propulsion Laboratory, California Institute of Technology and supported by Grant No. 1R01-CA20668-04 awarded by the National Cancer Institute, DHEW, and Grant No. 70-1859 awarded by the Defense Advanced Research Project Agency.

[13] Polyacrolein Microspheres

By Shlomo Margel

Several types of hydrophilic acrylate microspheres containing on their surface a variety of functional groups, such as carboxylate, hydroxyl, amide, or pyridine groups, have been synthesized previously.[1] The functional groups were used to bind proteins covalently to the microspheres by means of a series of reactions.[2] The last step of the microspheres derivatization technique, prior to protein binding, is a reaction with glutaraldehyde, designed to introduce reactive aldehyde groups on the surface of the microspheres. In order to simplify the derivatization procedure, polyglutaraldehyde microspheres were formed.[3] However, polyglutaraldehyde microspheres possess several disadvantages, i.e., low yield of formation [approximately 3% (w/w)], instability of microspheres with diameters larger than 0.7 μm, and instability of microspheres with magnetic properties in PBS solution. Recently, polyacrolein (PA) microspheres were synthesized.[4-6] These microspheres do not have the above disadvantages and they can be considered as more advanced polyaldehyde microspheres.

"Naked" PA Microspheres

Synthesis. PA microspheres were prepared by aqueous polymerization of acrolein in the presence of an appropriate surfactant. The micro-

[1] A. Rembaum, S. P. S. Yen, and W. Volkson, *Chem. Technol.* **8,** 182 (1978).
[2] R. S. Molday, W. J. Dreyer, A. Rembaum, and S. P. S. Yen, *J. Cell Biol.* **64,** 75 (1975).
[3] S. Margel, S. Zisblatt, and A. Rembaum, *J. Immunol. Methods* **28,** 341 (1979).
[4] S. Margel, U. Beitler, and M. Offarim, *Immunol. Commun.* **10**(7), 567 (1981).
[5] S. Margel, U. Beitler, and M. Offarim, *J. Cell Sci.* **56,** 157 (1982).
[6] A. Rembaum, R. C. K. Yen, D. H. Kempner, and J. Ugelstad, *J. Immunol. Methods* **52**(3), 341 (1982).

spheres were prepared by polymerization under alkaline conditions. For a discussion of polymerization by irradiation, see this volume [12]. The procedure for the formation of alkaline polymerized microspheres is detailed below.

"Alkaline" Microspheres. NaOH (0.2 N) was added dropwise to an aqueous solution containing 8% (w/v) acrolein and 0.5% (w/v) of the surfactant sodium hydrogen sulfite–polyglutaraldehyde conjugate,[5,7] until a pH of 10.5 was reached. The reaction continued for 2 hr, and the mixture was then dialyzed extensively against distilled water and centrifuged through water four times at 2000 g for 20 min. The PA microspheres of average diameter 0.1 μm, as determined by scanning electron microscopy (SEM), could be redispersed easily in PBS or in distilled water. The diameter of the microspheres could be controlled in a predictable manner by varying the concentration of the surfactant or acrolein or the pH of the polymerization reaction. The surfactant sodium hydrogen sulfite–polyglutaraldehyde conjugate[7] was prepared by the reaction of polyglutaraldehyde (5.0 g) with sodium bisulfite (12.5 g in 30 ml H_2O). The reaction continued until all the polyglutaraldehyde was dissolved. The solution was dialyzed extensively against H_2O and then lyophilized. The obtained surfactant was specifically designed for stabilizing the "alkaline" microspheres, since it contains both electrostatic and appropriate steric stabilizing groups. Many other surfactants, e.g., poly(ethylene oxide), poly(vinyl alcohol), Tween 20, sodium dodecyl sulfate, do not stabilize these microspheres. The microspheres produced via the alkaline mechanism can be prepared with extremely uniform diameter, ranging from 0.04 up to ~8 μm (Fig. 1a–d).

PA microspheres with fluorescent or magnetic properties were synthesized by carrying out the above described polymerization procedures in the presence of appropriate fluorochromic or ferrofluidic compounds, respectively.

Binding Capacity. The PA microspheres covalently bind through their aldehyde groups various amino ligands, e.g., proteins, drugs, enzymes, and antibodies, to form the Schiff base products. The reaction between the microspheres and the ligands is a single-step reaction which can also be achieved under physiological pH. Proteins bound to the microspheres do not leach into the solution due to the polyvalent Schiff base bonds formed. However, leaching of ligands bound to the microspheres through a single amino group can be prevented by a further reduction of the Schiff base bond with sodium borohydride or sodium cyanoborohydride[8] to produce a single C–N bond [reaction (1)].

[7] S. Margel and E. Weisel, *J. Polym. Sci.* **22,** 145 (1984).
[8] R. F. Borch, M. D. Bernstein, and H. D. Dupont, *J. Am. Chem. Soc.* **93,** 2897 (1971).

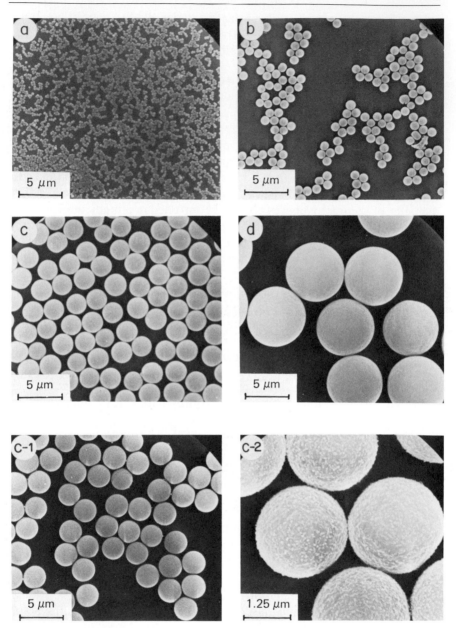

Fig. 1. SEM photomicrographs of "alkaline" PA microspheres of various diameters (a–d) and of hybrido PA microspheres (c-1 and c-2).

TABLE I
ALDEHYDE CONTENT OF PA MICROSPHERES[a]

Microsphere	% N in the oxime product	Aldehydes (mmol)/ microspheres (g)
"Alkaline"	4.0	2.9
"Irradiated"	13.0	9.3

[a] Fifty milligrams of microspheres in 5 ml H_2O was stirred at room temperature for 24 hr with 500 mg of hydroxylamine hydrochloride. The microspheres were then washed several times by centrifugation and dried under vacuum at 60°.

$$RCHO + R'NH_2 \rightleftharpoons RCH{=}NR' \xrightarrow{NaBH_4} RCH_2{-}NHR' \qquad (1)$$

The presence of aldehyde groups was ascertained by several methods. The infrared spectra indicate absorption bands at 1720 cm^{-1} and 2740 cm^{-1} due to the stretching of the nonconjugate aldehyde and the CH of aldehyde group, respectively. Additional evidence for the presence of aldehyde groups was shown by means of the 2,4-dinitrophenylhydrazine reaction which yielded yellow microspheres. Quantitatively, the aldehyde content of the PA microspheres was determined as the percentage of nitrogen resulting from the oxime prepared by the heterogeneous reaction of the microspheres with aqueous hydroxylamine hydrochloride solution.[9]

Under alkaline conditions acrolein probably polymerizes mainly through its aldehyde group. Therefore a relatively lower aldehyde content is expected in the final product. Table I does indeed show that the aldehyde content of the "irradiated" microspheres is 3.2 times higher than the aldehyde content of the "alkaline" microspheres: 9.3 mmol aldehydes compared with 2.9 mmol aldehydes for each gram of microspheres, respectively. The higher aldehyde content of the "irradiated" microspheres explains their greater ligand binding capacity as illustrated from their reaction with goat immunoglobulin (Fig. 2).

Hybrido PA Microspheres.[7] Hybrido microspheres were formed by grafting microspheres of approximately 0.1 μm diameter, formed by irradiation, onto the surface of PA microspheres obtained under alkaline conditions. An aqueous solution (4 ml) containing 4.5% (v/v) acrolein, 0.5% (w/v) poly(ethylene oxide), and 100 mg "alkaline" microspheres of 2.5 μm diameter was deaerated with argon. The stirred solution was then irradiated with a cobalt source (0.75 Mrad). The PA grafted microspheres

[9] P. J. Borchert, *Kunststoffe* **51**(3), 137 (1961).

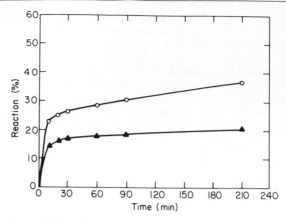

FIG. 2. Rate of reaction between PA microspheres and goat immunoglobulin: (○) "irradiated" microspheres: (▲) "alkaline" microspheres. One hundred milligrams of microspheres in 20 ml PBS was shaken at room temperature with 22 mg goat immunoglobulin. Samples were taken at intervals for the determination of unbound antibody.

formed were washed free of homo-PA microspheres by centrifugation through water four times at 500 g for 10 min. Figure 1c, c-1, and c-2 are SEM photomicrographs of hybrido microspheres; (c) shows the 2.5 μm "alkaline" microspheres, while (c-1) and (c-2) show the hybrido microspheres; (c-2) is a higher magnification. The grafted "irradiated" microspheres provide a greater surface area and thereby a higher aldehyde group content on the surface of the "alkaline" PA microspheres. Therefore the hybrido microspheres have greater binding capacity to amino

TABLE II
BINDING CAPACITY OF THE "ALKALINE" AND HYBRIDO PA
MICROSPHERES TOWARD VARIOUS AMINO LIGANDS[a]

Microspheres	Ligand	Binding capacity (mg/g microspheres)
"Alkaline"	BSA	25
Hybrido	BSA	31
"Alkaline"	Rabbit IgG	10
Hybrido	Rabbit IgG	30
"Alkaline"	Desferrioxamine[b]	25
Hybrido	Desferrioxamine	40

[a] One hundred milligrams microspheres in 3 ml PBS was shaken at room temperature for 24 hr with excess amino ligands.
[b] An iron chelating drug.

ligands than the "alkaline" microspheres, as shown experimentally (Table II).

Agarose-Polyacrolein Microsphere Beads (APAMB)[10]

Synthesis. APAMB were prepared by encapsulating PA microspheres of approximately 0.15 μm diameter within an agarose matrix. Briefly, a solution containing 1 g agarose and 25 ml of PA microspheres aqueous solution, 4% (w/v), was heated to 95° until the agarose gel was melted into a clear solution. The temperature was decreased to 70° and the solution was then poured into a stirred, 300 rpm, vessel containing peanut oil at 70°. Ten minutes later, the solution was cooled with ice. The beads formed were freed of the oil by several extractions with ether. Ether was then removed by evaporation. The diameter of the beads obtained ranged from 50 to 250 μm. Fractions containing beads of diameters 50–150 μm and 150–250 μm were obtained by passing the beads through appropriate sieves. Larger beads can be prepared with the same preparative procedure but at a lower stirring rate of the peanut oil. Figure 3a is a light micrograph of the APAMB, while Fig. 3b is a cross section of the APAMB, showing the "naked" microspheres uniformly encapsulated in the agarose matrix.

Cross-linking of the APAMB was accomplished by interacting the beads with divinylsulfone. One milliliter APAMB was added to an aqueous solution containing 2 ml of 0.5 M sodium carbonate buffer at pH 11.0 and 20 μl divinylsulfone. The reaction was allowed to proceed for 1 hr at 40°. The APAMB were then washed free of divinylsulfone by repeated decantation with water. A spacer arm of the polylysine–glutaraldehyde can be bound to the APAMB by interacting the beads first with polylysine (1 g beads and 5 mg polylysine in 2 ml water) and then with glutaraldehyde (1 g beads and 0.1 ml glutaraldehyde 50% in 2 ml water).[10]

Properties of the APAMB. APAMB fulfill the essential requirements of an effective immunoadsorbent. Agarose in a concentration of 4% (w/v) provides the mechanical strength, specificity, porosity, and biocompatibility of the beads. The microspheres encapsulated in the agarose matrix are used for the covalent binding of the ligands. The cross-linking of the beads do not effect the reactivity of the aldehyde groups and enable the beads to be autoclaved for sterilization. The binding capacity of the APAMB toward various amino ligands, e.g., proteins (BSA), antibodies (goat anti-mouse Ig, goat IgG, and monoclonal anti-Thy 1.2), lectins (concanavalin A and soybean agglutinin), and hormones (bovine insulin), is

[10] S. Margel and M. Offarim, *Anal. Biochem.* **128**, 342 (1983).

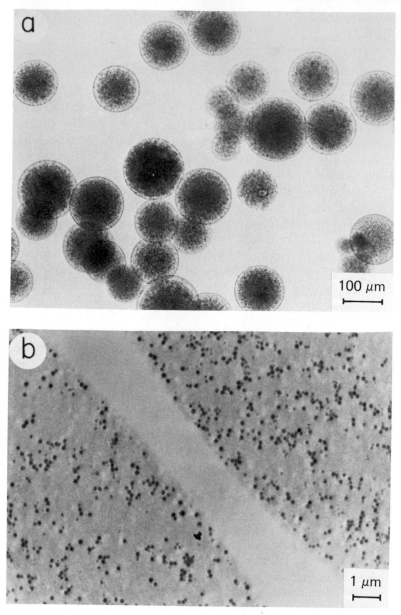

FIG. 3. (a) A light micrograph of the APAMB. (b) A transmission electron micrograph of a cross section of the APAMB.

TABLE III
BINDING CAPACITY OF THE APAMB TOWARD VARIOUS AMINO LIGANDS[a]

Ligand	Spacer	Binding capacity (mg/g beads)
BSA[b]	—	2.2
BSA[b]	Polylysine–glutaraldehyde	14.8
Goat IgG	—	5.0
Goat IgG[b]	Polylysine–glutaraldehyde	9.0
Soybean agglutinin	Polylysine–glutaraldehyde	5.9
Concanavalin A	—	2.4
Concanavalin A	Polylysine–glutaraldehyde	11.5
Goat anti-mouse Ig[b]	—	2.4
Goat anti-mouse Ig[b]	Polylysine–glutaraldehyde	12.4
Anti-Thy 1.2	—	5.0
Bovine insulin	—	7.2
Bovine insulin	Polylysine–glutaraldehyde	22.0

[a] One gram of the APAMB was shaken with excess quantities of the various ligands in 5 ml PBS for 12 hr at room temperature.
[b] Similar binding capacities were obtained for the reaction of cross-linked APAMB and the corresponding amino ligands.

shown in Table III. Steric requirements may explain the significantly increased binding capacity of the APAMB containing the spacer arm polylysine–glutaraldehyde. The beads preserved their physical and mechanical properties after coupling of the ligands. Nephelometric measurements did not indicate any release of microspheres from the agarose matrix during perfusion of saline through the cross-linked APAMB at a flow rate of 150 ml/min. Leakage of proteins bound to the APAMB into the supernatant fluid under basic, acidic, and physiological conditions was not detected by using the method of Lowry et al.[11] for protein determination. Furthermore, radioactivity in the supernatant of a PBS solution containing radioactive BSA (^{131}I-labeled) bound to APAMB was not detected during the radioactive lifetime.

Applications of the PA Microspheres

The potential use of the "naked" and the encapsulated microspheres was presented in several previous publications. A brief description follows.

[11] O. F. Lowry, N. J. Rosebrough, A. L. Farr, and R. J. Randall, *J. Biol. Chem.* **193**, 265 (1951).

Cell Labeling.[5] Different subpopulations of cells were resolved according to their different surface antigens by labeling the cells with immunomicrospheres, i.e., microspheres to which antibody is covalently bound. The high specificity of the "naked" microspheres toward cells was demonstrated with fluorescent microspheres of 0.1 μm average diameter by the following systems:

1. Labeling of human red blood cells (conjugated to rabbit anti-human red blood cells) with microspheres derivatized with goat anti-rabbit IgG.
2. Labeling of mouse B splenocytes with goat anti-mouse Ig derivatized microspheres.
3. Labeling of rat basophilic leukemia cells with the anti-allergic drug disodium chromoglycate derivatized microspheres.

Under optimal conditions, in all systems the experimental cells fluoresced under the microscope and the control cells did not, indicating the high specificity of the labeling.

Affinity Chromatography.[10] The removal and purification of antibodies from the serum of immunized animals were achieved with APAMB with diameters ranging between 50 and 150 μm. The beads conjugated to the desired protein (immunobeads) were washed successively with PBS, eluting medium (0.2 M glycine–HCl buffer at pH 2.4), and again with PBS. The immune serum was then passed through a column filled with the appropriate immunobeads at a flow rate of 1 ml/min (5–10 ml serum for each 1 ml of beads). The immunobeads were then washed several times with PBS. Adsorbed antibodies were then eluted with 0.2 M glycine–HCl buffer solution at pH 2.4, neutralized with NaOH, dialyzed against PBS, and then analyzed by polyacrylamide gel electrophoresis. The immunobeads, after the treatment with glycine–HCl buffer, were washed several times with PBS and stored cold in presence of sodium azide (0.05%) until reused. The binding capacity of some of the immunobeads to the appropriate antibodies and the resultant eluting data are given in Table IV. In all cases antibodies were not detected after the adsorption step, as determined by the ring test with the antisera. Antibodies eluted from the column and analyzed by polyacrylamide gel electrophoresis were found to contain only IgG. Normal rabbit serum was passed through a column containing irrelevant immunobeads. The beads were then treated as described for the isolation of antibodies. Under these conditions, proteins were not eluted from the beads, indicating that nonspecific absorption of proteins onto the beads did not occur. The immunobeads have also been used repeatedly during a period of a year without any significant loss of their antibody binding capacity.

TABLE IV
ISOLATION OF ANTIBODIES WITH IMMUNO-APAMB[a]

Spacer	Antigen	Antigen bound (mg)	Antiserum used	Antibody bound (mg)	Antibody eluted (mg)
—	BSA	2.2	Rabbit	2.4	2.4
Polylysine–glutaraldehyde	BSA	14.8	Rabbit	16.5	17.0
Polylysine–glutaraldehyde	DNP–BSA	9.2	Rabbit	8.5	8.4
Polylysine–glutaraldehyde	Rabbit IgG	10.0	Goat	12.0	12.0
—	Mouse IgG	4.5	Rabbit	4.6	4.5

[a] Data given per milliliter APAMB.

Cell Fractionation.[12] An effective separation of B and T mouse spleen cells was achieved with APAMB with diameters ranging between 150 and 250 μm conjugated to goat anti-mouse Ig or to anti-Thy 1.2. The immunobeads (1 ml) were packed in a siliconized Pasteur pipet plugged with a glass wool. The beads were washed first with PBS to remove sodium azide and then with Hank's solution containing 5% horse serum (HS). Washed, viable cells ($1–2 \times 10^7$) suspended in 1–2 ml of Hank's + 5% HS were filtered through the immunobeads at room temperature, at a rate of 1–3 drops/min. The column was then rinsed with Hank's + 5% HS until the eluate became cell free. Nonadsorbed cells were recovered by centrifugation at 500 g for 15 min and then resuspended in 1–2 ml of Hank's + 5% HS, and the adsorbed cells were then recovered by gently stirring the beads with a Pasteur pipet. The supernatant containing the eluated cells was centrifuged at 500 g for 15 min, and the cells were then resuspended in 0.5 ml of Hank's + 5% HS. Mouse B splenocytes were detected with rhodamine-conjugated goat anti-mouse Ig. Mouse T splenocytes were detected with FITC conjugated to anti-Thy 1.2. The immunobeads were washed with PBS and then stored in PBS + 0.05% (w/v) sodium azide until reused. Quantitative data describing the efficiency of the cell fractionation through the various immunobeads are shown in Table V. In the control experiment, mouse splenocytes were passed through a column containing goat anti-rabbit IgG derivatized beads. The composition of the B and T cells remained almost constant: 53 and 43% of B and T cells, respectively, before fractionation compared to 51 and 45% after fractionation. A very pure population of T cells (97–98%) was obtained in the nonadsorbed cell fraction passed through the goat anti-mouse Ig derivatized beads. However, the cells eluted from these immunobeads formed

[12] S. Margel, M. Offarim, and Z. Eshhar, *J. Cell Sci.* **62,** 149 (1983).

TABLE V
CELL FRACTIONATION OF MOUSE SPLENOCYTES WITH IMMUNO-APAMB[a]

	Percentage of cells					
			Adsorbed cells		Non adsorbed cells	
Bound antibody	B cells	T cells	B	T	B	T
Goat anti-rabbit IgG (control)	53	43			51	45
Goat anti-mouse Ig	60	40	—[b]	6	0	98
Goat anti-mouse Ig	58	40	—[b]	6	0	97
Anti-Thy 1.2	57	40	40	54	86	13
Anti-Thy 1.2	59	40	37	54	84	14

[a] Mouse splenocytes ($1-2 \times 10^7$ cells) were passed at a flow rate of 1–3 drops/min through a column containing beads derivatized with either goat anti-mouse IG or anti-Thy 1.2.
[b] Small aggregates of cells made cell counting difficult.

small aggregates that made cell counting difficult. The efficiency of the separation obtained with the anti-Thy 1.2 derivatized beads as shown in Table V is also satisfactory, although it is not as efficient as with the goat anti-mouse Ig-conjugated beads. In all the experiments described in Table

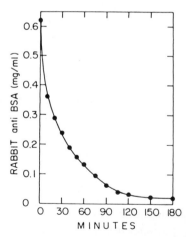

FIG. 4. Kinetics of the removal of rabbit anti-BSA from immunized rabbit. Immunized rabbit blood was circulated at a flow rate of 12 ml/min through a column containing 30 g of APAMB derivatized with BSA.

V the viability of the fractionated cells was unaffected and the recovery was between 80 and 100%. Further research involving the removal of human T cells for bone marrow transplantation purposes is underway in our laboratories.

Hemoperfusion, Removal of Specific Antibodies.[13,14] In the standard experimental procedure a blood pump cycled the arterial immunized blood through the immuno-APAMB column (1 mm average diameter) and then through a bubble trap and thereafter returned the blood to the jugular vein. In the model system the APAMB were covalently bound to BSA, and circulating anti-BSA were adsorbed onto the immunobeads from whole blood of an immunized animal. In practice, any antigen may be covalently bound to the beads and the appropriate antibody (in autoimmune disease) may be removed. A kinetics study of the adsorption *in vivo* of anti-BSA from whole blood of an immunized rabbit is shown in Fig. 4; 62% of the anti-BSA was removed in 30 min, 80% in 60 min, and 95% in 120 min. The APAMB are also biocompatible; there are negligible decreases during the hemoperfusion procedure in red blood cells, white blood cells, and platelets. Electrolytes and other soluble components also are minimally affected.

Other Applications. PA microspheres may also be used in many other clinical applications such as diagnostic tests, phagocytosis, drug delivery, and enzyme immobilization. Research dealing with some of these aspects is proceeding in our laboratories.

[13] S. Marcus, M. Offarim, and S. Margel, *Biomater., Med. Dev., Artif. Organs* **10,** 151 (1982).
[14] L. Marcus, A. Mashiah, M. Dalit, and S. Margel, *J. Biomed. Matter. Res.* (submitted for publication).

[14] Poly(vinylpyridine) Microspheres

By A. Schwartz and A. Rembaum

Introduction

Much time and effort have been directed toward understanding the mechanisms of formation, nucleation, and growth of microbeads when the diameter of the particles is less 1 μm. This understanding has aided in controlling the specific size and uniformity of the resulting materials.

The most widely known microbeads are made of polystyrene, a relatively hydrophobic, hard, inert material. They are highly spherical and will agglomerate in water if they are not in the presence of a surfactant. Formed by emulsion polymerization, nucleated within micelles, they grow in size to only about 0.5 μm. These can be used as seeds and swollen with additional monomer to undergo subsequent growth through multiple polymerizations, to finally reach sizes of 5–10 μm.[1]

Other types of microbeads based on acrylates and methacrylates can be formed by thermochemical treatment or treatment with ionizing radiation.[2-4] These microbeads are relatively hydrophilic and soft, usually contain reactive groups, and can be grown to a size of ~1.5 μm in one step.[5-10]

A third type of microbead, formed from vinylpyridine monomers, lies somewhere between the other two in that it is somewhat hydrophilic, and has limited reactivity. The size range of that can be obtained in one step is larger than the others, up to about 15 μm. The ability of vinylpyridine to copolymerize with a wide range of other monomers allows the introduction of a number of different functionalities to the resulting microbeads.

Vinylpyridine Monomers

The monomers of vinylpyridine may be considered analogs of styrene. Both molecules are vinyl derivatives of aromatic ring structures, but vinylpyridine has a nitrogen substituted in place of the carbon atom.

Styrene   4-Vinylpyridine

The presence of the nitrogen atom in the aromatic ring alters many of the chemical and physical aspects of the monomer, as well as the polymer, relative to styrene.

[1] J. Ugelstad, K. H. Kaggerud, F. K. Hansen, and A. Berge, *Makromol. Chem.* **180**, 137 (1979).

[2] A. Rembaum, S. P. S. Yen, E. Cheong, S. Wallace, R. S. Molday, I. L. Gordon, and W. J. Dreyer, *Macromolecules* **9**, 328 (1976).

[3] A. Rembaum and S. Margel, *Br. Polym. J.* **10**, 275 (1978).

[4] A. Rembaum, S. P. S. Yen, and R. S. Molday, *J. Macromol. Sci., Chem.* **A13**(5), 603 (1979).

[5] A. Rembaum, A. Gupta, and W. Volksen, U.S. Patent 4,170,685 (1973).

[6] A. Rembaum and W. Volksen, U.S. Patent 4,123,396 (1978).

[7] S. P. S. Yen and A. Rembaum, U.S. Patent 4,157,323 (1979).

[8] A. Rembaum, S. P. S. Yen, and W. Volksen, *CHEMTECH* March, p. 182 (1978).

[9] A. Rembaum and W. J. Dreyer, *Science* **208**, 364 (1980).

[10] A. J. K. Smolka, S. Margel, B. H. Nerren, and A. Rembaum, *Biochim. Biophys. Acta* **588**, 246 (1979).

TABLE I

COMPARISON OF MODELS AND MONOMERS[a]

Substance	Density (g/ml)	Boiling point (°C)	Refractive index
Benzene	0.8787	80.1	1.50108
Pyridine	0.9782	115–116	1.50920
Styrene	0.9059	145–146	1.54630
Vinylpyridine	0.9750	62–65 (15 mm Hg)	1.5500

[a] M. Windholz, ed., "The Merck Index," 10th ed. Merck & Co., Rahway, New Jersey, 1983.

The simple pyridine molecule can serve as a model to predict, to some degree, the chemical and physical properties of the polymer. The physical properties listed in Table I allow comparison of pyridine with its analog benzene as well as its monomer vinylpyridine and styrene. The nitrogen contained in the aromatic ring lends a polar nature to the molecule and tends to increase the density, boiling point, and refractive index. Pyridine resembles a highly deactivated benzene derivative in that it resists electrophilic substitution and undergoes nitration, sulfonation, and halogenation only under very vigorous conditions, e.g., 300°. The Friedel–Crafts reaction cannot be carried out under any conditions. However, nucleophilic substitution can take place in the 2- and 4-position, and amination, for example, can be carried out with sodium amide and moderate heating.

The basicity of pyridine ($K_b = 2.3 \times 10^{-9}$) is less than aliphatic amines ($K_b \sim 10^{-4}$). However, like other amines it can react with alkyl halides to form quaternary ammonium salts. Such a reaction product may find biological uses in the form of polymeric microbeads, as will be discussed later.

One other useful reaction is the catalytic reduction of the aromatic pyridine ring with hydrogen to an aliphatic heterocyclic. This would increase the basicity of the molecule to that of a secondary amine ($K_b = 2 \times 10^{-3}$).

Microbead Formation

The majority of microbead synthesis is carried out by emulsion polymerization in which monomer-swollen micelles are formed by surfactants. Polymerization is usually initiated by free radicals generated by thermal decomposition of specific chemicals. The best example of this process is the formation of styrene latex at 50° using sodium dodecyl sulfate as the surfactant and 2,2'-azobis(2-methylpropionitrile) as the free radical generator. Exclusion of oxygen by vacuum or nitrogen displacement allows the

TABLE II
POLY(VINYLPYRIDINE) MONOMERS

Monomer	Structure	N-Position
2-Vinylpyridine		Ortho
2-Methyl-5-vinylpyridine		Meta
4-Vinylpyridine		Para

free radicals to reach the necessary concentration for polymerization to proceed. The resulting latex must then be dialyzed to remove the residue initiator and surfactants.

Although microbead formation from vinylpyridine may be carried out through similar emulsion polymerization schemes, a far cleaner and easier procedure exists which eliminates the need for removal of the surfactants and chemical initiator. This procedure involves the use of polymeric stabilizers, co-solvents, and initiation by high-energy radiation.

The polymeric stabilizers, which may comprise anywhere from 0.5 to 5% by weight of the polymerization mixture, are usually high molecular weight nonionic aqueous soluble molecules. Poly(ethylene oxide), of molecular weight between 50,000 and 5,000,000, has been found to be the best polymer to stabilize the system, although polyoxyethylene sorbitan ester derivatives such as monolaurate (Tween 20) and trioleate (Tween 85) are also satisfactory as stabilizers. The polymeric stabilizers are usually dissolved in the appropriate amount of distilled water prior to addition of the monomers, which seems to help solubilize them.

The 2-vinylpyridine isomer and 2-methyl-5-vinylpyridine were successfully used to synthesize microbeads (Table II). The formed polymers comprise the three possible positions, ortho, meta, and para, for nitrogen atoms relative to the backbone of polymer chains. Such positions may have subtle effects on the chemical and physical properties of the resulting microbeads. For example, 2-vinylpyridine has in general been found to produce smaller microbeads and resists agglomeration better even in the absence of cross-linking and suspension agents.

The vinylpyridine monomers contain inhibitors e.g., 0.1% wt. t-butylcatechol (as received from Aldrich, Inc.). This may be removed quickly and easily by suspending 2.5 wt% Amberlite IR-45a in the monomer for a few minutes, and then filtering the monomer. Although this procedure removes the inhibitor, the monomer still has a light yellow-to-brown cast to it depending on the degree of degradation it has undergone. Such amine compounds are subject to oxidation; this can be retarded by storage in a cold and dark place. A better method of monomer purification is vacuum distillation over KOH pellets in the dark. The monomer is then obtained without discoloration.

Characterization of Microbeads

The physical and chemical characteristics include the average size (diameter or volume), distribution (coefficient of variation on diameter or size), density, opacity, hydrophilicity, reactivity (or inertness), and solubility in various solvents. The majority of these factors can be inferred from just knowing the average diameter, coefficient of variation, and the chemical composition, and consequently, how they will perform in a specific application.

The methods and assumptions concerning the measurement of average size and coefficient of variation are worth additional consideration. One of the most widely used methods of determining size is by making transmission or scanning electron micrographs of the beads and comparing them to micrographs of calibration standards, such as those supplied by the National Bureau of Standards (Etched Grid Lines in Brass, NBS 484a). The size distribution of the microbeads may be derived from these microbead diameters, provided that a large enough sample of microbeads (~1000) is measured to ensure statistical significance. A major drawback to this method is that microbeads that have undergone coalescence with two, three, or more beads during polymerization to form doublets, triplets, etc. are included in the distribution statistics and do not reflect the true distribution uniformity of singlet population. Knowing the percentage of doublets and triplets in the population, one can mathematically extract the singlet distribution. A more direct approach would be to make measurements and perform statistics on the separate populations. This method would be more in line with automated particle flow measurement methods. One of the most sensitive, accurate, and rapid methods of deter-

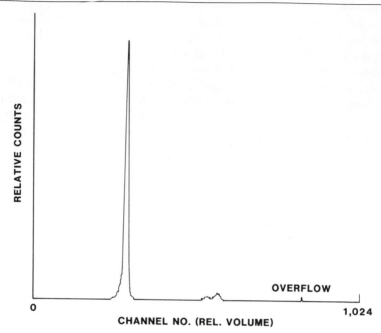

FIG. 1. Linear volume histogram from a UF100/ND100 (Becton Dickinson) of a 6.1-μm polystyrene microbead (Dow Diagnostics). The coefficient of variation on diameter was 0.9% for the singlet peak and the sample contained 8% doublets.

mining size and distribution of microbead suspension employs computerized instruments which measure electronic volume, as shown in Fig. 1. The sensitivity is derived from measuring the volume of the microbeads, a cubic function of the diameter. The accuracy is limited to the ability to calibrate the reference microbeads used to set up the instrument as long as the proper sized orifice is used. Again, this calibration is usually accomplished with electron microscopy. Also, the environment may play a role. In the dry state, under high vacuum in the electron microscope, there may be shrinkage of the microbeads in an aqueous suspension. This is especially true of microbeads made of hydrophilic materials throughout. Critical point drying has been found to help, but some shrinkage has still been observed.

With proper programming, the distribution of the singlet microbead population and the population of multiple coalesced microbeads can be determined with an electric volume sensing instrument. Conversion of the coefficient of variation for the volume (CV_V) to that of diameter (CV_D) can be calculated as shown in the following example:

$$CV_V = 5.0\%$$
$$0.05 + 1 = 1.05$$
$$\sqrt[3]{1.050} = 1.0163 - 1 = 0.016 \times 100$$
$$CV_D \sim 1.6\%$$

Factors Affecting Size and Distribution

The simplest formulation for vinylpyridine microbeads is to dissolve about 3% of freshly distilled 4-vinylpyridine in distilled water containing 0.1% poly(ethylene oxide), MW 100,000. After bubbling nitrogen gas through the solution for several minutes the container is sealed and exposed to γ radiation for a dose rate of 0.2 Mrad/hr, for a total of 2 Mrad. The resulting microbeads, shown in Fig. 2a, are about 0.2 μm, with a nonuniform size distribution.

The major factors controlling the size of microbeads are the concentration of monomers and the concentration of co-solvent. As shown in Table III, the size of the microbeads decreases with decreasing monomer concentration. Such microbeads are shown in Fig. 2a. Those microbeads represented in Table III contained 10% by monomer weight bisacrylamide, a hydrophilic cross-linker. Ethyleneglycol dimethacrylate, trimethylolpropane trimethacrylate, or hexahydro-1,3,5-triacryloyl-s-triazine are other cross-linkers which may be used that tend to make the microbeads more hydrophilic. The addition of ionic surfactants, e.g., sodium dodecyl sulfate, has been found to decrease the resultant microbead size.

TABLE III

SIZE OF PVP MICROBEADS AS A FUNCTION OF MONOMER CONCENTRATION

Concentration of 4-VP[a] (wt/wt%)	Ratio of BAM[b] to 4-VP (wt/wt%)	Monomer concentration (wt%)	Size[c] of beads (Å)
0.225	1:9	0.25	900
0.45	1:9	0.50	950
0.90	1:9	1.00	1300
1.35	1:9	1.50	1600
1.80	1:9	2.00	1600
2.25	1:9	2.50	1600
2.70	1:9	3.00	2500

[a] Vacuum distilled, and polymerized in degassed solution.
[b] Bisacrylamide.
[c] Size determined by SEM.

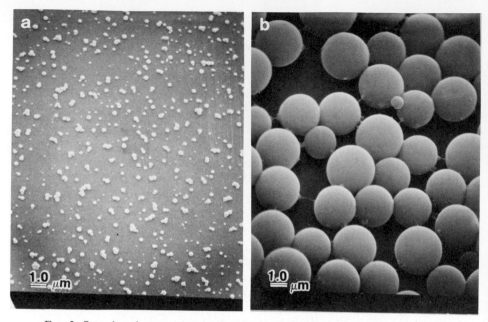

FIG. 2. Scanning electron micrographs of poly(vinylpyridine) microbeads polymerized (a) with no co-solvent and (b) with 15% acetone as co-solvent.

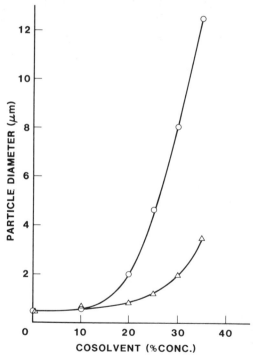

FIG. 3. Graph of PVP microbead size as a function of co-solvent concentration: (O) acetone; (△) methanol.

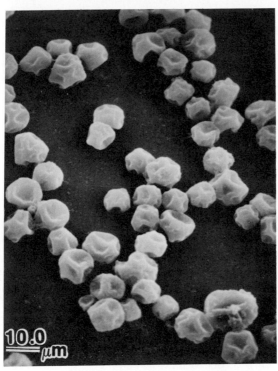

FIG. 4. Scanning electron micrograph of PVP microbeads which were formed in 35% acetone co-solvent. The "deflated" appearance suggests a hollow structure.

Two co-solvents which have been used successfully are acetone and methanol. The solubility of vinylpyridine monomers is much higher in these solvents than in water. This allows more of the monomers to dissolve in a co-solvent system and thus yield larger microbeads, as seen in Fig. 2b. There is an approximate exponential growth function with co-solvent concentration, as shown in Fig. 3, up to about 35%. Beyond this point, the solubility of the monomers is so great that gel formation is favored over bead formation, resulting in a clear homogeneous rigid polymer in the reaction tube when stirring is not employed.

Just prior to the gelation point, an instability in the system is indicated by a change in structure of the microbeads. As seen in Fig. 4, the large microbeads appear as "partially deflated basketballs" suggesting a hollow structure. Such a structure may lend itself to encapsulation applications with various compounds which could undergo diffusion-controlled release from the inner reservoir through the polymer membrane over an extended period of time.

TABLE IV
MICROBEAD SIZE AS A FUNCTION OF MOLECULAR
WEIGHT OF THE STABILIZER

PEO[a] molecular weight	Microbead size (Å)
4,000,000	2140
600,000	2330
300,000	2500
100,000	2650

[a] PEO, Poly(ethylene oxide).

The molecular weight of poly(alkylene oxide) polyether used as stabilizers can affect the size of the microbeads which are formed during irradiation. As shown in Table IV, increasing the molecular weight of the stabilizer poly(ethylene oxide) from 100,000 to 4,000,000 increases the microbead diameter from 2140 to 2650 Å, respectively.

As mentioned before, vinylpyridine is a basic molecule and results in a solution pH of approximately 9. When polymerized, the resulting microbeads have a distribution with a coefficient of variation of diameter of ~15%. However, if the pH is adjusted to 7.0–7.1 prior to irradiation, the coefficient of variation of diameter may be reduced below 3% as seen in Fig. 5. Reducing the pH below 7 tends to increase the distribution, and if it is reduced below 5, gel formation occurs.

Uniformity of the microbeads can also be increased by clarification of the monomer solution prior to degassing and irradiation. This may be accomplished by either filtration or centrifugation, which tends to remove undissolved stabilizer that may act as heterogeneous nuclei for the forming of microbeads. Using this additional step, coefficients of variation have been measured as low as 2.5% of diameter. The use of methylvinylpyridine in place of vinylpyridine was also found to yield very uniform microbeads, $CV_D \sim 2-3\%$.

The spherical shape and firmness of the microbeads can be improved by the use of increased radiation dose. Although the bead forms with as little as 0.5 Mrad, they are soft and tend to agglomerate at this stage. Continued radiation to about 4 Mrad has been found to reduce the tendency of sticking together and they appear less deformable and more spherical in shape. The addition of about 10% of the hydrophilic crosslinker bisacrylamide has also been found to reduce agglomeration and improve the spherical shape of the microbeads. Ethyleneglycol dimethacrylate used in place of the bisacrylamide had the same effect; however, the microbeads produced tend to be smaller in size.

Proteins can be coupled directly with poly(vinylpyridine) using CNBr

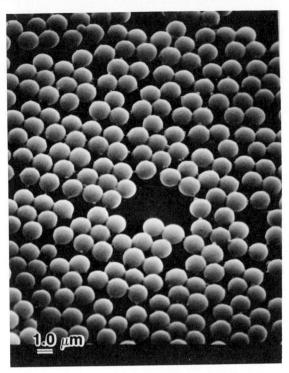

FIG. 5. Scanning electron micrograph of highly uniform PVP microbeads ($CV_D \sim 3\%$).

in a nonaqueous medium to activate the aromatic nitrogen, before exposing it to water then to a protein in aqueous solution.

This can be carried out with pyridine microbeads provided they are strongly cross-linked to prevent dissolution during the exposure of the microbeads to the nonaqueous solvent.[11]

Metallic Poly(vinylpyridine) Microbeads

The pyridine moiety of the polymer not only introduces a polar component to the microbead but also has the characteristic of complexing

[11] F. Pittner, T. Miron, G. Pittner, and M. Wilchek, *Enzyme Eng.* **5**, 447 (1980).

various ionic metals with its aromatic nitrogen. The transition metal salts can be adsorbed and complexed to poly(vinylpyridine) microbeads in almost a quantitative yield. Owing to the slightly hydrophobic nature of vinylpyridine, addition of ethanol to the microbead suspension appears to enhance diffusion of the metallic ions into the interior of the microbead.

Microbeads suspended in $CoCl_2 \cdot 6H_2O$ solution become blue, and suspended in $FeCl_3$ they take on a green appearance. In general, metal compounds from the periodic groups IVa, VIa, VIIa, and VIII in the forms of oxides, salts, and acids can be complexed by the aromatic nitrogen in the pyridine ring.

The metal containing acids react easily with poly(vinylpyridine) to form a strong quaternary bond. However, metal salts do not appear to quaternize the pyridyl groups, but form a strong ionic bond. Addition of water-miscible solvents such as the lower alcohols provide better interaction of the salts with the pyridine polymer. Heating has also been observed to enhance the interaction. Such solvents and heating most likely act to aid diffusion of the salt throughout the polymer microbeads. With stoichiometric uptakes of these heavy metal salts, the weight of the microbeads may increase by 50%.

In addition to the transition metal salts and acids, the noble metal acids such as chloroauric acid ($HAuCl_4 \cdot 3H_2O$) and chloroplatinic acid ($HPtCl_6 \cdot 6H_2O$) can quaternize the aromatic pyridine nitrogen to form a stable complex. The noble metal complexes can be reduced by reducing agents such as sodium borohydride to the zero valence of the metal, and the microbeads become black in appearance.

Particulate metals and metal oxides may also be incorporated into poly(vinylpyridine) microbeads during polymerization. This is accomplished simply by adding the particulates to the monomer solution prior to degassing and polymerization. Obviously the particle must be small relative to the bead size, usually 50–200 Å. An interesting application of the synthesis methodology is incorporation of colloidal magnetite which provides the microbeads with magnetic properties.

Poly(vinylpyridine) Copolymer Microbeads

Formation of microbeads from vinylpyridine has a number of advantages over other monomers. The major advantage is control over size and

TABLE V
COPOLYMER MICROBEADS OF VINYLPYRIDINE

Monomer	Functionality introduced into the microbead
Methacrylic acid	$-C\overset{\displaystyle O}{\underset{\displaystyle H}{\big\langle}}$
Acrylic acid	$-C\overset{\displaystyle O}{\underset{\displaystyle H}{\big\langle}}$
2-Hydroxyethyl methacrylate	$-CH_2-OH$
Allylamine	$-CH_2-NH_2$
Allyl mercaptan	$-CH_2-SH$
Acrylamide	$-C\overset{\displaystyle O}{\underset{\displaystyle NH_2}{\big\langle}}$

distribution. In one polymerization step, highly uniform microbeads can be formed over a range from a few hundred angstroms to 15 μm. With other monomers such as styrene, it is very difficult to form microbeads in the 1–15 μm range and must be done by multiple overcoating steps or seed swelling techniques.

Vinylpyridine microbeads also have the advantage of being able to incorporate metals through complexing with its aromatic nitrogen. The ring itself can undergo chemical modification to introduce other functional groups, e.g., amines.

However, through copolymerization with other monomers, an even wider range of functional microbeads may be formed which retain the advantage of high uniformity and control of size. Table V lists a number of copolymer microbeads which are possible to synthesize. The amount and properties of comonomers used with vinylpyridine will affect the resulting microbeads, e.g., hydrophobic monomers will tend to reduce the size; however, if the comonomer concentration is kept to about 10% or less, the microbead formation is not greatly affected in most cases.

The advantage of copolymerization of vinylpyridine with other monomers during microbead formation is that it not only changes the physical properties of the microbead, e.g., hydrophilicity, but also introduces functional chemical groups onto the surface for conjugation with other molecules. This is particularly important for conjugation of biological active molecules such as enzymes, antibodies, and protein dyes.

Copolymerization with acrylamide or monomers containing carboxyl groups such as methacrylic acid and acrylic acid allows conjugation of the acid group to an amine of the protein via carbodiimide activation. This results in a stable covalent amide linkage between the microbead and the amine-containing molecule of interest.

If the comonomers contain hydroxyl groups, e.g., hydroxyethyl methacrylate, then the microbeads can be activated with CNBr and form a stable covalent bond through the amine of the conjugated molecule.

A self-activated microbead can be prepared by copolymerization with acrolein. However, it appears that the hydrophilicity and properties of this comonomer are so different that a phase separation occurs and small microbeads are formed on the surface of the larger vinylpyridine (Fig. 6). Such a composite microbead has several advantages in that it will readily react with molecules containing primary amines under relatively mild conditions (pH 8–9). In addition, the large number of small microbeads covering the major bead will considerably increase the surface area.

Poly(vinylpyridine) Microbead Applications

The range of size and chemical functionality with which vinylpyridine may be formed into microbeads permit a wide spectrum of applications which include instrument calibration standards, cell labeling and separation immunoassays, and catalytic and enzyme supports.

Calibration Standards

In the field of particle analysis, instrumentation in many ways rivals theory and practice in its sophistication. At present, instruments of high sensitivity are capable of collecting, storing, analyzing, and correlating vast amounts of data. Simultaneous collection of data related to electronic volume sizing, forward and 90° light scattering, and multifluorescent emissions can now be accomplished. Proper correlation and analysis of these parameters can yield a wealth of information on the particles or cells under investigation.

This sophisticated capability requires that these instruments, particle counters and flow cytometers, have highly uniform and stable standards on which they can be calibrated.

The uniformity and size with which the poly(vinylpyridine) can be formed, together with the ability to incorporate fluorescent dyes, make these microbeads a useful standard, especially for flow cytometers. Ideal standards for a flow cytometer would be microbeads which are the same size as the cells being measured, i.e., generally platelets to macrophages,

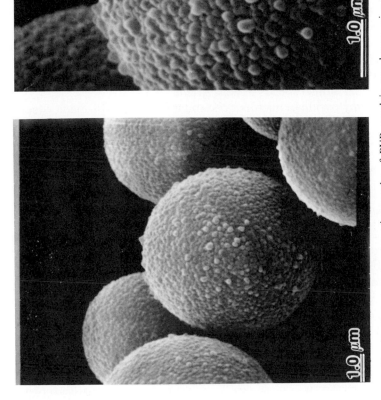

Fig. 6. Scanning electron micrographs of PVP–acrolein copolymer microbeads. The larger bead, presumably mostly poly(vinylpyridine), appears covered with smaller (~0.1 μm) acrolein beads.

which cover the size range from ~1.5–12 μm. For cells that are to be labeled with fluorescent dyes, the microbead standards would also have to contain the same dyes. These dyes could be conjugated to an allyl or vinyl moiety and the fluorescent monomer could be copolymerized. An example of such a fluorescent monomer would be conjugating isothiocyanate with allylamine to form allylfluorescein.

An alternative route would be to copolymerize a monomer with a primary amine into the bead, and then treat the microbeads with the reactive fluorescent dye. This would result in the dye being only on the surface of the microbead. For quantitative work, this latter method is extremely important because the environment of the dye affects the adsorption and emission of these dyes. As seen in Fig. 7, fluorescein when polymerized into a hydrophobic material exhibits a shift in its excitation spectrum toward the blue end and is observed with a corresponding shift and broadening of the emission spectrum toward the red. Dye attached only to the surface is in contact with an aqueous environment and gives the same spectra as the dye in solution. Under a fluorescent microscope using a 485-nm short-pass excitation filter and a 515-nm long-pass emission filter, a difference in these beads is quite evident, with internally incorporated fluorescein microbeads appearing yellow, while surface-fluoresceinated microbeads appear green.

Poly(vinylpyridine) microbead standards can also be made which contain two different fluorescent dyes that would show up in different regions of the spectrum, i.e., fluorescein in the green and tetramethylrhodamine in the red. These microbeads can be useful to set two photomultipliers simultaneously so as to compensate for spectral overlap.

Cell Labeling and Separation

In biological systems, there are various sets and subsets of cells which have specific functions. The classic examples are the formed elements in whole blood, which include platelets, erythrocytes, lymphocytes, monocytes, macrophages, etc. Within the set of lymphocytes, subsets of T cells and B cells can be distinguished, and even within these subsets specific subsets of cells can be identified which have specific immunological functions. To identify these specific cells, much work has been carried out with immunomarkers. Electron-dense immuno-specific conjugated mate-

---

FIG. 7. Excitation (a) and emission (b) spectra of fluorescein in solution (solid line), copolymerized within a hydrophobic poly(methyl methacrylate) microbead (---) and conjugated to the surface of a poly(methyl methacrylate) microbead (————). All spectra were taken of isotonic (pH 7.2) solutions and suspensions.

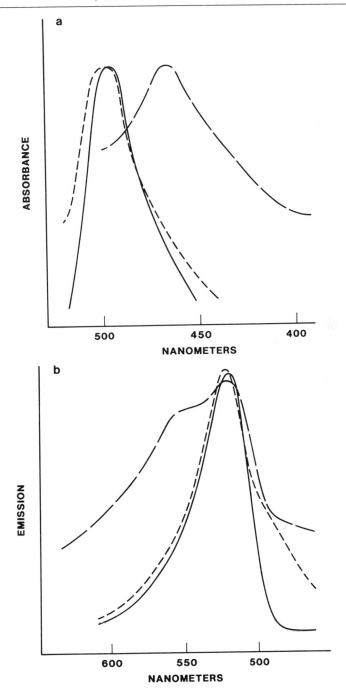

rials, e.g., ferritin,[12] can be observed with electron microscopy and can be used to mark whole cells.

The use of microbeads as cell markers can be thought of as an amplifying system with which the electron density (as studied by electron microscopy) or fluorescence (as measured by flow cytometery) can be increased by factors of 1000 or more. This can be extremely important for cell subsets which have very few antigens on their surface, because the number of soluble antibodies carrying the marker that attaches to the cell could be below the detection limit of the eye or instrument.

Many different types of microbeads have been employed to label cells. Initially, hydrophobic beads made from polystyrene were used. These can absorb antibodies but are plagued by nonspecific interaction with other cells which are not members of the subset of interest. Hydrophilic microbeads made from hydrophilic methacrylates such as 2-hydroxyethyl methacrylate acid have been found to be very useful for cell labeling in that they do not suffer from nonspecific interactions.

With hydrophilic microbeads, the antibodies have to be chemically conjugated to the beads rather than be dependent on passive adsorption. Like the hydrophilic microbeads, poly(vinylpyridine) microbeads can be formed as copolymers which contain functional groups to be conjugated to antibodies. These microbeads have the advantages of being able to incorporate fluorescent dyes and/or electron-dense materials, as well as having a wide size range.

Poly(vinylpyridine) microbeads have an advantage over other beads for use in electron microscopy in that they are able to actively complex the heavy metal salts such as uranyl acetate and lead citrate which are used in biological tissues. An example of this is shown in Fig. 8, in which poly(vinylpyridine) microbeads have been ingested by a macrophage. It is also possible to use the poly(vinylpyridine) microbeads which have already been complexed with electron-dense material, e.g., gold.

With large fluorescent microbeads, it would be necessary to attach only one to a cell to be able to distinguish a labeled cell by either size or fluorescence. Of course, assurance of the nonspecific interaction of such microbeads with other cells would have to be demonstrated with control cell populations.

Advantage of the large microbeads, 2–10 μm, can be taken when cell separation of subpopulations is required. Although flow cytometers can sort subpopulations of cells labeled with soluble fluorescent antibodies at rates of several thousand per second, it still takes over an hour of constant running to get a million cells if the particular subpopulation comprised less than 10% of the sample.

[12] G. Roth, *J. Immunol. Methods* **18,** 1 (1977).

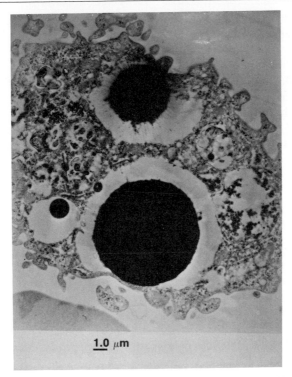

FIG. 8. Transmission electron micrograph of PVP microbeads ingested by a macrophage. Section stained with uranyl acetate and lead citrate.

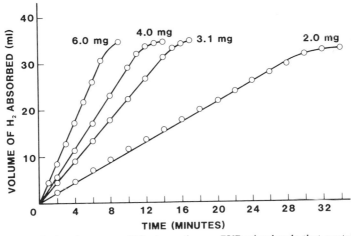

FIG. 9. Graph showing rates of H_2 absorption on PVP microbeads that contain atomic platinum.

Faster separation can be accomplished using larger, more dense beads by direct physical methods such as centrifugation, or by magnetic fields in the case of poly(vinylpyridine) microbeads which contain magnetic particles. A specific subpopulation of microbeads could be separated by low-speed centrifugation if they were attached to poly(vinylpyridine) microbeads because such beads have a higher density, $\rho \sim 1.3$, than the cells, or to comparable microbeads such as polystyrene, $\rho \sim 1.05$.

With magnetite incorporated into such beads, separation of cellular subsets is made even more direct by placing a magnet on the side of the vessel during gentle stirring. The cells labeled with the magnetic microbeads are held against the side while the remaining cells and suspending medium may be poured off.

Catalytic and Enzymatic Supports

One of the limiting aspects of catalytic reactions on solid supports is the availability of the catalysts to the reactants. Maximizing the surface area of the catalyst would naturally aid in the reaction rate.

Incorporating atomic platinum into poly(vinylpyridine) microbead by complexing the aromatic nitrogen with chloroplatinic acid followed by reduction to atomic platinum has been shown to catalytically aid in the hydrogenation of the model compound 1-hexene. Poly(vinylpyridine) microspheres containing 20% by weight of platinum were found to hydrogenate 1-hexene dissolved in methanol in a 100% yield at room temperature and pressure. This is surprising since the presence of the pyridine nitrogen usually poisons platinum catalysts. Figure 9 shows that the rate of hydrogenation is dependent on the amount of platinum relative to olefin.[6]

Poly(vinylpyridine) microbeads may also act as enzyme supports or biological catalysts. Poly(vinylpyridine) polymers were conjugated to trypsin, and were found to have activity toward the model substrate (sodium) benzoylarginine ethyl ester close to that of soluble trypsin.[11]

Acknowledgment

This report represents one phase of research carried out at the Jet Propulsion Laboratory, California Institute of Technology, under Contract NAS7-918, sponsored by the National Aeronautics and Space Administration, and by Grant No. 1R01-CA20668-04 awarded by the National Cancer Institute, DHEW, and Grant No. 70-1859 awarded by the Defense Advanced Research Project Agency.

[15] Artificial Cells Containing Multienzyme Systems

By Thomas Ming Swi Chang

Introduction

An artificial cell is a spherical ultrathin membrane of cellular dimensions enveloping biologically active material.[1,2] The enveloping membrane of each artificial cell separates the contents of the artificial cells from the external environment. A typical artificial cell has an ultrathin membrane of 200 Å thickness, an equivalent pore radius of about 18 Å, and a large surface/volume relationship (2.5 cm^2 in 10 ml of 20-μm diameter microcapsules). This allows for an extremely rapid equilibration of external permeant molecules at a rate which is 400 times faster than that of a standard hemodialysis machine. This way proteins, enzymes, and other macromolecules or particulate matter can be retained within the artificial cells to act on external permeant molecules which equilibrate rapidly into the artificial cells. In addition to this standard system, artificial cells can also be made with other variations in membrane composition, permeability, and other characteristics. It is possible to enclose almost any combination of enzymes, multienzyme systems, cofactor-regenerating enzyme systems, cell extracts, whole cells, proteins, adsorbents, magnetic materials, and multicompartmental systems.

Most enzymes function in an intracellular environment in the form of multienzyme systems. Enzyme systems immobilized within the intracellular environment carry out their function by acting sequentially on substrates. If one were to ignore this natural scheme and inject heterogeneous enzymes in free solution into the body, hypersensitivity reactions, immunological reactions, and rapid removal and inactivation could result. Free enzymes in solution cannot be kept at the sites where the action is desired. Enzymes in free solution are not stable, especially at a body temperature of 37°. Furthermore, multienzyme systems and those for cofactor recycling require the enzymes and substrates to be in close proximity, preferably in an intracellular environment. Unlike enzymes immobilized to solid support, enzymes immobilized in artificial cells remain in solution, and thus the problems of diffusion of substrates and cofactors are much less than other types of immobilized multienzyme

[1] T. M. S. Chang, *Science* **146**, 524 (1964).
[2] T. M. S. Chang, "Artificial Cells." Thomas, Springfield, Illinois, 1972.

METHODS IN ENZYMOLOGY, VOL. 112

systems. Finally, there are fewer problems with the amount or number of different enzyme systems that can be enclosed together into one artificial cell. Artificial cells are being used in detoxifiers, artificial kidneys, artificial livers, immunosorbents, blood substitutes, drug carriers, cell cultures, and hybrid organs. Detailed reviews in the area of artificial cells are available.[2-6]

Methods of Preparation

As emphasized earlier,[2] the basic methods used for the preparation of artificial cells are, in fact, physical examples for demonstrating the principle of artificial cells. Many new physical systems are being developed to demonstrate the same principle.[2-6] A typical procedure is described to demonstrate one of the many possible approaches.

Example of Procedure of Preparation

Reagents

Hemoglobin solution: 15 g of hemoglobin (Sigma bovine hemoglobin type 1, 2× crystallized, dialyzed, and lyophilized) is dissolved in 100 ml of distilled water, then filtered through Whatman No. 42 filter paper. The final concentration of hemoglobin in the filtrate is adjusted to 10 g/100 ml.

Enzyme solution: Enzymes (single, multiple, insolubilized, cell extract or other material) are dissolved or suspended in the hemoglobin solution. Suitable adjustment is required to maintain a final concentration of 10 g of hemoglobin per 100 ml. Also, a minimum pH of 8.5 should be maintained by the use of Tris buffer.

Organic solution: 100 ml of ether (analytical grade) is saturated with water by shaking with distilled water in a separating funnel, then discarding the water layer.

Cellulose nitrate solution: This is prepared by evaporating to completely dry 100 ml of collodion (USP, 4 g of cellulose nitrate in a 100-ml mixture of one part alcohol and three parts ether) into a thin

[3] T. M. S. Chang, "Biomedical Applications of Immobilized Enzymes and Proteins," Vols. 1 and 2. Plenum, New York, 1977.

[4] T. M. S. Chang, *in* "Drug Carriers in Biology and Medicine" (G. Gregoriadis, ed.), p. 271. Academic Press, New York, 1979.

[5] T. M. S. Chang, *in* "Methods in Enzymology" (K. Mosbach, ed.), Vol. 44, p. 201. Academic Press, New York, 1976.

[6] T. M. S. Chang, "Microencapsulation Including Artificial Cells." Humana Press, New York, 1984.

sheet, then dissolving to its original volume by using a solution consisting of 82.5 ml of ether and 17.5 ml of alcohol. This exact composition is important for membrane formation.

n-Butyl benzoate (Eastman)

Tween-20 solutions: 50% Tween 20 solution is prepared by dissolving 50 ml of Tween 20, an oil/water emulsifying agent (Atlas Powder Company, Canada Ltd., Montreal, Quebec, Canada) in an equal volume of distilled water. 1% Tween 20 solution is prepared by dissolving 1 ml of Tween 20 in 99 ml of water. The pH in both solutions should be adjusted using buffer to 7.

Magnetic stirrer: It is important to have a magnetic stirrer with sufficient power to give the speed (rpm) stated, especially when stirring the very viscous Tween 20 mixture. The Jumbo Magnetic Stirrer (Fisher Scientific Company, Montreal, Quebec, Canada) is used in the procedures described below. Also needed is a 4-cm magnetic stirring bar and a 150-ml glass beaker with an internal diameter of less than 6 cm.

Procedure. All procedures are carried out at 4°. To a 150-ml glass beaker containing 2.5 ml of Tris-buffered hemoglobin solution (containing the enzymes to be included), add 25 ml of the organic solution. The mixture is immediately stirred with the magnetic stirrer at a speed setting of 5 (1200 rpm). After stirring for 5 sec, 25 ml of the cellulose nitrate solution is added and stirring is continued for another 60 sec. The beaker is then covered and allowed to stand unstirred at 4° for 45 min.

The microcapsules should be completely sedimented by the end of 45 min. Decant most of the supernatant. Immediately add 30 ml of the *n*-butyl benzoate to the beaker and stir at a speed setting of 5 for 30 sec. The beaker is then allowed to stand uncovered and unstirred at 4° for 30 min.

The final step is to transfer the microcncapsulated enzyme from the organic liquid phase into an aqueous phase. First, butyl benzoate supernatant should be removed completely. In the case where microcapsules greater than 50 μm in diameter are prepared, this can be done readily if the microcapsules have sedimented completely by 30 min. If the microcapsules have not sedimented completely, then centrifugation at 350 g for 5 min is required.

After removal of supernatant, 25 ml of the 50% Tween 20 solution is added. The suspension is immediately dispersed by stirring with the Jumbo Magnetic Stirrer at a speed setting of 5 for 30 sec, and then 25 ml of water is added. After a further 30 sec of stirring, the suspension is further diluted with 200 ml of water. The slightly turbid supernatant may now be removed by centrifugation of the suspension at 350 g for 5 min.

The microencapsulated enzymes so obtained are washed repeatedly in

1% Tween 20 solution until no further leakage of hemoglobin takes place and the smell of butyl benzoate is no longer detectable. The removal of n-butyl benzoate with 1% Tween 20 solution is very important because, unless it is completely removed, butyl benzoate may affect the permeability characteristics of the semipermeable microcapsules.

The final preparation is suspended in a solution containing 0.9 g sodium chloride per 100 ml of water or in a suitable buffer solution. This is not a difficult procedure if the exact concentrations, timing, material, and other steps are followed exactly as described. Omissions or modification, even extremely minor, may result in a great deal of problems requiring extensive readjustment. It is suggested that the exact procedure be followed at first. Modifications should be carried out after the procedure has been repeated well enough to result in good artificial cells.

Methods of Forming Artificial Cell Membranes

Spherical ultrathin polymer membranes have been formed using emulsification followed by interfacial polymerization.[1,2,5,7] Numerous chemical reactions available for interfacial polymerization have been adapted for the same purpose.[8–11] Multiple-compartment membrane systems consisting of smaller artificial cells enveloped within larger artificial cells have also been developed.[2,7,12,13] Another approach is to use a secondary emulsion using silastics, cellulose acetate, and other polymers.[2,7,12,13] Liquid membranes can also be used in the secondary emulsion approach.[3] Biological and biodegradable membranes can also be used to microencapsulate enzymes and other biologically active materials. Protein has been used to form spherical ultrathin cross-linked protein membranes.[1,2,7,12] In this approach if the cross-linking reaction is allowed to continue the cross-linking of protein will proceed deeper into the microspheres until the whole artificial cell is cross-linked.[1,2,3,12] In the same way, artificial cells containing protein can be treated with cross-linking agents such as glutaraldehyde to form solid cross-linked protein microspheres.[14] Artificial

[7] Y. T. Yu and T. M. S. Chang, J. Enzyme Microb. Technol. **4,** 327 (1982).

[8] T. Mori, T. Tosa, and I. Chibata, Biochim. Biophys. Acta **321,** 653 (1973).

[9] M. Shiba, S. Tomioka, M. Koishi, and T. Kondo, Chem. Pharm. Bull. **18,** 803 (1970).

[10] R. B. Aisina, N. F. Kazanskata, E. V. Lukasheva, and V. Berezin, Biokhimiya **41,** 1656 (1976).

[11] T. M. S. Chang, F. C. MacIntosh, and S. G. Mason, Can. J. Physiol. Pharmacol. **44,** 115 (1966).

[12] T. M. S. Chang, Ph.D. Thesis, McGill University, Montreal (1965).

[13] T. M. S. Chang, Trans. Am. Soc. Artif. Intern. Organs **12,** 13 (1966).

[14] T. M. S. Chang, Biochem. Biophys. Res. Commun. **44,** 1531 (1971).

cells with spherical ultrathin lipid membrane have been prepared,[15] using a slight modification of the original technique for forming artificial cells.[1] In order to strengthen the weak lipid membrane, lipids have been complexed into spherical ultrathin cross-linked protein membrane or into the standard spherical ultrathin polymer membrane.[2,16,17] Another approach to strengthen the lipid membrane has been the use of multilamellar lipid membrane system of liposomes.[18,19] This system has more recently been modified to a spherical ultrathin lipid membrane,[20] similar to the earlier artificial cell systems.[1,15]

Methods of Enclosing Contents

Almost any biologically active material can be enclosed within artificial cells. The following have been enclosed: enzymes; multienzyme systems; cell extracts and other proteins[1-6]; granules of enzyme systems and proteins[2]; and combined enzyme and adsorbent system.[2,3,5] Artificial cells containing biological cells[2,11] are now receiving renewed attention because of applications in biotechnology. Magnetic materials have also been included within artificial cells to allow external magnetic fields to direct the movements of the artificial cells.[13] Other materials include radioisotope-labeled enzymes and proteins, insolubilized enzymes, multienzyme systems with cofactor recycling systems, antigens, antibodies, vaccines, and hormones.

Methods of Microencapsulation of Enzymes and Multienzyme Systems

By selecting one of the many updated methods available,[2,5,7] and by the proper adjustment of pH and other factors, most enzymes tested so far can be successfully enclosed within artificial cells. The 10 g/dl hemoglobin solution present in the standard artificial cells gives an intracellular environment somewhat comparable to red blood cells. This way, the enzymes enclosed in the artificial cells are stabilized by the high concentration of protein. Further stabilization can be obtained by cross-linking the hemoglobin with glutaraldehyde.[14] In most cases except for those treated with glutaraldehyde, the enzyme retains about 90% of its original activity if the artificial cells are disrupted to release the enzymes for

[15] P. Mueller and D. O. Rudin, *J. Theor. Biol.* **18,** 222 (1968).
[16] T. M. S. Chang, *Fed. Proc., Fed. Am. Soc. Exp. Biol.* **28,** 461 (1969).
[17] A. M. Rosenthal and T. M. S. Chang, *J. Membr. Sci.* **6**(3), 329 (1980).
[18] G. Gregoriadis, P. D. Leathwood, and B. E. Schnure, *FEBS Lett.* **14,** 95 (1971).
[19] G. Sessa and G. Weissman, *J. Biol. Chem.* **245,** 3295 (1970).
[20] G. Gregoriadis, "Drug Carriers in Biology and Medicine." Academic Press, New York, 1979.

analysis. However, because of the high concentrations of enzymes in the artificial cells and the membrane restriction to free diffusion of substrates, the assayed activity of the artificial cells containing the enzymes is usually about 30% of that of the enzymes in free solution. Hemoglobin can be purified to remove contaminating enzymes before use.[21] Other macromolecules can also be used as the carrier.[22,23]

Cofactor Recycling in Artificial Cells Containing Multienzyme Systems

Most metabolic functions are carried out in cells by complex multienzyme systems with cofactor requirements. As a result, research has been carried out here for the microencapsulation of multienzyme systems with cofactor regeneration. Artificial cells containing hexokinase and pyruvate kinase could recycle ATP for the continuous conversion of glucose into glucose-6-phosphate and phosphoenol pyruvate into pyruvate; similarly artificial cells containing alcohol dehydrogenase and malic dehydrogenase can recycle NADH making use of NAD^+.[24,25] A multienzyme system consisting of urease, glutamate dehydrogenase, and glucose-6-phosphate dehydrogenase, all within each artificial cell, has been used to convert urea into amino acids.[26] In this approach, urea is first converted into ammonia. Ammonia in the presence of α-ketoglutarate and NADPH is converted into an amino acid, glutamate. Glucose-6-phosphate dehydrogenase in the same artificial cell serves to recycle the cofactor NADPH. In order to allow for the use of blood glucose instead of glucose-6-phosphate, a new system of artificial cells containing urease, glutamine dehydrogenase, and glucose dehydrogenase has been developed using glucose as energy to recycle the cofactor.[27] Further research has resulted in an artificial cell system which can function using glucose at concentrations normally present in the blood.[27] Further development of this approach has been carried out to convert glutamate formed from urea or ammonia into other amino acids. The first system tested involved the conversion of urea into ammonia then to glutamic acid and then to alanine.[28] This was done using artificial cells each containing a multienzyme system of urease, glutamate dehydrogenase, alcohol dehydrogenase, and aminotransferase

[21] J. Grunwald and T. M. S. Chang, *J. Appl. Biochem.* **1,** 104 (1979).

[22] J. Grunwald and T. M. S. Chang, *Biochem. Biophys. Res. Commun.* **81**(2), 565 (1978).

[23] J. Grunwald and T. M. S. Chang, *J. Mol. Catal.* **11,** 83 (1981).

[24] J. Campbell and T. M. S. Chang, *Biochim. Biophys. Acta* **397,** 101 (1975).

[25] J. Campbell and T. M. S. Chang, *Biochem. Biophys. Res. Commun.* **69,** 562 (1976).

[26] J. Cousineau and T. M. S. Chang, *Biochem. Biophys. Res. Commun.* **79**(1), 24 (1977).

[27] T. M. S. Chang and C. Malouf, *Artif. Organs* **3**(1), 38 (1979).

[28] T. M. S. Chang, C. Malouf, and E. Resurreccion, *Artif. Organs* **3,** S284 (1979b).

(transaminase). Artificial cells containing multienzyme systems have also been studied for galactose conversion.[29]

The above studies demonstrate the feasibility of using multienzyme systems in artificial cells for the conversion of substrate which could not be done previously with single enzyme systems. This has been facilitated by the ability of multienzyme systems in artificial cells to recycle the required cofactors. This way, a very low external concentration of cofactor is required to carry out these types of reactions. However, for *in vivo* applications, although it is possible to introduce external cofactors for recycling, it would be much more desirable to retain the cofactors within the artificial cells so that high concentrations facilitate continuous recycling. One approach is to link cofactors to dextran to form soluble macromolecules for inclusion within the artificial cells.[22,23] For example, artificial cells formed with nylon–poly(ethyleneimine) membranes can be made to contain ethyl alcohol dehydrogenase, malate dehydrogenase, and a soluble dextran-NAD^+. In the presence of the substrate ethanol and oxaloacetate, dextran-NAD^+ was successfully recycled within the artificial cells by the sequential reactions of the included enzymes. The stability of the microcapsules can be further improved by cross-linking the microcapsules with glutaraldehyde but this treatment causes a considerable decrease in the recycling activity. The microcapsules are suitable for use in continuous flow reactors without the need to supply external cofactors.

Lipid–polymer membrane artificial cells can immobilize free cofactors in solution together with urease, glutamate dehydrogenase, alcohol dehydrogenase, and α-ketoglutarate.[7] This way, external urea or ammonia entering the artificial cells can be converted into amino acids.[7] NAD^+ is recycled in the presence of external alcohol.

Applications in Medicine, Industry, and Biology

Artificial cells containing enzymes and proteins have been used in a number of experimental and therapeutic conditions.[2–4,6,7] Artificial cells with red blood cell hemolysate have been tested for use as red blood cell substitutes.[1,2,9,11,30] Of the different approaches, the cross-linking hemoglobin approach[1,2,30] is now being developed to form smaller soluble polyhemoglobin blood substitutes with promising results.[31,32] Artificial cells

[29] T. M. S. Chang and N. Kuntarian, *Enzyme Eng.* **4**, 193 (1978).
[30] T. M. S. Chang, *Trans. Am. Soc. Artif. Intern. Organs* **26**, 354 (1980).
[31] P. E. Keipert, J. Minkowitz, and T. M. S. Chang, *Int. J. Artif. Organs* **5**, 383 (1982).
[32] P. E. Keipert and T. M. S. Chang, *Trans. Am. Soc. Artif. Intern. Organs* **29**, 329 (1983).

containing urease have been used as a model immobilized enzyme system for experimental therapy.[1,2,11,12] The basic results obtained pave the way for other types of enzyme replacement therapy. Thus, artificial cells have been used successfully in experimental enzyme therapy in hereditary enzyme deficiency of catalase in acatalasemic mice.[33,34] Repeated injections did not result in the production of immunological reactions to the heterogeneous enzyme in the artificial cells.[34] Additionally, artificial cells containing tyrosinase have been used in extracorporeal hemoperfusion to lower tyrosine levels in rats[35-37]; asparaginase for tumor suppression[38-40]; and phenylalanine ammonia lyase for phenylketonuria.[41]

Artificial cells have been used for the construction of artificial kidneys, artificial livers, and detoxifiers.[2,13,41-44] Ten milliliters of microencapsulated urease in an extracorporeal hemoperfusion chamber lowered the systemic blood urea of dogs by 50% within 45 min.[13] The ammonium formed was removed by microencapsulated ammonia adsorbent,[13] in the hemoperfusion chamber. This principle of urea removal using urease and ammonia adsorbent was later adapted into the Redy system for urea removal.[45] Oral administration of artificial cells containing urease and an ammonia adsorbent system has been used to remove urea diffusing into the intestinal tract in rats.[46-48] This basic research has now been developed to a stage for clinical assessment. Studies have been carried out to

[33] T. M. S. Chang and M. J. Poznansky, *Nature (London)* **218**, 243 (1968a).

[34] M. J. Poznansky and T. M. S. Chang, *Biochim. Biophys. Acta* **334**, 103 (1974).

[35] C. D. Shu and T. M. S. Chang, *Int. J. Artif. Organs* **3**(5), 287 (1980).

[36] C. D. Shu and T. M. S. Chang, *Int. J. Artif. Organs* **4**, 82 (1981).

[37] Z. Q. Shi and T. M. S. Chang, *Trans. Am. Soc. Artif. Intern. Organs* **28**, 205 (1982).

[38] T. M. S. Chang, *Nature (London)* **229**, 117 (1971b).

[39] T. M. S. Chang, *Enzyme* **14**(2), 95 (1973).

[40] E. D. Siu Chong and T. M. S. Chang, *Enzyme* **18**, 218 (1974).

[41] L. Bourget and T. M. S. Chang, *J. Appl. Biochem. Biotechnol.* **10**, 57 (1984).

[42] V. Bonomini and T. M. S. Chang, eds., "Hemoperfusion," Contrib. Nephrol. Ser. Karger, Basel, 1982.

[43] S. Sideman and T. M. S. Chang, eds., "Hemoperfusion: I. Artificial Kidney and Liver Support and Detoxification." Hemisphere, Washington, D.C., 1980.

[44] E. Piskin and T. M. S. Chang, eds., "Past, Present and Future of Artificial Organs." Meteksan Publisher, Ankara, Turkey, 1983.

[45] A. Gordon, M. A. Greenbaum, L. B. Marantz, M. S. McArthur, and M. D. Maxwell, *Trans. Am. Soc. Artif. Intern. Organs* **15**, 347 (1969).

[46] T. M. S. Chang and S. K. Loa, *Physiologist* **13**, 70 (1970).

[47] T. M. S. Chang, *Kidney Int.* **10**, S218 (1976b).

[48] D. L. Gardner, R. D. Falb, B. C. Kim, and D. C. Emmerling, *Trans. Am. Soc. Artif. Intern. Organs* **17**, 239 (1971).

use artificial cells containing tyrosinase for hemoperfusion in galactos-
amine-induced fulminant hepatic failure in rats. In these studies it was
found that hemoperfusion through tyrosinase artificial cells resulted in a
significant lowering of tyrosine in the systemic circulation.[35–37]

Artificial cells offer the potential for a variety of applications, many of
which are currently under investigation.

Section II

Drug Conjugates

[16] Preparation of Antibody–Toxin Conjugates

By A. J. Cumber, J. A. Forrester, B. M. J. Foxwell, W. C. J. Ross, and P. E. Thorpe

Introduction

Cell-type-specific cytotoxic agents have been constructed in several laboratories by linking highly potent toxins or their subunits to antibody molecules. This field has been reviewed by Olsnes and Pihl[1] and by Thorpe *et al.*[2] The toxins most widely used have been diphtheria toxin (the exotoxin from *Corynebacterium diphtheriae*) and the plant toxins ricin (from *Ricinus communis*) and abrin (from *Abrus precatorius*).

Abrin, ricin, and diphtheria toxin are composed of two polypeptide subunits, A and B, joined by a disulfide bond. All three toxins have similar modes of cytotoxic action.[3] They bind by means of recognition sites on the B chain to components of the plasma membrane of virtually all cell types in sensitive species of animal: abrin and ricin recognize galactose-terminating glycoproteins and glycolipids whereas the receptors for diphtheria toxin have not yet been identified. The A chain then penetrates (or is transported across) the plasma membrane or the membrane of an endocytotic vesicle and kills the cell by inactivating its machinery for protein synthesis: Diphtheria toxin A chain catalyzes the transfer of the ADP-ribosyl moiety of NAD^+ to the elongation factor EF-2, whereas the A chains of the plant toxins damage the 60 S subunit of ribosomes directly.

There are two ways of constructing antibody–toxin conjugates with cell-type specificity. The first is to link the antibody by a disulfide bond to the isolated A chain moiety or alternatively to one of the single-chain plant peptides, such as gelonin from *Gelonium multiflorum*, whose damaging action on eukaryotic ribosomes is apparently identical to that of the A chains of abrin and ricin. The second is to link the intact toxin to the antibody and block the cell recognition site on the B chain to prevent the conjugate from binding to and killing cells nonspecifically.

---

[1] S. Olsnes and A. Pihl, *Pharmacol. Ther.* **15**, 355 (1982).

[2] P. E. Thorpe, D. C. Edwards, W. C. J. Ross, and A. J. S. Davies, *in* "Monoclonal Antibodies in Clinical Medicine" (J. Fabre and A. McMichael, eds.), p. 167. Academic Press, London, 1982.

[3] S. Olsnes and A. Pihl, *in* "Molecular Action of Toxins and Viruses" (P. Cohen and S. van Heyningen, eds.), p. 51. Elsevier/North-Holland Biomedical Press, Amsterdam, 1982.

METHODS IN ENZYMOLOGY, VOL. 112

In this chapter we describe methods for preparing these two types of conjugate and for characterizing them physicochemically and biologically. We also relate the advantages and limitations of A chain conjugates to those of intact toxin conjugates as selective cytotoxic agents *in vitro* and as chemotherapeutic agents in animals.

Purification of Toxins

Before any purification procedures or other manipulations of toxins are contemplated, the safety procedures outlined in the Appendix should be carefully studied.

Ricin

This galactose-binding toxic lectin may be readily purified by a two-stage chromatographic process based on the method of Nicolson and Blaustein.[4] Galactose-binding proteins are first isolated on an agarose matrix and are then subject to a gel filtration step.

1. Untoasted castor bean cake (pomace) may be obtained from castor bean processors. It should be further defatted by extraction three times with 5 volumes (v/w) 40–60° petroleum ether. The air-dried material (500 g) is then extracted overnight by stirring it in 4 liters phosphate-buffered saline (PBS: 0.15 M NaCl, 0.01 M phosphate, pH 7.4). The supernatant is partially clarified by filtration through nylon gauze followed by centrifugation at 1500 g for 1 hr. To produce a clear solution suitable for chromatographic procedures, the supernatant is subjected to ammonium sulfate precipitation at 4°, the fraction precipitating between 40 and 60% saturation being collected by centrifugation (1500 g, 1 hr), redissolved in about 500 ml PBS, and dialyzed against three changes of 6 liters PBS.

2. This solution, clarified by further centrifugation if necessary, is then applied to a column (bed volume ~800 ml) of Sepharose 4B running in PBS. The gel should have been pretreated with 1 M propionic acid at room temperature for at least 2–3 weeks to enhance its binding capacity for lectins. The column should be jacketed and run at a temperature of <10° to optimize lectin binding. After application of the sample, the column is washed with at least 4 bed volumes of PBS. Some sources of castor beans, especially in our experience those of Chinese origin, contain two species of toxins differing in isoelectric point. The prolonged elution of the column ensures that the species with higher pI (~8.2), which binds

[4] G. L. Nicolson and J. Blaustein, *Biochim. Biophys. Acta* **266**, 543 (1972).

only feebly to Sepharose, is washed through together with all nonbinding proteins. When the UV absorbance (280 nm) of the eluate has reached a stable, low value (<0.1), the strongly binding lectins are displaced from the column with 100 mM galactose in PBS. The toxin and lectin elute together as a single, sharp peak.

3. This mixture is resolved by gel filtration on Sephacryl S-200, also running in PBS. The sample size should be restricted to 3–4% of the bed volume of the column, under which conditions the lectin of molecular weight 120,000 and the toxin of molecular weight 60,000 are fully resolved. The material recovered from the affinity column (step 2) may be concentrated by ultrafiltration (Amicon PM10 membrane) to ~20 mg/ml before applying to the gel filtration medium. Some sources of beans contain quantities of a material which has a molecular weight about 90,000 and is both toxic and a hemagglutinin. Toxin fractions should be selected so as to avoid contamination with this material. Five hundred grams of defatted pomace should yield about 1250 mg toxin, $E_{1\,cm}^{1\%}$ 11.8, at 280 nm. Sterile solutions of holotoxin may be stored at 4° for at least 12 months without detectable loss of activity or deep frozen (−30°) for several years.

Abrin

Abrin may be purified from seeds of *Abrus precatorius* by essentially similar methodology.

1. The seeds (500 g), which should be bright scarlet, are ground (e.g., in a coffee grinder) before extracting overnight with 4 liters of PBS at 4°. The yield may be increased by reextracting the pellet of softened stroma from the first centrifugation with a further quantity (2 liters) of PBS for a further 4–6 hr at room temperature after homogenizing in a Waring blender (2 min).

2. The combined extracts are then subjected to ammonium sulfate precipitation, and the material precipitating between 40 and 70% saturation is collected.

3. Other procedures are identical to those for ricin, but it may be noted that material of intermediate molecular weight (i.e., 90,000) is not seen in the galactose-binding proteins of *Abrus* extracts. Five hundred grams of ground beans will yield about 800 mg of toxin, $E_{1\,cm}^{1\%}$ 15.9, at 280 nm.

Diphtheria Toxin

Partially purified diphtheria toxin is available from Connaught Laboratories, Willowdale, Ontario, Canada. It may be purified to homogeneity by the method of Collier and Kandel,[5] upon which the following experimental procedure is based.

1. A dialyzed solution of 500 mg crude diphtheria toxin in 0.01 M sodium phosphate buffer, pH 7.0, is applied to a column (2.6 × 34 cm) of DEAE-cellulose equilibrated in the same buffer.

2. The column is washed at 100 ml/hr with a sequence of sodium phosphate buffers (pH 7.0) of increasing molarity as follows: 300 ml of 0.01 M buffer, 300 ml of 0.05 M buffer, 600 ml of 0.1 M buffer, 300 ml of 0.25 M buffer. The toxin elutes as a major peak with a trailing shoulder during the 0.1 M buffer step. The toxin may be identified by means of its precipitation reaction with horse antitoxin in Ouchterlony immunodiffusion assays.

3. The toxin solution is concentrated by ultrafiltration (Amicon PM10 membrane) to 25 ml and applied to a column (5 × 76 cm) of Sephadex G-100 (Superfine grade) equilibrated in 0.05 M sodium borate buffer, pH 8.5. Elution with the same buffer removes the toxin as a major component (~90%) of approximate molecular weight 65,000. A minor component (~10%) of approximate MW 130,000, probably dimeric toxin, may be seen.

4. The toxin should be concentrated to 10 mg/ml ($E_{1\,cm}^{1\%}$, 12.3 at 280 nm) and stored in aliquots at −30°.

Isolation of Toxin A Chains

Methods for preparing toxin A chains involve reductive cleavage of the disulfide bond followed by separation of the subunits. Special measures are necessary to ensure removal of all traces of holotoxin from the desired products. We describe three methods, the first applicable to both abrin and ricin, the second for ricin, and the third for diphtheria toxin.

Abrin or Ricin

The A chains may be conveniently prepared by cleaving the disulfide bond while the toxins are immobilized on an agarose matrix via the binding site of the B chain.

Experimental Procedure

1. Toxin (100–150 mg) is applied to a column of acid-treated Sepharose 4B, bed volume 25 ml. This represents about half the maximal loading of the gel. Sample and column should be equilibrated with 0.1 M

[5] R. J. Collier and J. Kandel, *J. Biol. Chem.* **246,** 1496 (1971).

phosphate buffer, pH 8.0, containing 0.1 mM EDTA. Immediately after the sample has been run onto the column, it is followed by 15 ml of the running buffer containing 5% 2-mercaptoethanol. The flow is then stopped and the column allowed to stand for 2 hr at room temperature.

2. Liberated A chain is then eluted with the phosphate buffer, B chain and residual holotoxin being retained on the column. The eluted peak is dialyzed against three changes of 0.1 M phosphate buffer, pH 8.0, containing 0.1 mM each of EDTA and dithiothreitol to guard against oxidation of the sulfhydryl group.

3. Traces of B chain or holotoxin should be removed by absorption twice with asialofetuin immobilized on Sepharose: approximately 2 mg asialofetuin per milliliter of gel, 1 ml of settled gel per 20 ml of A chain solution. Yield of recovered A chain should be between 75 and 95% of theoretical.

4. Free A chains should not be frozen (freezing tends to cause loss of enzymatic function) but rather stored at 4°. They are particularly prone to surface denaturation, and manipulations such as concentration by ultrafiltration should be avoided where possible. The addition of 20% glycerol is recommended to stabilize samples before transport. In general, it is desirable to prepare A chains as and when required.

Ricin

Alternative Method. This method for isolating both polypeptide chains of ricin derives from the procedure described by Olsnes and Pihl.[6]

1. Ricin at a concentration of 5 mg/ml in 0.1 M Tris-HCl, pH 8.5, is made 0.5 M with respect to galactose and 2% to 2-mercaptoethanol. The pH is adjusted to 8.5 and the sample incubated overnight at room temperature followed by 2–3 hr at 37°.

2. The sample is then applied to a column of DEAE-Sepharose equilibrated with 0.1 M Tris-HCl buffer, pH 8.5. One milliliter of Sepharose is used for each 5 mg of ricin being processed.

3. The column is washed with 0.1 M Tris-HCl buffer, pH 8.5, until all the unbound material—essentially pure A chain—has been eluted.

4. The column is then washed with 0.1 M galactose in 0.1 M Tris-HCl, pH 8.5, which removes only bound ricin.

5. Finally the column is washed with 0.1 M Tris-HCl, pH 8.5, containing 0.1 M galactose and 0.5 M NaCl to elute B chain.

6. The A chain is dialyzed into PBS and the B chain into 5 mM phosphate buffer, pH 6.5, containing 0.1 M galactose.

[6] S. Olsnes and A. Pihl, *Biochemistry* **12**, 3121 (1973).

7. To remove any contaminating ricin from the A chain, the material is passed down a 2-ml asialofetuin–Sepharose column (2–3 mg asialofetuin per milliliter of gel) or through a 2-ml column of anti-B-chain antibodies immobilized on Sepharose. The A chain is then filtered through a 0.22 μm membrane and stored at 4°.

8. The purity of the B chain fraction may be increased by passing it down a CM-Sepharose column equilibrated with 5 mM phosphate, pH 6.5, containing 0.1 M galactose—approximately 10 ml of gel per 25 mg of protein. Alternatively (or additionally), a column of anti-A-chain antibodies immobilized on Sepharose can be used.

9. The B chain is then dialyzed into PBS containing 0.1 M galactose, filtered through a 0.22-μm pore size membrane and stored at 4°.

Diphtheria Toxin

The following method is a modification of that described by Michel and Dirkx.[7]

1. To 1 ml of a 25-mg/ml solution of purified diphtheria toxin in 0.05 M borate buffer, pH 8.5, is added 1 ml of a 2-μg/ml solution of trypsin in the same buffer containing 2 mM EDTA.

2. The mixture is incubated for 2 hr at 25° to cleave any single-chain toxin proteolytically to its nicked (i.e., A and B chain) form; 500 μl of a 10-μg/ml solution of soybean trypsin inhibitor is added to terminate the reaction.

3. The nicked diphtheria toxin is reduced by adding 500 μl of 0.6 M dithiothreitol and incubating for 90 min at 37°. Liberated B chain sediments as a dense, white precipitate.

4. The reduced toxin is centrifuged and the clear supernatant applied to a column (2.6 × 50 cm) of Sephadex G-100 (Superfine grade) equilibrated with 0.05 M borate buffer, pH 8.5, containing 1 mM EDTA and 0.1 M 2-mercaptoethanol. Elution with the same buffer solution removes the A chain as the major peak (MW ~22,000).

5. The A chain is concentrated to 5 ml by ultrafiltration (Amicon UM05 or YM5 membrane) and further purified by gel filtration on a column (2.6 × 80 cm) of Sephadex G-75 (Superfine grade) equilibrated with the same buffer.

6. The A chain is then heated to 80° for 10 min to destroy any residual traces of toxin or B chain.[8]

[7] A. Michel and J. Dirkx, *Biochim. Biophys. Acta* **365,** 15 (1974).
[8] G. Kukor, M. Solotorovsky, and R. J. Kuchler, *J. Bacteriol.* **115,** 277 (1973).

Radiolabeling of Toxins and A Chains

Tracer quantities of radiolabeled toxins and A chains are incorporated into all the synthetic procedures described later (see Coupling of Antibodies and Intact Toxins; Coupling of Antibodies and Toxin A Chains) to permit determination of the composition of the conjugates obtained. Toxins may be readily labeled with iodine isotopes using either the chloramine-T[9] or, more conveniently, the Iodo-Gen[10] method. Labeled A chains should always be prepared by taking a quantity of iodinated holotoxin and subjecting it to the cleavage procedures described above on a suitably reduced scale. Attempts to iodinate free A chains under oxidizing conditions invariably lead to destruction of the sulfhydryl function and, therefore, loss of ability to take part in coupling reactions (see Coupling of Antibodies and Toxin A Chains). To ensure homogeneity with respect to sulfhydryl function, A chain mixed with radiolabeled A chain may be bound to a column of activated thiopropyl-Sepharose 6B and eluted with reducing agent immediately before use. The reducing agent must obviously be removed, e.g., by passage through G-25 Sephadex, before using the material in a coupling reaction (see Coupling of Antibodies and Toxin A Chains).

Coupling of Antibodies and Intact Toxins

The problem for the chemist is to prepare antibody–toxin conjugates of defined composition that satisfy the following requirements.

1. The recognitional specificity of the antibody should not be impaired.
2. The membrane penetration qualities of the toxin and the capacity of the A chain to damage ribosomes should be unaltered.
3. Cross-links between the A and B chains of the toxins should be avoided since it is probably necessary for the two subunits to separate for cytotoxicity to be expressed.
4. Homopolymer formation in antibody or toxin should be avoided.
5. When the conjugate is intended for use in animals, the linkage must also be able to withstand the degradative influences (e.g., enzymes) that prevail in blood and extracellular fluids.

The first coupling agents used to link antibodies and toxins were homobifunctional reagents (e.g., glutaraldehyde or diethyl malonimidate)

---

[9] P. J. McConahey and F. J. Dixon, this series, Vol. 70 [11].
[10] P. J. Fraker and J. C. Speck, Jr., *Biochem. Biophys. Res. Commun.* **80,** 859 (1978).

having two identical reactive groups. These were simple to use since they had only to be added to a mixture of the two proteins, but they suffered from a tendency to produce intramolecular cross-links in the toxin and to generate homopolymers. They have been superseded by heterobifunctional agents in which the two reactive groups are different and which react with amino acids in proteins under different conditions. These reagents permit far greater control over the coupling reaction and much reduce the risk of forming intra- and intermolecular cross-links. In the following paragraphs we describe the use of three such heterobifunctional reagents.

N-Hydroxysuccinimidyl Ester of Chlorambucil

A heterobifunctional reagent satisfying the above requirements is N-hydroxysuccinimidyl ester of 4-[bis(2-chloroethyl)amino]benzenebutanoic acid (chlorambucil). The activated ester moiety of the coupling agent is first reacted with amino groups in the immunoglobulin at low temperature at which the N-chloroethylamine groups are relatively unreactive. An excess of toxin is then added to the substituted immunoglobulin and the temperature raised to promote reaction of the bis-2-chloroethylamine groups with amino acid side chains. The principal product is an antibody–toxin conjugate, although some antibody–antibody polymer is also formed. When carried out at pH 9, the bis-2-chloroethylamine groups react preferentially with amine groups of lysine side chains in the toxin with the probable formation of a piperazine linkage.[11] The conjugate would then have the structure

$$
\text{Ig—NH} \cdot \text{OC(CH}_2)_3 \cdot \text{C}_6\text{H}_4 \cdot \text{N} \overset{\text{CH}_2 \cdot \text{CH}_2}{\underset{\text{CH}_2 \cdot \text{CH}_2}{\diagup \diagdown}} \text{N—Toxin}
$$

The N-hydroxysuccinimidyl ester of chlorambucil is prepared by condensing chlorambucil with N-hydroxysuccinimide using dicyclohexyl carbodiimide in tetrahydrofuran solution and is crystallized from ether–light petroleum as flattened needles, m.p. 82–84°.

Experimental Procedure

1. To 30 mg immunoglobulin at a concentration of 10 mg/ml in borate–saline buffer (0.05 M borate, 0.3 M NaCl, 0.5% n-butanol, pH 9.0) is added 1 mg of the N-hydroxysuccinimidyl ester of chlorambucil dissolved in 0.5 ml dry dimethyl sulfoxide.

[11] P. E. Thorpe and W. C. J. Ross, *Immunol. Rev.* **62,** 119 (1982).

2. After stirring on ice for 1 hr, the solution is applied to a jacketed column (1.6 × 36 cm) of Sephadex G-25F equilibrated with borate–saline buffer and maintained at 4°.

3. The column is washed with borate–saline buffer and the derivatized globulin, which elutes first, is added immediately to an ice-cold solution of 50 mg toxin in 5 ml of the same buffer.

4. This solution (~20 ml) is concentrated to 5.0 ml in a cooled Amicon ultrafiltration cell using a PM10 membrane. The concentrated solution is then allowed to warm to room temperature and left for 48 hr to react before resolution by gel filtration (see Purification and Storage of Conjugates).

We have used this procedure to prepare conjugates in which diphtheria toxin,[11] abrin,[12] ricin,[13] and the ribosome-inactivating protein gelonin[11] have been coupled to various antibodies. The above conditions result in the introduction of six to eight molecules of chlorambucil per molecule of immunoglobulin; and, when used with a 4- to 5-fold molar excess of toxin, it is usual to obtain yields of 5–10% of 1 : 1 antibody : toxin conjugate based upon the amount of antibody used.

N-Succinimidyl-3-(2-pyridyldithio)propionate

An alternative approach which overcomes the problem of homopolymer formation is to introduce different reactive groups into both the toxin and the antibody such that the predominant reaction when the two substituted proteins are mixed is between the introduced functional groups. A reagent which permits the introduction of two such functional groups into proteins is N-succinimidyl-3-(2-pyridyldithio)propionate (SPDP), first described by Carlsson et al.[14]

2-Pyridyl disulfide groups are introduced into one of the proteins to be coupled and in the other thiol groups are introduced by pyridyldithiolation and subsequent reduction with dithiothreitol. On mixing, the two substituted proteins react to form a conjugate in which the linkage is a disulfide bond. The conjugates have the general structure

$$\text{Ig—NH} \cdot \text{OC} \cdot \text{CH}_2\text{CH}_2 \cdot \text{SS} \cdot \text{CH}_2\text{CH}_2\text{CO} \cdot \text{NH—Toxin}$$

We prefer to introduce the sulfhydryl group into the immunoglobulin component rather than into the toxin to avoid exposing the latter to reduc-

[12] D. C. Edwards, A. Smith, W. C. J. Ross, A. J. Cumber, P. E. Thorpe, and A. J. S. Davies, *Experientia* **37**, 256 (1981).

[13] P. E. Thorpe, D. W. Mason, A. N. F. Brown, S. J. Simmonds, W. C. J. Ross, A. J. Cumber, and J. A. Forrester, *Nature* (*London*) **297**, 84 (1981).

[14] J. Carlsson, H. Drevin, and R. Axén, *Biochem. J.* **173**, 723 (1978).

ing conditions that could result in separation of its A and B subunits. In addition, the risk of forming toxin polymers, by disulfide bond formation, is avoided.

Experimental Procedure

1. To 10 mg of toxin in 1.0 ml borate–saline buffer is added 150 μg SPDP in 25 μl dry dimethylformamide.

2. After stirring for 30 min at room temperature, the solution is applied to a column (1.6 × 22 cm) of Sephadex G-25F equilibrated with phosphate–saline buffer (0.1 M phosphate, 0.1 M NaCl, 1 mM EDTA, 0.02% NaN$_3$, pH 7.5). The derivatized toxin is collected in a volume of 10–12 ml.

3. To a solution of 25 mg of immunoglobulin in 2.5 ml borate–saline buffer is added 185 μg SPDP in 30 μl dry dimethylformamide.

4. After stirring at room temperature for 30 min, the solution is applied to a column (1.6 × 36 cm) of G-25F Sephadex equilibrated with acetate–saline buffer (0.1 M acetate, 0.1 M NaCl, 1 mM EDTA, 0.02% NaN$_3$, pH 4.5). The derivatized antibody is collected in a volume of 10–12 ml.

5. Ten milliliters of this solution is concentrated to 2.5 ml by ultrafiltration (Amicon PM10 membrane) and 22 mg dithiothreitol in 500 μl acetate buffer added.

6. After stirring for 30 min at room temperature, the solution is applied to a column (1.6 × 36 cm) of Sephadex G-25F equilibrated in nitrogen-flushed phosphate–saline buffer. The emergent protein peak is immediately added to 10 ml of the derivatized toxin.

7. The reaction mixture is then concentrated to 5 ml by ultrafiltration and left at room temperature for 18 hr before resolution by gel filtration (see Purification and Storage of Conjugates).

We have used this method to couple ricin,[15] abrin,[16] and gelonin[17] to various antibodies.

The degree of substitution of either protein after the primary derivatization may be readily assayed by the method of Carlsson et al.[14] We introduce approximately two reactive groups per molecule into both the antibody and the toxin and use equivalent amounts of each component in the coupling reaction. This usually yields 20–40% of the 1 : 1 antibody : toxin conjugate.

---

[15] W. C. J. Ross, A. J. Cumber, and P. E. Thorpe, unpublished results.
[16] D. C. Edwards, W. C. J. Ross, A. J. Cumber, D. P. McIntosh, A. Smith, P. E. Thorpe, A. Brown, R. H. Williams, and A. J. S. Davies, *Biochim. Biophys. Acta* **717**, 272 (1982).
[17] P. E. Thorpe, A..N. F. Brown, W. C. J. Ross, A. J. Cumber, S. I. Detre, D. C. Edwards, A. J. S. Davies, and F. Stirpe, *Eur. J. Biochem.* **116**, 447 (1981).

N-Hydroxysuccinimidyl Ester of Iodoacetic Acid

This method relies on the selective reaction of iodoacetyl groups introduced into the toxin with thiol groups introduced into the immunoglobulin. Iodoacetylation of the toxin is achieved by means of the *N*-hydroxysuccinimidyl ester of iodoacetic acid first described by Rector *et al.*[18] Thiolation of the immunoglobulin is accomplished by means of the SPDP reagent as described above. As with the SPDP reagent, this method avoids homopolymer formation. The conjugate has the structure

$$\text{Ig—NH} \cdot \text{OC} \cdot \text{CH}_2\text{CH}_2 \cdot \text{S} \cdot \text{CH}_2 \cdot \text{CO} \cdot \text{NH—Toxin}$$

This differs from the product obtained with the SPDP method in that it contains a central sulfide (or thioether) linkage rather than a disulfide bond. The thioether bond is stable under reducing conditions where the disulfide bond is not. Importantly, the linkage of ricin to antibodies by this method produces impairment of the galactose binding sites of the B-chains of the toxin in the majority of conjugate molecules formed. This markedly reduces the tendency of the conjugates to bind to and kill cells nonspecifically.

Experimental Procedure

1. To a solution of 20 mg ricin in 1.5 ml borate–saline buffer is added 300 μg of the *N*-hydroxysuccinimidyl ester of iodoacetic acid in 200 μl of dry dimethylformamide.

2. After stirring for 30 min at room temperature, the solution is applied to a column (1.6 × 22 cm) of Sephadex G-25F equilibrated with phosphate–saline buffer. The derivatized ricin is collected in a volume of 10–12 ml.

3. Immunoglobulin (45 mg) is derivatized with SPDP [see *N*-Succinimidyl-3-(2-pyridyldithio)propionate, Steps 3 and 4] and then concentrate to 3.0 ml. This intermediate is reduced by the addition of 22 mg of dithiothreitol in 500 μl acetate buffer.

4. After stirring for 30 min at room temperature, the solution is applied to a column (1.6 × 20 cm) of Sephadex G-25F equilibrated with nitrogen-flushed phosphate–saline buffer. The protein that elutes at the void volume is added immediately to 11 ml of the derivatized ricin solution.

5. The reaction mixture is then concentrated to 5.0 ml by ultrafiltration (Amicon PM10 membrane) and left at room temperature for 18 hr.

6. *N*-Ethylmaleimide (1 mg) dissolved in dimethylformamide is added to inactivate any remaining free sulfhydryl groups and the reaction mix-

[18] E. S. Rector, R. J. Schwenk, K. S. Tse, and A. H. Sehon, *J. Immunol. Methods* **24**, 321 (1978).

ture stirred gently for a further hour before resolution by gel filtration (see Purification and Storage of Conjugates).

We have used this procedure to couple ricin to several antibodies.[19] The number of iodoacetyl groups introduced into the ricin molecule can be established by a procedure described elsewhere.[19] As with the SPDP method, we introduce about two reactive groups into each protein and use an equivalent amount of each component in the coupling reaction. The yields of 1 : 1 antibody : toxin conjugate are usually 20–40%.

Coupling of Antibodies and Toxin A Chains

A Chains of toxin contain a free thiol group which can react with a 2-pyridyl disulfide-substituted antibody molecule to form a disulfide linkage. These conjugates have the general structure

$$\text{Ig—NH} \cdot \text{CO} \cdot \text{CH}_2 \cdot \text{CH}_2 \cdot \text{SS—A-chain}$$

Experimental Procedure

1. To 26 mg antibody at a concentration of 10 mg/ml in phosphate–saline buffer is added 220 μg SPDP in 100 μl dry dimethylformamide. After stirring for 30 min at room temperature, the solution is applied to a column (1.6 × 30 cm) of Sephadex G-25F equilibrated in the same buffer; the derivatized antibody elutes in the void volume and is collected in about 10–12 ml.

2. The product is concentrated to about 3.5 ml by ultrafiltration (Amicon PM10 membrane). To it is added 15 mg of toxin A chain in 7 ml of nitrogen-flushed phosphate–saline buffer. The solution is gently mixed for 10 min and then left at room temperature for 18 hr before resolution by gel filtration (see Purification and Storage of Conjugates).

We have used this procedure to link antibodies to the A chains of abrin,[20] ricin,[21] and diphtheria toxin.[11] We routinely attach about two pyridyl disulfide groups to the antibody and allow the product to react with a 2- to 3-fold molar excess of the A chain. It is usual to obtain yields of up to 50% of the 1 : 1 conjugate based upon the immunoglobulin used.

[19] P. E. Thorpe, W. C. J. Ross, A. N. F. Brown, C. Meyers, A. J. Cumber, B. M. J. Foxwell, and J. A. Forrester, *Eur. J. Biochem.* **140,** 63 (1984).

[20] J. A. Forrester, D. P. McIntosh, A. J. Cumber, W. C. J. Ross, and G. D. Parnell, *Cancer Drug Delivery* **1,** 283 (1984).

[21] D. P. McIntosh, D. C. Edwards, A. J. Cumber, G. D. Parnell, C. J. Dean, W. C. J. Ross, and J. A. Forrester, *FEBS Lett.* **164,** 17 (1983).

Although several alternative methods for the preparation of antibody–A chain conjugates have been described, the SPDP method is probably the easiest and most reproducible method to use. Moreover, it is probable that the A chain has to be released from the antibody in order to diffuse to ribosomes and inactivate them, and conjugates made without disulfide linkages are usually inactive.

Purification and Storage of Conjugates

The objective is to obtain from the unresolved reaction mixtures the component comprising one molecule of antibody and one of toxin since this simple conjugate is generally the most active in the biological test systems. The reaction mixtures contain, in addition to the 1 : 1 antibody : toxin conjugate, polymerized conjugate, more highly substituted antibody (2 : 1, 3 : 1, etc. toxin molecules per single antibody molecule), polymerized antibody, and unreacted antibody and toxin. The removal of unreacted toxin and polymeric materials is readily achieved by gel filtration on polydextran gels. The unreacted holotoxins elute later than would be predicted from their molecular weight, probably because they have a weak affinity for the carbohydrate residues of the gels. The A chains are of sufficiently low molecular weight that they, too, are clearly separated from the conjugates.

For preparations on the scale described above, columns of Sephadex G-200 Superfine or of the faster running Sephacryl S-300 (3.2 × 80 cm) provide adequate separation for holotoxin conjugates. For resolution of A chain conjugates, Sephacryl S-200 columns (2.6 × 80 cm) give satisfactory separations. The fractions containing the 1 : 1 conjugate are identified from the elution volume and by the radioactivity associated with the toxin moiety.

Further purification of holotoxin conjugates can be achieved by rerunning the product of MW 210,000–215,000 (1 : 1 antibody : toxin) on the gel filtration column, but this is not worthwhile in the case of the A chain conjugates in view of the small difference in molecular weight between the conjugate and the antibody.

Purification of conjugates can also be accomplished by means of the binding of the antibody[22] or the toxin moiety[22,23] to appropriate affinity adsorbents. Immobilized staphylococcal protein A affords a means of removing noncoupled toxins from mixtures, provided that the antibody is of an appropriate class to bind to this adsorbent. Similarly, antibody–ricin

[22] L. L. Houston and R. C. Nowinsky, *Cancer Res.* **41**, 3913 (1981).

[23] D. G. Gilliland, R. J. Collier, J. M. Moehring, and T. J. Moehring, *Proc. Natl. Acad. Sci. U.S.A.* **75**, 5319 (1978).

and antibody–abrin conjugates can be adsorbed to Sepharose and non-coupled antibody and antibody–antibody polymers washed away; however, this method cannot be used with conjugates of ricin prepared by the iodoacetyl method because these show a marked reduction in their ability to bind to galactose residues. Conjugates prepared by the other methods show no comparable impairment of galactose-binding capacity.

When an intact ricin conjugate prepared by the iodoacetyl method is fractionated on an acid-treated Sepharose column (1.6 × 20 cm), typically 60–80% of the conjugate fails to bind and elutes without application of galactose. When further fractionated upon a column of asialofetuin-Sepharose, about 20% of the conjugate fails to bind. Fluorescence-activated cell sorter (FACS) analyses have revealed that the blocked ricin conjugate that passes through the Sepharose column has a markedly diminished capacity to bind nonspecifically to cells and that which passes through the asialofetuin-Sepharose column has no detectable tendency to bind nonspecifically. The fractions of 1 : 1 conjugate isolated as described above (i.e., that which binds strongly to Sepharose and that which binds neither to Sepharose nor to asialofetuin) differ markedly in their toxicity toward cells *in vitro* (see Conjugates with A Chains versus Conjugates with Intact Toxins).

The reduction in the galactose-binding capacity of the ricin moiety appears to be due to obstruction by the antibody moiety itself. The evidence for this is that iodoacetylated ricin (an intermediate in the preparation of the conjugate) displays unaltered binding to Sepharose, thereby showing that the impairment in the galactose-binding capacity of the conjugate is not due directly to derivatization of the toxin but that it is a consequence of completing the linkage to the antibody. We have termed the fractions with impaired galactose recognition "blocked" ricin conjugates.

Purified conjugates from any of the procedures described may be stored at 4° for several months after filter sterilization without detectable loss of activity. Alternatively, samples may be rapidly frozen in a solid CO_2–ethanol mixture or in liquid nitrogen and stored at −70° or below. Conjugates recovered from the frozen state manifest unchanged activity, and this procedure is probably the most suitable for long-term storage.

Physicochemical Characterization

The elution volumes of the various fractions from gel filtration columns are compared with those of the starting materials and with marker proteins of known molecular weight to obtain an initial indication of the molecular weight of the fractions. More precise information can be obtained by analyzing them electrophoretically on polyacrylamide gels. The use of gradient gels (2.5–28%; Universal Scientific Ltd.) run in the presence of sodium dodecyl sulfate affords clear separation of all molecular species present.

Comparison with the starting materials run simultaneously on gel slabs identifies free toxin, A chain, or unreacted immunoglobulin in the conjugate preparation. Holotoxin conjugates purified as described above appear predominantly as a single band of a molecular weight corresponding to a 1 : 1 composition; with A chain conjugates, the principal band corresponding to the 1 : 1 component is often seen to be accompanied by bands of higher molecular weight, a finding which indicates the presence of conjugates with higher ratios of A chain to globulin.

The average composition of the conjugates may be determined from measurement of the optical density at 280 nm and from the radioactivity. The concentration of the toxin (or A chain) component can then be calculated since the specific activity of the starting material is known. The concentration of immunoglobulin is estimated from the optical density of the conjugate after subtracting the optical density calculated as attributable to the toxin component. The ratio of antibody to toxin in the conjugate fractions together with their molecular weight enable the investigator to determine the molecular composition of the conjugate.

Biological Evaluation

The "quality control" checks routinely performed on newly synthesized conjugates are the following tests.

1. Inhibition of protein synthesis in a cell-free system: This test is to check that the A chain moiety has not suffered damage during the conjugation procedure. The conjugate is treated with 0.1 M dithiothreitol for 2 hr at 37° and the capacity of the liberated A chain to inhibit protein synthesis in a reticulocyte lysate is assayed according to either Roberts and Stewart[24] or Olsnes.[25]

2. Antigen-binding capacity: This test is to check that the antibody moiety has not suffered damage during the conjugation procedure. Of the many assays available, one of the best is to measure the ability of the conjugate to compete with radiolabeled native conjugate for the antigens

[24] K. Roberts and T. S. Stewart, *Biochemistry* **18**, 2615 (1979).
[25] S. Olsnes, this series, Vol. 50 [35].

on the target cell surface.[26,27] Others have used ELISA assays[28] and indirect immunofluorescence assays.[19,29]

3. Cytotoxicity: The usual assays are to treat target cells with the conjugate at various concentration *in vitro,* culture them for a further 24- or 48-hr period, and then measure their viability by determining the rate at which they incorporate [^{3}H]- or [^{14}C]leucine into protein (e.g., Thorpe *et al.*[17]) or [^{3}H]thymidine into DNA,[30] proliferate,[31] or form colonies in clonogenic assays.[26] The specificity of cytotoxic effect can be checked by comparing the potency of effect on target cells with that on cells lacking the appropriate antigen or upon target cells coated with saturating amounts of native antibody. Alternatively, comparison may be made with the effects of a control conjugate prepared from an antibody of irrelevant specificity.

Conjugates with A Chains versus Conjugates with Intact Toxins as Selective Cytotoxic Agents

A Chain conjugates are effectively free from cytotoxic effect upon cells lacking the appropriate antigens but the toxicity that they exert upon target cells in tissue culture systems is variable and presently unpredictable. Several levels of variability can be defined. First, some types of antigen appear not to allow the A chain moiety of a conjugate to penetrate the cytosol and kill the cell. An example is the rat helper T-lymphocyte antigen recognized by the monoclonal antibody W3/25: This antibody coupled to A chain is without cytotoxic effect upon W3/25-expressing normal and leukemic cells.[11] Second, different cell types expressing the same antigen can have very different sensitivities to the same conjugate: For example, resting mouse T lymphocytes are highly susceptible to monoclonal anti-Thy 1.1 antibody coupled to gelonin whereas T lymphoblasts and T lymphoma cells are resistant.[11,17] Third, different A chains linked to the same antibody can elicit cytotoxic effects of markedly different potency: For example, diphtheria toxin A chain coupled to anti-

[26] P. Casellas, J. P. Brown, O. Gros, P. Gros, I. Hellström, F. K. Jansen, P. Poncelet, R. Ronucci, H. Vidal, and K. E. Hellström, *Int. J. Cancer* **30,** 437 (1982).

[27] T. W. Griffin, L. R. Haynes, and J. A. DeMartino, *J. Natl. Cancer Inst.* **69,** 799 (1982).

[28] T. F. Bumol, Q. C. Wang, R. A. Reisfeld, and N. O. Kaplan, *Proc. Natl. Acad. Sci. U.S.A.* **80,** 529 (1983).

[29] W. C. J. Ross, P. E. Thorpe, A. J. Cumber, D. C. Edwards, C. A. Hinson, and A. J. S. Davies, *Eur. J. Biochem.* **104,** 381 (1980).

[30] D. A. Vallera, R. J. Youle, D. M. Neville, C. C. B. Soderling, and J. H. Kersey, *Transplantation* **36,** 73 (1983).

[31] M. Seto, N. I. Umemoto, M. Saito, Y. Masuho, T. Hara, and T. Takahashi, *Cancer Res.* **42,** 5209 (1982).

Thy 1.1 antibody is virtually nontoxic to Thy 1.1-expressing splenic T lymphocytes, whereas conjugates of the same antibody and abrin A chain, ricin A chain, or gelonin are outstandingly cytotoxic.[11,17] These discordant results are presently unexplained, but they presumably reflect differences in the pathway of entry of conjugates into cells, the physico-chemical properties of antigen and A chain, resistance of the conjugate to proteolysis, and so on.

By contrast, conjugates constructed using whole ricin and abrin have so far always exhibited a powerful cytotoxic effect upon cells bearing the target antigen. This generalization also seems to apply to conjugates with intact diphtheria toxin when used against cells of species naturally sensitive to the toxin. The reason for the greater effectiveness of the holotoxin conjugates most probably resides in the ability of the B chain moiety to assist A chain entry. This is supported by the findings of Youle and Neville,[32] Vitetta et al.,[33] and McIntosh et al.[21] that the cytotoxicity of ricin A chain conjugates is markedly potentiated and accelerated in the presence of free B chain.

The biggest problem with intact toxin conjugates is that they lack complete specificity of cytotoxic effect because they are able to bind to nontarget cells by means of the cell recognition site(s) of the B chain. With intact abrin and ricin conjugates, this nonspecific toxicity can be attenuated in vitro by including high concentrations of the competing sugars, galactose, or lactose.[13]

The recently developed iodoacetyl method (see above) for preparing antibody–ricin conjugates with impaired galactose recognition offers a further means of avoiding nonspecific toxicity in vitro. Importantly, such conjugates appear to retain the entry-promoting function of the B chain. A W3/25-blocked ricin conjugate was found to be more toxic than ricin itself to W3/25-expressing rat leukemia cells, in sharp contrast to the A chain conjugate, which was not cytotoxic.[19] It is possible that the property of the B chain that is required for A chain entry resides in hydrophobic domains which, when brought into close proximity to the cell membrane by the antibody–antigen interaction, insert into the membrane to form a channel through which the A chain can pass.

In whole-animal test systems, the A chain conjugates have the advantage over intact toxin conjugates of being about 100 times less poisonous. The LD_{50} (iv) of ricin A chain conjugates in mice is about 12 mg/kg with respect to the toxin moiety as opposed to 40 μg/kg for nonblocked ricin conjugates and 160 μg/kg for blocked ricin conjugates.[34] The reason why

[32] R. J. Youle and D. M. Neville, J. Biol. Chem. 257, 1598 (1982).
[33] E. S. Vitetta, W. Cushley, and J. W. Uhr, Proc. Natl. Acad. Sci. U.S.A. 80, 6332 (1983).
[34] P. E. Thorpe, unpublished results.

blocked ricin conjugates are poisonous is unknown. One possibility indicated by recent experiments[35] is that the mannose-terminating oligosaccharide side chains of ricin are recognized by the reticuloendothelial system, thereby leading to rapid uptake and cleavage of the conjugate with subsequent release of free toxin (LD_{50} = 3 μg/kg). A potential disadvantage of A chain conjugates is that the disulfide bond used to form them may be prone to cleavage in animals by reduction or disulfide exchange.[16] Intact toxin conjugates can be constructed using nondisulfide linkages since the naturally stable cystine residue joining the A and B subunits is retained; therefore, they may be more stable in animals.

Appendix

Safety Considerations

A very small quantity (less than 100 μg) of a toxin could prove fatal if it were to be introduced systemically (e.g., through cuts or by inhalation of aerosols). The following safeguards, in operation in the authors' laboratories, are designed to minimize the likelihood of such accidents occurring or, if they do, from having dangerous consequences.

1. Plastic laboratory supplies rather than glassware should be used wherever practical when handling toxins. In particular, glass pipets (including Pasteur pipets) should not be used. Glass chromatography columns should only be used if they are enclosed within a plastic water jacket.

2. Pressurized ultrafiltration equipment should be operated and exhausted within an efficient fume cupboard.

3. Toxin solutions should be centrifuged in sealed thick-walled tubes inside sealed rotors to guard against the eventuality of a tube fracturing during centrifugation. After centrifuging, the entire rotor assembly should be removed and opened inside a fume cupboard.

4. Grinding of seeds should be performed in sealed grinding units inside a fume cupboard. (*N.B. Ricinus communis* also contains a very potent allergen which is distinct from the toxin.)

5. The use of dry powders, e.g., freeze-dried preparations of the toxins, should be avoided wherever possible.

6. Toxins can accumulate on solid surfaces with which they have been in contact: Beware, for instance, of the membranes of ultrafiltration cells.

7. When sterilizing toxin solutions by filtration through 0.22- or 0.45-μm pore size filter units, the operator should wear a face mask and pro-

[35] P. E. Thorpe and B. M. J. Foxwell, unpublished results.

tective clothing and use the filter apparatus inside a fume cupboard to guard against the risk of the syringe containing the toxin solution detaching from the filter unit and squirting its contents into the face of the operator.

8. The usual precautions when handling other toxic substances apply, of course, to toxins. These include wearing gloves and protective clothing, using spill trays, and displaying "Toxic" notices on vessels containing toxic solutions.

9. Clearly marked areas (preferably separate laboratories) should be set aside for the handling of toxins. Toxin solutions should be stored in locked containers and freezers, and records should be kept of the materials stored.

10. Toxin-contaminated apparatus and waste solutions can be made safe by incineration, autoclaving, or by treating for at least 24 hr with 10% Chloros solution, as appropriate.

11. There is no known remedy for toxin poisoning other than to administer antitoxin antibodies intravenously immediately after an accident in which a dangerous quantity of toxin is believed to have been introduced into the bloodstream.[36] Horse antidiphtheria toxin for administration to humans is commercially available; antibodies to abrin, ricin, and other toxins are not yet commercially available in a form suitable for use in humans. Workers proposing to handle diphtheria toxin should have their blood antitoxin levels measured and immunization contemplated if these prove to be low. Toxins that come into contact with unbroken skin are not absorbed and can be simply washed off. The toxicity of the toxins by the oral route is probably much lower than when administered intravenously (in mice, this difference is at least 50-fold for abrin[37]).

[36] B. M. J. Foxwell, S. I. Detre, and P. E. Thorpe, *J. Clin. Exp. Immunol.* (in press).
[37] P. E. Thorpe, unpublished results.

[17] Protein Conjugates of Fungal Toxins

By HEINZ FAULSTICH and LUIGI FIUME

Nearly all fatal cases of poisoning by mushrooms in man are due to three *Amanita* species (*A. phalloides, A. verna,* and *A. virosa*) which contain two families of toxic cyclopeptides, amatoxins and phallotoxins. Their structures and physical properties have been elucidated by

T. Wieland and colleagues.[1] Amatoxins—the main component is α-amanitin (αA)—are responsible for liver necrosis in human poisoning; they act by inhibiting RNA polymerases II.[2-4] Phallotoxins—the main component is phalloidin (PD)—do not play a role in human poisoning but cause hemorrhagic necrosis of the liver after parenteral administration to animals. At the molecular level, they bind to polymeric or oligomeric actin, thus stabilizing the filamentous form of this protein.[5] Conjugation of amatoxins and phallotoxins to proteins was performed for two purposes: (1) to obtain a selective killing of cells which display high uptake of the protein attached to the toxin[6-10]; (2) to induce the production of toxin-specific antibodies in animals or cultured hybrid cells. Such antibodies provided the tool for detection of the toxins in biological fluids by radioimmunoassays.[11-13] Another aim of these experiments was to obtain a serum useful in the treatment of human *Amanita* poisoning.[14-17] Preliminary results, however, have shown that an amatoxin-binding serum enhances rather than decreases the toxicity of α-amanitin.[18]

For protein coupling we used toxins with a native carboxylic function, such as β-amanitin (βA) and phallacidin (PC) or chemically modified toxins with carboxylic groups introduced by spacer moieties. Toxin derivatives with functional amino groups can also be coupled to proteins, as proved by the use of αA-N (Table I). The toxins and toxin derivatives used in the preparation of protein conjugates are shown in Fig. 1. Coupling was achieved either by water-soluble carbodiimide or by activation as

[1] T. Wieland and H. Faulstich, *CRC Crit. Rev. Biochem.* **5,** 185 (1978).

[2] F. Stirpe and L. Fiume, *Biochem. J.* **105,** 779 (1967).

[3] C. Kedinger, M. Gniazdowski, J. L. Mandel, Jr., F. Gissinger, and P. Chambon, *Biochem. Biophys. Res. Commun.* **38,** 165 (1970).

[4] T. J. Lindell, F. Weinberg, P. W. Morris, R. G. Roeder, and W. J. Rutter, *Science* **105,** 779 (1970).

[5] A. M. Lengsfeld, I. Löw, T. Wieland, P. Dancker, and W. Hasselbach, *Proc. Natl. Acad. Sci. U.S.A.* **71,** 2803 (1974).

[6] G. Barbanti-Brodano and L. Fiume, *Nature (London) New Biol.* **243,** 281 (1973).

[7] L. Fiume and G. Barbanti-Brodano, *Experientia* **30,** 76 (1974).

[8] E. Bonetti, M. Derenzini, and L. Fiume, *Virchows Arch. B* **16,** 71 (1974).

[9] R. S. Hencin and J. R. Preston, *Mol. Pharmacol.* **16,** 961 (1979).

[10] M. T. B. Davis and J. F. Preston, *Science* **213,** 1385 (1981).

[11] L. Fiume, C. Busi, G. Campadelli-Fiume, and C. Franceschi, *Experientia* **31,** 1233 (1975).

[12] H. Faulstich, H. Trischmann, and S. Zobeley, *FEBS Lett.* **56,** 312 (1975).

[13] H. Faulstich, S. Zobeley, and H. Trischmann, *Toxicon* **20,** 913 (1982).

[14] W. Boehringer, Thesis, Universität Frankfurt am Main (1959).

[15] C. Cessi and L. Fiume, *Toxicon* **6,** 309 (1969).

[16] H. Faulstich and H. Trischmann, *Hoppe-Seyler's Z. Physiol. Chem.* **354,** 1395 (1973).

[17] K. Kirchner and H. Faulstich, *Monoclonal Antibodies to Amatoxins, Fed. Eur. Biochem. Soc. Meet., 1983* p. 274 (1983).

[18] L. Fiume and C. Busi, unpublished data.

mixed anhydrides. The chemical procedures and the properties of the conjugates are described below.

Coupling of Amatoxins by Carbodiimides

In a typical preparation, 10 mg of βA (11 μmol) and 20 mg of albumin (0.29 μmol) are allowed to react with 5 mg of 1-ethyl-3-(dimethylamino-

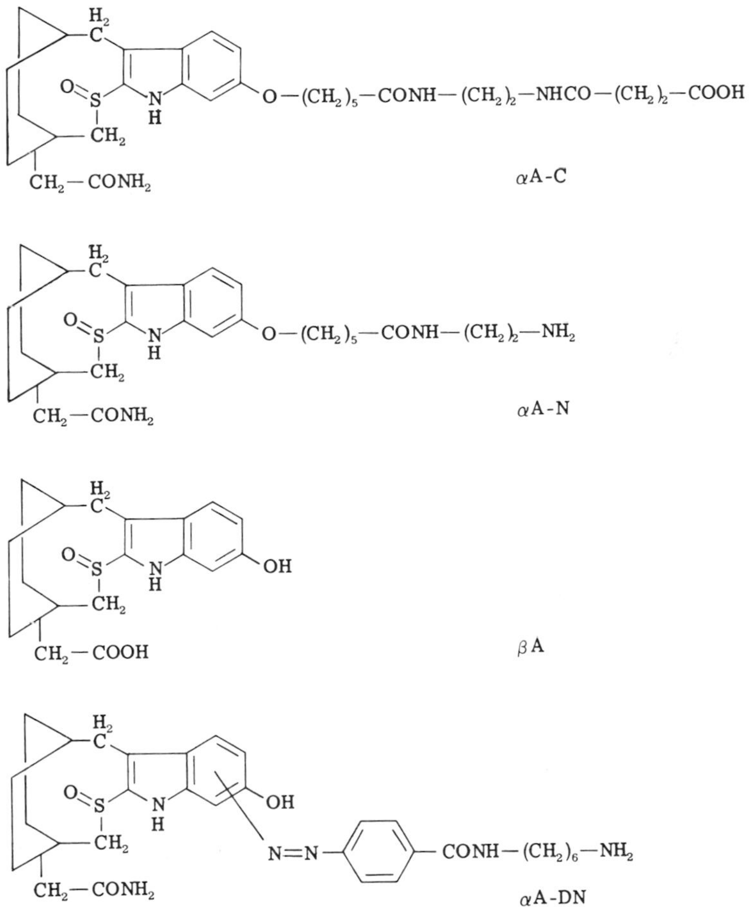

FIG. 1. Native amatoxins and phallotoxins and chemically modified toxins of both with functional groups and spacer moieties that have been used so far to prepare conjugates of fungal toxins with various proteins or poly(amino acids). αA-C, O-[5({[(succinoyl-amino)ethyl]amino}carbonyl)pent-1-yl]-α-amanitin; αA-N, O-[5{[(aminoethyl)amino]car-bonyl}pent-1-yl]-α-amanitin; βA, β-amanitin; αA-DN, 4-{6-[(1-aminohexylamino)car-bonyl]phenyldiazo}-α-amanitin; PC, phallacidin; PD-C, 1-aminomethyldithiolanophalloidin.

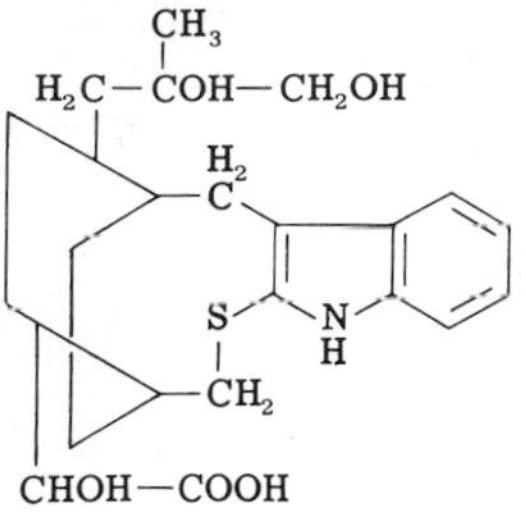

PC

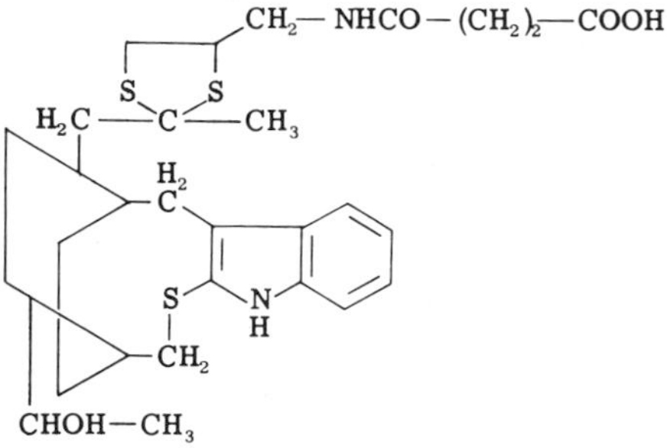

PD-C

FIG. 1. (*continued*)

propyl)carbodiimide (ECDI) (26 μmol) or with 8 mg of 1-cyclohexyl-3-[2-morpholinyl-(4)-ethyl]carbodiimide (MCDI) (18.9 μmol) in 1 ml of water at room temperature for 24 hr. The conjugated material is separated from free βA and from urea derived from carbodiimide by gel filtration through a 1.2 × 100-cm column of Sephadex G-75 equilibrated and eluted with 0.9% NaCl. The conjugate may be kept frozen or may be lyophilized after dialysis against water.

The molar ratio of βA to protein in the conjugate is calculated by measuring protein concentration according to Lowry *et al.*[19] and determining amanitin concentration from the extinction at 305 nm (ε, 1.5×10^4 M^{-1} cm^{-1}). The correction for light scattering is performed according to Leach and Scheraga[20] by measuring the apparent absorption of the conjugate in the 360- to 440-nm region in which neither albumin nor amanitin absorb and extrapolating back to the lower wavelengths. Data of amatoxin–protein conjugates prepared by carbodiimide coupling are listed in Table I.

[19] O. H. Lowry, N. J. Rosebrough, A. L. Farr, and R. J. Randall, *J. Biol. Chem.* **193,** 256 (1951).
[20] S. J. Leach and H. A. Scheraga, *J. Am. Chem. Soc.* **82,** 4790 (1960).

TABLE I
CARBODIIMIDE COUPLING OF AMATOXINS TO VARIOUS PROTEINS

Compound	Ratio	Carbodiimide used	Increase in βA toxicity after coupling to proteins (free βA = 1)[a]	Remarks	Reference
Bovine serum albumin–(βA)	0.6	MCDI	6	—	26
Bovine serum albumin–(βA)	1.6	ECDI	50–100	—	24, 25
Bovine serum albumin–(αA)	0.2	ECDI	—	Only conjugates of type II can be formed	27
Bovine serum albumin–(αA-N)	3.0	ECDI	—	Coupled at pH 5.5	22
Bovine γ-globulin–(βA)	2.4	ECDI	—	—	27
Calf asialofetuin–(βA)	5.7	ECDI	9	—	28
Calf asialofetuin–(αA)	0.2	ECDI	70	Only conjugates of type II can be formed	28

[a] The LD_{50} (ip) of free βA for mouse is 0.5 mg/kg body weight.

Two types of conjugates can be expected to be formed when βA is coupled to proteins by means of carbodiimides[21] (Fig. 2). In type I βA is linked to proteins by an amide bond; in type II, by an activated ester bond. Only type II can be formed in coupling αA, whose carboxylic group is amide substituted. The contribution of type II conjugates to the albumin–(βA) is 13–20% as estimated by the following two experiments. When αA was used for coupling, the ratio toxin : albumin was 0.2 instead of 1.6, as for βA (Table I). When albumin–(βA) (ratio 1.6) was incubated with hydroxylamine for several hours (a reaction that converts activated esters to hydroxamic acids under cleavage of the toxin), the molar ratio of toxin to protein decreased to 1.3.[22]

Carbodiimide coupling may give rise to some unwanted side reactions. For example, urea moieties can be introduced into the protein by an intramolecular acyl transfer in the intermediate O-acylisourea moieties.[23]

[21] H. Faulstich, H. Trischmann, and S. Zobeley, in "Amanita Toxins and Poisoning" (H. Faulstich, B. Kommerell, and T. Wieland, eds.), p. 37. Verlag Gerhard Witzstrock, New York, 1980.

[22] H. Faulstich, unpublished data.

[23] K. L. Carraway and D. E. Koshland, in this series, Vol. 25, p. 616.

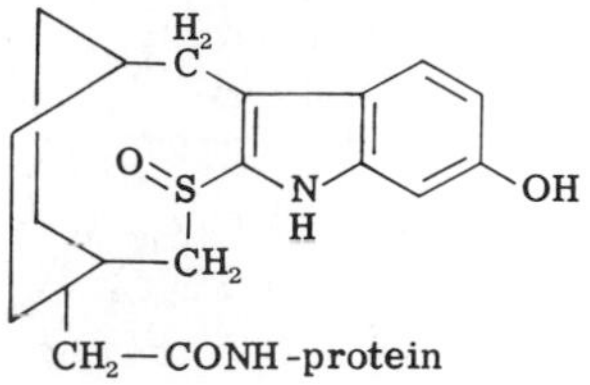

Fig. 2. Two types of linkages as expected from coupling reactions of β-amanitin (βA) with proteins, using carbodiimides. Type I: amide bond, formed from the carboxylic group of the toxin and probably ε-amino groups of lysine residues in the protein. Type II: activated ester bond, formed from a carboxylic group of the protein with the 6'-hydroxy group of the indole moiety in βA or αA.

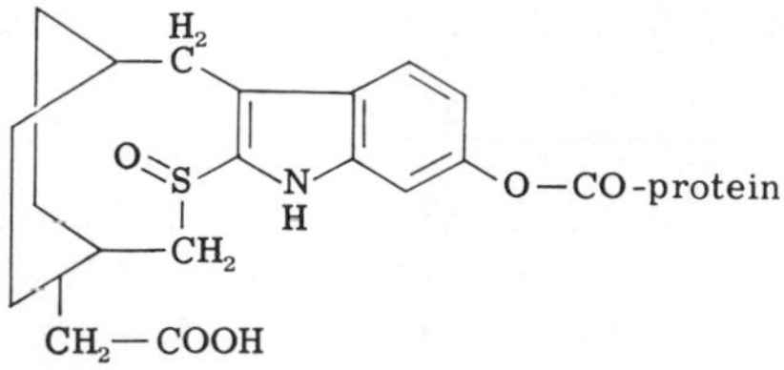

Furthermore, oligomerization of the protein may occur. In fact, SDS–PAGE of albumin–(βA) conjugates shows that only about a 30% portion of the conjugates has a molecular weight corresponding to the monomeric form of the protein; the larger part is made up by molecular species corresponding to albumin oligomers.[24] The molar ratio of amanitin to albumin is higher in conjugates of monomeric or dimeric albumin than in larger ones.

Biological Properties of Conjugates Prepared
  by Carbodiimide Coupling

Conjugates of βA with albumin are very hepatotoxic (Table I). In fact, the toxicity of βA increases 50–100 times when coupling is performed with ECDI[24,25] and 6–8 times when the conjugate is prepared with MCDI.[26,27] The reason why MCDI conjugates are less toxic than those obtained with ECDI is unknown. Conjugates of type I (Fig. 2) appear to be several times less toxic than conjugates of type II. Actually, the toxicity of βA increased 9 times by coupling it to desialylated fetuin, whereas the toxicity of αA (which in an analogous reaction allows the formation of type II conjugate only) increased 70 times.[28]

After coupling to albumin, βA changes its original target (hepatocytes) and produces the typical nuclear lesions[29,30] in sinusoidal liver cells,[24]

[24] M. Derenzini, L. Fiume, V. Marinozzi, A. Mattioli, L. Montanaro, and S. Sperti, *Lab. Invest.* **29,** 150 (1973).
[25] E. Bonetti, M. Derenzini, and L. Fiume, *Arch. Toxicol.* **35,** 69 (1976).
[26] L. Fiume, G. Campadelli-Fiume, and T. Wieland, *Nature (London), New Biol.* **230,** 219 (1971).
[27] L. Fiume and G. Campadelli-Fiume, unpublished data.
[28] L. Fiume and Derenzini, unpublished data.
[29] L. Fiume and R. Laschi, *Sperimentale* **115,** 288 (1965).
[30] V. Marinozzi and L. Fiume, *Exp. Cell Res.* **67,** 311 (1971).

which are very active in endocytosis of proteins. Since neither free αA nor βA enter into these cells, damage produced in them by albumin–(βA) conjugate represents the first example of a toxin forced to penetrate into cells by conjugation with a protein carrier.

Damage to sinusoidal cells may be produced either by the conjugate itself or through release of free amanitin into the cell. The first possibility cannot be discarded because amanitin bound to albumin still inhibits RNA polymerase II in a cell-free system, although the inhibitory effect decreases 3–5 times after conjugation.[24,26] The second possibility appears more probable since albumin after penetration into the cells is digested by lysosomal enzymes.[31] It is thus likely that the bonds linking the amanitin to albumin are split within lysosomes and the free toxin inhibits RNA polymerase II after crossing the lysosomal membrane. According to this second mechanism of action, the higher toxicity of type II conjugates may be due to a more rapid splitting of the ester bond linking amanitin to albumin in comparison to the amide bond in type I conjugate.

Amanitin–albumin conjugates obtained by direct carbodiimide coupling selectively damage cells which display a high albumin uptake.[6–8] Concentrations of conjugate that have no effect *in vitro* on lymphocyte viability and on their capacity to undergo blast transformation kill 100% mouse peritoneal macrophages.[6] Table II shows the higher sensitivity of macrophages to albumin–(βA) compared with the sensitivity of several cells of different origins and characteristics.

The conjugates prepared with ECDI are too toxic to be used as immunogens. Antibodies binding amatoxins were obtained by administering the less toxic conjugates prepared with MCDI to rats.[11] These rodents are several times more resistant to amatoxins,[32] as well as to their albumin conjugates, than are other animals. The antibodies were used to detect amatoxins in biological fluids by radioimmunoassay.[33,34]

Coupling of Amatoxins via Mixed Anhydrides

Activation of carboxylic groups in N-protected α-amino acids, as mixed anhydrides of carbonic acid monoesters, is a well-established method of modern peptide chemistry.[35] When performed using *N*-dimethyl-

[31] E. R. Unanue, *Adv. Immunol.* **15**, 95 (1972).

[32] O. Wieland, H. E. Fisher, and M. Reiter, *Naunyn-Schmiedebergs Arch. Exp. Pathol. Pharmakol.* **215**, 75 (1952).

[33] C. Busi, L. Fiume, D. Costantino, M. Borroni, G. Ambrosino, A. Olivotto, and D. Bernardini, *Nouv. Presse Med.* **6**, 2855 (1977).

[34] C. Busi, L. Fiume, D. Costantino, M. Langer, and F. Vesconi, N. Engl. J. Med. **300**, 800 (1979).

[35] E. Wünsch, *in* Houben-Weyl, "Methoden der Organischen Chemie" (E. Wünsch, ed.), Vol. 15, Part 2, p. 171. Thieme, Stuttgart, 1974.

TABLE II

TOXICITY OF ALBUMIN–(βA) AND αA FOR DIFFERENT CELLS[a]

Cell type	Albumin–(βA)		αA	
	CPE_{25}	CPE_{100}	CPE_{25}	CPE_{100}
Macrophages	2 (0.05)	30 (0.8)	2.5	10
HeLa	>250 (6.7)		3	10
KB	>250 (6.7)		5	>10
HEp-2	>250 (6.7)		10	>10
RTC	150 (4.2)	>250 (6.7)	5	>10
MDBK	>250 (6.7)		10	>10
MKS-Bu 100	250 (6.7)	>250 (6.7)	5	>10
HEF-SV	125 (3.4)	>250 (6.7)	2.5	>10
HEF	62 (1.7)	>250 (6.7)	5	>10
BHK	250 (6.7)	>250 (6.7)	2.5	10
VERO	>250 (6.7)		10	>10
CV-1	250 (6.7)	>250 (6.7)	—	—
MEF	>250 (6.7)		—	—

[a] CPE_{25} and CPE_{100} correspond to concentrations (μg/ml) of albumin–(βA) or αA killing 25 or 100% of the cells, respectively. The amount of amanitin contained in the conjugate (μg) is shown in parentheses. HeLa, KB, HEp-2, Neoplastic cells from human carcinomata; RTC, neoplastic cells from a methylcholanthrene-induced sarcoma in Fisher rats; MDBK, Madin–Darby bovine kidney cell line; MKS-Bu 100, mouse embryo fibroblasts transformed by simian virus 40; HEF-SV, human embryo fibroblasts transformed by simian virus 40; HEF, normal human embryo fibroblasts; BHK, baby hamster kidney cells; VERO, CV-1, green monkey kidney cells; MEF, primary mouse embryo fibroblasts. [Reproduced from *Experientia* (Fiume and Barbanti-Brodano[7]), with permission.]

formamide as solvent, the method is well suited for the activation of carboxylic groups in side chains of cyclic peptides.

In a typical preparation, 8 mg of βA or 10 mg of αA-C[36] (about 9 μmol) is dissolved in 0.2 ml of dimethylformamide, purified from water and dimethylamine by fractionated distillation as described in ref. 37. After cooling to $-15°$, 0.1 ml of a cold *N*-methylmorpholine solution (10 μmol) in dimethylformamide (0.112 ml in 10 ml) is added, followed by the addition of 0.1 ml of a cold solution of chloroethylcarbonate (10 μmol) in dimethylformamide (0.095 ml in 10 ml). The mixed anhydride is formed

[36] H. Faulstich, H. Trischmann, T. Wieland, and E. Wulf, *Biochemistry* **20**, 6498 (1981).
[37] "Organicum," 5th ed., p. 623. VEB Dtsch. Verlag der Wiss., Berlin, 1965.

during a 15-min reaction at $-15°$. The anhydride solution is added to a solution of 32 mg (0.46 μmol) of bovine serum albumin in 1.2 ml of water, adjusted to pH 7.8 with 0.1 N NaOH, with gentle magnetic stirring. The mixture is left for 3 hr to come to room temperature and is then applied to a 2.4 × 80-cm column of Sephadex G-75 equilibrated and eluted with 0.04% NH_4HCO_3. The protein elutes in the first peak (in the case of bovine serum albumin, usually in two peaks) and is lyophilized. During the lyophilization, most of the buffer salt is removed as a result of its volatility. Residual NH_4 salts may be removed, if necessary, by dialysis against NaCl solution. The second peak (third peak) contains free ama-toxins (βA, αA-C) which may be reused after purification on ion-ex-change columns, e.g., DEAE-Sephadex, eluted with acetic acid. The mo-lar ratio of amatoxin to protein can be calculated from the extinction values E_{310} and E_{280}, using the molar extinction coefficients of the amatox-ins (ε_{310}, 13,500 M^{-1} cm^{-1}; $\varepsilon_{280} = 0.55 \times \varepsilon_{310}$) and of the various proteins (ε_{280} protein) as follows:

$$\text{ratio} = \frac{\varepsilon_{310} \times (\varepsilon_{280} \text{ protein})}{13,500 \times (E_{280} - 0.55E_{310})}$$

As shown in Table III, most of the conjugates prepared by the method of mixed anhydrides have high molar ratios of toxin to protein. However, this cannot be attributed to the coupling procedure because in the case of the albumin–(βA) conjugate the ratio is only 1. Rather, the high molar ratios appear to be due to the use of the αA-C derivative, which by its 14-atom spacer exhibits higher reactivity toward proteins. In addition, the ratios of these conjugates are not changed during incubation with hydrox-ylamine, a finding indicating that all linkages between toxin and protein are of the amide type.

Biological Properties of Amatoxin Conjugates Prepared via Mixed Anhydrides

The capacity of αA-C to inhibit RNA polymerase II is five times lower than that of naturally occurring αA[36]; after coupling to albumin, fetuin, histone, or concanavalin A, it decreases to less than 1% (Table III). In contrast to RNA polymerase II, amanitin-specific IgG molecules recog-nize αA-C in protein conjugates rather well. In a solid-phase radioim-munoassay, where αA-C conjugated to various proteins competes for [^{3}H]amatoxin, 4–14% of the αA-C moieties are available for the amanitin-specific IgG molecules. In the particular case of human im-munoglobulins, the availability of αA-C is even 40%, probably as a result of the high flexibility of the carrier protein.

TABLE III

MIXED ANHYDRIDE COUPLING OF AMATOXINS TO VARIOUS PROTEINS, POLYLYSINE HYDROBROMIDE, AND POLYORNITHINE HYDROBROMIDE

Compound	Ratio	$\varepsilon_{280} \times 10^{-3,a}$	Inhibition (%) of RNA-polymerase II[b] (free αA = 100)	Binding capacity to amatoxin-specific antibodies (%)[c]	Remarks
Bovine serum albumin–(αA-C)	8.0	40.0	0.2	14	—
Bovine serum albumin–(βA)	1.0	40.0	—	—	—
Calf fetuin–(αA-C)	3.7	21.4	0.7	12	—
Hen egg lysozyme–(αA-C)	1.4	36.0	—	3	—
Calf histone (IV)–(αA-C)	4.3	15.0	0.06	4	Increases RNA synthesis by 100% at 2.5×10^{-7} M (histone effect)[d] and decreases by 50% at 7.5×10^{-6} M (amatoxin effect)
Human γ-globulin–(αA-C)	5.5	128.0	—	40	—
Bovine γ-globulin–(βA)	1.0	128.0	—	—	—
Ovalbumin–(αA-C)	2.5	32.6	—	4	—
Concanavalin A–(αA-C)	11.6	100.4[e]	0.9	10	—
Concanavalin A–(αA-C)	2.6	100.4	—	8	—

					Remarks
Ferritin–(αA-C)	1.0	—	—	(10)	The ratio was estimated from a radioimmunoassay assuming 10% availability of the α-amanitin molecules attached to the protein
Polylysine · HBr(30K)–(αA-C)	11.5	—	—	—	The ratio was determined spectrophotometrically and related on M_r 30 K
Polylysine · HBr(165K)–(βA)	39.0	—	—	—	In this compound the pK value of the phenolic OH of β-amanitin was decreased from 9.5 to 7.2
Polyornithine·HBr(30K)–(αA-C)	9.3	—	—	—	The ratio was determined spectrophotometrically and related on M_r 30 K
αA	—	$0.55 \times \varepsilon_{310}$	100	100	—

[a] Molar extinction coefficients (M^{-1} cm^{-1}) used for calculation of ratios.

[b] Inhibition of incorporation of [³H]UTP into RNA catalyzed by RNA polymerase II of *Drosophila melanogaster* embryos (αA = 100%).

[c] Competitive binding assay for O-[³H]methyldehydroxymethyl-αA with amatoxin-specific IgG covalently attached to nylon surfaces (αA = 100%).[13]

[d] B. D. Hall, M. Brzezinska, C. P. Hollenberg, and L. D. Schultz, *in* "Molecular Cytogenetics" (B. A. Hamkalo and J. Papaconstantinou, eds.), p. 217. Plenum, New York, 1973.

[e] Related on M_r 110K.

TABLE IV
PHALLOTOXIN COUPLING TO BOVINE SERUM ALBUMIN BY VARIOUS METHODS

Compound	Ratio	Coupling methods	Remarks	Reference
Bovine serum albumin–(PC)	2.7	ECDI	No affinity to actin	42
Bovine serum albumin–(PD-C)	9.6	MA	Affinity to actin ~5%	41
Bovine serum albumin–(PD-C)	2.4	ECDI, MCDI	Affinity to actin ~5%	41
Bovine serum albumin–(PD)	2.3	ECDI	—	38

Also the amatoxin conjugates prepared by the mixed anhydride method exhibit higher toxicity *in vivo* than do free amatoxins (up to 10-fold), although not so high as the conjugates prepared by ECDI coupling. Particularly, fetuin conjugates with αA-C have been used to raise antibodies in rabbits[13] or to induce immune response in rats, whose spleen cells after fusion with mouse myeloma cells have been cloned in order to obtain monoclonal antibodies to amatoxins.[17]

Phallotoxins Coupled to Proteins

Using either carbodiimide or mixed anhydride coupling, four different phallotoxin conjugates were prepared with the procedures described for amatoxins (Table IV). In the conjugate albumin–(PD) prepared with ECDI, most probably ester linkages are formed.[38] Phallacidin (PC) is the most abundant phallotoxin with a native carboxylic group in a side chain, which allows the formation of an amide bond. However, the carboxylic group is very close to the peptide backbone, and thus its coupling can affect the biological activity of the toxin. Therefore, a functional group has been introduced into the dihydroxy-L-leucine side chain of phalloidin (PD-C)[39] (Fig. 1), which has been shown to be least involved in the biological activity.[40] Phallotoxins are assumed to have their binding site buried in between the protein units of filamentous actin. Accordingly, conjugates with short spacers can be expected to be inactive on actin. In fact, there was no actin binding of albumin–(PC), which, under the assumption that PC is bound to the ε-amino group of a lysine residue, has only a 7-atom spacer between the backbones of peptide and protein.

[38] G. Barbanti-Brodano, M. Derenzini, and L. Fiume, *Nature (London)* **248**, 63 (1974).
[39] T. Wieland, A. Deboben, and H. Faulstich, *Liebig's Ann. Chem.* p. 416 (1980).
[40] H. Faulstich, *in* "Chemistry of Peptides and Proteins" (W. Voelter, E. Wuensh, Yu. Ovchinnikov, and V. Ivanov, eds.), Vol. 1, p. 279. de Gruyter, Berlin, 1982.

TABLE V

EFFECT OF PD AND ALBUMIN–(PD) ON MOUSE
PERITONEAL MACROPHAGES[a,b]

PD		Albumin–(PD)	
$\mu g/ml^c$	Percentage of dead cells	$\mu g/ml^c$	Percentage of dead cells
100	25	400 (12)[d]	100
50	10	200 (6)	100
25	0	100 (3)	75
12.5	0	50 (1.5)	75

[a] Reproduced from ref. 38, with permission.

[b] Mouse peritoneal macrophages were incubated at 37° in 0.5 ml of Eagle's basal medium (BME) containing PD or conjugate. After 15 hr, 2 ml of BME with 10% fetal bovine serum was added; and after a further 9 hr of incubation, the number of dead cells was determined by staining with trypan blue. In the albumin–(PD) experiment, most of the cells were detached from the glass at the end of the incubation period.

[c] Concentration during the first 15-hr incubation.

[d] The amount of PD contained in albumin–(PD) is listed within parentheses.

There is, however, residual (5%) affinity to actin in conjugates of PD-C where the corresponding spacer between the backbones has a length of 15 atoms.[41]

Bovine serum albumin–(PD), like amatoxin–albumin conjugates, is toxic to cells with high protein uptake *in vitro* and *in vivo*[38] (Table V). The lesions are similar to those found in hepatocytes after injection of free PD. The damage by conjugate is most likely caused by PD released free inside the cells by rapid hydrolysis of the ester bonds linking toxin and protein. On the other hand, albumin–(PC) does not display *in vivo* toxicity, even when the amount of coupled toxin injected is 2.4-fold the LD_{50} of free PC.[42]

Acknowledgments

The authors wish to thank Dr. Elisabeth Wulf for performing some of the inhibition assays of RNA synthesis. We gratefully acknowledge the excellent technical assistance of Mr. H. Trischmann and Miss Suse Zobeley.

[41] A. Deboben, Thesis, University of Heidelberg (1979).
[42] T. Wieland and A. Buku, *FEBS Lett.* **4**, 341 (1969).

[18] Transferrin Receptor as a Target for Antibody–Drug Conjugates

By DERRICK L. DOMINGO and IAN S. TROWBRIDGE

Introduction

Monoclonal antibodies linked to the A subunits of plant and bacterial toxin such as ricin or diphtheria toxin have been shown to be selectively cytotoxic against tumor cells.[1-4] Such conjugates have theoretical advantages in that their specificity resides exclusively in the antibody moiety and that the toxin A subunit is extremely cytotoxic, killing cells by inactivating protein synthesis. However, successful covalent attachment of the toxin A chain to antibody does not ensure that the resulting antibody–toxin conjugates will be cytotoxic.[5,6] Although the reasons for this are not fully understood, several factors may be involved including (1) the inability of the antibody, after chemical modification, to bind to its target antigen on the cell surface, (2) inefficient internalization of the antibody–toxin conjugate, or (3) failure of the toxic subunit to be released into the cytoplasm in an active form capable of inhibiting protein synthesis.

As the production of a potent and specifically cytotoxic antibody–ricin A conjugate requires that a variety of complex conditions are satisfied, it is advantageous to study antibody–ricin A conjugates that are prepared against well-characterized cell surface molecules. Cell surface receptors with known functions, such as the transferrin receptor, may offer additional benefits. For example, the transferrin receptor normally translocates transferrin-bound iron across the cell membrane and thus might be expected to facilitate the entry of bound antibody–toxin conjugates into the cell. Furthermore, as the transferrin receptor is abundantly expressed on some tumor cells, relative to most normal tissues, it is a target of considerable interest from the viewpoint of cancer treatment.[5] This chap-

---

[1] R. J. Youle and S. M. Neville, *Proc. Natl. Acad. Sci. U.S.A.* **77**, 5483 (1980).

[2] D. G. Gilliland, Z. Steplewski, R. J. Collier, K. F. Mitchell, T. H. Chang, and H. Koprowski, *Proc. Natl. Acad. Sci. U.S.A.* **77**, 4539 (1980).

[3] H. E. Blythman, P. Casellas, D. Gros, P. Gros, F. Jansen, F. Paolucci, B. Pau, and H. Vidal, *Nature (London)* **290**, 145 (1981).

[4] I. S. Trowbridge and D. L. Domingo, *Nature (London)* **294**, 171 (1981).

[5] I. S. Trowbridge and D. L. Domingo, *Cancer Surv.* **1**, 543 (1982).

[6] D. W. Mason, P. E. Thorpe, and W. C. J. Ross, *Cancer Surv.* **1**, 389 (1982).

ter will describe the preparation and properties of antibody–ricin A conjugates that are specific for the transferrin receptors of human and murine cells.

Preparation of Ricin Toxin

Ricin toxin was purified from locally obtained castor beans, as described by Nicolson and Blaustein.[7]

1. Decorticated castor beans (100 g) are ground in 1 liter PBS (0.15 M NaCl, 0.01 M sodium phosphate buffer, pH 7.3) in a Waring blender. Caution: This step should be performed with extreme care in a fume hood as the aerosols produced are very toxic. After grinding, the aerosols should be allowed to settle for at least 10 min before filtering out the grounds through cheesecloth.

2. Centrifuge the extract in 500-ml bottles at 10,000 g for 30 min in a Sorvall RC-2 centrifuge. The homogenate will separate into an aqueous layer and a lipid layer. The aqueous layer is decanted through 9 × × gauge nylon cloth and the centrifugation step repeated.

3. Add ammonium sulfate to give a 60% (w/v) saturated solution and collect the precipitate by centrifugation at 10,000 g for 30 min. Dissolve the precipitate in 100 ml of PBS and dialyze exhaustively against PBS.

4. Load the dialyzed extract onto a column (30 × 5 cm) of Sepharose 6B (Pharmacia, Uppsala, Sweden) and elute all unbound protein with PBS. Ricin is then eluted with 600 ml of 0.2 M galactose (or lactose) in PBS; collect 10 ml fractions. The yield of ricin ($E_{280\,nm}^{0.1\%} = 1.18$)[8] is 500–1000 mg.

5. Castor beans contain approximately equal amounts of ricin toxin (M_r 6 × 10^4) and ricin agglutinin (M_r 1.2 × 10^5). The toxin is separated from the agglutinin by chromatography on a Sephacryl S-200 (Pharmacia, Uppsala, Sweden) column (65 × 3.5 cm) equilibrated with PBS. Fractions (5 ml) are collected, and ricin toxin elutes as the second of two well-resolved protein peaks.

Isolation of Ricin A Chain

The isolation of the toxic subunit (A chain) from the binding subunit (B chain) of ricin toxin is performed as described by Olsnes and Pihl.[9]

[7] G. Nicolson and J. Blaustein, *Biochim. Biophys. Acta* **266**, 543 (1972).
[8] S. Olsnes, E. Saltvedt, and A. Pihl, *J. Biol. Chem.* **249**, 803 (1974).
[9] S. Olsnes and A. Pihl, *Biochemistry* **12**, 3121 (1973).

1. Ricin toxin (50 mg) in 3 ml of 0.5 M lactose, 0.1 M Tris–HCl, pH 8.5, is reduced by addition of 2-mercaptoethanol to 5% (v/v), followed by incubation for 16 hr at room temperature with stirring.

2. The reduced ricin is then chromatographed on a DEAE-cellulose (DE-52; Whatman, Maidstone, England) column (25 × 1.0 cm) equilibrated in 0.1 M Tris-HCl, pH 8.5. The partially purified ricin A chain elutes slightly before 2-mercaptoethanol, the absorbance of which at 280 nm may interfere with the detection of the toxin A chain. Fractions containing the ricin A chain can be identified by electrophoresis on a 10% SDS–polyacrylamide gel. Ricin B-chain ($E_{280\,nm}^{0.1\%} = 1.49$)[10] can be eluted from the column with 0.1 M NaCl in the same buffer.

3. Fractions containing ricin A chain are pooled and dialyzed against 0.01 M lactose, 0.1% 2-mercaptoethanol, 5 mM sodium phosphate buffer, pH 6.5, and chromatographed on a CM-cellulose (CM-52; Whatman, Maidstone, England) column (9.0 × 1.0 cm) equilibrated in the same buffer. After elution of unbound protein, a linear salt gradient (0–0.15 M NaCl in the same buffer, 40 ml total) is applied to the column and 1.0 ml fractions collected. As shown by SDS–polyacrylamide gel electrophoresis, pure ricin A ($E_{280\,nm}^{0.1\%} = 0.765$)[10] is found in the last of three protein peaks obtained. The yield of ricin A chain is 10–20%. Ricin A chain can be stored in elution buffer for at least a week at 4° but should not be frozen.

Chemical Modification of the Anti-Transferrin Receptor Antibody

Ricin A-chain–antibody conjugates are prepared using the cross-linking agent N-succinimidyl 3-(2-pyridyldithio)propionate (SPDP; Pharmacia, Uppsala, Sweden). SPDP is a heterobifunctional reagent that introduces aliphatic thiol groups while avoiding intramolecular cross-linking and the formation of homoconjugates.[11] It has been our experience that about 50% of the SPDP added to the reaction mixture couples to the antibody. For the human anti-transferrin receptor antibody designated B3/25 (a murine IgG$_1$, ref. 12) and the murine anti-transferrin receptor antibody designated RI7 217 (a rat IgG$_{2a}$, refs. 5,13), we used a 6-fold molar excess of SPDP to introduce an average of three 2-pyridyl disulfide groups per molecule of antibody. The procedure for preparing SPDP-derivatized antibody is as follows:

1. A 20-mM stock solution of SPDP is made up in absolute ethanol.
2. Ten microliters of stock SPDP is added with stirring to 5 mg anti-

[10] S. Olsnes, K. Refsnes, T. Christensen, and A. Pihl, *Biochim. Biophys. Acta* **405**, 1 (1975).
[11] J. Carlsson, H. Drevin, and R. Axén, *Biochem. J.* **173**, 723 (1978).
[12] I. S. Trowbridge and M. B. Omary, *Proc. Natl. Acad. Sci. U.S.A.* **78**, 3039 (1981).
[13] J. Lesley, D. L. Domingo, R. Schulte, and I. S. Trowbridge, *Exp. Cell Res.* **150**, 400 (1984).

body in 1 ml of PBS at room temperature, and the reaction mixture is incubated for 30 min.

3. The excess SPDP is separated from the antibody by gel filtration using a Sephadex G-25 Superfine (Pharmacia) column (15 × 1.8 cm) equilibrated in PBS.

Coupling of Ricin A Chain to Anti-Transferrin Receptor Antibody

1. Incubate ricin A chain (4–5 mg in 1 ml of CM-cellulose elution buffer) with fresh 0.1% 2-mercaptoethanol for 30 min at room temperature to ensure complete reduction, then separate ricin A chain from the reducing agent by filtration on a Sephadex G-25 column (15 × 1.8 cm) in PBS.

2. Slowly add ricin A (~3 mg in 4 ml of PBS) with stirring to 5 mg of SPDP-derivatized antibody (3 mol of pyridyl disulfide per mole of antibody) in 4 ml of PBS at 4° (molar ratio of ricin A chain to antibody of 3). After 5 min on ice, incubate mixture overnight at 37°.

3. Concentrate the mixture to 1–2 ml using an Amicon ultrafiltration apparatus or by vacuum dialysis, and separate the conjugate from unreacted ricin A chain by gel filtration on a Sephacryl S-200 column (120 × 1.5 cm) equilibrated in PBS.

4. Antibody–ricin A conjugates are analyzed under nonreducing conditions on a 5% SDS–polyacrylamide gel. As shown in Fig. 1, antibody conjugates containing one to three ricin A chains per antibody molecule are obtained under the reaction conditions described. The preparations contain variable amounts of unreacted antibody (~25%), but this does not appear to significantly interfere with the cytotoxicity of the conjugates.

Problems Encountered

A frequent problem, minimized by the protocol described above, is the precipitation of high-molecular-weight complexes during the conjugation reaction. Modification of IgG monoclonal antibodies with three 2-pyridyl disulfide groups gives the highest yield of conjugates containing one to two ricin A chains per antibody molecule without the formation of insoluble products. Using molar ratios of ricin A chain to antibody greater than 3 : 1 does not increase the yield of antibody–ricin A conjugates. The ionic strength of the buffer did not influence the conjugation of IgG anti-transferrin receptor antibodies, but increasing the salt concentration to 0.3 M NaCl reduced the amount of insoluble complexes formed during the coupling of ricin A to IgM antibodies such as RI7 208, a rat anti-murine transferrin receptor antibody.[14] In general, the coupling of ricin A-chain to IgM antibodies has been less reproducible than to IgG antibodies.

[14] I. S. Trowbridge, J. Lesley, and R. Schulte, *J. Cell. Physiol.* **112,** 402 (1982).

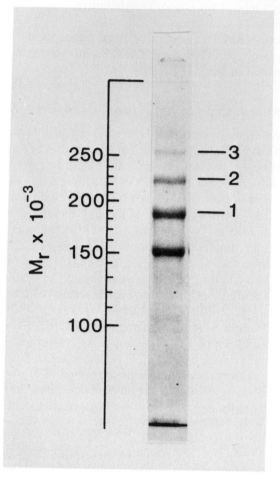

FIG. 1. Analysis of anti-transferrin receptor antibody–ricin A conjugate by SDS–poly-acrylamide gel electrophoresis. B3/25 monoclonal antibody against the human transferrin receptor was conjugated to ricin A using the SPDP method described in the text. The figure shows that on a 5% SDS–polyacrylamide gel run under nonreducing conditions, antibody conjugates with one to three ricin A chains per antibody molecule can be detected.

Storage and Stability of Antibody–Ricin A Conjugates

Antibody–ricin A conjugates are filter-sterilized immediately after preparation and stored at 4°. Most preparations have been stored for up to a year under these conditions without loss of activity. Although it is also possible to store conjugates frozen at −20°, freeze–thawing results in partial loss of activity.

In Vitro Assays for the Cytotoxic Activity of Antibody–Ricin A Chain Conjugates

Inhibition of Protein Synthesis

As a preliminary to cytotoxic testing, confirmation that the antibody–ricin A chain conjugate binds to target cells as effectively as native antibody may be obtained by quantitative fluorescence analysis on the cell sorter.[14] However, the major concern is the potency and specificity of the cytotoxic activity of the anti-transferrin receptor antibody–ricin A conjugates. This can be determined by measuring the inhibition of protein synthesis in cells exposed to the conjugates. The human T leukemic cell line CCRF-CEM[15] and the mouse T lymphoma cell line BW5147[16] are our standard test cells. Both cell lines express about 50,000 transferrin receptors per cell.

1. Cells (2×10^5 BW5147 cells in Dulbecco's modified Eagle's medium supplemented with 10% horse serum or 5×10^5 CCRF-CEM cells in RPMI 1640 supplemented with 10% fetal calf serum) are plated out in triplicate 35-mm petri dishes (Falcon 1008) in 2 ml tissue culture medium containing the desired concentration of antibody–ricin A conjugate. (In a typical experiment, 3-fold dilutions of conjugate over a concentration range of 10^{-8} to 10^{-11} M of antibody–ricin A conjugate are used.) The cells are then grown in the presence of the conjugate for 2 days.

2. [^{3}H]Leucine (10 μCi) is added to each dish and the cells incubated for an additional 2 hr.

3. The cells are harvested, centrifuged, and resuspended in 1 ml of 10% trichloroacetic acid.

4. After 15 min on ice, trichloroacetic acid-precipitable material is collected by filtration on Whatman GF/C glass fiber filters, which are then washed with acetone, dried, and counted in a scintillation counter.

From these data, the concentration of antibody–ricin A conjugate required to inhibit protein synthesis by 50% (IC_{50}) can be calculated. The table shows the IC_{50} values for two anti-transferrin receptor antibody–ricin A conjugates prepared with either B3/25 monoclonal antibody[12] specific for the human transferrin receptor or RI7 217 monoclonal antibody[13] specific for the murine transferrin receptor. Several controls, the results of which are also given in the table, should be performed.

[15] G. E. Foley, H. Lazarus, S. Farber, B. G. Uzman, B. A. Boone, and R. E. McCarthy, *Cancer* **18**, 522 (1965).
[16] R. Hyman and V. Stallings, *Immunogenetics* **6**, 447 (1974).

SPECIFICITY AND POTENCY OF ANTI-TRANSFERRIN RECEPTOR
ANTIBODY–RICIN A CONJUGATES DETERMINED BY *in Vitro*
INHIBITION OF PROTEIN SYNTHESIS

Addition to cells	$IC_{50}{}^a$	
	Mouse lymphoma BW5147	Human leukemia CCRF-CEM
B3/25^b − ricin A	$>1 \times 10^{-8}$	1×10^{-10}
B3/25 antibody	n.d.^{c}	$>1 \times 10^{-8}$
B3/25 + ricin A	n.d.	$>1 \times 10^{-8}$
RI7 217^d − ricin A	4×10^{-11}	$>1 \times 10^{-8}$
RI7 217 antibody	$>1 \times 10^{-8}$	n.d.
RI7 217 + ricin A	$>1 \times 10^{-8}$	n.d.
Ricin A chain	$>1 \times 10^{-8}$	$>1 \times 10^{-8}$
Ricin toxin	3×10^{-10}	3×10^{-11}

^{a} The concentration (M) required for 50% inhibition of protein synthesis, determined as described in the text.
^{b} Anti-human transferrin receptor antibody.
^{c} n.d., Not determined.
^{d} Anti-murine transferrin receptor antibody.

1. Antibody–ricin A conjugates should be tested on cells that do not express the target antigen. As the monoclonal antibodies against the mouse and human transferrin receptors do not cross-react between the two species, the specificity of killing can be established by testing anti-human transferrin receptor antibody–ricin A conjugates on mouse BW5147 cells and anti-murine transferrin receptor antibody–ricin A conjugates on human CCRF-CEM cells.

2. It should be demonstrated that the cytotoxicity of the antibody–ricin A conjugates cannot be produced by either antibody or ricin A alone or by a mixture of the two proteins.

3. Finally, a comparison of the IC_{50} value for ricin toxin itself tested on the same cells with that of the antibody–ricin A conjugate will provide a measure of the relative potency of the conjugate.

From the results in the table, it can be seen that antibody–ricin A conjugates directed against the transferrin receptor give potent and specific killing *in vitro*. This appears to be a quite general finding, as highly cytotoxic conjugates have been prepared with at least four different anti-transferrin receptor antibodies using either ricin A, diphtheria toxin fragment A[4], or *Pseudomonas* exotoxin.[17]

[17] D. J. P. Fitzgerald, I. S. Trowbridge, I. Pastan, and M. C. Willingham, *Proc. Natl. Acad. Sci. U.S.A.* **80,** 4134 (1983).

Clonal Growth Assay

An important question for many potential uses of cytotoxic antibody conjugates including tumor therapy is the frequency with which cells escape killing. The inhibition of protein synthesis assay would not detect a small subpopulation of resistant cells in cultures treated with antibody–ricin A conjugates, particularly as dying cells contribute a background of residual protein synthesis (usually about 10% of the control in the 2-day assay). However, cells escaping the cytotoxic effects of conjugates can be detected at a frequency as low as 10^{-6} by measuring the cloning efficiency of cells exposed to a high concentration of antibody–ricin A conjugate.

For anti-transferrin receptor–ricin A conjugates, BW5147 cells and CCRF-CEM cells are plated out at a cell density of 1×10^6 cells/ml in four, replicate, 0.2-ml wells of a 96-well microtiter plate (Falcon 3042) in the presence of $10^{-8} M$ antibody–ricin A conjugate. Twofold dilutions are made so that the final dilution gives 1 cell/ml. The plates are observed for 3–4 weeks for growth. As for the inhibition of protein synthesis assay, nonspecific effects of the conjugates on cloning efficiency can be detected by testing anti-human transferrin receptor antibody–ricin A conjugates on mouse BW5147 cells and anti-murine transferrin receptor antibody–ricin A conjugates on human CCRF-CEM cells.

The results of such clonal growth assays with three different anti-transferrin receptor antibodies conjugated to ricin A showed that no growth was observed even in wells containing 1×10^6 cells/ml. This is consistent with the view that because transferrin receptors are essential for cell growth, cells cannot readily escape the cytotoxic effects of anti-receptor antibody–ricin A conjugates by either antigenic modulation[18] or genetic loss mechanisms.[19] Whether mutants can be obtained with structurally altered receptors that retain function but lack the antigenic site for a particular anti-receptor antibody is not known.

Applications

Monoclonal antibody–ricin A conjugates directed against the transferrin receptor have been used in a variety of studies. Initially, it was hoped that such conjugates would give potent and specific killing *in vivo*. Although intravenous administration of B3/25 anti-human transferrin receptor antibody–ricin A conjugate inhibited the growth of a human melanoma cell line, M21, at a subcutaneous site in athymic nude mice,[4] antibody alone was equally effective. Similar effects on the growth of M21 cells have been reported using an antibody–ricin conjugate against a mela-

---

[18] L. Old, E. Stockert, E. Boyse, and J. H. Kim, *J. Exp. Med.* **127,** 523 (1968).
[19] R. Hyman, K. Cunningham, and V. Stallings, *Immunogenetics* **10,** 261 (1980).

noma-associated antigen,[20] and, in general, antibody–ricin A conjugates have proved to be much less effective *in vivo* than *in vitro*. It is believed that both rapid clearance of conjugates from the circulation by the reticuloendothelial system and instability of the disulfide bond linking ricin A to the antibody contribute to the relative ineffectiveness of antibody–ricin A conjugates *in vivo*. The problems associated with the *in vivo* use of antibody–ricin A conjugates remain to be resolved.

Anti-transferrin receptor antibody–ricin A conjugates have proved to be useful tools *in vitro*. For example, we were interested in determining the distribution of transferrin receptors on hematopoietic stem cells in mouse bone marrow. One reason for this was to identify normal cells in proliferating tissues that may be at risk if exposed to cytotoxic antitransferrin receptor antibody reagents during treatment. Myeloid (CFUc), erythroid (CFUe and BFUe), and pluripotent stem cells (CFUs) occur in low frequency in the bone marrow and can only be detected by the appropriate colony-forming assays. It could be shown that RI7 217 monoclonal antibody against the murine transferrin receptor coupled to ricin A almost completely inhibited the formation of myeloid and erythroid colonies when the conjugate was present throughout the culture period.[13] It was not possible, however, to distinguish from this data whether colony formation was abolished because transferrin receptors were induced on the progenitor stem cells during their proliferation *in vitro* or because the stem cells normally express transferrin receptors in bone marrow. This question was resolved by briefly preincubating bone marrow cells with RI7 217 antibody–ricin A conjugate and then washing away unbound conjugate before carrying out stem cell assays. It was known that almost maximal inhibition of protein synthesis in BW5147 cells was observed after a 60-min incubation of the cells with 10^{-8} M RI7 217 antibody–ricin A conjugate.[13] The results of the pulse experiments with bone marrow cells showed that whereas greater than 90% of erythroid precursors (CFUe) were killed by the conjugate, myeloid progenitor cells (CFUc) and pluripotent stem cells (CFUs) were unaffected. Thus, it could be concluded that only CFUe express sufficient numbers of transferrin receptors to be sensitive to anti-transferrin receptor antibody–ricin A conjugates.

Antibody–toxin conjugates directed against the transferrin receptor have also been useful in developing methods of enhancing the cytotoxicity of antibody conjugates. Fitzgerald *et al.*[17] prepared cytotoxic conjugates by coupling ricin A chain or *Pseudomonas* exotoxin to antibodies

[20] T. F. Bumol, Q. C. Wang, R. A. Reisfeld, and N. O. Kaplan, *Proc. Natl. Acad. Sci. U.S.A.* **80,** 529 (1982).

against the human transferrin receptor. These conjugates were shown to be toxic for human KB cells at concentrations ranging from 10^{-9} to 10^{-10} M if cells were exposed to the antibody–toxin conjugates for 16–18 hr. However, a brief 1-hr exposure to the anti-transferrin receptor conjugates was not sufficient to inhibit protein synthesis unless cells were treated in the presence of adenovirus. Addition of the virus alone had no effect on protein synthesis but markedly increased the cytotoxicity of the antibody conjugates. Thus, the IC_{50} values for a 1-hr incubation with antibody–toxin conjugates in the presence of adenovirus were similar to those for a 16- to 18-hr incubation with the same conjugates alone. The enhancement of cytotoxicity by adenovirus is probably related to the fact that virus and conjugate share a common transport pathway within the cell. Adenovirus enters cells by receptor-mediated endocytosis and is transported into endosomes. The virus apparently escapes into the cytoplasm by disrupting the endosome membrane. It is known that anti-transferrin receptor antibodies bound to the receptor enter cells via coated pits and also localize in endosomes.[17,21] It is probable that receptor-bound conjugates and adenovirus become localized in the same endosomes and that the virus consequently facilitates the movement of the antibody conjugate into the cytoplasm.

Concluding Remarks

The transferrin receptor offers a useful model system with which to analyze the requirements for developing effective cytotoxic monoclonal antibody conjugates. From the practical viewpoint of therapy, the receptor as a target antigen offers the potential advantages of being essential for tumor cell growth and functioning as a membrane transport molecule. These advantages must be weighed against the possibility that cytotoxic anti-transferrin receptor antibody conjugates may also damage normal tissues. In this respect, it is noteworthy that mouse pluripotent stem cells (CFUs) are spared by anti-transferrin receptor–antibody ricin A conjugates. At present, there are no monoclonal antibodies available that are completely specific for a particular tumor. However, selectivity for different tumors may be increased by using combinations of monoclonal antibodies. It is in this context that anti-transferrin receptor monoclonal antibody conjugates may be of practical therapeutic importance.

[21] C. R. Hopkins and I. S. Trowbridge, *J. Cell Biol.* **97**, 508 (1983).

[19] Lectins as Carriers: Preparation and Purification of a Concanavalin A–Trypsin Conjugate

By W. Thomas Shier

Lectins are a group of proteins with the characteristic ability to bind saccharide moieties at two or more sites on the molecule.[1] This structure endows lectins with the ability to agglutinate red blood cells and cultured cells by cross-linking cell-surface saccharide structures. The majority of the lectins which have been characterized are derived from plant seeds, and they have been extensively studied under the name of phytohemagglutinins. Extensive use has been made of lectins as laboratory tools.[2] Some lectins are used in blood typing. A variety of them have been used as experimental tools to investigate biological roles of cell-surface glycoproteins. Some lectins (e.g., wheat germ agglutinin, concanavalin A) specifically agglutinate tumor cells but not normal control cells under defined experimental conditions, thereby providing valuable tools for investigating biochemical changes associated with oncogenic transformation. Most lectins are mitogenic with either T or B lymphocytes and they have found extensive use as tools to study the mechanism of mitogenic stimulation of lymphocytes, to define lymphocyte subclasses, and to fractionate lymphocyte populations.

The best-characterized lectin is concanavalin A (Con A).[3] It was first crystallized by J. B. Sumner in 1919 from jack beans (*Canavalia ensiformis*). Con A exists at physiological pH as a tetramer of identical subunits of molecular weight 26,000. Each monomer contains one saccharide-binding site with a specificity for α-D-mannosyl or α-D-glucosyl moieties. It is readily purified by affinity chromatography on Sephadex (cross-linked dextrans), followed by crystallization. Typically 2–3 g of Con A is obtained from 100 g of jack beans. As a result of the inexpensive source and simple purification procedure, Con A is readily available commercially and relatively inexpensive (about $45 per gram).

Work on lectins as carriers of drugs and enzymes was initiated before the advent of monoclonal antibodies with the aim of developing site-specific mechanisms of targeting drugs and other agents.[4] Monoclonal

[1] N. Sharon and H. Lis, *Science* **177**, 949 (1972).
[2] H. Bittiger and H. P. Schnebli, eds., "Concanavalin A as a Tool." Wiley, London, 1976.
[3] G. M. Edelman, B. A. Cunningham, G. N. Reeke, J. W. Becker, M. J. Waxdal, and J. L. Wang, *Proc. Natl. Acad. Sci. U.S.A.* **69**, 2580 (1972).
[4] W. T. Shier, J. T. Trotter, and D. T. Astudillo, *Int. J. Cancer* **18**, 672 (1976).

METHODS IN ENZYMOLOGY, VOL. 112

antibodies can be found with higher specificities than lectins, a property making them more useful for many of the applications as drug and enzyme carriers originally envisaged for lectins, particularly research applications. However, there have been production problems with monoclonal antibodies which have limited their availability in the quantities needed for many commercial applications. Future developments, such as the transfer of the genes for monoclonal antibodies to *Escherichia coli* or another microbial host, may lead to production of larger amounts. However, it is unlikely that any of these methods will result in the production of monoclonal antibodies as inexpensive and abundant as Con A. Consequently, industrial applications, such as reversibly immobilized enzymes (discussed below) are more likely to use lectins than monoclonal antibodies.

Lectins have several additional advantages over monoclonal antibodies as carriers of drugs and enzymes. Lectins are readily purified by simple affinity chromatography techniques that are useful both in the initial isolation of the lectin and in purifying conjugates containing lectins. An additional advantage of lectins is that they can tolerate higher coupling frequencies without loss of binding activity than can antibodies.

Lectins also possess the unusual property of being retained for prolonged periods at the site of injection in a variety of tissues.[4] ^{125}I-Labeled lectins differ in the period for retention in mouse footpads—from as low as 4-fold longer (for 90% loss) than any typical globular protein tested (in the case of ^{125}I-labeled soybean agglutinin) to as high as 158-fold longer (in the case of ^{125}I-labeled *Ulex europaeus* agglutinin). Prolonged retention at the site of injection is assumed to be related to the observations of Edelson and Cohn[5] and Goldman and Raz[6] that lectins are interiorized inside vesicles in the cytoplasm of cells. The presence of the lectin somehow modifies the vesicles so that fusion with lysosomes is inhibited. As a result, the lectin persists for extended periods within the cells although not in actual contact with the cytoplasm. When ^{125}I-labeled Con A is broken down in cultured cells, it is released as fully hydrolyzed fragments (iodotyrosine), a finding consistent with the rate of degradation being determined by the rate of fusion of lectin-containing vesicles with lysosomes.[7] As discussed below, this property may allow medical applications involving slow release of drugs.

In addition to lower specificity, lectins have an additional disadvantage for biomedical applications relative to monoclonal antibodies in that

[5] P. J. Edelson and Z. A. Cohn, *J. Exp. Med.* **140,** 1364 (1974).
[6] R. Goldman and A. Raz, *Exp. Cell Res.* **96,** 373 (1975).
[7] W. T. Shier, *in* "Drug Carriers in Biology and Medicine" (G. Gregoriadis, ed.), p. 43. Academic Press, New York, 1979.

many of them are toxic and are potent inducers of inflammation.[8] Consequently, in some cases the toxicity of the lectin–drug conjugate may exceed the toxicity of the drug alone. While this is not a problem for cytotoxic drugs administered by intranodal injection,[9] it does limit most applications of lectins as drug carriers.

Materials

Concanavalin A (twice crystallized in saturated sodium chloride) was obtained from Miles Laboratories. Crystalline bovine trypsin, glutaraldehyde, p-tosyl-L-arginine methyl ester, azocasein, cyanogen bromide, and α-methyl-D-mannoside were obtained from Sigma. Sephadex G-200 and Sepharose 4B were obtained from Pharmacia. BioGel P-2 resin was obtained from Bio-Rad Laboratories. Sheep red blood cells were obtained in Alsevier's solution from Flow Laboratories. Benzamidine hydrochloride, p-aminobenzamidine dihydrochloride, N-benzyloxycarbonyl-6-amidocaproic acid, and 1-cyclohexyl-3-(2-morpholinoethyl)carbodiimide metho-p-toluenesulfonate were purchased from Aldrich.

Methods

Assays

Trypsin. During preparation and purification of the conjugate, trypsin activity was determined spectrophotometrically in a Gilford recording spectrophotometer using tosyl-L-arginine methyl ester (TAME) as substrate and crystalline bovine trypsin as standard. The assay procedure is essentially that given in "Worthington Enzyme Manual."[10] Trypsin activity was monitored by the change in absorbance at 247 nm using 1 mM tosyl-L-arginine methyl ester, 40 mM Tris-HCl buffer, pH 8.1, and 3 to 7 μg of crystalline trypsin or equivalent in 1-ml cuvettes. The amount of substrate hydrolyzed was determined by monitoring the absorbance at 247 nm for 3 min. The amount of trypsin activity was estimated by interpolation from a standard curve.

Trypsin activity was also determined in the purified conjugate using azocasein (sulfanilamide–azocasein) as substrate and monitoring production of peptides soluble in 5% (w/v) aqueous trichloroacetic acid. Azocasein (0.5 ml of 1 mg/ml) in 40 mM Tris-HCl, pH 8.1, was incubated at 25°

---

[8] W. T. Shier, J. T. Trotter, and C. L. Reading, *Proc. Soc. Exp. Biol. Med.* **146,** 590 (1974).

[9] T. Kitao and K. Hattori, *Nature (London)* **265,** 81 (1977).

[10] L. A. Decker, ed., "Worthington Enzyme Manual," p. 221. Worthington Biochemical Corp., Freehold, New Jersey.

with 0.5 to 1.5 μg of crystalline trypsin or equivalent for 10 min, at which time the reactions were stopped by addition of 0.5 ml of 10% (w/v) aqueous trichloroacetic acid. The precipitate was sedimented in a clinical centrifuge, and the absorbance of the supernatant fluid measured at 340 nm. The amount of trypsin activity was estimated by interpolation from a standard curve.

Hemagglutination. Concanavalin A activity in preparations of the Con A–trypsin conjugate was estimated by hemagglutination assays in triplicate in a Takasty microtiter apparatus (Microtiter, Cooke Engineering Co., Alexandria, VA). Assays were carried out with 25-μl aliquots of phosphate-buffered saline (0.14 M NaCl, 0.01 M potassium phosphate, pH 7.4) per well as diluent for 2-fold dilutions of 25-μl test samples followed by 25 μl of a suspension of 2×10^8 sheep red blood cells per milliliter. Because proteolytic enzyme treatment increases the agglutinability of cells, the sheep red blood cells were pretreated with trypsin (0.05%) in phosphate-buffered saline for 30 min at 25°, then washed in the same buffer. The Con A content of Con A–trypsin preparations was estimated by comparison of serial dilution end points (i.e., maximum concentration that does not cause agglutination) with standard concentrations of unconjugated Con A.

Radiolabeled Lectin Retention Assay. A characteristic of lectins and lectin–enzyme conjugates which has been detected and studied in this laboratory[4] is prolonged retention at the site of injection in tissues. This property is readily measured for retention of ^{125}I-labeled lectins following injection in the footpads of mice. The amount of radiolabel retained at the site of injection can be determined at various times by sacrificing the animal and excising and counting the foot in a gamma counter. Taking samples at frequent intervals from a group of mice treated in parallel allows the estimation of the time for 90% elimination from the mouse foot by interpolation of plots of radioactivity retained versus time after injection. By this method the time for 90% elimination of 25 μg of ^{125}I-labeled trypsin was estimated to be 2.56 ± 0.13 hr. When the amount of sample is limited, prolonged retention of a lectin or lectin–enzyme conjugate can be demonstrated using a single time point, typically at 24 or 48 hr. However, the relative increase in retention time cannot be estimated from a single point because the time courses of elimination appear to be complex functions at later time periods.[4]

Preparation of a Con A–Trypsin

Coupling Reaction. All operations were carried out at 0–4°. Con A (100 mg) in 2 ml of saturated NaCl was diluted to 10 ml with 0.5 M

NaCl buffered with 0.2 M Tris-acetate, pH 5 (TAS), containing crystalline bovine trypsin (100 mg) and 10 mM benzamidine hydrochloride to prevent trypsin from attacking the Con A during the coupling reaction. Cold 25% aqueous glutaraldehyde (0.4 ml) was slowly added with efficient stirring and the mixture allowed to stand at 0° for 90 min. The coupling reaction was terminated by applying the reaction mixture directly to the Con A affinity column (100 ml of Sephadex G-200), washing with 10 ml of TAS, 20 ml of 1% sodium metabisulfite, and an additional 200 ml of TAS.

During the development of the coupling technique, a variety of coupling agents were tested. Several (e.g., water-soluble carbodiimides) were effective coupling agents, but they resulted in severe loss of enzyme activity. Only glutaraldehyde served as an effective coupling agent without significantly inactivating any of the enzymes being studied. This presumably reflects the oligomeric or polymeric nature of the actual reactive coupling species with glutaraldehyde, which tends to keep the reactive species out of the catalytic region of the enzymes.

Preparation of Affinity Columns

Concanavalin Affinity Column. Affinity purification on Con A-containing species was carried out on columns of swelled Sephadex G-200 in TAS. The column was washed with 2 bed volumes of TAS buffer, then eluted with 1.5 bed volumes of 0.2 M α-methylmannoside in TAS.

Trypsin Affinity Column. Trypsin-containing species were purified on an affinity column prepared essentially by the method of Sampaio et al.,[11] in which p-(ε-aminocaproylamido)benzamidine was coupled to cyanogen bromide-activated Sepharose 4B. The ligand was prepared in two steps: p-Aminobenzamidine · 2HCl (5 g), N-benzyloxycarbonyl-6-amidocaproic acid (6.65 g), and 1-cyclohexyl-3-(2-morpholinoethyl)carbodiimide metho-p-toluenesulfonate (11.65 g) were dissolved in a mixture of pyridine (10 ml) and acetonitrile (150 ml) and stirred overnight. The solvent was evaporated and the residue dissolved in ethyl acetate : butanol (1 : 1). The solution was extracted with 1 N hydrochloric acid, water, 5% aqueous $NaHCO_3$, and water. The organic layer was dried over anhydrous sodium sulfate, concentrated under reduced pressure, and triturated with ether to produce 6 g of product. The benzyloxycarbonyl-protecting group was removed by heating at 90–95° for 30 min in 40 ml of trifluoroacetic acid. The trifluoroacetic acid was evaporated under reduced pressure, and the residue dissolved in an excess of methanol saturated with anhydrous HCl. The mixture was evaporated under reduced pressure; the residue was washed with ether and crystallized by dissolving in a mini-

[11] C. Sampaio, S. C. Wong, and E. Shaw, *Arch. Biochem. Biophys.* **165,** 133 (1974).

mum of methanol and bringing the mixture to the incipient cloud point by addition of ether. This procedure yields 1.9 g of p-(ε-aminocaproyl-amido)benzamidine · 2HCl, m.p. 284° (lit.[11] 284–285.5°).

Sepharose 4B (100 ml) was activated with cyanogen bromide according to the method of March et al.,[12] suspended in 70 ml of 0.1 M NaHCO$_3$ buffer, pH 10, and mixed with 35 ml of an aqueous solution containing 1 g of p-(ε-aminocaproylamido)benzamidine adjusted to pH 10 with aqueous NaOH. The mixture was maintained at 4° with occasional remixing for 24 hr, washed with water on a sintered glass funnel, and stored until used in 1 M NaCl containing 0.02% NaN$_3$.

Purification

The coupling reaction produces a random mixture of Con A–trypsin conjugates (potentially with more than one Con A and/or trypsin moiety) plus Con A, aggregated Con A, trypsin, aggregated trypsin, and unreacted glutaraldehyde oligomers. The initial purification step employed a Con A affinity column (Sephadex G-200) to retain only functional Con A-containing species. After the column was washed, the bound species were eluted with 150 ml of 0.2 M α-methyl-D-mannoside and the eluate passed onto a coupled trypsin affinity column consisting of 2 ml of p-aminobenzamidine coupled to Sepharose 4B as described. After the sample was applied, the columns were uncoupled and the trypsin affinity column washed with 20 ml of TAS to ensure removal of any species not containing a trypsin moiety with a functional benzamidine-binding site. The column was then eluted with 5 ml of 1 M benzamidine hydrochloride in TAS, followed by 2 ml of TAS. The preparations were stored in this form frozen in 1-ml aliquots. The aliquots were individually freed of benzamidine immediately before use. Removal of benzamidine was accomplished by applying the 1-ml aliquot to a column of BioGel P-2 resin (120 × 0.9 cm) in 0.15 M NaCl containing 0.02 M Tris acetate, pH 5.0, and eluting in the same buffer. The fractions in the excluded volume were combined and assayed for protein by the method of Lowry.[13]

Characterization

The procedure described above typically yields Con A–trypsin preparations containing 5–9 mg of total protein. The trypsin content determined by the esterase activity with tosyl-L-arginine methyl ester is 2–4 mg, or

---

[12] S. C. March, I. Parikh, and P. Cuatrecasas, *Anal. Biochem.* **60**, 149 (1974).

[13] O. H. Lowry, N. J. Rosebrough, A. L. Farr, and R. J. Randall, *J. Biol. Chem.* **193**, 265 (1951).

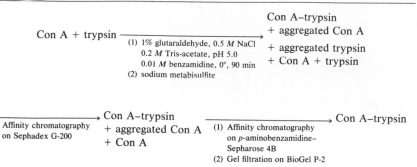

FIG. 1. Scheme for the preparation and purification of Con A–trypsin.

about 40% of the protein content. The Con A content determined by hemagglutination of trypsinized sheep red blood cells is typically 120–150% of protein content. The anomalously high Con A content presumably reflects the presence of (Con A)$_2$–trypsin or higher order species in the preparation. Aggregated lectins are more effective at agglutination of cells than those with normal valence or subnormal valence.[14]

Electrophoresis in polyacrylamide gels containing sodium dodecyl sulfate is of limited value in characterization of conjugates containing Con A because commercially available Con A contains a high percentage of fragmented molecules.[15] As a result, gel electrophoresis of crystalline Con A preparations contain multiple bands representing fragments from functional molecules. A method has been published[16] for preparing Con A free of fragmented molecules, but it was not practical in our hands. Electrophoresis of Con A–trypsin preparations in polyacrylamide gradient slab gels containing 10 to 15% polyacrylamide and 0.1% sodium dodecyl sulfate contained no band migrating with trypsin in parallel tracks, but they contained bands (1) migrating with free Con A subunit and (2) migrating at the expected molecular weight for a trypsin–Con A subunit conjugate. Free Con A subunits are expected to be present because the trypsin moiety in Con A–trypsin is expected to be bound to only one subunit of the Con A tetramer, thereby leaving the other three free.

The purification scheme (Fig. 1) and characterization of the conjugate described above demonstrate the presence of Con A in the conjugate by

[14] R. Lotan, H. Lis, A. Rosenwasser, A. Novogrodsky, and N. Sharon, *Biochem. Biophys. Res. Commun.* **55,** 1347 (1973).

[15] J. L. Wang, B. A. Cunningham, and G. M. Edelman, *Proc. Natl. Acad. Sci. U.S.A.* **68,** 1130 (1971).

[16] B. A. Cunningham, J. L. Wang, M. N. Pflumm, and G. M. Edelman, *Biochemistry* **11,** 3233 (1972).

affinity purification and by hemagglutination. The presence of Con A was demonstrated in two additional ways. The conjugate was prepared using trypsin labeled with ^{125}I by the method of McConahey and Dixon[17] and unlabeled Con A. The resulting conjugate containing 42,800 cpm/mg protein was used to demonstrate prolonged retention of the conjugated trypsin in mouse footpads. At 48 hr after injection of 25 μg of ^{125}I-labeled trypsin free or of an equal amount conjugated to Con A, the extent of retention was $0.38 \pm 0.01\%$ for free trypsin and $6.27 \pm 0.55\%$ for trypsin bound to Con A. The presence of Con A in the conjugate was also indicated by the demonstration that the conjugate is mitogenic with mouse T lymphocytes (*vide infra*).

The purification scheme and characterization procedures described above demonstrate the presence of trypsin in the conjugate by affinity chromatography and by the esterase activity characteristic of trypsin with tosyl-L-arginine methyl ester as substrate. Con A–trypsin was also assayed for proteolytic activity with a macromolecular substrate azocasein. The coupling reaction that forms the conjugate may occur such that the active site of the enzyme is partially blocked by the lectin. This can prevent the enzyme from utilizing macromolecular substrates such as azocasein but not low molecular weight substrates such as tosyl-L-arginine methyl ester. A comparison of proteolytic activities with these two substrates indicated that 50–80% of the proteolytic activity observed with the small substrate is retained with macromolecular substrates.

The presence of trypsin in the conjugate was also indicated by the demonstration of mitogenicity of the conjugate with B lymphocytes and fibroblasts (*vide infra*).

Results

The use of lectins as carriers of drugs or enzymes has been limited. Con A–L-asparaginase has been prepared in this laboratory and tested for effectiveness against an L-asparaginase-sensitive tumor.[4] The conjugate was totally ineffective even at dose levels of conjugated L-asparaginase activity sufficient to cause complete regression of the tumor if administered in the unconjugated form. The results are consistent with the Con A–L-asparaginase being interiorized in vesicles where it was ineffective at removing L-asparagine from the protein synthesis apparatus of the tumor cells. More success has been reported by Kitao and Hattori,[9] who coupled the cytotoxic drug daunomycin to Con A. They observed that the conjugate was more effective than free daunomycin at protecting mice from L1210 leukemia cells or Ehrlich ascites carcinoma cells.

[17] P. J. McConahey and F. A. Dixon, *Int. Arch. Allergy Appl. Immunol.* **29,** 185 (1966).

Con A–trypsin was prepared in this laboratory to provide an experimental tool to investigate interrelationships in the modes of action of various mitogens. Insulin, proteolytic enzymes, and Con A all exert mitogenic activity with selected cell types, and the results of a variety of studies suggest possible interrelationships between their mechanisms of action in some cell types. The following are some examples: (1) mitogenic stimulation of lymphocytes induces the appearance of insulin receptors[18]; (2) Con A competitively prevents the binding of insulin to insulin receptors, a finding indicating that Con A interacts directly with the insulin receptor[19]; (3) trypsin has insulin-like activity *in vivo*[20] and *in vitro;* (4) on some cells, insulin binding sites are exquisitely sensitive to proteolytic action[21]; and (5) trypsin induces the appearance of lectin receptor sites on the surface of fibroblasts and other cell types.[22]

The rationale behind this approach was that the Con A–trypsin conjugate retains within a single molecular species various activities that are mitogenic in different cell systems, and a comparison of its mitogenic activities with insulin and with the individual components of the conjugate, Con A and trypsin, might reveal previously undetected interrelationships between the various mitogenic activities. Mitogenic activities were compared in cultures of murine B lymphocytes (which respond to mitogenic stimulation by trypsin[23]) and T lymphocytes (which respond to mitogenic stimulation by Con A[23]), and rat embryo fibroblasts and 3T3-4a mouse fibroblasts (which respond to mitogenic stimulation by insulin and trypsin). Studies on B and T lymphocytes indicated that the Con A–trypsin conjugate retains within a single molecular species mitogenic activity for B and T lymphocytes. The results of studies with fibroblasts supported the following conclusions: (1) coupling trypsin to Con A lowers the optimal concentration for stimulation of DNA synthesis in susceptible cells, consistent with the concept that the concanavalin A moiety functions to concentrate the conjugated trypsin at the cell surface; (2) optimal mitogenic stimulation of fibroblasts with Con A–trypsin occurred at concentrations lower than the concentrations which cause detachment from the substratum, a finding consistent with the density of Con A receptors on 3T3-4a cells being higher near the sites for trypsin-induced mitogenesis than near the sites for trypsin-induced detachment from the substratum; and (3) the mitogenic activities of insulin at a saturating concentration and

[18] U. Krug, F. Krug, and P. Cuatrecasas, *Proc. Natl. Acad. Sci. U.S.A.* **69,** 2604 (1972).
[19] P. Cuatrecasas and G. P. E. Tell, *Proc. Natl. Acad. Sci. U.S.A.* **70,** 485 (1973).
[20] P. Rieser, *Am. J. Med.* **40,** 759 (1966).
[21] P. Cuatrecasas, *J. Biol. Chem.* **246,** 6522 (1971).
[22] M. Burger, *Proc. Natl. Acad. Sci. U.S.A.* **62,** 994 (1969).
[23] J. Watson, R. Epstein, I. Nakoinz, and P. Ralph, *J. Immunol.* **110,** 43 (1973).

of trypsin are additive across the mitogenic concentration range of trypsin activity in the form of free trypsin or Con A–trypsin, a finding consistent with the concept that trypsin and insulin stimulate mitosis by independent mechanisms.

Lectin–enzyme conjugates are potentially useful in situations in which the lectin concentrates the enzyme at its site of action and prevents its loss by dilution or flow of fluids. An example of this approach is given by the work of Barker et al.,[24] who prepared a conjugate of Con A and dextranase for evaluation as an agent to remove dental plaque. The rationale was that the Con A would bind to glucosyl moieties in the dextran component of dental plaque, thereby concentrating the conjugated dextranase at its intended site of action and preventing loss of the enzyme in the flow of saliva. Studies on the effectiveness of this conjugate have not been reported. Because proteolytic enzymes such as trypsin are also effective at dissolving plaque,[25] the Con A–trypsin conjugate was tested for ability to dissolve plaque in vitro. The conjugate was not more effective than free trypsin (W. T. Shier, unpublished results).

Discussion

Lectins as carriers of enzymes and drugs have several advantages over monoclonal antibodies. These include (1) the commercial availability of most lectins in greater quantities and at lower cost; (2) the fact that both the lectin and conjugates containing it are readily purified on readily available, reusable affinity columns; and (3) prolonged retention of lectins at the site of injection in tissues. These advantages can be expected to lead to applications of lectins as carriers in situations in which the principal disadvantages of lectins (i.e., lower specificity, higher toxicity) are not of major importance.

An example is provided by commercial biotechnology applications involving reversibly immobilized enzymes. Use of enzyme technology is often made economically feasible by immobilizing enzymes and passing the substrate over them. This approach permits reuse of the enzyme and eliminates or makes trivial removal of enzyme from the product. Immobilized enzyme technology cannot be used with insoluble substrates because of the requirement for diffusion to bring the substrate and the active site of the enzyme together. Many of the most promising industrial enzymes (e.g., pulanase, other proteases, cellulase) are limited in this manner. Lectin–enzyme conjugates offer an approach to circumventing this

[24] S. A. Barker, A. G. Giblin, and C. J. Gray, Carbohydr. Res. 36, 23 (1974).
[25] E. A. Sweeney and J. H. Shaw, J. Dent. Res. 44, 973 (1965).

problem in the form of reversibly immobilized enzymes, which can be recovered after digestion of a substrate by filtration through an affinity packing. This is technically not much more difficult than removing a bead-immobilized enzyme, provided a synthetic affinity material is available on a noncompressible support to provide high filtration speeds and extensive reusability of the affinity packing. In contrast, monoclonal antibodies usually require an immobilized protein for affinity purification; affinity packings made with immobilized proteins usually cannot be extensively reused.

The property of prolonged retention of lectins at the site of injection in tissues has potential applications for controlled release of drugs, although the toxicity and immunogenicity generally associated with lectins have limited its potential application in this area. The technique is potentially applicable to drugs which can be attached to a lectin but are not inactivated by lysosomal enzymes. The problem of not being able to attach many drug molecules directly to a lectin molecule could be addressed by attaching drugs to polycationic carrier molecules which would subsequently be attached to the lectin. Lectins of low toxicity may be identified and their tendency to be rapidly interiorized in tissues may prevent them from inducing anaphylaxis, which constitutes the greatest danger from hypersensitivity to a pharmaceutical agent.

Acknowledgments

This work was supported in part by Grant No. CA16123 from the National Cancer Institute, United States Public Health Service. Some of the experiments described here were carried out in the Cell Biology Laboratory, Salk Institute, San Diego, California. Studies on mitogenesis in rat embryo fibroblasts and 3T3-4a fibroblasts were carried out by Drs. H. J. Ristow and D. Paul. Studies on mitogenesis in murine B and T lymphocytes were carried out by Dr. J. D. Watson. Studies on plaque dissolution were carried out by Dr. D. L. Williams. I thank these investigators for their contributions.

[20] Hormone–Drug Conjugates

By J. M. VARGA

The impetus for the preparation of hormone–drug conjugates was the finding that biologically active peptides such as melanotropin (MSH),[1] epidermal growth factor (EGF),[2] and insulin[3] are internalized by their target cells. It was hoped that the peptide would act as a discriminator for the drug and make it "cell specific." In addition, the receptor-mediated endocytosis of the hormone would be harnessed so that the drug would be delivered directly into the cytoplasm or the nucleus where it would exert its toxic effects. It was assumed that the properties of a conjugate would be the simple sum of the properties of its components. While the results obtained with some conjugates have satisfied our expectations, others have not. In this chapter a few examples of conjugates of hormones with daunomycin and ouabain are discussed. Chemically, this group of molecules consists of peptides substituted with polycyclic compounds. Toxic derivatives of steroid hormones[4] and conjugates of hormones with the catalytic[5] or binding[6] subunits of lectins or bacterial toxins are beyond the scope of this chapter.

Melanotropin–Daunomycin

We prepared a conjugate of melanotropin and daunomycin[7] using a procedure developed by Hurwitz et al.[8] for the covalent binding of daunomycin to antibodies. The free amino groups of porcine βMSH (N-terminal Asp and the ε-amino groups of Lys-6 and Lys-17) were substituted with an oxidized derivative of daunomycin as shown in Fig. 1.

Procedure. Daunomycin (10 mg, 19 μmol, Sigma Co.) was dissolved in 0.75 ml phosphate-buffered saline, pH 7.4 (PBS). Sodium periodate

[1] J. M. Varga, G. Moellmann, P. Fritsch, E. Godawska, and A. B. Lerner, *Proc. Natl. Acad. Sci. U.S.A.* **73**, 559 (1976).

[2] G. Carpenter and S. Cohen, *J. Cell Biol.* **71**, 159 (1976).

[3] I. D. Goldfine, G. J. Smith, K. Y. Wong, and A. L. Jones, *Proc. Natl. Acad. Sci. U.S.A.* **74**, 1368 (1977).

[4] M. E. Wall, G. S. Abernathy, F. I. Carroll, D. J. Taylor, *J. Med. Chem.* **12**, 810 (1969).

[5] T-M. Chang, A. Dazord, and D. M. Neville, Jr., *J. Biol. Chem.* **252**, 1515 (1977).

[6] R. A. Roth, B. A. Maddux, K. Y. Wong, Y. Iwamoto, and I. D. Goldfine, *J. Biol. Chem.* **256**, 5350 (1981).

[7] J. M. Varga, N. Asato, S. Lande, and A. B. Lerner, *Nature (London)* **267**, 56 (1977).

[8] E. Hurwitz, R. Levy, R. Maron, M. Wilchek, R. Arnon, and M. Sela, *Cancer Res.* **35**, 1175 (1975).

METHODS IN ENZYMOLOGY, VOL. 112

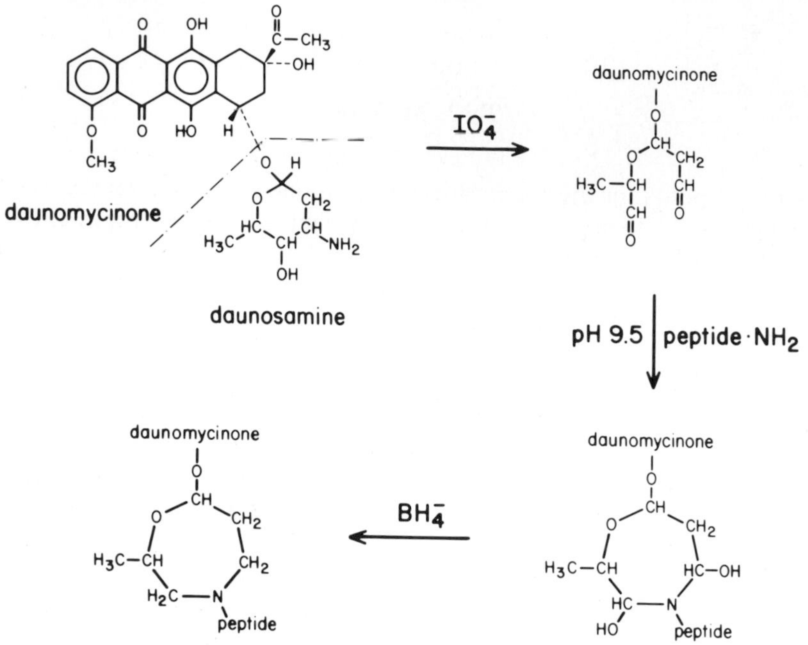

FIG. 1. Synthesis of β-melanotropin–daunomycin.

(5.8 mg, 27 μmol) was added, the pH was adjusted to 9.5 with 1 M NaOH, and the reaction mixture was incubated in the dark for 1 hr at room temperature. Glycerol was added to give a final concentration of 50 mM to consume excess periodate. HPLC-purified porcine β-melanotropin (15 mg, 8 μmol) containing 1 μCi of [125I]-labeled βMSH[9] was added in 0.5 ml of 0.15 M K$_2$CO$_3$ (pH 9.5). The mixture was incubated at 22° for 1 hr. The Schiff base was reduced with NaBH$_4$ (0.3 mg/ml) at 37° for 2 hr. The reaction mixture was applied to a Sephadex G-25 (90 × 1 cm) column and eluted with PBS. The appearance of red color in the void volume coincident with the major peak of [125I]βMSH indicated the presence of the peptide–toxin conjugate. The molar ratio of daunomycin to MSH was calculated to be 2.2 to 1. The conjugate was freeze-dried and hydrolyzed with 6 N HCl at 110° for 18 hr, and the amino acid composition was determined in parallel with the analysis of the hydrolysate of a mixture of daunomycin and unconjugated MSH. Approximately 50% of the lysine and aspartic acid residues were not recoverable in the conjugate, a finding

[9] D. T. Lambert, C. Stachelek, J. M. Varga, and A. B. Lerner, *J. Biol. Chem.* **257**, 8211 (1982).

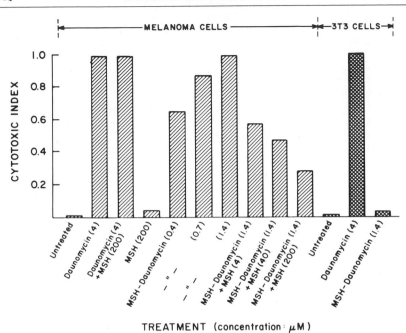

TREATMENT (concentration: μM)

FIG. 2. Cytotoxicity of melanotropin–daunomycin. Melanoma and 3T3 cells were cultured in the presence of substances indicated in the figure. Cytotoxic index[7] was determined after 3 days of cultivation. (From ref. 10, reproduced with permission.)

which indicated that the oxidized product of daunomycin (D) had reacted with the N-terminal amino group of aspartic acid and the ε-aminolysyl groups (6 and 17) of MSH. Therefore, the following general formula was suggested for the composition of the conjugate: (D)-Asp-Glu-Gly-Pro-Tyr-Lys(-D)-Met-Glu-His-Phe-Arg-Trp-Gly-Ser-Pro-Pro-Lys(-D)-Asp. We have attempted to cleave the conjugate at the arginine residue in order to delineate the position of the substitution of lysines by daunomycin. Although the unsubstituted peptide is susceptible to proteolytic cleavage in two positions, between Lys and Met and between Arg and Trp, we have not been able to produce proteolytic fragments from the conjugate with trypsin or chymotrypsin.[7]

The MSH–daunomycin conjugate was more toxic to murine melanoma cells than free daunomycin but was not toxic to 3T3 cells. Melanoma cells could be protected from the toxicity of the conjugate by the addition of a large excess of melanotropin[10] (Fig. 2). When cells were

[10] J. M. Varga and N. Asato, in "Targeted Drugs" (E. P. Goldberg, ed.), p. 73. Wiley, New York, 1983.

PROTECTION OF MELANOMA CELLS
BY LYSOSOMOTROPHIC AMINES

Treatment[a]	Degree of protection (%)[b]
DM	0
DM + methylamine, 0.1 mM	60
DM + NH$_4$ acetate, 1 mM	84
DM + chloroquine, 0.1 M	62
DM + dansylcadaverine, 0.1 mM	37
DM + cystamine, 10 μM	20
DM + procaine, 1 mM	14
DM + bacitracin, 1 mM	8

[a] Melanoma cells were treated for 1 day with daunomycin[c] or melanotropin–daunomycin (DM, 1.5 μM) in the presence of the lysosomotropic amines listed. The amines listed in the table had no protective effect when free daunomycin was used as the cytotoxic agent.

[b] Degree of protection = $[(CI_{toxin} - CI_{toxin+amine})/CI_{toxin}] \times 100$. Cytotoxic index (CI) was determined as described in ref. 7.

exposed to daunomycin or melanotropin–daunomycin at 37°, the red fluorescence of daunomycin could be seen in the nuclei of melanoma cells by fluorescence microscopy. No nuclear localization of the drug could be seen in 3T3 cells.[7] Subsequent studies have shown that melanoma cells can be protected from the toxicity of the MSH–daunomycin conjugate by lysosomotropic amines such as methylamine, chloroquine, and ammonium acetate.[11] The amines, added in the same concentration, had no protective effect on the toxicity of free daunomycin (see table). These findings suggest that the conjugate has to be processed in the lysosomes in order to be toxic to melanoma cells. Indirect evidence for this suggestion was found by Wiesehahn et al.[12] who showed that daunomycin, when bound to MSH, does not bind to DNA. Accordingly, it is possible that the drug is split off from the peptide or that the conjugate is metabolized in the lysosomes before an active form of the drug reaches the nucleus.[11] It is also possible that the conjugate kills the cells by a different mechanism, for example, by acting within the cell membrane (see Comments).

The time dependence of cytotoxicity of daunomycin and melanotropin–daunomycin were compared, and the conjugate was found to act more slowly than the free toxin.[10] Since receptors for MSH are expressed

[11] J. M. Varga, N. Asato, G. Wiesehahn, and J. E. Hearst, in "Phenotypic Expression in Pigment Cells" (M. Seiji, ed.), p. 399. Univ. of Tokyo Press, Tokyo, 1981.
[12] G. Weisehahn, J. M. Varga, and J. E. Hearst, Nature (London) 292, 467 (1981).

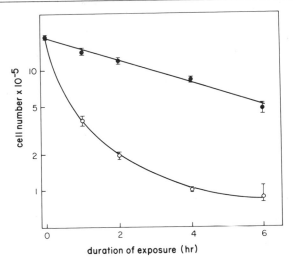

FIG. 3. Time dependence of cytotoxicity. Cells were exposed for a variable length of time to 1 μM daunomycin (○) or to melanotropin–daunomycin (●), and the number of viable cells was measured[7] after 2 days of cultivation. (From ref. 10, reproduced with permission.)

predominantly in the G_2 phase of the cell cycle,[13] cells must pass through the receptor-positive phase in order to be sensitive to the conjugate. The linear function, obtained when the number of surviving cells is plotted on a semilogarithmic scale (Fig. 3), is anticipated if killing of cells occurs in a cell cycle-dependent manner by melanotropin–daunomycin because of the following reason: the transition probability model of the cell cycle[14] assumes that the probability of transition from an indeterminate A state $(G_0 + G_1)$ to B phase $(S + G_2 + M)$ is constant (k). Thus, the probability of cells leaving A state can be described by the following equation: $dA/dt = kc^{-kt}$ where c is the number of cells in A state at 0 time. Therefore, if the cells surviving the treatment with melanotropin–daunomycin are the ones that remain in A phase $(G_0 + G_1)$, a linear function is expected if the logarithm of surviving cells is plotted against the time (t) of treatment. In comparison, a similar curve, obtained from an experiment when cells were treated with free daunomycin, is not linear (Fig. 3). In the presence of melanotropin–daunomycin, cells with large nuclei have accumulated.[7] The median cell volume in cultures treated with melanotropin–daunomycin was significantly larger than in cultures incubated with free daunomycin (Fig. 4).

13 J. M. Varga and P. Fritsch, in "Endocrinology" (V. H. I. James, ed.), Vol. 1, p. 492.
14 J. A. Smith and L. Martin, Proc. Natl. Acad. Sci. U.S.A. 70, 1263 (1973).

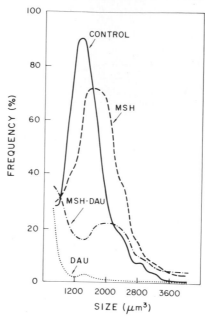

FIG. 4. Size of cells, treated with MSH, daunomycin, and MSH–daunomycin. Cells were grown without addition (control) or in the presence of 1 μM of MSH or daunomycin or MSH–daunomycin. The size of the cells was determined with a Coulter Channelyzer after 2 days of cultivation.

High-performance liquid chromatography (HPLC) has been exploited for the separation of amino-substituted derivatives of MSH.[15] Since βMSH has three free amino groups, three mono-, three bi-, and one trisubstituted conjugates are expected to be present after the reaction of oxidized daunomycin with MSH. When the conditions of reaction are as described above, several minor substituted peptides can be detected by HPLC (Fig. 5). Chemical identification of the purified derivatives and characterization of their hormonal and toxic properties are in progress.

Melanotropin–Ouabain

We have prepared a conjugate of melanotropin–ouabain[15] using a procedure similar to that used for the preparation of melanotropin–daunomycin. The rhamnose ring of ouabain was cleaved by periodate oxidation as described by Smith[16] and the dialdehyde was condensed with βMSH (Fig. 6).

[15] J. M. Varga, L. Airoldi, and G. Davila-Huerta, in "New Approaches in Liquid Chromatography" (H. Kalasz, ed.), p. 3. Elsevier, Amsterdam, 1984.
[16] T. W. Smith, J. Clin. Invest. **51**, 1583 (1972).

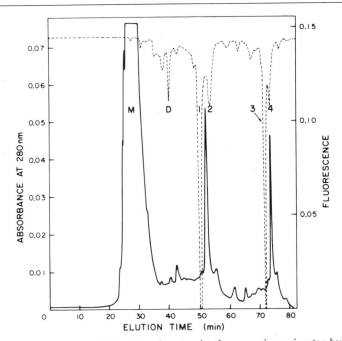

FIG. 5. Separation of individual β-melanotropin–daunomycin conjugates by reversed-phase HPLC. The liquid chromatograph consisted of a Model 420 Beckman solvent programmer (Berkeley, CA), two Model 110A Beckman pumps, a Beckman Model 210 sampling valve with a 100-μl sample loop, a Model 153 Beckman UV detector, and a Perkin-Elmer (Norwalk, CT) Model 650 LC fluorescence spectrophotometer. Data were recorded with a Kipp & Zonen BD 41 (Delft, Holland) dual channel recorder. The column effluent was monitored at 280 nm; fluorescence detection was carried out at 470 and 590 nm for excitation and emission, respectively. We used Waters (Milford, MA) μBondapak C_{18} (3.9 mm I.D. × 300 mm) columns. Gradient elution: Solvent A, aqueous solution of 0.05% trifluoroacetic acid (TF); Solvent B, 0.05% TFA in acetonitrile. The program for the gradient was 10 to 20% B in A in 5 min, from 20 to 30% B in A in 15 min, from 30 to 40% B in A in 35 min, from 40 to 60% B in A in 5 min. Continuous line shows OD_{280} nm; dashed line, fluorescence. Peaks: M, β-melanotropin derivatives, containing no daunomycin; D, oxidized daunomycin–ethanolamine derivatives; 2–4, anticipated β-melanotropin–daunomycin derivatives.

Procedure. To a solution of 10 mg ouabain (Sigma), containing 5 μCi of [³H]ouabain (New England Nuclear), 5 mg $NaIO_4$ was added in 0.5 ml PBS. The reaction was allowed to proceed at room temperature for 1 hr. Unreacted periodate was neutralized by addition of glycerol to a final concentration of 50 mM. Fifteen milligrams βMSH in 0.5 ml of K_2CO_3, pH 9.5, was added. After reaction for 1 hr at room temperature, the Schiff base was reduced by addition of 0.5 mg $NaBH_4$ in 0.1 ml water. The reaction mixture was kept at 37° for 2 hr, then passed through a BioGel

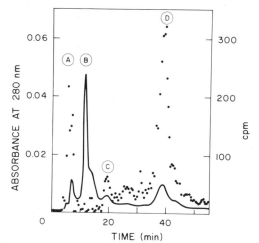

FIG. 6. Synthesis of β-melanotropin–ouabain.

FIG. 7. Separation of β-melanotropin–ouabain by reversed-phase HPLC. The same system was used as described in the legend for Fig. 2. Peak A, polymerized product of oxidized ouabain; peak B, native β-MSH; peaks C and D, βMSH–ouabain conjugates. Peak D was collected, freeze-dried, and analyzed for its ability to stimulate tyrosinase and its cytotoxicity. Continuous line, absorbance; points, cpm. (From ref. 15, reproduced with permission.)

P-2 column (20 × 0.8 cm) which had been equilibrated with water. The radioactive fraction appearing in the void volume was freeze-dried and then redissolved in water. Aliquots were injected into a Waters HPLC System operated in the isocratic mode with 24% acetonitrile in 0.1 M $NH_4H_2PO_4$, pH 8. Radioactivity was measured in a scintillation counter. The separation of the conjugate is shown in Fig. 7.

The ability of the major melanotropin–ouabain conjugate to stimulate the activity of tyrosinase in melanoma cells was 60% that of the unsubstituted peptide when measured at 0.1 μM concentration. Surprisingly, while free ouabain (1 μM) was highly toxic to murine melanoma cells, the conjugate had a potent growth stimulatory effect during the early phases of cell growth. This result suggests that receptors for MSH and the channels of the Na^+/K^+ pump may be in close proximity within the plasma membrane and that the two systems may interact.

Thyrotropin–Daunomycin

A conjugate of thyrotropin and daunomycin was prepared by Kaneko[17] using the method described for the preparation of MSH–daunomycin. Rabbit thyroid cells could be protected from the toxicity of the conjugate by a large excess of thyrotropin.[17]

Comments

The molecular basis for the use of peptide hormones as discriminators for delivery of drugs is their specific interaction with cell surface receptors. Since substitution of the peptide hormone with drugs may decrease its energy of binding to receptors, it is not surprising that in order to be toxic, the conjugate has to be present in higher concentrations than would be expected from the association constant of hormone binding. Melanotropin added in nanomolar concentrations causes half-maximal stimulation of tyrosinase in melanoma cells.[9] The dissociation constant of the binding of MSH to its receptor is also in this concentration range.[18] By contrast, melanotropin–daunomycin has to be added at concentrations two orders of magnitude higher to achieve 50% killing of cells.[7]

The hormone–drug conjugates used in earlier experiments were not chemically homogeneous but were mixtures of mono- and multisubstituted peptides and a variety of positional analogs.[7] With the availability of high resolution methods of separation, such as HPLC, homogeneous substituted peptides can now be obtained. When MSH is substituted with

[17] Y. Kaneko, *Horm. Metab. Res.* **13**, 110 (1981).
[18] A. DiPasquale, J. McGuire, and J. M. Varga, *Proc. Natl. Acad. Sci. U.S.A.* **74**, 601 (1977).

amino-reactive agents such as FITC or SPDP, the biological activity of the substituted peptide depends mostly on the degree and much less on the position of the substitution. For instance, mono-, bi, and tri-SPDP substituted MSH retain approximately 1/10, 1/100, and 1/1000 of the original biological activity of the hormone, respectively, while the activities of the three monosubstituted analogs are nearly identical.[19] The requirement for homogeneity of glycoprotein–drug conjugates[17] poses a more complex problem since a great number of free amino or carboxyl groups are available for substitution on the surface of these molecules. Clearly, sparse substitution of the protein cannot be a solution because at lower degrees of substitution, the presence of unsubstituted and also multisubstituted derivatives may compete for binding to receptors by the active molecular species. In this class of conjugates, specific sites must be identified on the molecule, perhaps in the carbohydrate moieties, before chemically defined, substituted hormones can be prepared.

So far our primary concern has been with the specific interaction of peptide hormones with receptors on the cell surface. However, recent studies of hormone action suggest that peptide hormones can be "recognized" not only by the receptors on the cell surface but also by additional cellular components. Interaction of peptide hormones with calmodulin[20] and other cytoplasmic[21] as well as nuclear components[22] has been suggested. It is possible that substitution of the hormone by a drug may alter binding to different cellular components in different ways. For example, substitution in a certain position might not alter binding to receptors on the cell surface but the same derivative may act as an antagonist of the hormone at some point within the cell and exert its toxic effect there. Because of this possibility, protection of the cell from the toxicity of the conjugate by a large excess of free hormone may be unrelated to competition of the hormone for receptors on the cell surface. Therefore, protection studies alone do not provide definitive proof for "receptor-mediated" drug targeting. Moreover, many or most of the peptide hormones are "recognized" and fragmented by specific peptidases present in the lysosomes.[23] Substitution of the hormone with a drug may prevent fragmentation of the conjugate which, in turn, may cancel its biological effect. β-Melanotropin contains two peptide linkages that are susceptible to proteolysis (see above). Presumably, biologically active fragments of in-

---

[19] G. Davila-Huerta and J. M. Varga, in "Chromatography, the State of Art" (H. Kalasz and L. S. Ettre, eds.). Elsevier, Amsterdam, 1984 (in press).

[20] D. A. Malencik and S. R. Anderson, *Biochemistry* **22,** 1995 (1983).

[21] B. I. Posner, Z. Josefsberg, and J. M. Bergeron, *J. Biol. Chem.* **254,** 12494 (1979).

[22] B. A. Yanker and E. M. Shooter, *Proc. Natl. Acad. Sci. U.S.A.* **76,** 1269 (1979).

[23] J. F. Caro and J. M. Amatruda, *J. Biol. Chem.* **255,** 10052 (1980).

ternalized MSH are generated by the splitting of one or both of the susceptible bonds. It is relevant to note that conjugates of MSH–daunomycin are resistant to proteolysis and they do not induce tyrosinase in melanoma cells.[7]

Anthracycline antibiotics, such as daunomycin, intercalate into DNA and inhibit transcription and replication.[24] Since toxicity of this drug has been equated with its ability to bind to DNA, it has been assumed that daunomycin has to be internalized to be toxic to target cells. Indeed, when daunomycin or MSH–daunomycin is added to melanoma cells in toxic concentrations, the red fluorescence of the anthracycline moiety can be detected in the nuclei of cells.[7] However, the coincidence of nuclear fluorescence and toxicity does not prove a cause-and-effect relationship. It is possible that the conjugate kills the cells by a different mechanism, for example, by acting in the cell membrane. Recent studies by Tritton and Yee[25] suggest that anthracycline antibiotics, bound to a solid support, can be cytotoxic without entering the cells. However, if daunomycin acts in the membrane of both melanoma and 3T3 cells, why should the conjugation of melanotropin to the drug cancel the toxic effect of daunomycin in only 3T3 cells?

Conjugates of drugs with peptide hormones are toxic derivatives that may be particularly useful for the selection of variant cells with defects in their receptors, in receptor-mediated endocytosis, or in as yet undefined intracellular processes that depend on hormonal stimulation. As modified drugs, the conjugates are novel pharmacological agents which may contribute to our understanding of the mechanism of drug action. Also, they may become important in the elucidation of basic rules of site-directed chemotherapy against certain forms of cancer.

Acknowledgments

These investigations were supported by Grant CA 26081 awarded by the National Cancer Institute, and by the National Foundation for Cancer Research. We thank Dr. Guadalupe Davila-Huerta for the separation of MSH–daunomycin derivatives by HPLC, Dr. Saul Lande for providing the βMSH, and Dr. Ann Korner for editing the manuscript.

[24] A. DiMarco and F. Arcamone, *Arzneim.-Forsch.* **25,** 368 (1975).
[25] T. R. Tritton and G. Yee, *Science* **217,** 248 (1982).

[21] Polylysine–Drug Conjugates

By LYLE J. ARNOLD, JR.

Introduction

In recent years numerous investigators have developed novel methods for reducing the toxic side effects associated with cancer chemotherapeutic agents and other potentially beneficial compounds. A number of these methods are disclosed in this volume. One method which has met with some success in the use of polymeric substances as carriers. Typically, the strategy has been to use a carrier which either releases the drug more slowly or "targets" the drug more selectively to the desired tissue.

One of the first polymeric supports used as a drug carrier involved complexes between the anthracyclines and DNA.[1] Subsequently, a large number of polymeric carriers have been evaluated. These fall into three basic categories: (1) drugs bound to carriers which are specific for membrane characteristics of target cells; these include surface charge, the fluidity of the membrane, surface carbohydrates, and transport properties;[1-12] (2) drugs attached to carriers which take advantage of the high rate of endocytosis in cancer cells;[1,10,13-24] (3) drugs that may be attached

[1] A. Trouet, D. Deprez-de Campeneere, and C. DeDuve, *Nature (London), New Biol.* **239**, 110 (1972).
[2] D. A. L. Davies and G. J. O'Neill, *Br. J. Cancer* **28**, Suppl. 1, 285 (1973).
[3] I. Fleschner, *Eur. J. Cancer* **9**, 741 (1973).
[4] T. Ghose, M. R. C. Path, and S. P. Nigam, *Cancer* **29**, 1398 (1972).
[5] T. Ghose, S. T. Norvell, A. Guclu, D. Cameron, A. Bodurtha, and A. S. MacDonald, *Br. Med. J.* **3**, 495 (1972).
[6] A. Guclu, T. Ghose, J. Tai, and M. Mammen, *Eur. J. Cancer* **12**, 95 (1976).
[7] F. L. Moolten and S. R. Cooperband, *Science* **169**, 68 (1970).
[8] T. Kitao and K. Hattori, *Nature (London)* **265**, 81 (1977).
[9] Y. Rahman, W. Kisielski, E. Buess, and E. Cerny, *Eur. J. Cancer* **11**, 883 (1975).
[10] A. Trouet, M. Masquelier, R. Baurain, and D. Deprez-De Campeneere, *Natl. Acad. Sci. U.S.A.* **79**, 626 (1982).
[11] H. J. P. Ryser, W. Shen, and F. B. Merk, *Life Sci.* **22**, 1253 (1978).
[12] L. J. Arnold, Jr., A. Dagan, and N. O. Kaplan, in "Targeted Drugs" (E. Goldberg, ed.), p. 89. Wiley, New York, 1983.
[13] I. Iliev, M. Georgieva, and V. Kabianov, *Russ. Chem. Rev. (Engl. Transl.)* **43**, 69 (1974).
[14] M. Szerke, R. Wade, and M. E. Whisson, *Neoplasma* **19**, 199 (1972).
[15] R. Wade, M. E. Whisson, and M. Szerkerke, *Nature (London)* **215**, 1303 (1967).
[16] N. G. L. Harding, *Ann. N.Y. Acad. Sci.* **186**, 270 (1971).

METHODS IN ENZYMOLOGY, VOL. 112

to polymeric carriers which exclude the drug from tissues where they produce toxic side effects.[12,17]

A polymeric carrier which has been studied by us and others is polylysine.[12,19,20,25–32] Studies over the last 30 years have indicated that polylysine interacts with many cells in both specific and unusual ways. Early studies indicated that polylysine possessed antiviral, antibacterial phage, and antibacterial activities.[26,33,34] At the same time, one early study indicated that polylysine had some activity against murine tumors.[35]

More recently studies have indicated that polylysine has an affinity for cancerous tissue[27] and is capable of enhancing the uptake into cells of macromolecules, such as albumin and peroxidase, which are coupled to it.[28,30,31] In addition, we have found that polylysine shows very specific effects with some types of tumor cells.[25] An additional desirable property of poly(L-lysine) is its fascile degradation by intracellular trypsin.

Consequently, polylysine shows many characteristics which suggest its use as a drug carrier: it has an affinity for at least some cancer cells, it is readily taken up by endocytosis, and it can be readily cleaved by trypsin once its conjugates enter cells.

---

[17] C. F. Chu and J. M. Whiteley, *Mol. Pharmacol.* **13,** 80 (1977).

[18] P. A. Kramer and T. Burnstein, *Life Sci.* **19,** 515 (1976).

[19] G. Atassi, M. Duarte-Karim, and N. J. Tagnon, *Eur. J. Cancer* **11,** 309 (1975).

[20] I. Brown and H. W. C. Ward, *Cancer Lett.* **2,** 227 (1977).

[21] G. Cornu, J. Michaux, G. Sokal, and A. Trouet, *Eur. J. Cancer* **10,** 695 (1974).

[22] C. DeDuve, T. DeBarsy, B. Poole, A. Trouet, P. Tulkens, and F. VanHoof, *Biochem. Pharmacol.* **23,** 2495 (1974).

[23] T. A. Marks and J. M. Venditti, *Cancer Res.* **36,** 496 (1976).

[24] M. Rozencweig, Y. Kenis, G. Atassi, M. Staquet, and M. Duarte-Karim, *Cancer Chem. Rep., Part 3* **6,** No. 2, 131 (1975).

[25] L. J. Arnold, Jr., A. Dagan, J. C. Gutheil, and N. O. Kaplan, *Proc. Natl. Acad. Sci. U.S.A.* **76,** 3246 (1979).

[26] C. Shahtin and E. Katchalski, *Arch. Biochem. Biophys.* **99,** 508 (1962).

[27] L. J. Anghileri, M. Heidbreder, and R. Mathes, *J. Nucl. Biol. Med.* **20,** 79 (1976).

[28] H. J. P. Ryser, *in* "Rehovot Symposium on Peptides" (E. R. Blout, F. A. Bovey, M. Goodman, and N. Lotan, eds.), p. 617. Wiley, New York, 1974.

[29] S. E. Kornguth and M. A. Stahmann, *Cancer Res.* **21,** 907 (1961).

[30] H. J. P. Ryser, W. C. Shen, and F. B. Mark, *Life Sci.* **22,** 1253 (1978).

[31] W. C. Shen and H. J. P. Ryser, *Proc. Natl. Acad. Sci. U.S.A.* **75,** 1872 (1978).

[32] W. C. Shen and H. J. P. Ryser, *Mol. Pharmacol.* **16,** 614 (1979).

[33] M. A. Stahmann, L. H. Graf, E. L. Patterson, J. C. Walker, and D. W. Watson, *J. Biol. Chem.* **189,** 45 (1951).

[34] L. Bichowski-Slomnicji, H. Berger, J. Kurtz, and F. Katchalski, *Arch. Biochem. Biophys.* **65,** 400 (1956).

[35] T. Richardson, J. Hodgett, A. Lindner, and M. A. Stahmann, *Proc. Soc. Exp. Biol. Med.* **101,** 382 (1959).

The practical use of polylysine as a drug carrier requires additional considerations: (1) toxicity; (2) retention of conjugate integrity until the conjugate has reached the target cell; and (3) the ability of the conjugate to be degraded to an active drug form once the conjugate has reached the target site.

In earlier studies[25] we found that polylysine alone possessed antineoplastic activity and toxic effects. This observation suggested that polylysine might not be useful as a drug carrier since one might be unable to attain efficacious doses of a polylysine conjugate because of the toxicity of the carrier alone. At the same time it might be difficult to determine if the beneficial effects of a conjugate are due to the conjugate or carrier alone.

These difficulties have been largely resolved since we found that the toxicity of polylysine and its neoplastic activity are highly dependent upon molecular weight and concentration. At high molecular weights and high concentrations, polylysine has pronounced effects. At low concentration and low molecular weight, polylysine has minimal effects.[12,25] Thus, low concentrations of low molecular weight lysine polymers are indicated as drug carriers.

The principal aim of this chapter is to describe methods for the synthesis of various drug conjugates with polylysine. At the same time this chapter provides data on the action of these conjugates with various cells in culture as well as in whole animals.

Methods

Synthesis of 6-Aminonicotinamide–Polylysine

The 6-aminonicotinamide–polylysine conjugates were formed by first synthesizing 6-aminonicotinamide-N^6-succinic acid (see Fig. 1). This derivative was typically produced by reacting 20 g (0.146 mol) 6-aminonicotinamide with 23 g (0.23 mol) succinic anhydride in 200 ml dimethyl sulfoxide (DMSO) for several hours at 100° followed by overnight incubation at 70°. The progress of the reaction was determined using thin-layer cellulose plates with an ascending solvent of ether : methanol : H_2O : concentrated NH_4OH (13 : 6 : 1 : 1). Once complete, the reaction mixture was poured into 300 ml ice-cold water. The precipitate was filtered and air dried. It was then ground with diethyl ether and refiltered. The final yield was 29 g (84%), and the structure was verified by ^1H NMR using a Varian 220 MHz spectrometer.

The 6-aminonicotinamide-N^6-succinate–polylysine conjugate was formed using 1-ethyl-3-(3-dimethylaminopropyl)carbodiimide. A typical reaction employed 400 mg (2.1 mmol) of 1-ethyl-3-(3-dimethylamino-

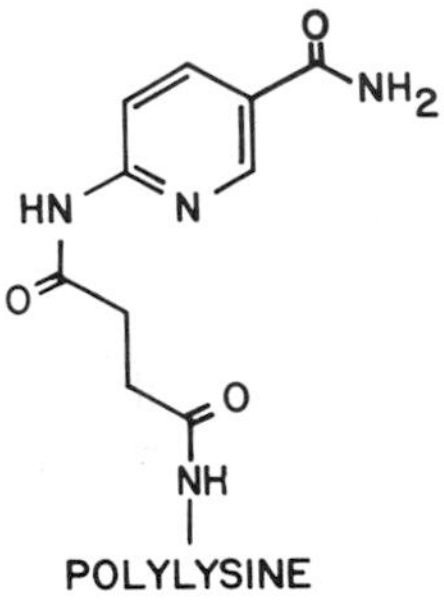

FIG. 1. The structure of the 6-aminonicotinamide conjugate.

propyl)carbodiimide, 200 mg (0.85 mmol) 6-aminonicotinamide-N^6-succinic acid, and 10 ml water. After heating the solution nearly to boiling, a small amount of K_2HPO_4 was added and the pH adjusted to 7.0 with NaOH upon cooling.

Two hundred milligrams of poly(L-lysine) HBr was then added and the mixture incubated for 1 hr. The pH was again adjusted to 7 with NaOH and reacted further for about 1 hr with occasional heating to 70–80°. The reaction mixture was then dialyzed and purified further using a 1.5 × 45-cm BioGel P-2 column with dilute HCl (pH 2) as the eluant. Using fluorescamine to detect polylysine it was shown that this column cleanly separated the polylysine conjugate from unreacted 6-aminonicotinamide-N^6-succinic acid. The weight ratio of attached 6-aminonicotinamide was determined to be 1 : 2.6 (w/w) (6-aminonicotinamide : poly(L-lysine)) from the weight of the pure conjugate and a calculated $\varepsilon_{283\ nm}^{1\%} = 118$ for 6-aminonicotinamide-N^6-succinic acid.

Alternatively, 6-aminonicotinamide-N^6-succinic acid was also coupled to polylysine with 1-ethyl-3-(3-dimethylaminopropyl)carbodiimide with N-hydroxysuccinimide as a catalyst in DMSO and pyridine.

Synthesis of Methotrexate–Polylysine Conjugates

The methotrexate–polylysine conjugate was formed by condensing the glutamate residue of methotrexate with the ε-amino groups of polylysine (see Fig. 2). Typically, 7 mg (0.015 mmol) methotrexate, 5 mg (0.043 mmol) N-hydroxysuccinimide, and 15 mg polylysine were dissolved in 2 ml of 10% pyridine in DMSO. After heating to dissolve any solid material, 15 mg (0.078 mmol) 1-ethyl-3-(3-dimethylaminopropyl)carbodiimide was added. The reaction mixture was incubated 5–30 min, followed by the addition of 5 volumes of water. The mixture was then applied to a 3-ml RP-2 (Merck) reverse-phase column. After washing with several column

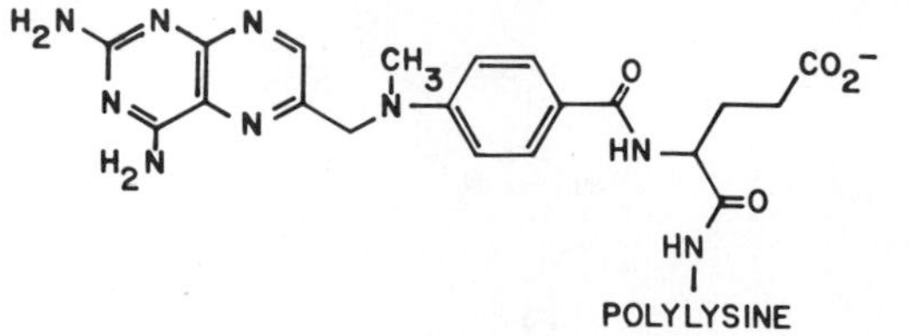

FIG. 2. The structure of the methotrexate–poly(L-lysine) conjugate. Coupling occurs primarily at the more reactive α-carboxyl of the glutamate residue, even though both the α and γ carboxyl groups are available for reaction.

volumes of water and dilute KOH (pH 11) to remove uncoupled methotrexate, the conjugate was eluted with dilute HCl (pH 2) and acetone (1:1, v/v).

The purity of the conjugate was determined by chromatographing the final product on thin-layer silica gel plates using an ascending solvent of methanol:water:NH$_4$OH (conc.) (40:10:1). This strongly alkaline solvent system was necessary to deprotonate the polylysine (R_f 0.0) and to prevent salt interactions from occurring which prevent uncomplexed methotrexate (R_f 0.9) from migrating. The coupling ratio was determined from the weight of the pure conjugate and the absorption of methotrexate ($\varepsilon_{307 \text{ nm}} = 19.7 \times 10^3$ at pH 1).

Synthesis of Doxorubicin– and Daunomycin–Polylysine Conjugates

The conjugates of the anthracyclines with polylysine were formed by using succinate to bridge between the amino group of the anthracycline and the ε-amino group of polylysine (see Fig. 3). Direct succinylation of the anthracycline was found to be somewhat difficult; however, the reaction of the anthracyclines with active succinate esters proceeded well. Typically, 14 mg (0.125 mmol) N-hydroxysuccinimide, 5 mg (0.05 mmol) succinic anhydride, and 96 mg (0.5 mmol) of 1-ethyl-3-(3-dimethylaminopropyl)carbodiimide were reacted for 20 min in 3 ml DMSO. Daunomycin hydrochloride (1.2 mg, 0.002 mmol), which was dissolved in 0.5 ml DMSO, was then added, and the mixture vortexed occasionally for 1 hr. The mixture was then cooled to 0°, followed by the addition of 2 ml chloroform and 2 ml cold 0.1 M K$_2$HPO$_4$, pH 7. The solution was mixed and the chloroform phase removed. The solution was then extracted a second time with 2 ml cold chloroform. The chloroform extracts were pooled and back-extracted once with 2 ml cold 0.1 M K$_2$HPO$_4$, pH 7, and three times with 2 ml cold water.

The chloroform phase was evaporated to leave the daunomycin succinate N-hydroxysuccinimide ester as a residue. This was then dissolved in 2 ml water, and 25 mg poly(L-lysine) · HBr (M_r 35,000) in 1 ml water was

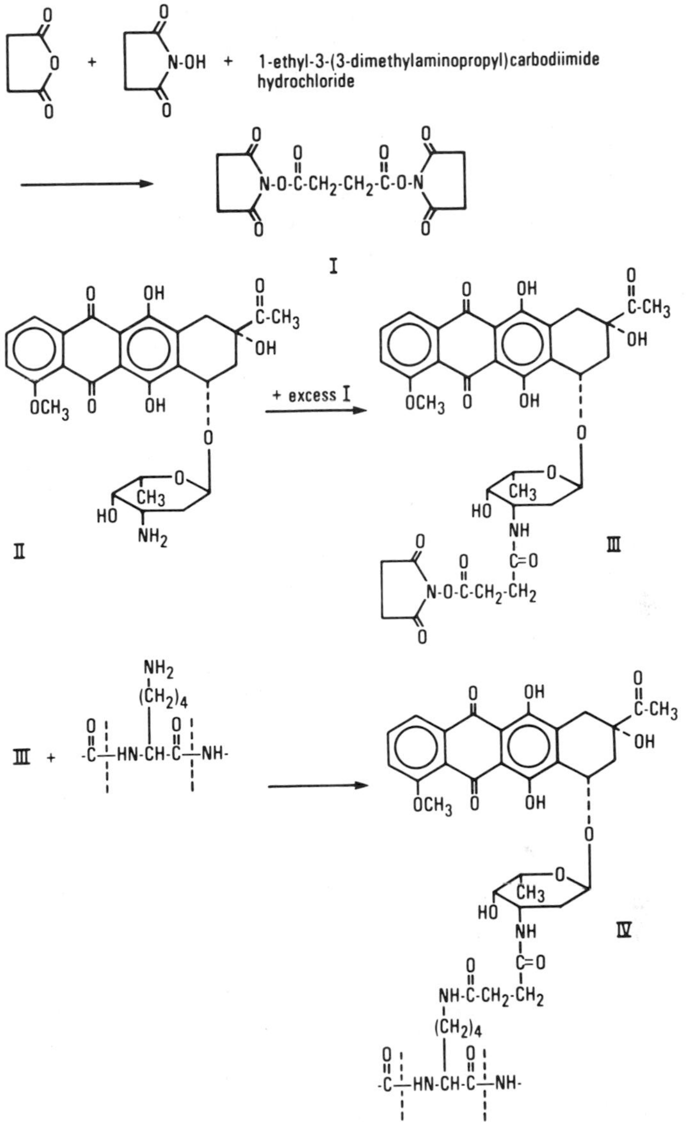

FIG. 3. The steps in the synthesis of the daunomycin conjugate. Taken from Arnold *et al.*[12] Reprinted with permission from "Targeted Drugs" (E. Goldberg, ed.), Wiley, New York, 1983.

added immediately. The reaction mixture was then vortexed continuously for 10 min and subsequently mixed occasionally for 2 hr. The mixture was then applied to a DE-52 column which had been washed with 10% K_2HPO_4, pH 7, followed by washing with water. The column was eluted

TABLE I

VIABILITY OF HeLa CELLS TREATED WITH VARIOUS CONJUGATES OF
DAUNOMYCIN–POLYLYSINE CONJUGATES USING DIFFERENT SPACER GROUPS[a]

		Percentage of controls		
Spacer		600 ng/ml	60 ng/ml	6 ng/ml
Succinic acid	HOOC(CH₂)₂COOH	24	98	—
Glutaric acid	HOOC(CH₂)₃COOH	11	91	—
Suberic acid	HOOC(CH₂)₆COOH	17	97	—
Dithiodiacetic acid	HOOCCH₂S—SCH₂COOH	10	94	—
Dithiodipropionic acid	HOOC(CH₂)₂S—S(CH₂)₂COOH	—	96	108
Oxalic acid	(COOH)₂	7	78	—
Phthalic acid	[benzene ring with two COOH groups (phthalic acid structure)]	7	92	—
Daunomycin (free)		2	9	102

[a] The coupling ratios are approximately 1 : 25 (w/w) daunomycin–poly(L-lysine) · HBr. The poly(L-lysine) is M_r 35,000.

with an isocratic gradient of 0.05 M K_2HPO_4, pH 7. The daunomycin–polylysine was the first to elute free of unconjugated daunomycin derivatives. The conjugate was dialyzed and lyophilized. The ratio of daunomycin to polylysine was determined to be 1 : 25 [w/w; daunomycin–poly(L-lysine) · HBr] by dissolving a known amount of the conjugate in water and determining the amount of bound anthracycline using $\varepsilon^{1\%}_{495\ nm}$ 196.

Using this same procedure, we conjugated daunomycin to M_r 13,000, 35,000, and 70,000 poly(L-lysine) and M_r 70,000 poly(D-lysine). Similarly, doxorubicin was conjugated to M_r 35,000 poly(L-lysine) . Using this procedure we also coupled daunomycin to poly(L-lysine) through several other diacid spacers including phthalic acid, oxalic acid, glutaric acid, saberic acid, dithiodiacetic acid, and dithiodipropionic acid (see Table I).

Cells and Culture. Cells were maintained in a 1 : 1 mixture of Ham's F-12 and Dulbecco's modified Eagle's media which was supplemented with 1.2 g sodium bicarbonate per liter and 15 mM HEPES as well as 192 units of penicillin, 200 μg streptomycin, and 25 μg of ampicillin per milliliter. In addition, 2.5% fetal calf serum and 5% horse serum (v/v) were added to the media. Cultures were grown in 100-mm plastic culture dishes (Falcon) in a humidified atmosphere at 37° in the presence of 5% CO_2. Stock cultures were passaged every 4–5 days.

Some studies were conducted in serum-free media. In these cases the same media was employed, with the exception that serum was replaced by the following hormones: insulin (5 μg/ml), hydrocortisone (36 μg/ml), transferrin (5 μg/ml), luteinizing hormone (NIH,B10; 2 μg/ml), and epidermal growth factor (40 ng/ml).

Experiments were started by washing stock plates with phosphate-buffered saline (PBS), trypsinizing the cells with 0.1% trypsin in 0.03% EDTA at 37°, and plating the cells into 35-mm dishes at a density of 10^4 cells/dish. The dishes were then allowed to stand overnight in the incubator to permit the cells to become attached before the addition of drug conjugates. Cells were counted when suspended using a model B Coulter counter.

Cell Growth Studies. Once cells had become attached, experiments were started by replacing the media with new media containing the poly-lysine-drug conjugate at the appropriate concentration. At the various times indicated, duplicate plates were taken from the incubator, washed with PBS, trypsinized, and counted.

Animal Studies. Mice of the strain indicated (see individual experiments) were injected intraperitoneally with 0.1 ml of either L1210 (10^5 cells) or Ehrlich ascites carcinoma (2×10^6–10^7 cells) on day 0. Beginning on day 1, injections of the appropriate conjugate in approximately 0.1 ml were given for 5 consecutive days. Animals were subsequently monitored for death as the result of tumor growth. Animals showing an increase of more than 300% in life span in comparison to controls were considered free of tumor (long-term survivors) and the experiment discontinued.

Toxicity studies were carried out as described above except that no tumor cells were injected and the mice were monitored for death due to the conjugate alone.

Results and Discussion

The activity of the conjugates were preliminarily evaluated using cultured HeLa cells. This was done by following the inhibition of cell growth with time and dose of the conjugate.

Anthracycline–Poly(L-Lysine) Conjugates

The activity of daunomycin attached to 35,000 M_r poly(L-lysine) through various spacers is shown in Table I. Table I includes two spacers which contain disulfide bonds. These were synthesized with the hope that daunomycin would be released very efficiently by cleavage of the disulfide once the conjugate entered the reducing environment of the cell. This strategy, however, did not significantly improve the efficacy of the conjugate. These daunomycin conjugates appear to have very similar activities regardless of the spacer moiety. Typically, the activity of the conjugate was 10–15% that of the free drug.

In an effort to prepare conjugates with higher activity compared to the free drug, we decided to couple daunomycin to poly(L-lysine) using the periodate oxidation procedure described by Hurwitz and co-workers.[36] This procedure involves treatment of daunomycin with periodate to generate aldehyde at the 2' and 3' carbon positions. The dialdehyde is then incubated with lysine residues to form Schiff bases, which are subsequently reduced with sodium borohydride.

Upon preparation of the conjugate by this procedure followed by purification on a DE-52 cellulose column, we found that the conjugate had no activity with HeLa cells in culture.

It might be argued that the complex does not enter the cells; however, from the range of substances that poly(L-lysine) is capable of transporting into cells,[30–32] this seems unlikely. In addition, we have found, employing fluorescence-labeled polylysine conjugates of daunomycin, that the conjugates are rapidly taken up by HeLa cells (L. J. Arnold Jr. and A. Dagan, unpublished results).

This result indicates that even though the activity of the daunomycin conjugates when prepared by our procedure is reduced compared to that of the free drug, the procedure is superior to the periodate oxidation procedure—at least for poly(L-lysine) polymers. A retention of only 15% activity in the conjugate compared to the free drug should not discourage *in vivo* experiments since selectivity provided by the carrier may still greatly increase the therapeutic index. That is to say that it is not of great concern if the activity of the bound drug is reduced as long as its selectivity for cancer cells is increased in comparison to normal cells. To date, we have tested daunomycin conjugates against two tumor types: Ehrlich ascites in BALB/c mice and L1210 leukemia in BDF$_1$ mice. In both cases, the activity of the conjugate was very low. This suggests that, although these conjugates retain activity in culture, additional factors are necessary for activity *in vivo*.

Interestingly, Trouet *et al.*[10] have examined the lysosomal cleavage of daunomycin from bovine serum albumin. The daunomycin was joined directly to a lysine residue in the albumin, using either a succinate bridge (as we have) or a longer spacer arm of one or more amino acids and succinate. Trouet *et al.* found in their studies that the longer spacer arms were cleaved readily by lysosomes; but when succinate alone was the bridging group, daunomycin was cleaved very slowly from albumin. The

---

[36] E. Hurwitz, R. Levy, R. Maron, M. Wilchek, R. Arnon, and M. Sela, *Cancer Res.* **35,** 1175 (1975).

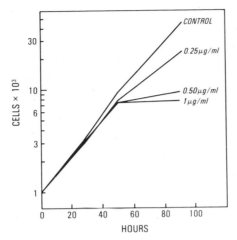

FIG. 4. The inhibition of the HeLa cell growth by the methotrexate conjugate.[12] The conjugate is 1 : 20 (w/w) methotrexate/poly(L-lysine) · HBr and the poly(L-lysine) used is M_r 3000. Note the lack of inhibition until the second day.

cleavage of daunomycin from albumin was paralleled by an increase in life span of animals inoculated with L1210. This study suggests that the activity of the poly(L-lysine)–anthracycline conjugates might be improved by using longer spacer arms.

Methotrexate–Poly(L-Lysine) Conjugates

The conjugates of methotrexate possess very good activity. The activity of a typical conjugate is shown in Fig. 4. Regardless of the concentration of the conjugate (this was also true for free methotrexate), there was little inhibition of cell growth up to 2 days. This appears to be due to a requirement to exhaust thymidine pools before either free methotrexate or the conjugates exert their effects. After this depletion period, the conjugate demonstrated a clear dose-dependent growth inhibition (see Fig. 4). It seems clear that this activity is due to the inhibition of dihydrofolate reductase by methotrexate, since it is possible to rescue the cells by adding exogenous thymidine or folic acid to the cultures.

An interesting observation was made when we compared the effects of free and bound methotrexate with cells grown in the presence or absence of serum. Figure 5 shows the effects of the methotrexate conjugate on the growth of HeLa S, a HeLa subline selected for growth in serum-free medium. Note that the presence of serum does not affect the cytotoxic activity of free methotrexate. In the presence and absence of serum,

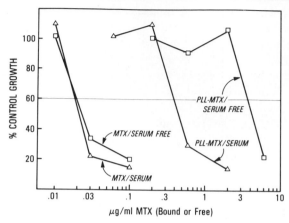

FIG. 5. The ID$_{50}$ of free methotrexate (MTX) and methotrexate conjugate (PLL-MTX) with HeLa S cells in serum-containing and serum-free media.[12] Cells were counted after 4 days of growth. Note that the ID$_{50}$ is actually at the 60% level. This is due to the fact that the cells grown for about 2 days, regardless of the level of treatment, i.e., high doses, still give 10% growth. The conjugate was the same as that in Fig. 4.

the ID$_{50}$ is approximately 0.02 μg/ml methotrexate. In contrast, there is a profound effect of serum on the activity of the conjugate. With serum, the conjugate has approximately 10 times the activity it has in serum-free medium. This is directly opposite the effect we find for poly(L-lysine) alone, which is normally far more toxic to cells grown in serum-free media.[25] As a result, this unusual effect of the conjugate does not appear to be directly related to the way in which polylysine alone reacts with these cells.

The major observation remains that HeLa S cells are much more sensitive to methotrexate–poly(L-lysine) when they are grown in the presence of serum. We have addressed several questions in an attempt to account for this finding. (1) Does albumin, which is present in serum-containing media, increase the toxicity of the conjugate? (2) Does trypsin, which is present in small amounts in serum, cleave the conjugate to more active forms? (3) Are there constituents of serum-free media (e.g., hydrocortisone) which block the activity of the conjugate? In exploring these questions, we have made the following observations.

1. The addition of purified albumin to serum-free media, to the same level as that present in serum-containing media (3 mg/ml), gave no increase in sensitivity of HeLa S cells to the methotrexate conjugate.

2. Incubation of methotrexate–poly(L-lysine) overnight in serum-free medium with trypsin added caused the complete loss of activity over the

dose range tested. At the same time, the addition of soybean trypsin inhibitor at a dose sufficient to block the degradation of poly(L-lysine) in serum-containing media had no effect on the activity of the methotrexate conjugate. In other words, although small amounts of trypsin are present in serum, they are insufficient to significantly affect the activity of the methotrexate conjugate.

Both of these experiments indicate that the active form of the methotrexate conjugate is the intact complex and not a degradation product. These results also indicate that the conjugate is not catabolized and activated externally by HeLa cells. Studies by Shen and Ryser support this conclusion.[32]

3. Serum-containing media supplemented with hydrocortisone to the level present in serum-free media did not give any increased resistance to the cytotoxicity of the methotrexate conjugate.

Up to this time we have not been able to determine the features which make cells grown in serum-containing media more than 10 times as sensitive to the methotrexate–poly(L-lysine) conjugate as the same cells grown in serum-free media. The factors which account for this curious phenomenon can probably best be determined by carrying out additional mixing experiments using substances known to be present in either serum-containing or serum-free media.

A determination of the components which enhance the sensitivity of HeLa S cells to the methotrexate conjugate might then make it possible to enhance the sensitivity of potential target cells or tissues even further and thus to improve site specificity by providing other key components at the target site.

In addition to using HeLa S cells, we have also determined the activity of the methotrexate conjugate with L1210 cells and Ehrlich ascites cells. The L1210 cell is very sensitive to methotrexate, and thus one might expect that the conjugate would be very active with these cells. Table II shows the ID_{50} values for L1210 cells and Ehrlich ascites carcinoma cells in serum-containing media. As expected, L1210 cells are much more sensitive to free methotrexate. In contrast, they are very resistant to the methotrexate conjugate. The activity of the conjugate with L1210 cells is essentially zero since it shows only about 1/1000 of the activity of the free drug. This low activity level in fact might be accounted for by trace contamination of the conjugate with free methotrexate. It should be noted that in our previous studies, poly(L-lysine) alone had little effect on increasing the life span of mice inoculated with L1210, in contrast to mice

TABLE II
ID_{50} OF METHOTREXATE–POLY(L-LYSINE) (MTX-PLL) WITH EHRLICH ASCITES AND L1210 LEUKEMIA CELLS[a]

	ID_{50} (ng/ml)	
Cell type	MTX (free)	MTX-PLL
Ehrlich ascites	21	450
L1210 leukemia	5.4	3200

[a] The conjugate was 1 : 20 (w/w) methotrexate–poly(L-lysine)·HBr. The poly(L-lysine) is M_r 3000. The concentration is expressed in terms of free or polylysine-conjugated methotrexate (MTX).

injected with Ehrlich ascites cells.[25] The results shown in Table II similarly indicate a very low interaction between L1210 cells and the methotrexate conjugate.

Regardless of the exact mechanisms which affect the activity of the methotrexate conjugates, it is clear that their activity is highly dependent upon cell type.

In summary, we have made the following observations.

1. Methotrexate conjugates have good activity with some cell types.

2. The activity of the methotrexate conjugate with cultured cells is dependent upon components present in the media. The methotrexate conjugate with HeLa S cells is very active in serum-containing media, but much less active in serum-free media.

3. The active form of the methotrexate conjugate appears to be the intact species and not a tryptic digestion product.

4. The activity of the methotrexate conjugate is very dependent upon cell type. It shows reasonable activity with Ehrlich ascites cells but virtually no activity with L1210 leukemia cells.

6-Aminonicotinamide–Poly(L-Lysine)

Figure 6 shows the activity of the 6-aminonicotinamide conjugate with HeLa cells in culture. At the concentrations of polylysine employed here, the polylysine carrier has no effect on cell growth.[25] Growth is substantially decreased, however, when the cells are treated with the 6-aminonicotinamide conjugate.

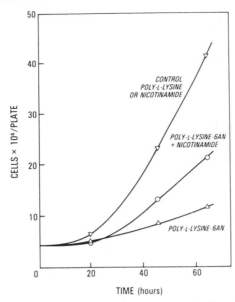

FIG. 6. The inhibition of HeLa cell growth by the 6-aminonicotinamide (6AN) conjugate.[14] The concentration of the conjugate was 30 μg/ml, and it had a coupling ratio of 6-aminonicotinamide : poly(L-lysine) of 1 : 2.5 (w/w). External nicotinamide was added at a concentration of 500 μg/ml.

To assure ourselves that the activity was associated with the drug conjugate, we added exogenous nicotinamide, which competes for intracellular sites with 6-aminonicotinamide. Under these conditions, much of the growth-inhibiting activity of the conjugate is eliminated. The residual growth inhibition may be due to the added nicotinamide since exogenous nicotinamide alone has a inhibitory effect on cell growth.

In addition, we examined the antineoplastic activity of the 6-aminonicotinamide–poly(L-lysine) conjugate against Ehrlich ascites tumors in white Swiss mice. Figure 7 shows the effect of the conjugate as well as the effect of free 6-aminonicotinamide on tumor-bearing mice. The free drug is known to have antineoplastic activity, but it also possesses a severe neurotoxicity. As a result, free 6-aminonicotinamide alone has never been effective as an antineoplastic agent.

Mice treated with free 6-aminonicotinamide alone died within 6–10 days from paralysis which is associated with central nervous system (CNS) toxicity. At the same time, animals given the tumor and no free drug or conjugate died from tumor growth at 25–40 days. Tumor-bearing animals treated with the 6-aminonicotinamide conjugate at a dose equiva-

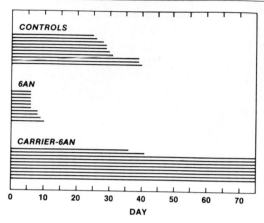

Fig. 7. The increase in life span of white Swiss mice inoculated intraperitoneally with Ehrlich ascites tumors and either 6-aminonicotinamide or conjugate.[12] Controls are animals inoculated with tumor only. The 6AN represents treatment with the free drug alone (8.5 mg/kg for 5 consecutive days). Carrier–6AN is 6-aminonicotinamide–poly(L-lysine) given at a concentration equivalent to that of the free drug. Note that the conjugate does not show the toxicity of the free drug. A portion of the antineoplastic activity of the conjugate may be due to the nature of the conjugate itself.

lent to the treatment with free 6-aminonicotinamide showed marked inhibition of tumor growth without evidence of the neurotoxicity of the free drug. In subsequent studies we were able to show that exogenous nicotinamide only partially blocked the antitumor effect of the conjugate. This result suggests that at least some of the activity of the conjugate may be due to the physical nature of the conjugate itself. Significantly we could not find any dose of the free drug which would give a marked increase in life span without producing severe CNS toxicity.

Conclusions

Polylysine is an effective drug carrier with various cell types. Some cells, however, such as L1210, appear to be resistant to the cytotoxic effects of drug–polylysine conjugates. The explanation for this phenomenon is probably associated with the fact that L1210 cells interact weakly with polylysine.[25]

Several drugs, including methotrexate, daunomycin, and 6-aminonicotinamide, have been attached to polylysine with a retention of drug activity. Methotrexate can be coupled directly to the ε-amino groups of polylysine via its glutamate moiety. Daunomycin can be coupled to polylysine using a number of spacer groups including succinate, and 6-aminonicotin-amide can be covalently linked to polylysine via a succinate spacer.

Generally, the conjugate possesses only 5–20% of the activity of the free drug. The activity of the conjugate, however, is strikingly different from that of the free drug. The activity of the conjugate is highly dependent upon the target cell type and other constituents present within the target cells. In some cases, the toxicity of the conjugate in animals is decreased relative to the free drug, while beneficial antineoplastic activity is retained. These studies indicate that polylysine has potential for "targeting" drugs to specific cell or tissue sites.

[22] Dextran and Inulin Conjugates as Drug Carriers

By LUIGI MOLTENI

Many drugs with good biopharmacological properties are limited in usefulness because of poor chemical stability, metabolism, or lack of organ tropism. Owing to their short half-life, the therapeutic use of most drugs is often limited by the need to give frequent doses to the patient, which causes considerable discomfort. Moreover, large amounts of drugs are required, and while most is converted into generally inactive derivatives or metabolites, only a portion of it is utilized. Therefore, the question is often raised as to how to increase the chemical and biological stability of a drug, that is to modify its half-life and to confer organ tropism.

An appealing solution to the problem consists of chemically coupling the drug to a carrier able to transfer the active substance, biologically unchanged, to the site of pharmaceutical utilization. Among the different compounds which can be employed as a possible carrier of pharmacological substances, polysaccharides, mainly dextran and inulin, seem to have some of the necessary qualifications. They react to form defined and stable compounds and of themselves lack any significant toxicity or organ or cellular tropism.

Dextrans are biosynthetic polymers consisting of linear chains of glucose in a 1 : 6 linkage. Dextrans with average molecular weights of 1000, 5000, 40,000, 75,000, 110,000, 700,000, and more are available. Inulin (isolated from *Dahlia* tubers) is a natural polysaccharide formed by fructose and glucose linear chains having an approximate average molecular weight of 5000. Dextrans 40,000, 70,000, and 110,000 are used in medicine

METHODS IN ENZYMOLOGY, VOL. 112

as blood expanders; inulin is employed in a diagnostic test for renal function.

Dextran and Inulin Conjugates

The direct esterification of dextran and inulin with carbonyl groups of drugs may be easily carried out in two ways: by reacting the polysaccharide with the acid chlorides in the presence of bases (for instance, alkaline hydroxides or pyridines) or by using carbodiimides, azides, or mixed anhydrides such as chloroacetic or trifluoroacetic anhydride. The literature does not report any drug ester of inulin.

Dextran 40,000 was esterified with acetylsalicylic acid[1] and nicotinic acid[2] and dextran 150,000 with acetylsalicylic acid.[3] Preliminary studies showed that the pharmacological active moiety was present in the polymer in high percentages and that the compounds had good pharmacological activities, characterized by long lifetimes. The conjugate acts as a slow-release system.

According to a private communication by E. Oradi, a general method of direct esterification of dextrans and inulin with acid chlorides is basically the same as described in the French Patent No. 1,604,123 to synthesize the dextran nicotinic acid ester. The acid chloride dissolved in 25-fold excess pyridine at 80° is added to a pyridine solution of dextran and the mixture warmed at 120° for 50 min. After cooling the solution is neutralized with 25% ammonium hydroxide, filtered, and precipitated with ethanol.

Examples of dextran and inulin conjugates using nonpolar bonds are widely reported in the literature of the last 20 years. These derivatives are obtained by activation of the hydroxyl groups of the polysaccharide followed by coupling with amine, hydroxyl, thiol, keto, or formyl moieties of the drug which is to be bound.

Activation of Dextran and Inulin

The methods of activation of dextran normally used are the following: oxidation of the polysaccharide by periodate to form polyaldehyde dextran; preparation of an azide dextran; activation by cyanogen halides or organic cyanates to form imidocarbonate dextrans and, finally, preparation of bromo- or chlorohydroxypropyl dextran, through a reaction with

[1] E. Oradi, unpublished data.
[2] French Patent No. 1,604,123 (1971).
[3] P. Papini *et al.*, *Ann. Chim.* **59**, 1943 (1969).

bromo- or chloropropyl epoxide. This last method has been the only one used to activate inulin. By all these methods it is possible to synthesize drug-linked polysaccharides as defined stable compounds with a suitable molecular weight. Oxidization of the hydroxyls of dextran with sodium periodate solutions leads to aldehyde functionalities that easily react with amine derivatives.

$$
\begin{array}{ccc}
x\text{—OH} & & x\text{=}O \\
x & & x \\
x & \xrightarrow{\text{NaIO}_4} & x \\
x & & x \\
x\text{—OH} & & x\text{=}O
\end{array}
$$

Many examples of dextran periodate activation for the preparation of protein conjugates are described in the literature. This straightforward reaction consists of dissolving dextran in 0.03 M NaIO$_4$ and keeping the solution at room temperature for 12 hr, in the dark. The reagents are calculated to provide 1 mol of NaIO$_4$ per mole of glucosidic residue in dextran, to obtain 100% oxidization. The polyaldehyde dextran, purified by dialysis and obtained as a dry powder by lyophilization, has to be used in a short time for the conjugation reaction. The activation of dextran by the azide system where the glucosidic hydroxyls are bound to –CH$_2$–CO–NH–NH$_2$ radicals is long, tedious, and complex. For these reasons it is not used very often.

$$
\begin{array}{ccccc}
x\text{—OH} & & x\text{—O—CH}_2\text{—COOH} & & x\text{—O—CH}_2\text{—CONH—NH}_2 \\
x & & x & & x \\
x & \xrightarrow{\text{Cl—CH}_2\text{—COOH}} & x & \xrightarrow{\text{H}_2\text{N—NH}_2} & x \\
x & & x & & x \\
x\text{—OH} & & x\text{—O—CH}_2\text{—COOH} & & x\text{—O—CH}_2\text{—CONH—NH}_2
\end{array}
$$

Most dextran conjugates are obtained by activation of dextran with cyanogen halides followed by formation of imidocarbonate dextrans.

$$
\begin{array}{ccc}
x\text{—OH} & & x\text{—O} \\
x & & x \\
x & \xrightarrow{\text{BrCN}} & x \quad \diagdown \\
x & & x \quad \diagup C\text{=}NH \\
x\text{—OH} & & x\text{—O}
\end{array}
$$

A rather similar activation by organic cyanates also gives imidocarbonate derivatives.

$$
\begin{array}{ccc}
x\text{—OH} & & x\text{—O} \\
x & & x \\
x & \xrightarrow{\text{RO—C}\equiv\text{N}} & x \quad \diagdown \\
x & & x \quad \diagup C\text{=}NH \\
x\text{—OH} & & x\text{—O}
\end{array}
$$

Therefore it can be considered as a variation of the activation by cyanogen halides.

The activation of dextran by CNBr has been described by R. Axen[4] and, in spite of its limitations, it is widely used. Dextran (5 g) is dissolved in 160 ml of water and the pH is raised to 11 by addition of NaOH, and portions of about half a gram of CNBr are added rapidly (total quantity of the reacting CNBr, 2.5 g). Throughout the reaction the pH is maintained at 11 by the addition of NaOH. The final cloudy solution is gel-filtered on a G-25 Sephadex column, and the eluted portions, which are dextrorotating in the polarimeter, are collected and used directly for the coupling reaction. Considerable interest has been generated by the preparation of activated dextran (and inulin) using chlorohydroxypropyl dextran. This procedure permits solid-phase activation and long-term storage. The activation of dextran and inulin with epihalohydrins to halohydroxypropyl polysaccharides can be performed in aqueous solution alone or in the presence of a catalyst. The catalyst allows shorter activation times and gives higher levels of activation.

$$
\begin{array}{c}
x{-}OH \\
x \\
x \\
x \\
x \\
x{-}OH
\end{array}
\quad
\xrightarrow[\displaystyle CH_2{-}CH{-}CH_2Hal]{\displaystyle \overset{O}{\diagup\,\diagdown}}
\quad
\begin{array}{c}
x{-}O{-}CH_2{-}CH({-}OH){-}CH_2Hal \\
x \\
x \\
x \\
x \\
x{-}O{-}CH_2{-}CH({-}OH){-}CH_2Hal
\end{array}
$$

In our experiments on the preparation of chlorohydroxypropyl dextran and chlorohydroxypropyl inulin, we found that either using the catalyst $Zn(BF_4)_2$ or carrying out the activation reaction under ultraviolet light (λ, 254 nm) illumination leads to equally well-activated polysaccharides. K. Bolewsky[5] first used $Zn(BF_4)_2$ to catalyze the reaction between dextran and epichlorohydrin.

In a 250-ml round flask equipped with a reflux condenser and a stirrer, 10 g dextran, 25 ml of water, 50 ml epichlorohydrin, and 10 ml of a 25% water solution of $Zn(BF_4)_2$ are mixed. This solution is warmed on a steam bath for 5 hr and then poured into 700 ml of acetone. The precipitate is filtered, washed with acetone, and dried. The dry crude product is dissolved in a minimum amount of water and precipitated again with acetone. This procedure is repeated twice in order to obtain pure chlorohydroxypropyl dextran. Ultraviolet light (λ, 254 nm), as an alternative catalyst, employs the same basic synthesis conditions (10 g of the polysaccharide, 25 ml of water, and 50 ml epichlorohydrin); however, the longer reaction time of 48 hr is used. These reactions produce chlorohy-

[4] R. Axen et al., Nature (London) **214,** 1302 (1967).
[5] K. Bolewsky et al., Ann. Pharm. Poznam **10,** 81 (1973).

droxypropyl dextran or chlorohydroxypropyl inulin with a chlorine content of 5–6%, a content corresponding to an activated hydroxyl every two or three glucosidic residues.

Conjugation Reactions

Primary amino groups of drugs easily react with iminocarbonate dextran (obtained via CNBr or organic cyanates activation). Since it is not possible to isolate the activated polysaccharides, the coupling has to follow activation immediately. Iminocarbonate dextrans are conjugated with drugs in 2% sodium borate ($Na_4B_2O_7 \cdot 10 H_2O$), pH 9.2, at 5° for 48 hr. The same mild conditions (pH 9.2, sodium borate, 5°, 48 hr) are used for conjugation of amino group-containing drugs to chlorohydroxylpropyl dextran and chlorohydroxypropyl inulin. The activated polysaccharides are dry powders, which may be stored for as long as several years. The reaction can also be used for the conjugation of primary alcohol and thiols. In these cases, the solution pH is raised to 11–13.

Chlorohydroxypropyl dextran and the drug to be linked (weight ratios of drug to activated dextran 1 : 1) are dissolved in a borate buffer solution at pH 9.2 and reacted at 5° for 48 hr. The conjugation of polyaldehyde dextran with the primary amino groups of drugs is usually carried out at room temperature in the dark for 12 hr in a phosphate buffer solution, pH 7.2. The resulting Schiff base is reduced with $NaBH_4$ at about 40° for 2 hr. Sometimes a chemical bridge consisting of an alkyl or arylalkyl chain is put between the drug and the activated polysaccharide, for pharmacological or chemical purposes.

The work of J. Pitha[6] on dextran–alprenolol conjugates shows that dextran can be sterically unfavorable for binding to the receptor. This hindrance can be reduced by inserting a linker between the polysaccharide and the beta blocker. From a private communication with M. Hashida, the *in vitro* cleavage rate of mitomycin from its esters with the alkylcarboxylic derivatives of dextran imidocarbonate becomes slower with increasing chain length of the spacer. Hashida prepared these derivatives by first coupling the activated dextran with ω-amino acids and, subsequently adding mitomycin using a carbodiimide-promoted reaction. In order to link a keto or aldehydic compound to dextran, an amino derivative of dextran has to be prepared.

For the synthesis of the conjugate of formylrifamycin SV with dextran, we first prepared a hydrazine derivative of the polysaccharide, reacting chlorohydroxylpropyl dextran with hydrazine. After reaction with

[6] J. Pitha *et al.*, *Proc. Natl. Acad. Sci. U.S.A.* **77**, 2219 (1980).

the drug, a hydrazone is formed which when reduced forms a hydrazine. Chlorohydroxypropyl dextran (chlorine content, 6%), dissolved in a 2% sodium borate buffer solution, is treated with an excess of $NH_2-NH_2 \cdot H_2O$ and the mixture is kept overnight at room temperature. After dialysis and lyophilization, a white powder is obtained (nitrogen content, 2.3%). The hydrazinehydroxypropyl dextran (1 g) is dissolved in 10 ml of 10% sodium acetate and 1 g of formyl drug dissolved in 10 ml of the same solvent is added. The solution is incubated for 2 hr at room temperature and is then reacted with sodium borohydride (50 mg) to reduce the hydrazone bond to a more stable hydrazine.

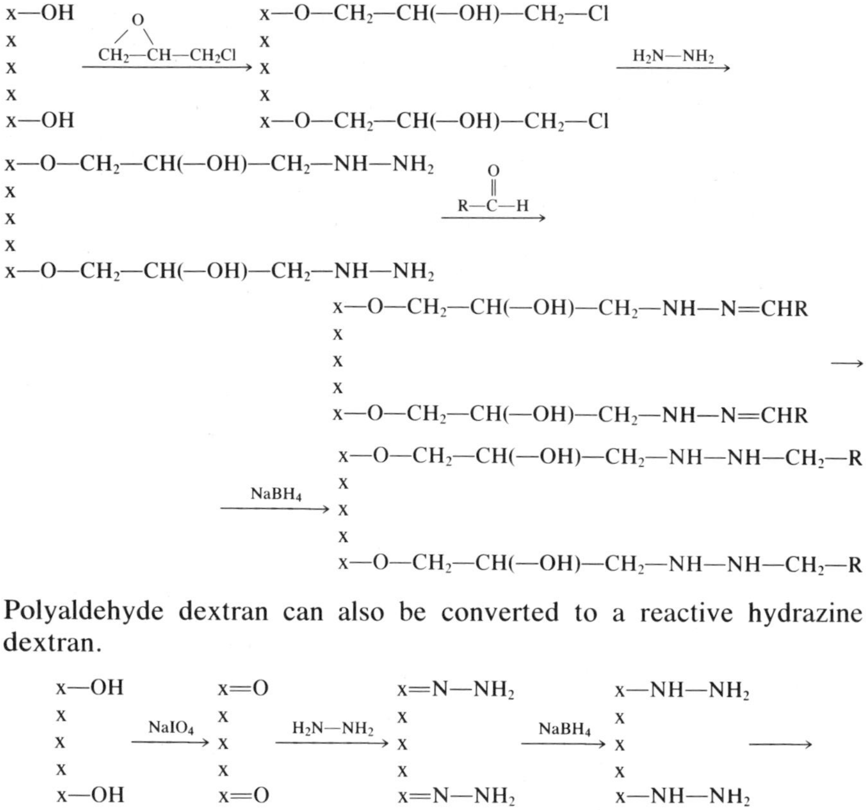

x—OH
x
x
x
x—OH

$\xrightarrow{\overset{\displaystyle O}{\overset{\diagup\diagdown}{CH_2-CH-CH_2Cl}}}$

x—O—CH₂—CH(—OH)—CH₂—Cl
x
x
x
x—O—CH₂—CH(—OH)—CH₂—Cl

$\xrightarrow{H_2N-NH_2}$

x—O—CH₂—CH(—OH)—CH₂—NH—NH₂
x
x
x
x—O—CH₂—CH(—OH)—CH₂—NH—NH₂

$\xrightarrow{R-\overset{\displaystyle O}{\overset{\|}{C}}-H}$

x—O—CH₂—CH(—OH)—CH₂—NH—N=CHR
x
x
x
x—O—CH₂—CH(—OH)—CH₂—NH—N=CHR

$\longrightarrow$

x—O—CH₂—CH(—OH)—CH₂—NH—NH—CH₂—R
x
x
x
x—O—CH₂—CH(—OH)—CH₂—NH—NH—CH₂—R

$\xrightarrow{NaBH_4}$

Polyaldehyde dextran can also be converted to a reactive hydrazine dextran.

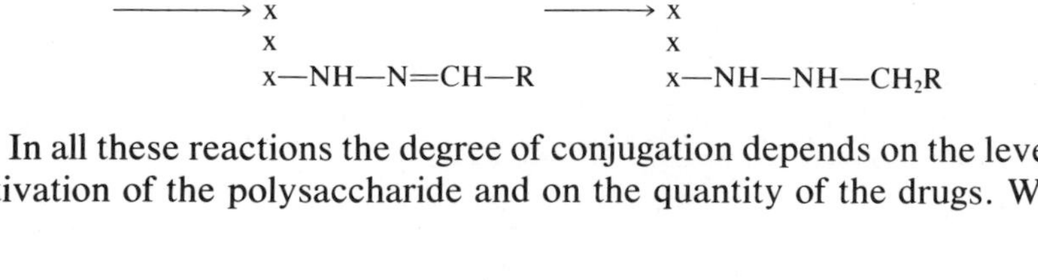

x—OH
x
x
x
x—OH

$\xrightarrow{NaIO_4}$

x=O
x
x
x
x=O

$\xrightarrow{H_2N-NH_2}$

x=N—NH₂
x
x
x
x=N—NH₂

$\xrightarrow{NaBH_4}$

x—NH—NH₂
x
x
x
x—NH—NH₂

$\longrightarrow$

x—NH—N=CH—R
x
x
x
x—NH—N=CH—R

$\xrightarrow{R-\overset{\displaystyle O}{\overset{/\!\!/}{C}}-H}$

x—NH—NH—CH₂R
x
x
x
x—NH—NH—CH₂R

$\xrightarrow{NaBH_4}$

In all these reactions the degree of conjugation depends on the level of activation of the polysaccharide and on the quantity of the drugs. When

STABILITY OF WATER-SOLUBLE DEXTRAN AND INULIN CONJUGATES

Drug	Conjugate	Test[a]	Analysis[b]	Results
CoB$_{12}$ vitamin	—	1	TLC	Unstable
CoB$_{12}$ vitamin	Dextran 70,000	1	TLC	Stable
Noradrenaline	—	1	TLC	Unstable
Noradrenaline	Dextran 70,000	1	TLC	Stable
S.O.D. enzyme	—	2	EA	−30%
S.O.D. enzyme	—	3	EA	−60%
S.O.D. enzyme	Dextran 70,000	2,3	EA	Stable
S.O.D. enzyme	Inulin	2,3	EA	Stable
Cysteine	—	4	TLC	Oxid. to cystine
Cysteine	Dextran 70,000	4	TLC	Stable
Cysteine	Inulin	4	TLC	Stable

[a] Test 1: 3 hr solar lamp illumination; room temperature. Test 2: 5 days in the dark; 4°. Test 3: 2 days in the dark; 38°. Test 4: 1 day solar illumination; room temperature.

[b] TLC, Thin-layer chromatography; EA, enzymatic activity.

highly activated dextran or inulin or small quantities of the drug to be conjugated (to obtain low conjugation) are used, it is necessary, after drug attachment, to react the partially activated polysaccharide with an excess of glycine, or another nucleophile, to prevent subsequent cross-linking reactions or adverse side reactions due to unreacted epoxy radicals present in the polymer. The conjugates may be purified from unreacted ligand and/or blocking reagent by gel filtration. Useful molecular sieve media are Sephadex or BioGel. The choice of the type of molecular sieve is dependent upon the molecular weight of the polysaccharide. Pure conjugates are obtained using suitable Sephadex or BioGel columns. Dialysis or the use of Amicon hollow fibers are alternative routes to molecular sieving and are better suited to industrial-scale preparations. Dextran and inulin conjugates may be obtained as dry powders from aqueous solutions by freeze-drying lyophilization or by evaporation of water–alcohol solutions *in vacuo*.

The conjugation of drugs to dextran and inulin increases drug stability to light, temperature, hydrolysis, and chemical agents (see table). The enhanced chemical stability of the drugs as well as the water solubility of these conjugates makes possible new pharmaceutical and medical applications.

Pharmacological Properties

Dextran and inulin are pharmacologically inert, and as a result the pharmacological properties of their conjugates are due to the linked drug. Consequently, functional groups responsible for, or involved in, the pharmacological activity of the drug cannot be permanently blocked.

Noradrenaline dextran (70,000 MW) does not exert any effect on blood pressure when injected intravenously in rabbits, and upon injection amphetamine dextran (70,000 MW) does not modify the activity of mice. The conjugates are pharmacologically inactive because the amino group of the parent drug is blocked.

On the other hand the anesthetic activity of novocaine–dextran (70,000 MW) in rabbits using the corneal reflex test seems to decrease only 5% in comparison to an equimolar dose of novocaine. The free amino groups of novocaine (diethylaminoethyl p-aminobenzoate) is not of particular importance for the pharmacological activity of the drug. Diethylaminoethyl benzoate, the novocaine molecule without the p-amino group, has the same anesthetic properties. Although the conjugates have the same basic pharmacological properties of the parent drugs, they differ in their potency, bioavailability, pharmacokinetics, and toxicity.

The potency of dextran and inulin conjugates, calculated on a molecular basis, depends on many factors. Different molecular weight dextrans do not impair conjugations but they lead to conjugates with different pharmacological activities. Large-sized dextrans seem to interfere with the affinity of macromolecular drugs for their receptors.

In investigating the effect on gastric acid secretions in the rat after administration of gastrin-like compounds according to the method of Ghosh,[7] it was found that 5 μg of pentagastrin linked to dextran (70,000 MW) injected intravenously was not active, while the smaller dextran (5000 MW) pentagastrine derivative at the same dosage retained the same activity on gastric secretion as the parent drug (Fig. 1).

The chemical and physical properties of the drug to be conjugated will affect the type of binding. While the linking reactions occur under conditions weak enough to guarantee chemical integrity of even labile drugs, the chosen conjugation reaction may be responsible for a certain loss of activity of the derivatives. The steric rigidity acquired by some dextran conjugates is sometimes responsible for a decreased pharmacological activity and may be diminished by inserting an inert chemical bridge between the drug and the polysaccharide. The potency of the conjugate is dependent on spacer length.[6] Finally, a certain loss of activity (calculated on a

[7] M. N. Ghosh, *Brit. J. Pharmacol.* **13**, 54 (1958).

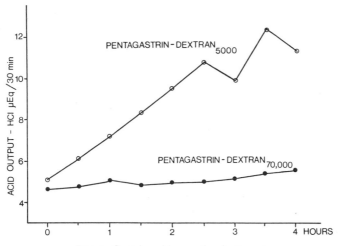

FIG. 1. Gastric acid secretion in the rat.

molecular basis) of substances such as proteins and enzymes can be due to the blocking of sites during the coupling reaction.

Dextrans and inulin are not absorbed following oral administration, and, since the bond between the polysaccharide and the linked drug is particularly stable (except the hydrazone and Schiff base), their conjugates are not active orally. Decreased activity of orally administered drugs can sometimes be achieved by conjugation with 5000 MW dextran (Fig. 2). On the other hand, the absorption of the conjugates following intramuscular administration is generally good and depends on the molecular size of the polymer.

The most striking pharmacological difference between dextran conjugates, inulin conjugates, and the parent unlinked drugs is their kinetic fate. While the two polysaccharides chemically react in the same way and confer to their conjugates the same stability and bioavailability properties, their kinetic distribution is completely different and, as a consequence, their conjugates are dominated by the kinetics of the carrier. The serum half-life of dextran derivatives is typically much higher than the parent drugs. The polysaccharide acts as a shield against rapid metabolic degradation of the drug and the larger molecular weight of the conjugate appears to retard tubular filtration by the kidneys. The conjugation of a drug to dextran gives it long-lasting pharmacological characteristics, depending in part on long circulatory lifetimes.

Dextran conjugates are temporarily retained by tissues of the reticuloendothelial system and accumulates mostly in the liver where they are

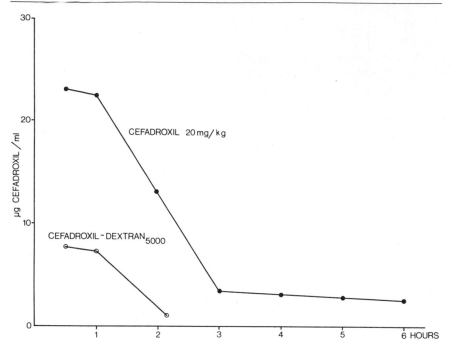

FIG. 2. Serum cefadroxil concentrations after oral administration to a rabbit.

metabolized by dextranases with a kinetic rate related to molecular size. With drugs conjugated to dextrans, it is consequently possible to achieve liver targeting. In studies on the chronic toxicity of CoB_{12}–dextran (70,000 MW, intramuscular administration in rats), an accumulation of the conjugate was found in the liver of the animals treated for a month. The presence of the CoB_{12}–dextran conjugate was determined by titration of the vitamin B_{12} content and by histology of the organ. However, the liver concentration of the polymer was reversible (Fig. 3).

In contrast, the main kinetic characteristic of inulin and inulin derivatives is their very short serum half-life and their rapid and complete urinary elimination. By inulin conjugation, drugs can be carried into the urinary tract without undergoing any metabolic degradation, showing the same kinetic fate as the polysaccharide itself. The molecular size of inulin conjugates is not large enough to prevent a rapid renal filtration and the urinary tract tropism of the carrier. When inulin and dextran of the same molecular weight (5000) are conjugated to the same drug, it is the reticuloendothelial tropism of the dextran which directs the dextran derivatives to the liver, while the inulin, by its very rapid renal filtration, gives its

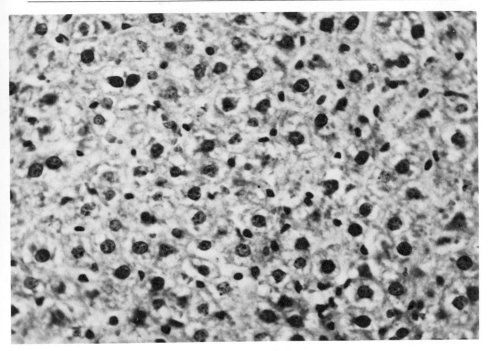

FIG. 3. Liver micrograph. Fixation in 10% in formalin; staining with hemotoxylin–eosin. The white spots correspond to CoB_{12}–dextran deposits. ($\times 80$)

conjugates a urinary tract tropism, even if the compounds are very similar in structure, molecular size, and chemical and biological stability.[8]

The kinetic profiles of some antibiotic conjugates of dextran (MW 5000) and inulin (MW 5000) are good examples. The conjugates of amoxicillin, ampicillin, gentamicin, amikacin, streptomycin, kanamycin, neomycin, limecyclin, colistin, cefadroxil, cephadrine, cefuroxime, cefoxitine, and rifamycin are alike in their kinetic fate: long serum lifetimes and liver accumulation of the dextran derivatives, contrasted by short serum lifetimes and good urinary elimination of the inulin conjugates (20–30% of the free antibiotic). The percentage of antibiotic in the conjugate ranges from 12 to 60%, and consequently the antimicrobial potency of the conjugate is unique for every compound. The percentage of incorporation of the drug depends on the chemical reactivity of the amino groups during coupling.

By intramuscular administration, the conjugates are well absorbed,

[8] L. Molteni, *in* "Optimization of Drug Delivery" (H. Bundgaard, A. B. Hansen, and H. Kofod ed.), p. 285, Alfred Benzon Symposium 17. Munksgaard, Copenhagen, 1982.

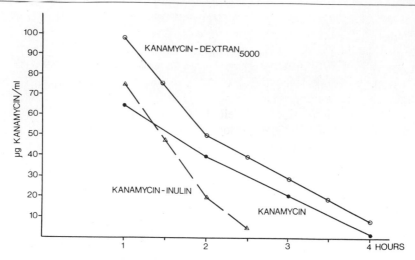

Fig. 4. Serum kanamycin concentration in rats after intramuscular administration (25 mg/kg).

and sometimes the carrier improves the bioavailability of the antibiotic, as exemplified by derivatives of kanamycin (Fig. 4).

The toxicity of the dextran and inulin derivatives is low and depends on the toxicity and percentage of the drug linked to the macromolecule. The immunogenicity of certain enzymes is decreased when complexed with dextran, where probably the binding of the polysaccharide sterically protects the foreign protein from recognition by the immune system.[9] The daunomycin–dextran conjugate is endowed with a markedly higher therapeutic index, as compared with the free drug, because of its low toxicity.[10] Additionally, research on dextran derivatives of β-adrenergic drugs shows that the conjugates are not able to penetrate the cell or reach the central nervous system and that some typical side effects of β-adrenergic agents may be eliminated when conjugated.[11]

Possible Therapeutic Application

The reduced toxicity of some toxic drugs when conjugated to dextran can result in safer and more effective therapy. The lack of oral bioavailability of dextran and inulin limits the therapeutic application of their conjugates to parenteral or topical administration. The chemical stability

[9] A. P. Kashkin et al., Khim. Farmatsert. Zh. 12, No. 7, 27 (1978).
[10] Levi-Schaffer et al., Cancer Treat. Rep. 66, No. 1, 107 (1982).
[11] J. Pitha, Abstr. Paper Ann. Chem. Soc. 181, Meeting ORPL 16 (1981).

and the good water solubility at neutral pH of the conjugates (even if the parent drug is unstable, poorly soluble, or soluble only at acid or alkaline pH) permit the use of these preparations as injectables or topical dermatologics as well as ophthalmic agents.

A topical delivery for the gastrointestinal tract can be achieved by the oral use of the nonabsorbable drug conjugates of dextran and inulin. Dextran and inulin derivatives are well absorbed by intramuscular administration and eventually useful in prolonging or shortening the serum half-lives of the coupled drugs. The intraarticular injection of dextran derivatives of antibiotics, antiinflammatory agents, or antiarthritic drugs permits a longer *in situ* time in order to reduce the frequency of injections.

On the other hand, some problems of immunogenicity rarely arise after infusion administration of high molecular weight dextrans. As a consequence, the use of their conjugates (intramuscular or intravenous) may involve the possible risk of some immunological reactions. For this reason, it is desirable to use in the conjugate preparation low molecular weight dextrans (1,000, 5,000, 10,000), which are also able to provide their conjugates with good chemical and biological stability and prolonged pharmacokinetics. It must also be kept in mind that during the conjugation reaction, some cross-linking occurs and that the molecular weight of the product conjugate is bigger than expected.

Conclusions

Many drugs conjugated to a polysaccharide carrier, such as dextran and inulin, exhibit improved chemical and biological stability as well as changes in their pharmacokinetics. The bond between the carrier and the drug can be an ester, and in this case the resulting conjugate acts as a slow-release system. The bond attaching the drug to the polymer may give the macroconjugates better chemical, chemicophysical, and biological properties as well as different pharmacokinetics.

In the case of some derivatives, the macromolecule acts by itself, without releasing the free drug. For this reason the functional groups related to the pharmacological activity of the drug should not be blocked during conjugation. Different molecular weight conjugates can be obtained with dextrans. The higher the molecular weight of the dextran derivative, the more they are characterized by long serum pharmacokinetics which favor uptake by the reticuloendothelial system. The inulin derivatives are rapidly excreted by kidneys without undergoing any metabolism, exerting their pharmacological activity directly in the urinary tract.

In conclusion, by esterification of a drug with dextran and inulin it is possible to produce a slow-release form of the drug. On the other hand, linking a drug or a biologically active substance by a nonpolar bond to dextran, it is possible to prolong its serum lifetime and pharmacobiological activity and to selectively reach the RES tissues, while the nonpolar bond to inulin gives the final conjugate a urinary tract tropism. In all these conjugates (by ester or nonpolar linkage), the polysaccharide carrier will improve the chemical and chemicophysical stability of the linked drug.

[23] Carrier Potential of Glycoproteins

By JUDITH L. BODMER and ROGER T. DEAN

Introduction

In a previous theoretical article,[1] one of us has outlined the wide variety of possible applications of glycoproteins in targeting drugs. It is now well appreciated that glycoproteins can be used to direct drugs to specific cell types, by virtue of the range of plasma membrane receptors for glycoproteins which characterize particular cell types. Some of these receptors are listed in the table. In general they mediate entry into cells by endocytosis and so are valuable in directing molecules to endosomes, lysosomes, and other structures with which such vesicles fuse. In addition, small permeable molecules may be deposited inside particular cells by such means and then, after hydrolysis of their carrier, be freed to diffuse throughout the cell, thus reaching cytosolic and other locations nonselectively. Glycoproteins have also found application as spacer molecules, thereby allowing larger amounts of a drug to be linked to antibodies (which were used to target the drugs) than is possible to antibody alone.[2]

Glycoproteins may also be used to direct liposomes, erythrocyte ghosts, or viral envelopes to specific cell types, with whose cell surface they fuse, thereby depositing their soluble contents into the interior of the cell. If the molecules so delivered are essentially membrane impermeable, then their effects can be restricted to the areas topologically included within the cytosolic space of the recipient cells.

[1] R. T. Dean, in "Drug Carriers in Biology and Medicine" (G. Gregoriadis, ed.), p. 71. Academic Press, New York, 1979.
[2] R. W. Baldwin et al., Bulletin du Cancer 70, 132 (1983).

METHODS IN ENZYMOLOGY, VOL. 112

SOME CELLULAR RECEPTORS FOR GLYCOPROTEINS

Receptor	Representative location	Reference for introduction
Serum glycoproteins recognized primarily by carbohydrate termini		
Galactose-terminated glycoproteins	Hepatocytes	3
Other serum glycoproteins		
Transferrin	Reticulocytes	4
α_2-Macroglobulin	Macrophages; fibroblasts	5
Lysosomal enzymes		
Mannose-terminated enzymes	Macrophages	6
Mannose 6-phosphate terminated enzymes	Fibroblasts	7
Hormones and growth factors		
TSH	Thyroid	8
Gonadotropins	Gonadal cells	8
Cellular growth factors, such as epidermal growth factor	Fibroblasts	8
Opsonins		
Complement components	Leukocytes	9
Antibodies	Leukocytes	9

It is less widely realized that glycoproteins might also be used to direct molecules selectively to other intracellular locations outside the confines of the vacuolar system and the cytosol. For instance, a particle containing on its surface a glycoprotein recognizing an externally disposed molecule on mitochondria might be loaded into the cytoplasm and then interact selectively with mitochondria. The loading might be achieved by entrapping the particle within unilamellar liposomes, which could be of a constitution permitting fusion with the cell surface and could themselves bear a surface glycoprotein directing them to the cell type of interest.[1]

Such possibilities remain only theoretical and will not be discussed in this primarily practical chapter. However, the first-mentioned general category of glycoprotein targeting—directing molecules to specific cells which can then endocytose them—has now received a significant amount of experimental attention, although no such targeting mechanism has achieved routine clinical use. The interest has been such that several of the glycoprotein–receptor systems now merit individual attention and are to be found in other parts of this volume. Their use in directing liposomes will be covered in the second part of this volume and thus is excluded from the present chapter. We will also exclude discussion of the fact that useful glycoprotein carriers need to be presented in such a way that they are both stable in the bloodstream and yet not highly immunogenic.

Thus, we present practical details for a necessarily arbitrary selection from the remaining glycoprotein–receptor systems which are being studied as drug carriers, primarily by endocytic routes (see table).[3-9] We mention more briefly their future application in directing viral coats containing drugs to chosen cells since this topic is not receiving substantial treatment elsewhere in the volume. It needs to be admitted at the outset that the inclusion of the word *potential* in the title of this chapter is entirely apposite; and hence we have resisted the temptation to elaborate on an excessive number of individual cases but rather have chosen a few which illustrate the general methods.

Glycoproteins Which Can Be Used to Target Drugs by Pinocytosis

Some of the major cell surface receptors for gycoproteins are listed in the table. Although it is not absolutely clear in all cases, it seems likely that in most, cell surface binding of glycoprotein ligand to its receptor can lead to pinocytosis of the ligand. Thus, most of these ligands have potential application as drug carriers, whether by conjugating a natural ligand or a synthetic analog of the ligand (e.g., a neoglycoprotein) to a drug. We have chosen to describe four systems which use receptors for carbohydrate moieties of serum proteins and lysosomal enzymes and for the serum proteinase inhibitors α_2-macroglobulin (A_2M) to carry drugs. All these systems, of course, require the purification of the glycoprotein, but this will not be discussed here. Rather, we describe the conjugation procedures and such limited information about the application of the product as is available.

Galactose-Terminated Fetuin as a Carrier for Pepstatin

Pepstatin is a group-specific inhibitor of aspartic proteinases and has some clinical applications already. Together with other more selective congeners, which might be made specific, for instance, for renin, the inhibitor has considerable interest. Although directing pepstatin to the liver is not difficult,[10] the first pepstatin–glycoprotein carrier which has been made, asialofetuin (galactose-terminated)–pepstatin, has been de-

[3] L. Fiume et al., FEBS Lett. **153**, 6 (1983).
[4] J.-N. Octave et al., Trends Biochem. Sci. **8**, 217 (1983).
[5] J. Kaplan and E. A. Keogh, Cell **24**, 925 (1981).
[6] C. Tietze et al., J. Cell Biol. **92**, 417 (1982).
[7] A. Gonzalex-Noriega et al., J. Cell Biol. **85**, 839 (1980).
[8] D. Evered and G. M. Collins, eds., Ciba Found. Symp. **92** (1982).
[9] S. C. Silverstein et al., Annu. Rev. Biochem. **46**, 669 (1977).
[10] R. T. Dean, Nature (London) **257**, 414 (1975).

signed to achieve selectivity toward hepatocytes. The procedure involves purification of the serum glycoprotein fetuin (from commercial samples) and its enzymatic (or chemical) desialylation—to expose terminal galactose residues. The conjugation has been developed by Furuno et al.,[11] who employed enzymatically desialyated fetuin.

The coupling of pepstatin to asialofetuin was through the ε-amino groups of the protein to the N-hydroxysuccinimide ester of pepstatin. Eighteen μmol of pepstatin is dissolved in 1 ml of dimethylformamide. Thirty-six μmol of 1-ethyl-3-(3-dimethylaminopropyl)carbodiimide and 36 μmol of N-hydroxysuccinimide are then added. After 2 hr at room temperature, the resultant mixture is added dropwise to 30 ml of 0.1 M sodium bicarbonate solution containing 3 μmol of asialofetuin. This mixture reacts for 2 hr at room temperature. Products can be separated on a column of LH-20 (3.5 × 50 cm), eluted with 0.9% NaCl. The conjugate elutes in the void volume while free pepstatin is retarded by the gel. The product contains approximately three pepstatin molecules per mole of protein, although the inhibitory action of pepstatin is somewhat reduced in the conjugate.

When rats were injected in the jugular vein with the resulting conjugate (3 mg/animal), the conjugate reached very similar specific activities in hepatocytes and nonparenchymal cells (as measured in isolated cells obtained from EGTA-perfused livers). In contrast, pepstatin conjugated to denatured bovine serum albumin was primarily present in the nonparenchymal cell fraction. When free pepstatin was administered to the animals, very little accumulated in the liver, a finding showing the effectiveness of the conjugate. Subcellular fractionation experiments showed that the conjugate reached lysosomes within 2 hr of presentation, but that after more prolonged periods (~6 hr), 80% or more of the pepstatin was lost into the bile.

Unfortunately, the conjugate did not substantially retard the degradation of fluorescein–asialofetuin by the livers of the same rats. However, pepstatin is not very effective in any circumstances in inhibiting the degradation of this molecule, so that it remains possible that active intralysosomal concentrations of the inhibitor were obtained. A further difficulty the authors observed was that apparently intact lysosomes lost pepstatin (presumably free pepstatin after hydrolysis of the carrier) when incubated in vitro, and while retaining some lysosomal enzymes. This is in spite of the fact that pepstatin presented externally to cell membranes or to lysosomal membranes does not seem to penetrate. This exit of pepstatin may be due to the acid pH within the lysosomes (contrasting with the neutral

[11] K. Furuno et al., J. Biochem. (Tokyo) 93, 249 (1983).

medium). Thus, the conjugate may load not only the lysosomes but also the cytosol of the recipient cells.

Asialofetuin as a Carrier for Trifluorothymidine

As reviewed in ref. 3, conjugates of inhibitors of DNA synthesis are of interest in viral and other diseases because the free inhibitors are often rather toxic and have widespread effects on DNA synthesis besides those on the appropriate target cells. Several possibly valuable conjugates have been developed: we shall describe the preparation of the trifluorothymidine (F_3T) conjugate of asialofetuin (ASF), as developed by Fiume et al.[12] Similar methods have been used to conjugate adenine 9-β-D-arabinofuranoside, which is less toxic.[3,13] The possible difficulty of immunogenicity may be avoided by the use of the neoglycoprotein, lactosaminated human albumin, which also binds to the hepatocyte galactose-glycoprotein receptor.[3]

The conjugation of F_3T to asialofetuin is performed by coupling the hydroxysuccinimide ester of F_3T-glutarate to ASF, as follows. F_3T (500 mg, 1.69 mmol) is made up in 20 ml pyridine. Glutaric anhydride (245 mg, 2.15 mmol) is added gradually over 2 hr with stirring, at 80°. The mixture is then kept at the same temperature overnight before being evaporated under vacuum. The residue is dissolved in 1.5% aqueous acetic acid and chromatographed in the same solvent on Sephadex LH-20 (200 × 3 cm); 20-ml fractions are collected. Under these circumstances, fractions 73–86 contain F_3T and fractions 87–104 contain F_3T-glutarate (~180 mg). The latter are lyophilized and seem to contain a mixture of 5'-mono- and 3',5'-diglutarate.

The glutarate mixture is dissolved in dimethylformamide (5 ml). N-Hydroxysuccinimide (79 mg, 0.69 mmol) and dicyclohexylcarbodiimide (142 mg, 0.69 mmol) are added gradually at 4°. After 72 hr at 4°, the precipitate is filtered, and the solution added over a period of 4 hr to a solution of 200 mg ASF in 20 ml water at pH 6–7, 4°. The reaction mixture is stirred at 4° overnight. After this, the mixture thoroughly is dialyzed against water at 4° and lyophilized. The product is dissolved in 10 ml of 0.05% ammonium bicarbonate and chromatographed in the same solvent on Sephadex G-75 (100 × 2 cm). Fractions of 5 ml are taken, and F_3T–ASF is recovered in fractions 9–24, with a F_3T : ASF molar ratio of about 8 : 1.

The product is recognized normally by the hepatic galactose-glycoprotein receptor; and its clearance is not inhibited by fetuin or denatured

[12] L. Fiume et al., FEBS Lett. **103**, 47 (1979).
[13] L. Fiume et al., FEBS Lett. **116**, 185 (1980).

bovine serum albumin. It has been tested as an inhibitor of the hepatitis caused by ectromelia virus—specifically of the enhanced DNA synthesis (mainly viral DNA) in hepatocytes which occurs after fairly prolonged infection. F_3T–ASF was much more effective in liver than F_3T. But it had an inhibitory effect also on DNA synthesis in bone marrow, an effect of comparable magnitude to that produced by F_3T alone. Thus, although selective for liver, the conjugate was nevertheless reaching other undesired target cells, possibly after release of F_3T from the conjugate by hydrolysis in hepatocytes. This problem might be avoided by use of inhibitors which are rather less membrane permeable.

α_2-Macroglobulin as a Carrier for the Toxin Ricin

Although ricin is itself a glycoprotein, binding sites for it (containing galactose) are relatively widely distributed among cell types, and, thus, targeting of the toxin is necessary if it is to be used for selective killing of a particular kind of cell, such as a tumor cell. In this section we describe targeting by means of conjugates with the proteinase inhibitor α_2-macroglobulin, whose receptors are less widely distributed. The succeeding section will deal with ricin targeting by incorporating a modified carbohydrate ligand into the toxin.

The toxic moiety of ricin, the A chain, is normally carried by the B chain, which binds to the cell surfaces and in an unknown manner facilitates entry of the A chain into the interior of the cell, where it inhibits protein synthesis. The B chain also seems to protect the A chain against extracellular proteolysis. To remove the directional influence of the B chain, it needs to be removed and replaced by a targeting molecule which will also prevent excessive breakdown. α_2-Macroglobulin (A_2M) has been chosen for this purpose.[14]

The preparations of A_2M and ricin used[14] will not be described. Ricin A chain was obtained by chromatofocusing. The following procedure is used for the conjugation of A_2M and ricin A chain,[14] using N-succinimidyl-3-(2-pyridyldithio)propionate. Human A_2M (4 mg/ml) is dissolved in 50 mM sodium phosphate buffer, pH 7.6, containing 200 mM NaCl. It is reacted with 46 μM N-succinimidyl-3-(2-pyridyldithio)propionate and after 20 min at room temperature separated from unreacted reagent by chromatography on Sephadex G-50. Ricin A chain (0.45 mg/ml) is then reacted with the modified A_2M (0.94 mg/ml) for 12 hr at 25° while being dialyzed against 250 mM sodium phosphate buffer, pH 7.6, also containing 200 mM NaCl. The resulting conjugate can be purified by gel filtration

[14] H. B. Martin and L. L. Houston, *Biochim. Biophys. Acta* **762**, 128 (1983).

of Sephadex G-100 in a phosphate-buffered saline at pH 6.5. After concentration by ultrafiltration, the product can be stored at −80°.

The conjugate was much more effective in inhibiting protein synthesis in cultured human fibroblasts than was free ricin A chain, though it was far less effective than native ricin. Inhibition required uptake; and from morphological studies, it seemed that the conjugate must enter by receptor-mediated endocytosis, and then at least the A chain must survive the endosomal or lysosomal acidity. The application of this conjugate has so far only been *in vitro*; further work will be needed to establish whether useful targeting (and stability) can be obtained *in vivo*. However, the *in vitro* studies are at least encouraging.

Mannose 6-Phosphate-Target Ricin

Several neoglycoproteins have been synthesized, at least partly with the aim of using the newly incorporated carbohydrate moiety to direct the recipient macromolecules to specific cells.[15] The main example we describe is the incorporation of mannose 6-phosphate (M6P) oligosaccharides into ricin[16]; although later we mention some particles whose carbohydrate moieties can similarly be chemically modified. Modification of synthetic soluble polymers, incorporating both drugs and carbohydrate directional ligands, is discussed elsewhere in this volume.

In the present example, M6P in monophosphopentamannose (MPM) is incorporated into the native ricin molecules. The MPM can be isolated from certain yeast strains and from some commercial mannan preparations. The conjugation of MPM to ricin can be achieved by reductive amination of the Schiff base between C-1 of the reducing terminal sugar and free amino groups on the protein.[16] A 0.2 M solution of the MPM, with ricin (15 mg/ml) and NaCNBH$_3$ (159 mM) in N,N-bis(2-hydroxyethyl)glycine (bicine), 50 mM, pH 9, is incubated for 24 hr at 37°. Unreacted materials are then removed by dialysis at 4° against 10 mM Tris-HCl, pH 7.5. Molar ratios (MPM : ricin) of about 7 : 1 are obtained.

This conjugate, of course, still contains the ricin B chain binding sites. *In vitro* the action of these can be suppressed by 100 mM lactose, and thus it could be shown that the M6P-containing ricin had a highly selective toxicity for fibroblasts (containing an M6P receptor) and much less effect on HeLa and other cells which do not bear this receptor, providing lactose was present. Unfortunately, the B chain seemed still to be important in the conjugate; and in parallel experiments,[16] M6P conjugates of the

[15] Y. C. Lee and R. T. Lee, in "The Glycoconjugates" (M. I. Horowitz, ed.), Vol. 4, p. 57. Academic Press, New York, 1982.
[16] R. J. Youle *et al., Proc. Natl. Acad. Sci. U.S.A.* **76,** 5559 (1979).

separated A chain of another toxin (diphtheria) were not toxic. Ricin A chain was not stable to the reductive amination procedure and so could not be used alone. Thus, since lactose could not be conveniently used with the ricin–MPM conjugate *in vivo*, this approach to cell-specific toxins remains, like most, at the potential stage. But there seems no clearly insuperable problem in the application.

Targeting of Phagocytosable Particles

While microcapsules and liposomes are excluded from this chapter since they are covered extensively elsewhere in these volumes, it may be worth noting that other phagocytosable particles, whether synthetic (latex or Sepharose) or organic (cell wall materials or even killed parasites), may on occasion be useful carriers or immunogens and may require targeting. Such particles are avidly engulfed by leukocytes—"professional" phagocytic cells. In addition, as noted in the table, opsonins produced by the host (including antibodies and complement) may enhance the efficiency of this process. However, such opsonins usually enhance the activity of most leukocytic phagocytes in endocytosing the opsonized particles.

Where selectivity of uptake is needed for a particular phagocyte (e.g., the macrophage), or for another cell type which has intrinsic, albeit slighter, phagocytic capacity (e.g., the fibroblast), other targeting methods are needed. And, indeed, some such targeting methods are used by certain organisms themselves, most notably parasites. The possible utility of a rigid particle for successive penetration of both the cell membrane and then the cytoplasmic face of an intracellular membrane has been mentioned in the introduction; thus, there probably will be future applications of this idea.

To illustrate a possible approach, we choose the targeting of particles to macrophages by means of the mannose-glycoprotein receptor (see table). There is evidence that zymosan binds to this receptor during phagocytosis and that this receptor interaction is sufficient to induce several kinds of macrophage activation (e.g., ref. 17 for discussion). Thus, many of the methods of linking mannose to polymers[15] might be used to prepare mannose-terminated inert particles to which drugs could then also be conjugated. Several such particles (containing the mannose termini) are commercially available, intended for affinity chromatography of lectins and glycosidases. Of these, mannose conjugated to 6% agarose (Pierce) does possess selective binding capacity for macrophages (our unpublished results).

[17] J. L. Bodmer and R. T. Dean, *Biochem. Biophys. Res. Commun.* **113**, 192 (1983).

Targeting by Means of Fusion at the Cell Surface

Glycoproteins have considerable potential for targeting molecules to cells when they are required to be deposited in the cell cytosol rather than in the lysosomal system. Normally this involves sequestration of the carried drug within a capsule through which it essentially cannot permeate. The capsule is then designed so that it can fuse with the surface of the target cell, depositing its contents within. Liposomes and capsules and several other such vehicles are discussed elsewhere in these volumes. Here we mention briefly the use of viruses for encapsulating and delivering drugs, targeted by glycoproteins.

Reconstituted envelopes of fusogenic animal viruses such as Sendai virus can be used to carry various molecules such as proteins or drugs. Besides incorporating molecules into the intraviral space, it is also possible to incorporate them into the surface membrane, providing that they are not excessively sensitive to proteolysis by viral coat enzymes. Thus, a glycoprotein could be incorporated into the surface to determine the specificity of binding of the reconstituted particle and, thus, to direct its contents to particular cell types.

There are many problems to be overcome before such a system will be practically applicable. For instance, many such viruses enter the cytosol after endocytosis,[8] although fusion (instead of endocytosis) may be encouraged in some cases by incubation at low pH (<6.0). However, such a regime is unlikely to be useful *in vivo*. Nevertheless, the paramyxoviruses seem to undergo fusion as their prime mode of entry, so these may be possible vehicles.

Future Developments: Targeting to Other Intracellular Sites

Glycoproteins are relatively rare in the interior of cells, particularly in the cytosol. But they are present to some extent on internal membranes, and they may be concerned with specific membrane–membrane interactions.[18] Thus, by using a two-layered structure as outlined in the introduction, investigators might exploit intracellular glycoprotein recognition to direct materials to particular organelles.

---

[18] R. C. Hughes, "Membrane Glycoproteins." Butterworth, London, 1976.

Section III

Prodrugs

[24] Theory and Practice of Prodrug Kinetics

By Robert E. Notari

Drug Problems: Dosage Form versus Biological

Prodrug Objectives

Prodrug goals and their corresponding kinetic properties have been reviewed elsewhere.[1,2] The theoretical requirements, feasibility for success, assessment of success, and the number of examples of successful prodrugs all vary with the specific goal. Some typical problems which prodrugs are designed to overcome are listed in Table I. The pharmaceutical problems have more readily succumbed to the prodrug approach than have the pharmacokinetic problems wherein the number of successful examples decreases roughly in the order of the problems listed.

There is one goal common to all prodrugs. After completion of the objective, each prodrug should quantitatively yield its drug. For example, if a prodrug improves on the taste of the drug, then rapid prodrug conversion should take place preferably within the gastrointestinal tract but no later than prodrug arrival in the blood. If drug is poorly absorbed following oral administration, its absorbable prodrug should rapidly reverse in the blood. If the drug is unstable as a powder, the stable prodrug may reverse upon dissolution but no later than its arrival in the blood. Intact prodrug circulating in the blood is subject to loss by routes which compete with drug formation. Also, distribution of prodrug throughout the body necessitates clinical toxicology and pharmacology studies on the prodrug itself. Thus, with the exception of target site enrichment or rate-limiting conversion for extended duration, circulating intact prodrug is contrary to the objectives.

Prodrug Evaluation

The kinetic requirements specific to each prodrug goal will define the evaluation procedure and the type of data which constitute success. Although drug bioavailability is paramount in all cases, the evaluation of a prodrug designed specifically for that task will differ from one designed to improve taste or to prolong duration. The prodrug designed solely to

[1] R. E. Notari, *Pharmacol. Ther.* **14**, 25 (1981).
[2] R. E. Notari, *in* "Optimization of Drug Delivery" (H. Bundgaard, A. B. Hansen, and H. Kofod, eds.), p. 117. Munksgaard, Copenhagen, 1982.

TABLE I
DRUG PROBLEMS POTENTIALLY ALLEVIATED BY
PRODRUG FORMATION

Pharmaceutical	Pharmacokinetic
Unpleasant taste	Poor bioavailability
Pain on injection	Short duration
Poor solubility	High first-pass metabolism
Instability	Toxicity/side effects
Slow dissolution rate	Nonspecificity

overcome a pharmaceutical problem must, in addition to that primary task, provide a drug plasma concentration time course that is bioequivalent to the reference. Conversely, bioequivalency represents failure in a prodrug designed to improve absorption. The bioavailable dose of a prodrug for prolonged action should be equal to or better than the drug itself, but the duration of the terminal phase of the drug plasma concentration time course should be significantly longer. Each unique prodrug task thereby imparts individual specifications on its own assessment.

Reversal Mechanisms

"Triggers"

It is necessary to control the starting point for reversal of prodrug to drug as well as the ensuing rate of conversion. The mechanism employed to "trigger" the reaction may be chemical or enzymatic. It may not be possible to actually start the conversion mechanism, but rather the rate of conversion may suddenly be greatly increased. Triggers may include a change in pH or the sudden introduction of water or enzymes. Success is limited more by the absence of a reliable means to trigger the reaction than by the ability to control the rate once it has begun.

Enzymatic Reversal

The conversion of these prodrugs to their corresponding drugs is dependent upon the presence of enzymes capable of metabolizing the prodrug–drug linkage. The trigger may rely on the absence of enzymes in the formulation and their known presence in the body or their difference in regional activity throughout the body. To the extent that enzymatic activity can be reliably predicted, reversal can be controlled. When the enzymes are ubiquitous, these prodrug applications become limited.

High interpatient variability may be anticipated unless there is excess bioactivity in most patients, i.e., esterases. High variability is suggested by the observation that highly metabolized drugs elicit more intersubject variability in elimination rates than those which do not undergo metabolism.[3]

In spite of this drawback, enzymatic reversal offers greater potential for control of the reversal site than does the chemical mechanism.

Chemical Reversal

Prodrug conversion rates by nonenzymatic mechanisms may show less intersubject variability than those which rely on metabolic routes. However, chemically reversed prodrugs present limited opportunity to gain an advantage. This is due to the lack of a "trigger." The most common reversal mechanism used is that of hydrolysis. Differential hydrolysis rates must arise from differing pH values. It is common to formulate liquid preparations to pH values compatible with the physiological pH of the site of administration. In this case reversal *in vivo* means prodrug instability *in vitro*.

To some extent the pH differentials within the body may be employed. For example, the stomach pH varies from approximately 1 to 4, the intestinal pH from 5 to 8, and the blood pH is 7.4. In most cases one would expect the chemical reversal rate in acid to exceed that at neutral pH. The opposite extreme would be represented by a prodrug subject only to hydroxyl-ion attack. While this is unlikely for the stomach pH region, the reversal rate at pH 7 would be 10^3 faster than at pH 4 for such a case. Such a prodrug could improve on the oral absorption of an acid-unstable drug provided that the prodrug is more acid stable than the drug but converts rapidly to drug in the intestines or blood. If the reversal rate in the intestines (pH 7) had a 10-min half-life, then in the extreme (10^3) difference, the half-life at pH 3 would be 167 hr. This provides a wide margin of differentiation, assuming that reversal represents the only mechanism for prodrug loss. In practice, the prodrug may also be likely to undergo the same degradation pathway as the drug unless the degrading moiety has been derivatized. In addition, the rate of loss will usually increase at the lower gastric limit of pH ~1.

In spite of these potential limitations, success has been realized in designing prodrugs which exhibit increased gastric stability and rapid reversal in the intestines and/or blood.[4] A prodrug designed for increased

[3] R. E. Notari, "Biopharmaceutics and Clinical Pharmacokinetics," 3rd ed., p. 290. Dekker, New York, 1980.
[4] A. Tsuji, E. Miyamoto, T. Terasaki, and T. Yamana, *J. Pharm. Sci.* **68**, 1259 (1979).

dry stability may chemically reverse in contact with water for injection.[5] Prodrugs can be designed to increase duration through controlled chemical hydrolysis of prodrug in the blood following parenteral administration.[6] Beyond these few successful examples, it is difficult to hypothesize many more problems, especially of the biological type, which can be overcome by a chemically reversing prodrug.

Mixed Mechanisms

Prodrugs may undergo both chemical and enzymatic reversal. On oral administration, the site for reversal may be the intestines or the blood or both. Often the mechanism and site are not precisely known. When the prodrug is designed to improve bioavailability (as in the case of drug instability in the stomach), then a mixture of mechanisms and sites is acceptable so long as drug bioavailability is improved. In many cases it is easier to demonstrate that a prodrug is successful than it is to prove the exact site and mechanism for reversal. While conversion is seldom solely enzymatic or chemical, it is common for the rate of one or the other to predominate, depending upon physiological conditions.

In Vitro/In Vivo Kinetics

Model-Independent versus Model-Dependent Assessments

It has been convenient to employ pharmacokinetic modeling in the development of prodrug theories.[1,7–9] Such an approach has produced useful generalities which can be defined in model-independent terms. In practice, it is prudent to develop theory by use of models and to evaluate prodrugs by model-independent assessments. Unless specific data are obtained to unequivocally define a particular model, the use of models in prodrug evaluation serves only to add a degree of speculation to the final conclusions.

The choice of an appropriate model using the time course for drug and prodrug concentration in plasma is often an arbitrary selection from kinetically equivalent models. The same may be said for the choice of a model following administration of the drug itself.[10] The time course for drug

[5] J. S. Wold, R. R. Joost, H. R. Black, and R. S. Griffith, J. Infect. Dis. 137, Suppl., S17 (1978).
[6] L. E. Kirsch and R. E. Notari, J. Pharm. Sci. 73, 728 (1984).
[7] P. R. Byron, R. E. Notari, and M.-Y. Huang, Int. J. Pharm. 1, 219 (1978).
[8] R. E. Notari, M.-Y. Huang, and P. R. Byron, Int. J. Pharm. 1, 233 (1978).
[9] V. J. Stella and K. J. Himmelstein, J. Med. Chem. 23, 1275 (1980).
[10] J. G. Wagner, J. Pharmacokinet. Biopharm. 3, 457 (1975).

concentration in plasma following a rapid intravenous injection of drug can often be described by

$$C = \sum_{i=1}^{n} C_i e^{-\lambda_i t} \qquad (1)$$

If Eq. (1) is found to be adequate, independent of the dose, the drug is said to behave according to linear kinetics. This implies that drug plasma concentrations are linearly related to dose and that the time course for C is the sum of one or more apparent first-order rate processes. To date, most drugs behaving by linear kinetics have been accommodated by a monoexponential ($n = 1$), biexponential ($n = 2$), or triexponential ($n = 3$) form of Eq. (1). These model-independent descriptions can be reliably employed to evaluate useful pharmacokinetic parameters such as half-life, clearance, bioavailability, and volume of distribution (Table II). If drug is administered by an extravascular route with apparent first-order input constant (k_a), then in theory Eq. (1) would require an additional term over that observed by rapid intravenous injection, $-[\sum_{i=1}^{n} C_i] e^{-k_a t}$. In practice, this term is often a hybrid with the first or last intravenous term so that data obtained by an extravascular route may not predict the n, C_i, or λ_i values to be observed by intravenous administration. Nonetheless, when the model-independent equations are operationally valid, they can be used to make predictions and assessments.

The further extension of these equations to compartmental interpretation serves only to introduce uncertainty if $n > 1$.[10] For example, when $n = 2$, the minimum model satisfying this equation is one with the sampled compartment (compartment 1) reversibly transferring drug to at least one other compartment (compartment 2), where one or both have exit constants. There are at least three possibilities

$$3 \leftarrow 1 \rightleftarrows 2 \qquad (2)$$
$$1 \rightleftarrows 2 \rightarrow 3 \qquad (3)$$
$$4 \leftarrow 1 \rightleftarrows 2 \rightarrow 3 \qquad (4)$$

where compartments 3 and 4 represent drug metabolized or eliminated from the body. Multiple exit constants would allow additional models, all of which provide biexponential ($n = 2$) loss for C when drug is rapidly introduced into compartment 1. Based upon data for drug concentration in plasma, these models are kinetically indistinguishable, yet model (2) is commonly employed.

When one model is arbitrarily selected out of several possibilities, Eq. (1) (describing the time course for compartment 1) remains operative, but predictions for compartment 2 and all calculations employing the individ-

TABLE II

CALCULATION OF PHARMACOKINETIC DESCRIPTIVE PARAMETERS
USING MODEL-INDEPENDENT EQUATIONS

Pharmacokinetic property	Equation	Application
Time course for drug concentration in plasma	$C = \sum_{i=1}^{n} C_i e^{-\lambda_i t}$	Following rapid intravenous administration
	$C = \sum_{i=1}^{n} C'_i e^{-\lambda_i t} - \left[\sum_{i=1}^{n} C'_i\right] e^{-k_a t}$	First-order input where $k_a > \lambda_1$
Biological half-life	$t_{1/2} = 0.693/\lambda_z$	Where λ_z is the smallest λ_i following intravenous administration
Area under the plasma concentration curve from time 0 to ∞	$AUC = \sum_{i=1}^{n} (C_i/\lambda_i)$	For $C = \sum_{i=1}^{n} C_i e^{-\lambda_i t}$
	$AUC = \sum_{i=1}^{n} (C'_i/\lambda_i) - \left(\left[\sum_{i=1}^{n} C'_i\right]/k_a\right)$	Following first-order input where $k_a > \lambda_1$
	AUC from trapezoidal rule	Directly from time-course data; no descriptive equation required
Bioavailability fraction	$f = AUC_{ev}/AUC_{iv}$	When AUC is proportional to dose, the extravascular (ev) bioavailability is related to the dose-adjusted AUC values
Total body clearance	$CL = fD/AUC$	Any route; following intravenous administration $f = 1$
Renal clearance	$CL_r = f(\text{total excreted})/AUC$	Any route; set $f = 1$ for intravenous administration
Apparent volume of distribution	$V_z = fD/\lambda_2 AUC$	Where λ_z is the smallest λ_i following intravenous administration

ual rate constants in the model become speculative. For example, both models (2) and (3) are adequately described by Eq. (1) with $n = 2$, but the physical meanings of the four constants differ, as shown in Table III. In addition, the time course for compartment 2 (which must be predicted) can differ considerably when model (2) is compared to model (3), as illustrated in Fig. 1. This has particular significance in understanding why prodrug conversion rate constant values derived from compartmental modeling have not been predictable from *in vitro* constants.

A rapid intravenous dose of prodrug may also result in data which are adequately described by Eq. (1) wherein the number of exponentials re-

TABLE III
COMPARISON OF THE EXPONENTIAL MULTIPLIERS (C_1, C_2) AND COEFFICIENTS (λ_1, λ_2)
WHEN DRUG ELIMINATION IS FROM COMPARTMENT 1, MODEL (2) COMPARED TO
ELIMINATION FROM COMPARTMENT 2, MODEL (3)

Parameter[a]	Model (2)	Model (3)
C_1	$\dfrac{\text{DOSE}(k_{21} - \lambda_1)}{V_1(\lambda_2 - \lambda_1)}$	$\dfrac{\text{DOSE}(k_{21} + k_{20} - \lambda_1)}{V_1(\lambda_2 - \lambda_1)}$
C_2	$\dfrac{\text{DOSE}(k_{21} - \lambda_2)}{V_1(\lambda_1 - \lambda_2)}$	$\dfrac{\text{DOSE}(k_{21} + k_{20} - \lambda_2)}{V_1(\lambda_1 - \lambda_2)}$
$\lambda_1 > \lambda_2$	$\dfrac{K1 \pm (K1^2 - 4K2)^{1/2}}{2}$	
$K1$	$(k_{12} + k_{21} + k_{10})$	$(k_{12} + k_{21} + k_{20})$
$K2$	$k_{21}k_{10}$	$k_{12}k_{20}$

[a] V_1 is the apparent volume of compartment 1 and k_{ij} is the first-order rate constant from compartment i to j.

quired to describe the prodrug (n') may or may not be the same as that required for the drug (n). Even when they are the same, there remains the uncertainty of the time courses for the compartments in the models for each species and the fact that the compartments are not anatomical pools

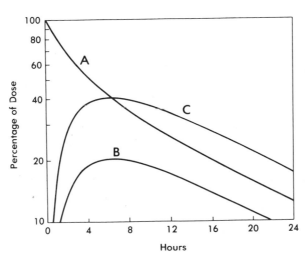

FIG. 1. Semilogarithmic plots illustrating how a single time course for compartment 1 (curve A) can predict different profiles for compartment 2 based upon its interpretation as model (2) (curve B) or model (3) (curve C). The time course for compartment 1 is 50% $e^{-\lambda_1 t}$ + 50% $e^{-\lambda_2 t}$ where $\lambda_1 = 0.34$ hr^{-1} and $\lambda_2 = 0.059$ hr^{-1}. The k_{12}, k_{21}, and k_{loss} ratios are 1, 2, 1 for model (2) and 2, 1, 1 for model (3).

but simply mathematical descriptions. Without definitive distribution data, one cannot assign either conversion sites or physiological meaning to a prodrug–drug compartmental model. This was borne out in a study specifically designed to test for the possibility of an *in vitro–in vivo* correlation.[6] The prodrug cyclocytidine was shown to convert to cytarabine by simple chemical hydrolysis. Each compound was described by Eq. (1), where $n' = n = 2$ following its rapid intravenous administration to rabbits. While the compartment 1 volume values were similar for both, the apparent volume of distribution for prodrug was roughly twice that of the drug. In spite of these apparent volumes, the model of best fit for drug and prodrug time courses following intravenous prodrug administration required conversion from both prodrug compartments 1 and 2 to compartment 1 of drug. While the conversion rate constant agreed with the *in vitro* hydrolysis rate constant, the model lacks physiological meaning. As discussed in that report, neither compartmental modeling nor physiological flow models can predict *in vivo* prodrug conversion rates from *in vitro* data but both are capable of retrospectively fitting *in vivo* data to the known *in vitro* result. This limits the value of *in vitro* testing to the screening of a few prodrug types.

In Vitro Screening

While *in vitro* testing may not be predictive, it is useful in a limited number of cases. In particular, *in vitro* screening is successful in the evaluation of prodrugs which are meant to undergo rapid enzymatic reversal in blood. It is also applicable to testing prodrugs designed to enhance stability wherein chemical and/or enzymatic reversal occurs prior to prodrug arrival in blood. In these examples, prodrugs are selected for *in vivo* testing when *in vitro* conversion rates are sufficiently rapid to ensure no significant plasma concentrations of intact prodrug. *In vitro* testing of prodrug esters of penicillins typify this approach.[11–13] These inactive esters must facilitate absorption and then rapidly hydrolyze in blood. Screening for *in vivo* testing was based on *in vitro* conversion rates with half-life values of a few minutes or less in human blood or diluted human blood together with sufficient stability in gastrointestinal fluids to allow absorption.

[11] B. Ekström, *Drugs Exp. Clin. Res.* **7**, 269 (1981).
[12] J. P. Clayton, M. Cole, S. W. Elson, H. Ferres, J. C. Hanson, L. W. Mizen, and R. Sutherland, *J. Med. Chem.* **19**, 1385 (1976).
[13] Y. Shiobara, A. Tachibana, H. Sasaki, T. Watanabe, and T. Sado, *J. Antibiot.* **27**, 665 (1974).

Drug Yields

Bioavailability

The bioavailability of drug from administered prodrug must be evaluated regardless of the route of prodrug administration. Even a rapid intravenous dose of prodrug can result in a reduction in bioavailability of drug if prodrug is lost to non-drug-producing routes. A model-independent approach should be used to assess the bioavailable dose which is defined as fD, where D is the administered dose and f is the fraction of that dose systemically available as drug. Calculations must be carried out on a molecular basis since the molecular weight of the drug and prodrug differ. The area under the time course for concentration in plasma (C in M) from time zero to infinity is called the area under the curve (AUC). In linear kinetics, the AUC is proportional to the bioavailable dose. The AUC value may be calculated graphically from the C versus t plot using the trapezoidal rule.[14] It may also be calculated from the equation of best fit. When the time course following an intravenous injection of drug is described by Eq. (1), the AUC may be calculated from

$$\text{AUC} = \sum_{i=1}^{n} (C_i/\lambda_i) \tag{5}$$

Thus, for a triexponential ($n = 3$) equation,

$$\text{AUC} = \frac{C_1}{\lambda_1} + \frac{C_2}{\lambda_2} + \frac{C_3}{\lambda_3} \tag{6}$$

In theory, the time course for plasma concentration of drug following the administration of its prodrug can exhibit up to five exponential terms when $n' = n = 2$ and administration is extravascular (Table IV).[1] In practice, it is more likely that the drug concentration in plasma will be described by an equation of two or three exponentials made up of constants which may or may not agree with the theoretical values. This is due to an inability to define all of the phases so that the operational equation contains hybrids of the theoretical components.[1] Regardless of the physical meaning (or lack thereof) for the constants, the equation describing the data can be used to calculate AUC from Eq. (5) provided it is of the general form of Eq. (1), as illustrated in Fig. 2.

Absolute Bioavailability

The fraction (or percentage) of the prodrug dose systemically available as drug constitutes absolute bioavailability. For linear kinetics, the

---

[14] R. E. Notari, "Biopharmaceutics and Clinical Pharmacokinetics," 3rd ed., p. 86. Dekker, New York, 1980.

TABLE IV

EXPONENTIAL COEFFICIENTS IN THE TIME-COURSE EQUATION FOR
CONCENTRATION OF DRUG IN PLASMA (C) FOLLOWING
ADMINISTRATION OF ITS PRODRUG[a]

Eq. (1)		Intravenous prodrug		Extravascular prodrug	
n'	n	Sum	Rate constants	Sum	Additional constants[b]
2	2	4	λ_1' λ_2' λ_1 λ_2	5	k'
1	1	2	λ_1' λ_1	3	k'
1	2	3	λ_1' λ_1 λ_2	4	k'
2	1	3	λ_1' λ_2' λ_1	4	k'

[a] For equations, see ref. 1.
[b] k' is the sum of k_a' plus any other competing first-order routes of prodrug loss from the site of administration.

bioavailable fraction (f) can be calculated from the ratio of the dose-adjusted AUC values for drug plasma concentration versus time following prodrug administration (AUC_{pd}) to that following an intravenous dose of drug itself (AUC_{iv}),

$$f = AUC_{pd}/AUC_{iv} \tag{7}$$

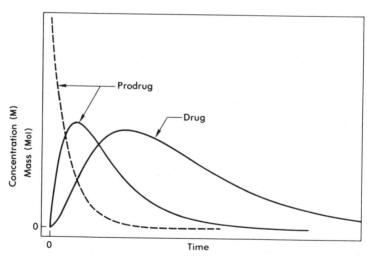

FIG. 2. The time course for prodrug remaining at the extravascular site of administration (−−), its concentration in plasma (——), and the resultant drug concentration which can be described by $C = (1.33e^{-4t} - 4.00e^{-2t} + 2.67e^{-t}) \times 10^{-4}\ M$. Using Eq. (5), AUC = [(1.33/4) − (4.00/2) + (2.67/1)] $\times 10^{-4}\ M$ − time = $1 \times 10^{-4}\ M$ − time.

assuming that clearance (CL) remains constant within the comparison. This relationship is based on the model-independent calculation for clearance

$$CL = fD/AUC \qquad (8)$$

when the ratio D/AUC is known to be constant.

Relative Bioavailability

Absolute bioavailability requires that the drug be administered by the intravenous route. This may not be feasible. Then, the bioavailability of drug from administered prodrug may be compared to an extravascular dose of a reference standard. The standard may be the drug product candidate for improvement, a solution of the drug, or both. According to Eq. (8), $(AUC)(CL) = fD$. When drug clearance is constant and AUC values are adjusted for dose, then the bioavailability of drug from prodrug (f_{pd}) relative to that from the reference standard (f_{ref}) may be calculated from

$$f_{pd}/f_{ref} = \text{relative bioavailability} = AUC_{pd}/AUC_{ref} \qquad (9)$$

where AUC values are calculated from drug plasma concentration time courses and neither f value is known.

Intentional Bioinequivalency

Bioequivalency requires identical drug plasma concentration time courses following equal doses of drug or prodrug. Since AUC values do not take the shapes of the curves into account, a relative bioavailability value of unity does not ensure bioequivalency.

The prodrug goals will dictate which comparisons are appropriate, and these will normally be obvious. A prodrug designed for increased shelf life must minimally be bioequivalent. One designed to increase absorption rate cannot be bioequivalent but should show a high peak drug plasma concentration (C_{max}) at a short time (t_{max}) without sacrificing bioavailable dose fD. Prodrugs designed to increase the duration of drug in the body must prolong the terminal drug concentration time course (by a process known as the "flip-flop" phenomenon[15]). In so doing, successful prodrugs cannot be bioequivalent. While it would be considered ideal to maintain a relative bioavailability ratio of unity in achieving this goal, it may not be possible. When duration can be increased with a significant resultant clinical advantage, then a compromise in f may be acceptable. However,

---

[15] P. R. Byron and R. E. Notari, *J. Pharm. Sci.* **65**, 1140 (1976).

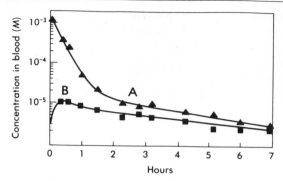

F<small>IG</small>. 3. Concentration time course for prodrug (A, cyclocytidine) and its drug (B, cytarabine) following an intravenous prodrug dose (3.8×10^{-4} mol/kg) to a rabbit. During the initial 15 min, the prodrug : drug ratio exceeds 100 so that prodrug controls to test conversion during the assay workup provided cytarabine HPLC peaks equal to those in the corresponding samples. After approximately 20 min cytarabine data could be corrected if necessary. Data from Table I in ref. 6.

the degree of compromise is the basis for debate since poor bioavailability can predispose dosage forms to higher interpatient variability.[3]

Assay Specificity

Conversion during Workup

The most common source of assay nonspecificity is prodrug conversion to drug during sample preparation for assay. Since a prodrug is expected to give rise to the drug during the experiment itself, its partial conversion during assay procedures has often gone undetected. Control experiments which reproduce the entire sampling and assay conditions must be conducted using prodrug and assaying for drug to determine the extent of conversion.

The significance of slight conversion will vary with the situation. Early samples in a pharmacokinetic study may contain predominantly prodrug so that only a small percentage conversion during the assay produces yields similar to the concentration of drug in the sampled fluid (Fig. 3). Thus, the prodrug concentration may be within the limits of acceptability for the assay while the drug concentration assay is totally incorrect. In cases where prodrug can be assayed but drug concentration is below the limit of the assay sensitivity, both the control and the sample will provide the same result for drug concentration since in both instances drug arises only from the procedure. Here, control experiments using the exact procedure and conditions starting with the prodrug concentration, as determined by the assay, can be used to correct the false positives obtained for the concentration of drug.[6] During a time-course experiment involving prodrug conversion to drug, the situation may eventually reverse so that drug concentration predominates over prodrug. A small percentage conversion of prodrug as a minor component will then become insignificant. The ratio of prodrug to drug wherein these conditions are satisfied can be calculated from the experimentally observed percentage conversion.

Ideally, assays should be specific for drug and also for prodrug. In practice an assay for drug before and after a treatment for total prodrug conversion to drug often serves as an alternative. This suffers from the increased variability associated with any difference assay. Mass balance should be applied to situations where the total is known, such as the case of a prodrug designed to increase shelf life. It cannot be applied to *in vivo* sampling since the total cannot be known in an open system. Neither the ability to assay both prodrug and drug nor to test for mass balance indicate conversion during the procedure. This can only be done by assaying for drug after applying the entire procedure to prodrug.

Effect of Nonspecificity on Pharmacokinetic Evaluation

Since the primary nonspecificity error stems from conversion during the sample treatment, the drug assays are prone to be overestimates of the actual concentrations. The degree to which this occurs is dependent upon the extent of conversion and the prodrug concentration in the sample. Obviously, evaluations such as drug bioavailability would indicate higher estimates than the true value. In addition, the prodrug plasma concentration time course will be that of an independent entity, as previously discussed relative to Table IV. Therefore, the shape of the apparent drug plasma concentration time course, which is the sum of the drug concentration from plasma and that from prodrug conversion in the workup, can be altered when it reflects the two separate time courses.[16] Such altered profiles will result in miscalculated pharmacokinetic parameters and incorrect assessments of prodrug success in goals such as duration and bioavailability.

Site Specificity

Entrapment versus Enrichment

Theoretically site specificity could be achieved by entrapment wherein the one-way delivery of prodrug within the target would be followed by the slow release of drug or where the released drug itself would then be

[16] R. E. Notari, in "Design of Biopharmaceutical Properties through Prodrugs and Analogs" (E. B. Roche, ed.), p. 82. Am. Pharm. Assoc., Washington, D.C., 1977.

trapped within the target. Practically, there are few examples of success—the work of Bodor[17] being a notable exception. Also, if the drug normally has difficulty entering the target site or if it accumulates in extraordinary amounts as a result of drug or prodrug entrapment, then questions regarding its pharmacology and toxicology must be readdressed. This is in contrast to one of the chief attractions for rapidly reversing prodrugs for which the pharmacology and toxicology of the drug are already known.

Site enrichment is the increase in all or some portion of the drug concentration time course within the target relative to that achieved following an equivalent bioavailable dose of the drug itself. In contrast to a simple increase in bioavailability, enrichment must provide increased target to plasma drug concentration ratios or increased target enrichment relative to other reference tissues such as known sites for toxicity. These changes in prodrug disposition, relative to that achieved by the drug itself, can be brought about by altering the partitioning behavior of prodrug, provided that sufficient conversion takes place within the target. Relative to the drug itself, enrichment may increase the AUC, peak height, or duration within the target.

Neither entrapment nor enrichment represent drug specificity per se. In both cases it would be necessary for prodrug to be preferentially taken up by the target or to release drug only within the target. These are not easily achieved in practice. It has been concluded, based on theory, that prodrugs cannot provide increased site specificity when applied to most currently used drugs.[9] Increased prodrug distribution will favor increased prodrug delivery to multiple sites wherein enzymatic activity can be available to varying degrees. Prodrugs can, nonetheless, alter the time course for drug disposition in a manner which favors the target site during a specified time period relative to that observed following administration of drug itself. The degree of drug specificity may thus be temporarily increased.

Feasibility for Enrichment

Most reports of successful drug delivery to target tissues by prodrugs are based on site enrichment rather than specificity. Reported conclusions have not always been compatible with what is kinetically feasible. For example, a prodrug cannot achieve enrichment if it is immediately and totally converted to drug at some pretarget site, such as the blood. The drug tissue to plasma ratio in such a case must be similar to that achieved

[17] N. Bodor, *in* "Design of Biopharmaceutical Properties through Prodrugs and Analogs" (E. B. Roche, ed.), p. 98. Am. Pharm. Assoc., Washington, D.C., 1977.

from an equal plasma concentration of drug following administration of the drug itself. It is also not feasible to achieve enrichment during steady-state dosing. In this case both drug and prodrug will each distribute in accordance with their individual steady-state distribution kinetics.

A rapidly distributing drug will not provide increased peak height or early enrichment by using prodrugs. Conversely, a drug which normally distributes slowly and is administered by intermittent dosing (not steady-state) might show target enrichment during the period of time normally preceding completion of its own distribution. In this case the therapeutic onset may be faster but the enrichment will be transient as drug distribution will eventually return to normal.

It is also possible to prolong the terminal concentration time course for drug within the target by slow prodrug input to the target or slow drug release within the target. Prodrug in either case behaves as a depot which can be used to prolong the therapeutic duration. The extended input of prodrug to the target is more readily achieved from an intramuscular depot injection than by the slow site delivery of systemically circulating prodrug. Here, the prodrug in the blood is more likely to be cleared by competing mechanisms with a resultant loss of available prodrug for tissue distribution.

While there are many examples of prodrugs which increase delivery of the drug to the target tissues or organs, few approach site specificity. In general, it has been easier to achieve improved formulation characteristics, stability, bioavailability, and depot injections for increased duration and the number of examples of these is extremely large.[18,19]

[18] A. A. Sinkula and S. H. Yalkowsky, *J. Pharm. Sci.* **64**, 181 (1975).
[19] R. E. Notari, *J. Pharm. Sci.* **62**, 865 (1973).

[25] Prodrug Lability Prediction through the Use of Substituent Effects

By M. CHARTON

A prodrug is a derivative of a known (parent) drug which is converted within the organism to the parent drug. The conversion usually involves some kind of cleavage, though some instances have been reported in which the prodrug is a precursor which reacts to form the drug in the organism. Cleavage may be hydrolytic, reductive, or oxidative, the former being the most frequently encountered. The design of prodrugs is therefore very different from that of parent drugs. The desired level of bioactivity is already present in the latter and requires no further modification. The function of a prodrug is to improve the pharmaceutical performance and overcome deficiencies in the physical properties of the parent drug. Thus, it may improve transport to a receptor, prevent drug loss before the receptor is reached, maintain a desired drug level in the organism, increase shelf life, improve taste, increase solubility, improve injectability, or make effective tableting possible. The design of the prodrug must achieve one or more of these results while ensuring that the parent drug forms in the appropriate place at a suitable rate. In order to accomplish this end, it is important to be able to predict the degree of drug formation in different parts of the organism as a function of molecular structure. This is generally referred to as prodrug lability. The term is used whether parent drug formation results from cleavage or from a precursor. It is important to be able to predict prodrug lability not only in the vicinity of the receptor at which the prodrug is to act but in all body fluids, tissues and membranes through which the prodrug may pass on its way to the target receptor.

As most prodrugs undergo parent drug formation by hydrolytic cleavage, this work will be largely devoted to this mode of formation. Prodrugs may be cleaved by simple chemical reactions occurring in any of the body fluids through which they pass in the course of their transport from the site of entry into the organism to the receptor site at which the parent drug is to act. Alternatively they may be cleaved by enzymatic reactions involving enzymes in body fluids, skin, membranes, and other components of the organism.

Principles of Correlation Analysis

The most effective, as well as the simplest, method of predicting chemical reactivity or biological activity as a function of molecular structure is correlation analysis (frequently termed linear free energy relationships).[1-3] Its application to the prediction of prodrug lability has been described.[4]

[1] J. Shorter, "Correlation Analysis in Organic Chemistry." Oxford Univ. Press (Clarendon), London and New York, 1973.

[2] M. Charton, *Chem. Tech.* (*Heidelberg*) **4**, 502 (1974); **5**, 245 (1975).

[3] J. Shorter, "Correlation Analysis of Organic Reactivity." Research Studies Press, Chichester, 1982.

[4] M. Charton, *in* "Design of Biopharmaceutical Properties through Prodrugs and Analogs" (E. B. Roche, ed.), p. 228. Am. Pharm. Assoc., Washington, D.C., 1977.

Consider a set of substances that has the general structure XGY. Y, the active site, is a group of atoms at which some quantifiable chemical reaction or biological activity takes place. X is a substituent which is varied from one member of the data set to another. G is a skeletal group to which X and Y are bonded. Both Y and G are held constant throughout the data set. For ordinary chemical reactions in solution, the variation of reactivity with change in X is a function of the electrical effects of the X group and frequently of its steric effects as well. It is convenient to dissect the electrical effect into two components, the localized (field and/or inductive) and delocalized (resonance) effects. Thus, in the most general case, chemical reactivity is described by the *LDS* equation

$$Q_X = L\sigma_{IX} + D\sigma_{DX} + Sv_X + h \qquad (1)$$

where Q_X is the logarithm of the rate or equilibrium constant of the X-substituted member of the data set. *L, D, S,* and *h* are coefficients determined by multiple linear regression analysis. The σ_{IX} and σ_{DX} constants are localized and delocalized electrical effect parameters, respectively.[5] The v_X constants are steric parameters[6] and are required when X and Y are in close proximity to each other (in van der Waals contact). When X and Y are well separated, no term in v is necessary and Eq. (1) simplifies to the *LD* equation

$$Q_X = L\sigma_{IX} + D\sigma_{DX} + h \qquad (2)$$

When X is bonded to a C atom of the skeletal group G which is itself involved in π bonding, the term in σ_D is required. When it is bonded to a C atom of G which is itself involved only in σ bonding, the term in σ_D is generally unnecessary and only localized electrical effects need be considered. When steric effects are absent, the data are described by the *L* equation,

$$Q_X = L\sigma_{IX} + h \qquad (3)$$

Depending on the nature of Y and the extent to which it is capable of conjugation with a π-bonded skeletal group, four different σ_D constants may be used. When Y is incapable of conjugation with G, best results are obtained with $\sigma_R°$ constants. When Y is a weak electron acceptor group capable of conjugation with G, the σ_R constants are appropriate. When it is a strong electron acceptor capable of conjugation with G (such as a carbenium ion), best fit generally results with the σ_R^+ constants. When Y

[5] M. Charton, *Prog. Phys. Org. Chem.* **13,** 119 (1981).
[6] M. Charton, *Top. Curr. Chem.* **114,** 57 (1983).

is capable of conjugation with G and has a lone pair on the atom directly bonded to G, the σ_R^- constants are required.

It is frequently convenient to use composite electrical effect parameters, particularly when studying meta- and para-substituted benzene data sets. In general, composite substituent constants can be defined by the expression

$$\sigma_T = \lambda\sigma_I + \delta\sigma_D \tag{4}$$

where λ and δ are coefficients which determine the composition of the composite. The composition is described by the equations

$$P_D = (\delta \cdot 100)/(\lambda + \delta) \tag{5}$$

and

$$P_L = 100 - P_D \tag{6}$$

A very useful type of composite substituent parameter[5] is the σ_n constant which is defined by the relationship

$$\sigma_n = \sigma_I + (P_D/P_L)\sigma_D \tag{7}$$

The σ_m and σ_p constants used in the Hammett equation,

$$Q_X = \rho\sigma_X + h \tag{8}$$

are special cases of σ_n with n (which represents the composition of the constant and is equal to P_D) having values of 28 and 50, respectively. The σ_n constants are used in the Hammett equation and in the $\rho\sigma$ equation. The latter is given by

$$Q_X = \rho\sigma_X + Sv_X + h \tag{9}$$

and is used when steric effects are present.

When two or more substituents are bonded to equivalent positions in a skeletal group the LDS equation [Eq. (1)] takes the form

$$Q_X = L \sum_{i=1}^{n} \sigma_{IX_i} + D \sum_{i=1}^{n} \sigma_{DX_i} + S \sum_{i=1}^{n} v_{X_i} + h \tag{10}$$

Thus, the substituent effects are assumed to be additive. For electrical effects this is usually a very good approximation. In the case of steric effects, it is often unjustified.

There is an alternative to the use of the v steric parameters (or any other steric parameters based on van der Waals radii) as a measure of steric effects. The simple branching (SB) equation, which takes the form[7-9] shown below, is a useful alternative.

$$Q_{Ak} = \sum_{i=1}^{m} a_i n_i + a_0 \tag{11}$$

In Eq. (11), n_i is the number of C atoms in an alkyl (Ak) group which bear the label $i + 1$. Consider the system (**I**) in which an alkyl group is bonded to some skeletal group. The Ak group is numbered beginning with the C atom bonded to G. As

$$n_i = \sum C^{i+1} \tag{12}$$

in (**I**), $n_1 = \Sigma C^2 = 3$, $n_2 = \Sigma C^3 = 4$, $n_3 = \Sigma C^4 = 3$, $n_4 = \Sigma C^5 = 3$. A major advantage of the SB equation is that in contrast to most correlation equations it is in the case of Ak groups independent of experimentally determined substituent parameters and the errors associated with them. The values of n_i are determined solely by counting the number of C atoms with the appropriate label and are not subject to error. The method can be extended to cycloalkyl groups (cAk) by calculating effective n_i values from experimental data. These $n_{i,ef}$ constants are experimentally determined substituent parameters and are of course subject to error. It follows, then, that although cAk groups can be included in a correlation with Eq. (11) this can be done only at the expense of the advantage of error-free parameters.

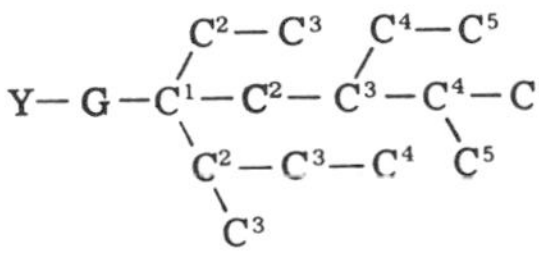

$$\begin{array}{c} \text{C}^2\text{—C}^3 \quad \text{C}^4\text{—C}^5 \\ / \qquad\quad / \\ \text{Y—G—C}^1\text{—C}^2\text{—C}^3\text{—C}^4\text{—C}^5 \\ \backslash \qquad\quad \backslash \\ \text{C}^2\text{—C}^3\text{—C}^4 \quad \text{C}^5 \\ \backslash \\ \text{C}^3 \end{array}$$

I

The steric effects of many substituents are conformationally dependent and therefore no one set of steric parameters will describe the behavior of such groups under all possible conditions. The SB equation is capable of modeling this variation in the steric effect at least to some extent. It suffers from the disadvantage that it assumes all branches at a

---

[7] M. Charton, *J. Org. Chem.* **43**, 3995 (1978).
[8] M. Charton, *Adv. Pharmacol. Res. Pract., Proc. Congr. Hung. Pharmacol. Soc., 3rd, 1979* p. 211 (1980).
[9] M. Charton, *J. Chem. Soc., Perkin Trans. 2* p. 97 (1983).

given C atom to be equivalent and does not discriminate among them. The expanded branching (XB) equation,[9]

$$Q_{Ak} = \sum_{i=1}^{m} \sum_{j=1}^{3} a_{ij} n_{ij} + a_{00} \tag{13}$$

does distinguish between successive branches at a given C atom. The n_{ij} are given by

$$n_{ij} = \sum C^{i+1,j} \tag{14}$$

Consider the system (**II**) which is identical to (**I**) except for the numbering: $n_{11} = n_{12} = n_{13} = 1$, $n_{21} = 3$, $n_{22} = 1$, $n_{23} = 0$, $n_{31} = 2$, $n_{32} = 1$, $n_{33} = 0$, $n_{41} = 2$, $n_{42} = 1$, $n_{43} = 0$. The XB equation is a more accurate model of steric effects than is the SB equation. It too has the advantage that for Ak groups the n_{ij} parameters are free of experimental error. The disadvantage of the XB equation is that it has three times as many independent variables as does the SB equation and therefore requires a correspondingly larger data set in order to give good results.

$$
\begin{array}{c}
C^{31}\!-\!C^{31} \quad C^{41}\!-\!C^{51} \\
\diagup \qquad\qquad \diagup \\
Y\!-\!G\!-\!\overset{\displaystyle}{C^1}\!-\!C^{22}\!-\!\overset{\displaystyle}{C^{31}}\!-\!C^{42}\!-\!C^{51} \\
\diagdown \qquad\qquad\qquad \diagdown \\
C^{23}\!-\!C^{31}\!-\!C^{41} \quad C^{52} \\
\diagdown \\
C^{32}
\end{array}
$$

II

The correlation of reaction rates for enzymatic reactions frequently requires the use of a transport parameter. The most general correlation equation then becomes

$$Q_X = L\sigma_{IX} + D\sigma_{DX} + Sv_X + T\tau_X + h \tag{15}$$

where τ is the transport parameter and T is its coefficient. The most commonly used transport parameters are log P and π.[10] P is a partition coefficient for the distribution between water and some organic solvent (usually n-octanol) of a member of the data set. π is defined by the equation

$$\pi_X = \log P_X - \log P_H \tag{16}$$

Log P, and therefore π as well, are composite parameters. Log P represents the difference in intermolecular forces between water–substrate and

[10] C. Hansch and A. J. Leo, "Substituent Constants for Correlation Analysis in Chemistry and Biology." Wiley, New York, 1979.

organic solvent–substrate. Such intermolecular forces include dipole–dipole (dd), dipole–induced dipole (di), induced dipole–induced dipole (ii), ion–dipole (Id), ion–induced dipole (Ii), hydrogen bonding (hb), and possibly charge transfer (ct) interactions. Thus, the equation

$$Q_X = A\alpha_X + L\sigma_{IX} + D\sigma_{DX} + Sv_X + H_1 n_{HX} + H_2 n_{nX} + Ii_X + B_0 \quad (17)$$

or relationships derived from it have been shown to correlate various types of transport parameters including $\log P$ and π.[11-13] Equation (17) can also be used to correlate bioactivities. The α parameter, defined by

$$\alpha_X = (MR_X - MR_H)/100 \quad (18)$$

where MR_X and MR_H are the group molar refractivities of X and H, respectively, is a measure of group polarizability. n_H is the number of OH or NH bonds in X, n_n is the number of lone pairs on O or N atoms in X, i is 1 for charged (ionic) and 0 for uncharged X, and the H, I, and B terms are the linear regression coefficients.

It is occasionally necessary in modeling cleavage rates in body fluids to introduce a term in τ^2 into Eq. (15), giving

$$Q_X = L\sigma_{IX} + D\sigma_{DX} + Sv_X + T_1\tau_X + T_2\tau_X^2 + h \quad (19)$$

Data sets are often encountered in which X is restricted to alkyl groups. It has been shown that the electrical effects of alkyl groups are constant.[5,14-16] For alkyl groups, α is a linear function of the number of C atoms in the group.

$$\alpha_{Ak} = b_1 n_C + b_0 \quad (20)$$

Finally, consider the case for alkyl groups $n_H = n_n = i = 0$. From Eqs. (19) and (20) it is clear that Q_{Ak} is a function only of n_C and v. When the steric effects are represented by the SB or XB equations [Eqs. (11) and (13), respectively], the former results in the alkyl bioactivity branching equation,

$$Q_{Ak} = \sum_{i=1}^{m} a_i n_i + a_C n_C + a^\dagger (n_C)^2 + a_0 \quad (21)$$

where $a^\dagger$ is the coefficient of $(n_C)^2$, and the latter in the expanded alkyl

[11] M. Charton and B. I. Charton, *J. Theor. Biol.* **99**, 629 (1982).

[12] M. Charton and B. I. Charton, *in* "Quantitative Approaches to Drug Design" (J. Dearden, ed.), p. 260. Elsevier, Amsterdam, 1983.

[13] M. Charton, *Abstr. Pap., 186th Meet. Am. Chem. Soc.* Pest. 92 (1983).

[14] M. Charton, *J. Am. Chem. Soc.* **99**, 5687 (1977).

[15] M. Charton, *J. Org. Chem.* **44**, 903 (1979).

[16] M. Charton and B. I. Charton, *J. Org. Chem.* **47**, 8 (1982).

bioactivity branching equation,

$$Q_{Ak} = \sum_{i=1}^{m} \sum_{j=1}^{3} a_{ij}n_{ij} + a_C n_C + a^{\dagger}(n_C)^2 + a_{00} \tag{22}$$

These relationships are very useful in modeling the bioactivities of alkyl-substituted data sets.[8,17]

Finally, it has recently been shown that data for a number of different organisms can be combined into a single data set represented by a single correlation equation on the condition that the same mechanism for the bioactivity occurs in all of the organisms.[18] The method can be generalized to all types of biocomponents such as different tissues or body fluids within an organism.[19] The basis of the method lies in defining the bioactivities observed for any one substituent in the overall data set as the biocomponent parameter. This technique can be applied to the combination of data sets involving factors such as dose size or biological test type as well. It can also be used to combine data sets of chemical reactivities obtained for the same reaction under different reaction conditions.

Reversible Groups

Functional groups which are capable of undergoing cleavage in an organism, either by ordinary chemical reactions or with enzymatic catalysis, are called reversible groups (Gr). In order to make a prodrug of the cleavage type a modifiable group (Gm) on the parent drug, Z must be converted to a reversible group which then becomes the active site Y when the cleavage reduction takes place (Y = Gr). A partial list of reversible groups is given in Table I.

The body fluids through which a drug may pass on its way to the receptor at which it is to act can determine the nature and extent of nonenzymatic hydrolysis by their pH. Values of the pH range for various body fluids are given in Table II. The extent of enzymatic hydrolysis will depend on the body fluids, tissues, and membranes with which the prodrug interacts as both the type and concentration of enzymes varies with the nature of the biocomponent. For the prediction of cleavage rate in body fluids, it is frequently useful to determine rates of cleavage for a set of compounds directly in the body fluid of interest, determine the correlation

[17] M. Charton *in* "Quantitative Approaches to Drug Design" (J. Dearden, ed.), p. 67. Elsevier, Amsterdam, 1983.

[18] M. Charton, *in* "Quantitative Approaches to Drug Design" (J. Dearden, ed.), p. 69. Elsevier, Amsterdam, 1983.

[19] M. Charton, unpublished results.

TABLE I
REVERSIBLE GROUPS

Modifiable group	R^a	Reversible group	Z^a
RC=O	H,Ak,Ar	$RC(OZ)_2$	Ak
	H	$HC(O_2CZ)_2$	Ak,Ar
	H,Ak,Ar	C=C—OZ	Ak
	H,Ak,Ar	$C=C—NZ^1Z^2$	Ak
	H	$CHCl(O_2CZ)$	Ak,Ar
	H,Ak,Ar	C=NZ	Ak,Ar,OMe
	H,Ak,Ar		H,CO_2Ak
	H,Ak,Ar	$C=C—O_2CZ$	Ak,Ar
CO_2H	—	CO_2Z	Ak,Ar
	—	C=O(SZ)	Ak,Ar
	—	$C=O(NZ^1Z^2)$	H,Ak,Ar,OH,NH_2
	—	$CO_2CHZ^1O_2CZ^2$	$Z^1 = H,Ph$
			$Z^2 = Ak,Ar$
	—	CO_2CHZCl	H,Ar
	—	$C(OZ)_3$	Ak
	—	CO_2CO_2Z	Ak,Ar
	—	CO_2CH_2Z	SMe,SOMe, SO_2Me
CONHR	H	$CONHCH_2NZ^1Z^2$	Ak,Ar
	H	$CONHCH_2OH$	—
OH	—	O_2CZ	Ak,Ar, . . .
	—	$(—O)_3CZ$	Ak,Ar, . . .
	—	OZ	Ak,Ar
	—	—O—C=C	—
	—	$(—O—)_2CZ^1Z^2$	H,Ak,Ar
	—	$O_2CNZ^1Z^2$	H,Ak,Ar
	—	O_2COZ	H,Ak,Ar
	—	O_2CO_2CZ	H,Ak,Ar
	—	$OSiZ_3$	Ak
	—	$(—O)_2CO$	—
	—	$(—O)_3PO$	—
	—	$(—O)_2SO_2$	—
NHR	H,Ak,Ar	NR(C=O)Z	Ak,Ar
	H,Ak,Ar	$NRCO_2Z$	Ak,Ar
	H,Ak,Ar	NRC=C	—
	H,Ak,Ar	NR(C=O)NHZ	Ak,Ar
	Ar	$NArCH_2N$	—
—NH	—	$—N—CH_2O_2CH$	Ak,Ar
SH	—	S(C=O)Z	Ak,Ar

^{a} Ak, Alkyl; Ar, aryl.

TABLE II
pH AND ENZYME CONTENT OF BODY FLUIDS[a]

Body fluid	pH range (adults)[b]	Hydrolytic enzymes
Whole blood		
Arterial	7.315–7.439 (nb)	Cholinesterase,
	7.386–7.462	aminopeptidase,
		pepsin, trypsin
Capillary	7.360–7.420 (m)	Same as arterial
	7.366–7.430 (w)	
Plasma		
Arterial	7.35–7.43	
Venous	7.378–7.418	
Cerebrospinal fluid		
Cisternal fluid	7.327–7.371	Esterases
Lumbar fluid	7.327–7.371	
Synovial fluid	7.31–7.64 (a)	
Saliva		
Parotid	5.1–6.25 (a)	
Submandibular	5.9–7.3 (a)	
Total saliva	6.40–8.24 (c)	Cholinesterases
	5.8–7.1 (a)	
Gastric juice	1.2–7.4 (a, nb)	
	0.9–7.7 (a, c)	
	1.92 (b, m)	
	2.59 (b, w)	
Pancreatic juice	7.5–8.8 (a)	Lipase, some peptidases
Bile	6.2–8.5 (a)	Esterases, phosphatases,
		lipases
Intestinal juice		
Duodenum	6.5–7.6	
Jejunum	6.3–7.3	
Ileum	7.6 (b)	
Colon	7.9–8.0	
Sweat	4–6.8 (a)	

[a] Data from "Scientific Tables," 7th Ed. (R. Diem and C. Lentner, eds.). Ciba-Geigy, Basle, 1970.
[b] pH values are 95% range for adults unless otherwise noted. a, Extreme range; b, mean value; m, men; w, women; c, children; nb, newborn.

equation for the set, and predict cleavage rates for compounds of particular interest from this correlation equation. The cleavage reaction may be enzymatic, nonenzymatic, or some combination of both. It was remarked earlier that prodrug design involves a combination of two components: (1) the incorporation of the desirable physical and chemical properties which are the reason for modifying the parent drug; and (2) designing the prodrug so as to generate the parent drug at an appropriate rate in the vicinity of the receptor site at which it is to act. The drug designer might begin by choosing some reversible groups which would serve to introduce the desired properties. The correlation equations for the appropriate reactions can then be used to predict whether the proposed prodrugs will undergo significant cleavage before reaching the site at which they are to act and whether they will form the parent drug at a suitable rate. When the path of the prodrug from site of entry to site of action is fully known, the lability of the prodrug is predictable provided the necessary correlation equations are known. Generally, this is not the case. Nevertheless, the correlation equations can be used to make qualitative estimates of lability which can be useful both in the choice of a reversible group and in the selection of the most suitable derivatives of a reversible group.

Examples of the Determination of Correlation Equations

The key to the prediction of prodrug lability as a function of substituent effects lies in the correlation equation. Examples of the determination of these equations are presented here.

Wagner, Grill, and Henschler[20] have reported hydrolysis rates for 3-(O-acyl) derivatives of etilefrine (**III**), in human blood (a), rat blood (b), and rat liver homogenate (c). The hydrolysis rate-determined derivatives with R = Ak were (alkyl group, and values for a, b, and c) Me, 57.5, 44, 182; Et, 47.5, 46.5, 226; Pr, 32, 25.4, 204; iPr, 34.5, 32, 220; Bu, 30.7, 34, 194; tBu, 5.6, 4.6, 196; 1-methylcyclohexyl, 0.4, —, 45.

$$\overset{R}{\underset{O=\overset{|}{C}-O}{}} \text{—CH—CH}_2\text{—}\overset{+}{\text{NH}}_2\text{Et}\ \ \text{Cl}^-$$
$$\underset{\text{OH}}{}$$

III

As only alkyl groups are included in the subset being studied, the alkyl bioactivity branching equation [Eq. (1)] can be used. The data sets are too small to give reliable results. They were, therefore combined, and the logarithms of the rates for Ak = Me were used to parameterize the biocomponents. Thus,

$$\omega \equiv \log \text{rate}_{\text{Me}} \tag{23}$$

the equation used for correlation was

$$Q_{\text{Ak}} = \sum_{i=1}^{m} a_i n_i + a_C n_C + a^{\dagger}(n_C)^2 + O\omega + a_0 \tag{24}$$

[20] J. Wagner, H. Grill, and P. Henschler, *J. Pharm. Sci.* **69**, 1243 (1980).

As 1-methylcyclohexyl is a cAk group for which no $n_{C,ef}$ and $n_{i,ef}$ values are available, these are estimated. $n_{C,ef}$ was assumed equal to that for cycloheptyl. $n_{1,ef}$ was obtained by adding 1 to the value for cyclohexyl giving 2.5, and $n_{2,ef}$ was assumed equal to that for cyclohexyl.

In order to avoid collinearity of n_C and $(n_C)^2$, the n_C values were rescaled as suggested by Goodford[21] and Berntsson[22] using the expression

$$\tau^* = \tau - (\tau_{max} + \tau_{min})/2 \tag{25}$$

Correlation with Eq. (24) was carried out using n_C^* and $(n_C^*)^2$ in place of n_C and $(n_C)^2$. The correlation matrix shows that n_C and n_1 are strongly co-linear. As Student's t-test gave greater significance for a_1 than for a_C, the correlation was repeated with the exclusion of n_C^* giving the correlation equation

$$Q_{Ak} = 1.77(\pm 0.245)\omega - 0.11(\pm 0.028)(n_C^*)^2 - 0.470(\pm 0.0747)n_1$$
$$- 0.448(\pm 0.14)n_2 - 0.631 \tag{26}$$

with $100R^2 = 85.33$; $s = 0.292$; $F = 21.82$; and $n = 20$.

The dependence on $(n_C^*)^2$ indicates enzymatic hydrolysis. Steric effects are the major factor in determining lability in all three biocomponents. Branching at C^1 and C^2 are of equal importance.

Wechter and co-workers[23] have studied 5'-acyl derivatives of aracytidine as potential prodrugs. They have studied rates of the hydrolysis by mouse plasma (a), human plasma (b), and human synovial fluid (c) and of the alkaline hydrolysis (d). For R = Ak the values obtained were [Ak, 10^3(a), 10(b), 10(c), (d)] 1-adamantyl, 139, 2.0, 1.5, 0.02; Et$_3$C, 0.3, —, —, —; tBu, 2.1, 0.21, —, 0.06; Et$_2$CH, 0.3, 0.2, —, 0.006; Hp, 321, 321, 95, 0.28; cHx, 116, 139, 82, 0.15; cBu, 87, 76, 6, —; iPr, 9.9, 7.6, 0.8, 0.34; Me, 1.7, 10.8, 0.07, 1.1; iBu, —, —, —, 0.093; Hpd, —, —, —, 0.27.

As no $n_{i,ef}$ values for the 1-adamantyl group are available, the alkyl bioactivity branching equation cannot be used. Values of v and α are available for all set members, however, thus making possible the use of the equation

$$Q_{Ak} = Sv_{Ak} + T_1\tau_{Ak} + T_2(\tau_{Ak})^2 + h \tag{27}$$

which is derived from Eq. (19). α^* was used as the transport parameter.

Correlation of logarithms of rate constants for hydrolysis by mouse plasma gave the equation

[21] P. J. Goodford, *Adv. Pharmacol. Chemother.* **11,** 51 (1973).
[22] P. Berntsson, *Acta Pharm. Suec.* **17,** 199 (1980).
[23] W. J. Wechter, M. A. Johnson, C. M. Hall, D. T. Warner, A. E. Berger, A. H. Wenzel, D. T. Gish, and G. L. Neil, *J. Med. Chem.* **18,** 339 (1975).

$$\log k_r = -2.14(\pm 0.309)v_{Ak} + 9.77(\pm 1.70)\alpha^*_{Ak}$$
$$- 2.05(\pm 12.7)(\alpha^*_{Ak})^2 + 3.26(\pm 0.399) \tag{28}$$

with $100R^2 = 91.51$; $F = 17.97$; $s = 0.438$; and $n = 9$.

As the coefficient of $(\alpha^*)^2$ is not significant, this parameter was dropped from the equation. The final result was

$$\log k_r = -2.13(\pm 0.281)v_{Ak} + 9.74(\pm 1.55)\alpha^*_{Ak} + 3.23(\pm 0.329) \tag{29}$$

with $100R^2 = 91.42$; $F = 32.17$; $s = 0.400$; and $n = 9$.

For hydrolysis by human plasma, the equation

$$\log k_r = -3.62(\pm 0.485)v_{Ak} + 7.19(\pm 1.67)\alpha^*_{Ak}$$
$$- 15.2(\pm 12.4)(\alpha^*_{Ak})^2 + 4.45(\pm 0.518) \tag{30}$$

with $100R^2 = 93.33$; $F = 18.66$; $s = 0.419$; and $n = 8$ was obtained. As a

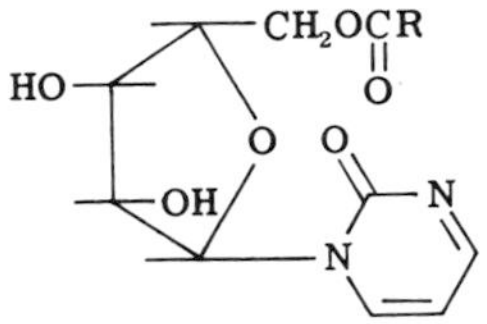

IV

t-test gave only a 70.0% confidence level for $T_2(\alpha^*)^2$ was excluded and the correlation repeated giving

$$\log k_r = -3.50(\pm 0.499)v_{Ak} + 6.86(\pm 1.73)\alpha^*_{Ak} + 4.17(\pm 0.488) \tag{31}$$

with $100R^2 = 90.81$; $F = 24.72$; $s = 0.439$; and $n = 8$.

The correlation equation obtained for human synovial fluid was

$$\log k_r = -1.99(\pm 1.41)v_{Ak} + 9.53(\pm 3.11)\alpha^*_{Ak}$$
$$- 48.0(\pm 20.4)(\alpha^*_{Ak})^2 + 2.81(\pm 1.01) \tag{32}$$

with $100R^2 = 91.92$; $F = 7.579$; $s = 0.547$; and $n = 6$. These results are *not* significant. They suggest, but do not prove, a dependence on steric effects and polarizability. Steric effects seem to be dominant in all three data sets. Hydrolysis in human synovial fluid may show a dependence on $(\alpha^*)^2$.

Correlation of rate constants for the alkaline hydrolysis of (**III**) with the equation

$$\log k_r = Sv_{Ak} + h \tag{33}$$

was successful, giving

$$\log k_{rAk} = -1.98(\pm 0.173)v_{Ak} + 0.938(\pm 0.174) \tag{34}$$

with $100R^2 = 94.93$; $s = 0.166$; and $n = 9$.

Clearly, lability of these prodrugs is a function of steric and, in the case of hydrolysis in body fluids, polarizability.

Attempts to combine the body fluid data into a single data set by correlation with the equation

$$\log k_r = Sv_{Ak} + A_1\alpha^*_{Ak} + A_2(a^*_{Ak})^2 + O\omega + h \tag{35}$$

were unsuccessful, thus suggesting that hydrolyses in the body fluids were proceeding by different paths.

Patel and Repta[24] proposed the use of enol esters as prodrugs. They have determined hydrolysis rates of these compounds (1-phenylvinyl alkanoates, $H_2C{=}CPhO_2CAk$) in human (a) and rat plasma (b), human and rat liver supernatant (c, d), and rat kidney (e). Their values are (Ak, a, b, c, d, e) Me, 26.1, 3.65, 1.84, 2.02, 0.44; Et, 53.3, 13.90, 4.18, 7.97, 0.70; Pr, 57.8, 9.90, 8.35, 6.48, 0.43; iPr, 6.30, 2.24, 2.42, 4.44, 0.30; tBu, —, 3.47, 1.04, 1.76, 0.18.

The data were combined into a single data set. As only simple alkyl groups are present, the alkyl bioactivity branching equation is a good choice for determining the correlation equation. Unfortunately for this particular small set of alkyl groups n_C is equal to $n_1 + n_2$. The correlation was therefore carried out with the equation

$$Q_{Ak} = a_1n_1 + a_2n_2 + a^\dagger(n_C)^2 + O\omega + a_0 \tag{36}$$

was defined as $\log k_{r,Me}$. The resulting correlation equation was

$$\log k_r = -0.104(\pm0.0505)n_1 + 0.101(\pm0.135)n_2 - 0.188(\pm0.0550)(n_C)^2$$
$$+ 1.02(\pm0.0890)\omega + 0.428(\pm0.121) \tag{37}$$

with $100R^2 = 89.31$; $F = 39.69$; $s = 0.239$; and $n = 24$.

As a_2 was not significant, n_2 was excluded and the correlation repeated giving

$$Q_{Ak} = -0.110(\pm0.0493)n_1 - 0.206(\pm0.0490)(n^*_C)^2$$
$$+ 1.02(\pm0.0879) + 0.474(\pm0.103) \tag{38}$$

with $100R^2 = 89.00$; $F = 53.93$; $s = 0.236$; and $n = 24$.

Borgman, Baldessarini, and Walton[25] have suggested diacyl derivatives as apomorphine prodrugs (V). They have determined for several of these compounds the duration of stereotyped behavior in the rat. Their results were obtained at three different dose levels, 50, 100, and 200 μmol/kg intraperitoneally. The values reported are (Ak, 50, 100, 200 μmol/kg) Me, 2.1, 2.3, 2.7; Et, 1.6, 1.9, 2.9; iPr, 2.5, 3.3, 4.9; tBu, 3.1, 5.1, 7.8.

[24] J. P. Patel and A. J. Repta, Int. J. Pharm. 9, 29 (1981).
[25] R. J. Borgman, R. J. Baldessarini, and K. G. Walton, J. Med. Chem. 19, 717 (1976).

The data were combined into a single set. As there are only 12 data points available, v rather than branching was chosen as a method of parameterizing steric effects. In this small set of alkyl groups, v is linear in n_C and $(n_C)^2$. Correlations were therefore carried out separately with the equations

$$Q_{Ak} = Sv_{Ak} + O\omega + h \tag{39}$$
$$A_{Ak} = a_C n_C^* + a^\dagger(n_C^*)^2 + O\omega + h \tag{40}$$

The correlation equation obtained with Eq. (39) is

$$\log t_{Ak} = 0.498(\pm 0.0734)v_{Ak} + 2.38(\pm 0.462)\omega - 0.786(\pm 0.181) \tag{41}$$

with $100R^2 = 88.96$; $F = 36.27$; $s = 0.0728$; and $n = 12$.
    That obtained from Eq. (40) is

$$\log t_{Ak} = 0.120(\pm 0.0185)n_C^* + 0.0545(\pm 0.0207)(n_C^*)^2$$
$$+ 2.38(\pm 0.455)\omega - 0.470(\pm 0.172) \tag{42}$$

with $100R^2 = 90.50$; $F = 25.39$; $s = 0.0717$; and $n = 12$. It is not possible in

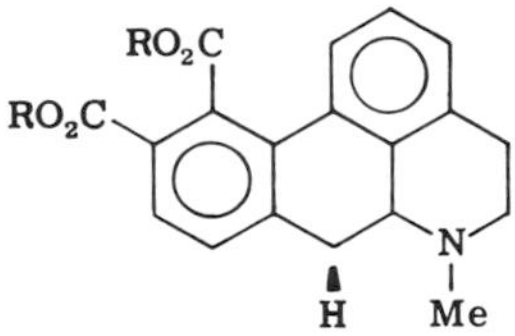

V

this data set to determine whether prodrug activity is dependent only on steric effects or only on polarizability or on some combination of the two.
    Bundgaard, Hansen, and Larsen[26] have proposed the use of esters (**VI**) of malonuric acid as prodrugs. These compounds react to form the barbituric acids according to the equation

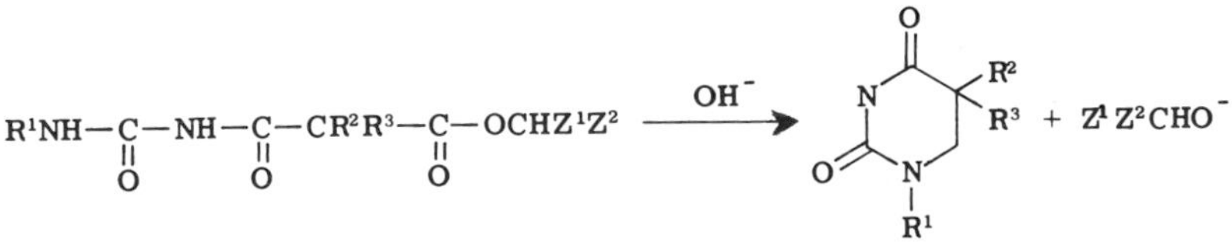

VI

This is an example of *in situ* formation of parent drugs as opposed to cleavage. Rate constants for the reaction were determined in buffer (pH 7.4, 0.5 M) at 24 and 37°. The values obtained are (Z^1, Z^2, $10^4 k_r$ at 24°

[26] H. Bundgaard, A. B. Hansen, and C. Larsen, *Int. J. Pharm.* **3**, 341 (1979).

and at 37°) H, H, 5.5, 2.5; Me, H, 2.3, 1.0; Me, Me, 0.29, 0.12; Ph, H, 4.4, 1.08; OMe, H, 8.5, 3.8; MeO_2C, H, 10.2, 5.1.

The data were combined into a single data set and correlated with the equation

$$Q_X = L\Sigma\sigma_{IZ} + Sv_X + O + h \tag{43}$$

where $X = OCHZ^1X^2$. Estimated v values were used for OCH_2CCl_3 (0.70), OCH_2OMe (0.45), and OCH_2CO_2Me (0.57). The resulting correlation equation was

$$\log 10^4 k_{rX} = 2.79(\pm0.423)\Sigma\sigma_{IZ} - 2.29(\pm0.416)v_X$$
$$+ 1.04(\pm0.355) + 0.577(\pm0.299) \tag{44}$$

with $100R^2 = 90.60$; $F = 25.71$; $s = 0.209$; and $n = 12$.

It seems fairly certain that the rate of formation of barbituric acids depends on the localized electrical effect of Z and the steric effect of X.

Differt and co-workers[27] have studied the hydrolysis of carbonates of salicylic acids (VII) in phosphate buffers at pH 7.4 and at pH 12 and in 2%

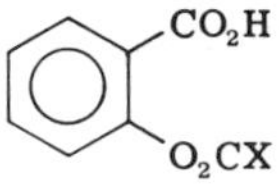

X = ZCH₂O

VII

human plasma (a, b, and c, respectively). These compounds were of interest as substitutes for aspirin. The values for $t_{1/2}$ are (X, a, b, c) EtO, 41, 3.8, 13.3; BuO, 29, 4.5, 11.7; HxO, 15, 4.8, 8.2; Cl_3CCH_2O, 0.53, 4.1, 0.35. The number of compounds studied is too small to permit conclusions to be drawn. The application of correlation analysis can nevertheless reveal trends. The hydrolysis rate constants were correlated with the equation

$$Q_X = L\sigma_{IZ} + Sv_X + h \tag{45}$$

The correlation equation for pH 7.4 is

$$\log k_{rX} = -3.45(\pm0.737)\sigma_{IZ} - 2.89(\pm1.50)v_X + 2.99(\pm0.846) \tag{46}$$

with $100R^2 = 99.07$; $F = 53.20$; $s = 0.145$; and $n = 4$.

That for pH 12 is

$$\log k_{rX} = -0.366(\pm0.0183)\sigma_{IZ} + 0.769(\pm0.0373)v_X + 0.206(\pm0.0210) \tag{47}$$

[27] L. W. Differt, H. C. Caldwell, T. Ellison, G. M. Irwin, D. E. Rivard, and J. V. Swintosky, *J. Pharm. Sci.* **57,** 828 (1968).

with $100R^2 = 99.78$; $F = 231.0$; $s = 0.00359$; and $n = 4$. The rate constants in human plasma were correlated with the equation

$$Q_X = L\sigma_{IZ} + A\alpha_Z + h \tag{48}$$

giving

$$\log k_{rX} = -3.87(\pm 0.133)\sigma_{IZ} - 1.13(\pm 0.306)\alpha_Z + 1.15(\pm 0.0486) \tag{49}$$

with $100R^2 = 99.90$; $F = 520.5$; $s = 0.0403$; and $n = 4$.

The chemical hydrolysis is probably a function of electrical and steric effects. The enzymatic hydrolysis which is the probable path in human plasma seems to depend upon electrical and polarizability effects. The data available are not sufficient to permit the drawing of any definitive conclusions.

Beckett, Taylor, and Garrod[28] have proposed trialkylsilyl ether (**VIII**) derivatives of methylephedrine, ephedrine, and norephedrine as prodrugs. They have reported rate constants for the hydrolysis of these compounds at pHs 4.0 and 7.4. Their values are (Ak^1, Ak^2, Ak^3, 10^4k_r at pH 4.0 and at pH 7.4) Me, Me, Me, 2100, 66; Et, Me, Me, 680, 20; Pr, Me, Me, 720, 23; Bu, Me, Me, 540, 13; Et, Et, Me, 290, —; Pr, Pr, Me, 310, 5.8; Bu, Bu, Me, 150, —; Et, Et, Et, 90, 1.5; Pr, Pr, Pr, 110, —; Bu, Bu, Bu, 36, —.

The data were correlated with the SB equation [Eq. (11)] considering SiAk^{1}Ak^{2}Ak^3 as an alkyl group of the type in (**I**) with C^1 = Si. As for all of the data points, $n_1 = 3$, it was excluded from the correlation. Only n_2, n_3, and n_4 were used as variables. The correlation equations obtained were (for pH 4.0)

$$\log k_{rX} = -0.446(\pm 0.0121)n_2 + 0.0246(\pm 0.0118)n_3$$
$$- 0.158(\pm 0.0118)n_4 + 3.31(\pm 0.0206) \tag{50}$$

with $100R^2 = 99.76$; $F = 832.7$; $s = 0.0312$; and $n = 10$; and (for pH 7.4)

$$\log k_{rX} = -0.556(\pm 0.0185)n_2 + 0.0252(\pm 0.0236)n_3$$
$$- 0.203(\pm 0.0471)n_4 + 1.85(\pm 0.0327) \tag{51}$$

with $100R^2 = 99.79$; $F = 310.1$; $s = 0.0413$; and $n = 6$.

As the same pattern of steric dependence on branching is observed at both pH 4.0 and pH 7.4, it is very likely that the hydrolysis mechanism is the same at both pH values. The data sets were therefore combined into a single set and correlated with the equation

$$Q_X = a_2n_2 + a_3n_3 + a_4n_4 + O\omega + a_0 \tag{52}$$

[28] A. H. Beckett, D. C. Taylor, and J. W. Gorrod, *J. Pharm. Pharmacol.* **27**, 588 (1975).

defining ω as log k_{rMe}. The resulting correlation equation was

$$\log k_{rX} = -0.491(\pm 0.0191)n_2 + 0.0366(\pm 0.0204)n_3$$
$$-0.154(\pm 0.0230)n_4 + 1.09(\pm 0.0238)\omega - 0.232(\pm 0.0669) \quad (53)$$

with $100R^2 = 99.56$; $F = 623.7$; $s = 0.0660$; and $n = 16$.

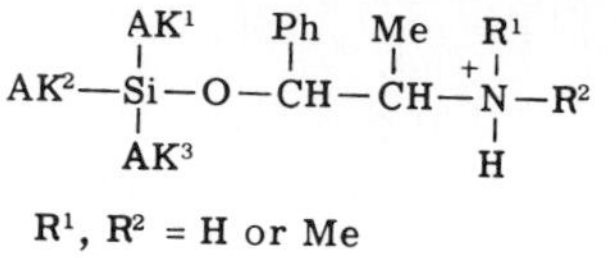

$$\begin{array}{ccccc} & AK^1 & Ph & Me & R^1 \\ & | & | & | & \overset{+}{|} \\ AK^2 - & Si-O-CH-CH-\overset{}{N} & -R^2 \\ & | & & & | \\ & AK^3 & & & H \end{array}$$

R^1, R^2 = H or Me

VIII

Finally, Bundgaard and Johansen[29] have proposed Mannich bases of amides as prodrugs. They have determined rate constants for the hydrolysis of 4-substituted N-piperidinomethylbenzamides (**IX**). Their values are (X, k_r) H, 0.051; NO_2, 0.17; Cl, 0.070; OMe, 0.035. The Hammett equation is suitable for the correlation of these data, and the resulting correlation equation is

$$\log k_{rX} = 0.658(\pm 0.0210)\sigma_{pX} - 1.28(\pm 0.0089) \quad (54)$$

with $100R^2 = 99.80$; $F = 978.7$; $s = 0.0162$; and $n = 4$.

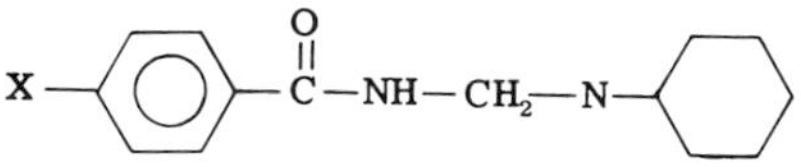

$$X - \underset{}{\bigcirc} - \overset{\overset{O}{\|}}{C} - NH - CH_2 - N - \bigcirc$$

IX

The electrical effect constants used in all of these correlations were taken from ref. 5; the v steric constants are from ref. 6.

[29] H. Bundgaard and M. Johansen. *Arch. Pharm. Chem. Sci. Ed.* **8**, 29 (1980).

[26] Alteration of Drug Metabolism by the Use of Prodrugs

By SIDNEY D. NELSON

Metabolism studies are an integral part of the testing procedures that are applied to evaluate the safety and efficacy of drugs, food additives, pesticides, and other industrial chemicals. Such studies generally provide a better understanding of a drug's mode of action, toxicity, and interactions with other drugs, and sometimes they provide insight into biochem-

ical reaction mechanisms that are involved in biotransformations and toxicities. Hopefully, the insight that is provided by such studies can be used
to increase the efficacy of a drug. In general terms, drug efficacy can be
increased by increasing its potency, selectivity, and duration of action or
by decreasing the probability of undesirable toxic reactions.

By definition, prodrugs must be transformed chemically and/or enzymatically to active drug *in vivo*. Several other chapters in this section
describe how prodrug structure can alter the physicochemical characteristics of a drug, and thereby rates of absorption, distribution, and elimination. Many examples have been presented which illustrate transforma-
tions of prodrugs to active drug by specific metabolic reactions. This
chapter will focus on the effects of prodrug structure on pathways of
metabolism which involve the formation of highly reactive and sometimes
toxic products.

Background

Studies on Acetanilide Analgesic-Antipyretics

Two widely used acetanilide analgesic-antipyretics are acetaminophen
(**I**) and phenacetin (**II**). Although phenacetin may have some therapeutic
activity as the parent structure, most of its activity has been attributed to
its oxidative deethylation product, acetaminophen. As such it is a fortui-
tous prodrug.

However, alterations in metabolism caused by this structural modification are dramatic.[1,2] A significantly greater proportion of a dose of
phenacetin is hydrolyzed at the amide linkage, and subsequent oxidation
of the resulting *p*-phenetidine is thought to yield metabolites that are
responsible for methemoglobinemia, hemolysis, and kidney toxicities. In
addition, phenacetin is oxidized to reactive metabolites by N-hydroxyl-

[1] J. A. Hinson, *Environ. Health Perspect.* **49**, 71 (1983).
[2] S. D. Nelson, *J. Med. Chem.* **25**, 753 (1982).

ation and probably arene oxide formation. In contrast, acetaminophen primarily is conjugated at the phenolic oxygen, and a smaller proportion of the dose is oxidized to a reactive quinoneimine. In therapeutic doses this metabolite is primarily conjugated with glutathione. In overdose situations this metabolite causes severe hepatic necrosis. Interestingly, hepatic necrosis has never been documented in man even after large doses of phenacetin. Thus, ethylation of the phenolic group of acetaminophen dramatically changes the metabolism and toxicity of the drug.

Three other prodrugs of acetaminophen have been described in the literature; a tetrahydropyranyl ether (**III**),[3] an ethoxyethyl ether (**IV**),[4] and a mutual prodrug ester of aspirin (**V**).[5] The two phenoxy ethers were prepared to decrease the bitter taste characteristics of acetaminophen. Both are hydrolyzed rapidly in the gastric juice to yield acetaminophen. Therefore, systemic metabolism should be the same as if acetaminophen were administered.

On the other hand, benorylate (**V**) is not hydrolyzed in the gastric juice and is more slowly absorbed than either acetaminophen or aspirin. However, on passage through the gut, portal circulation, and liver, benorylate is extensively hydrolyzed to aspirin and acetaminophen such that presystemic metabolism yields plasma levels of acetaminophen and aspirin comparable to the same doses of the individual drug moieties. This is consistent with therapeutic results showing that doses of aspirin and acetaminophen, which are equivalent to what is anticipated from doses of benorylate, produce the same degree of analgesic and antipyretic response[6] and of antiinflammatory activity.[7] The major advantage of the

[3] A. J. Repta and J. Hack, *J. Pharm. Sci.* **62,** 1892 (1973).
[4] A. Hussain, P. Kulkarni, and D. Perrier, *J. Pharm. Sci.* **67,** 545 (1978).
[5] A. Robertson, J. P. Glynn, and A. K. Watson, *Xenobiotica* **2,** 339 (1972).
[6] J. Weill, R. Gaillon, C. Rendu, and C. Lejeune, *Therapie,* **23,** 541 (1968).
[7] D. L. Beales, H. C. Burry, and R. Graham, *Br. Med. J.* **2,** 483 (1972).

drug appears to be as a prodrug of aspirin that can be used to treat chronic inflammatory diseases at a decreased dosing rate and without attendant irritation of the gastric mucosa.[8]

A major criticism of the studies that were reported on benorylate is the omission of toxicological data. Because both aspirin and acetaminophen can cause toxicity and because animal models are available for the toxicities, benorylate should have been examined in the appropriate animal models for toxicological effects. Since the introduction of benorylate, two reports of fatality in children due to centrilobular hepatic necrosis, a common feature of acetaminophen toxicity, have appeared.[9,10] These might have been avoided if proper toxicological studies in animals had been carried out and reported.

Studies on Antiinflammatory Drugs

Prodrugs of aspirin which do not cause gastric irritation but revert upon hydrolysis to aspirin rather than the salicylic acid derivative have been prepared (*vide infra* and ref. 11). As has been demonstrated,[12] careful analytical studies must be carried out to ensure that metabolites are correctly identified. The phenylalanine derivative of aspirin (**VI**) was proposed as a prodrug of aspirin based on titrimetric studies.[13] However, HPLC analysis of metabolites revealed that aspirin (**VII**) was not a product; rather, salicyl phenylalanine (**VIII**) was.[12] Thus, subtle changes in

[8] D. N. Craft, J. H. P. Cuddigan, and C. Sweetland, *Br. Med. J.* **3**, 545 (1972).
[9] M. Sacher and H. Thaler, *Lancet* **1**, 481 (1977).
[10] D. N. K. Symon, E. S. Gray, O. J. Hanmer, and G. Russell, *Lancet* **2**, 1153 (1982).
[11] J. E. Truelove, A. A. Hussain, and H. B. Kostenbauder, *J. Pharm. Sci.* **69**, 231 (1980).
[12] Z. Muhi-Eldeen and A. Hussain, *J. Pharm. Sci.* **72**, 1093 (1983).
[13] P. K. Banerjee and G. L. Amidon, *J. Pharm. Sci.* **70**, 1307 (1981).

acyl linkage can markedly affect rates of hydrolysis. Unfortunately, all of the above-mentioned studies were carried out with peptidases *in vitro,* and no studies were reported *in vivo* where pH differences could effect the results.

The importance of carrying out metabolism studies *in vivo* in more than one species recently was demonstrated for a cinnamate ester antiinflammatory prodrug (**IX**).[14] This compound was rapidly hydrolyzed and reduced to the 2-phenylbutyric acid metabolite (**X**) in all species examined, but loss of the one carbon fragment to generate the active antiinflammatory 2-phenylpropionic acid metabolite (**XI**) did not take place in mouse, baboon, or man.

$$HC{\equiv}C-\underset{H}{\overset{CH_3}{C}}-O-\langle\bigcirc\rangle-\underset{H}{\overset{CH_3}{C}}=CCOOC_2H_5 \longrightarrow \longrightarrow HC{\equiv}C-\underset{H}{\overset{CH_3}{C}}-O-\langle\bigcirc\rangle-\underset{H}{\overset{CH_3}{C}}-CH_2-COOH$$

IX   **X**

$$HC{\equiv}C-\underset{H}{\overset{CH_3}{C}}-O-\langle\bigcirc\rangle-\underset{H}{\overset{CH_3}{C}}-COOH$$

XI

The ethinyl group itself has been employed as a precursor to the acetic acid moiety in biphenyl antiinflammatory agents.[15] The prodrug, 4′-ethynyl-2-fluorbiphenyl (**XII**), is converted in the rat to (2-fluoro-4′-biphenyl)acetic acid (**XIII**), an active antiinflammatory compound.

$$\langle\bigcirc\rangle_F{-}\langle\bigcirc\rangle-C{\equiv}CH \xrightarrow{P-450} \langle\bigcirc\rangle_F{-}\langle\bigcirc\rangle-CH_2-\overset{O}{\overset{\|}{C}}-OH$$

XII   **XIII**

Based on extensive mechanistic studies, it has been postulated that terminal acetylenes are oxidized by cytochrome P-450 to reactive intermediates that can suicide inactivate the enzyme as well as hydrolyze to acetic acid derivatives.[16] Therefore, structural modifications that employ the ethinyl group may yield compounds that inhibit cytochrome P-450 which could result in drug interactions as has been observed with some contraceptive steroids.[17]

[14] T. R. Marten, R. J. Ruane, P. B. East, and S. L. Malcolm, *Xenobiotica* **13,** 1 (1983).
[15] H. R. Sullivan, P. Roffey, and R. E. McMahon, *Drug Metab. Dispos.* **7,** 76 (1979).
[16] P. R. Ortiz de Montellano and K. L. Kunze, *Arch. Biochem. Biophys.* **209,** 710 (1981).
[17] K. Einarsson, J. L. E. Ericsson, J. A. Gustafsson, J. A. Sjövall, J. Zietz, and E. Zietz, *Biochim. Biophys. Acta* **369,** 278 (1974).

Miscellaneous Examples

Acylation is one of the most common reactions used in the design of prodrugs. Unfortunately, in some cases such a structural modification might lead to a more toxic product. For example, acetylation of isoniazid (**XIV**) generates acetylisoniazid (**XV**) which hydrolyzes preferentially at the isonicotinoyl carbonyl group to yield acetylhydrazine (**XVI**), a hepatotoxin.[18] The effect of other acyl groups on the toxicity of hydrazines has not been investigated. However, alkylation of hydrazines yields derivatives that are usually more toxic.[18,19]

XIV   XV   XVI

In at least one case, acylation has been found to decrease the incidence of toxic reaction to a drug. Procainamide (**XVII**) is a widely used antiarrhythmic that can induce a lupus reaction in patients. Acetylation of the amine function protects against this undesired reaction.[20] However, *N*-acetylprocainamide (**XVIII**) is hydrolyzed only minimally to procainamide in man.[21] Although it is not a prodrug, it does have substantial antiarrhythmic activity of its own.[22] Thus, structural modification in this instance altered both the metabolic disposition and toxicity of the drug without significantly changing its spectrum of therapeutic activity.

XVII R = H

XVIII R = CH$_3$—C(=O)

General Considerations in Experimental Design

In the development of a prodrug, it is desirable to examine the effects of prodrug structure on metabolism to determine whether compounds are

[18] J. R. Mitchell, H. J. Zimmerman, K. G. Ishak, U. P. Thorgeirsson, J. A. Timbrell, W. R. Snodgrass, and S. D. Nelson, *Ann. Intern. Med.* **84,** 181 (1976).

[19] H. Druckrey, *Xenobiotica* **3,** 271 (1973).

[20] D. M. Roden, S. B. Reele, S. B. Higgins, R. Smith, J. A. Oates, and R. L. Woosley, *Am. J. Cardiol.* **46,** 463 (1980).

[21] J. M. Strong, J. S. Dutcher, W. K. Lee, and A. J. Atkinson, Jr., *J. Pharmacokinet. Biopharm.* **3,** 223 (1975).

[22] D. E. Drayer, M. M. Reidenberg, and R. W. Sevy, *Proc. Soc. Exp. Biol. Med.* **146,** 358 (1974).

formed with undesirable therapeutic or toxic properties. Similarly, when a prodrug is developed from a compound with documented therapeutic and toxicological properties, the effects of prodrug structure on both of these properties should be investigated. Two basic requirements for such investigations are choosing an appropriate animal model and accurate methods for metabolite analysis.

Animal Models

Animal models for therapeutic activity and toxicity of a prodrug should derive from those already established by studies with parent drug structures. Thus, it would seem reasonable that any prodrug of acetaminophen would be compared to the parent drug with regards not only to time and dose requirements for analgesic-antipyretic activities, but also to time and dose requirements for hepatic and renal toxicity, inasmuch as animal models for these have been developed. Presumably, any unusual activity or toxicity of the prodrug in animals will be documented in testing carried out on the drug prior to application for investigational new drug status to the FDA. For example, methemoglobinemia and hemolysis are toxicological reactions of phenacetin not common to acetaminophen. This would indicate some unusual property of the prodrug which would have to be further investigated by comparative metabolism studies.

Metabolism Studies

It is not the purpose of this chapter to review all of the methods that are used in metabolism studies. Suffice it to say that they should be rigorous enough to establish the major routes of biotransformation in various animal species including man. As has been pointed out with the cinnamate ester antiinflammatory prodrug (**IX**), there can be considerable species variation in metabolism. Therefore, it is important to find an animal species that metabolizes the drug in a manner similar to that of man.

The rationale for the development of many prodrugs is improvement of presystemic pharmaceutical characteristics, i.e., taste, absorption, etc. For these drugs it is important to establish the contribution of prodrug structure to presystemic extraction and metabolism of the drug. Design of such studies is covered in this volume (see [28]). In contrast, the development of a prodrug to target a drug to a particular tissue will require both sampling the systemic circulation and the target tissue for drug and metabolites.

Additional studies need to be carried out if the prodrug or one of its metabolites forms reactive metabolites. Often these biotransformation products are so reactive that undetectable amounts reach the general circulation. Therefore, radiolabeled analogs of the prodrug are required to monitor covalent binding of the metabolites to tissue macromolecules. Methods and precautions for carrying out such studies have been published.[23] Several examples of structural elements that are present in drugs and form reactive metabolites have been cited and discussed.[2]

Drug Interactions

Finally, the structural modification to drug structure that is imparted by development of a prodrug needs to be assessed in regards to potential drug interactions. For example, the mutual prodrug benorylate (**V**) has the potential for drug interactions between its two hydrolysis products, aspirin and acetaminophen, both pharmacokinetically and pharmacodynamically. Metabolite profiles of each in the presence of the other need to be compared with those generated by the prodrug. Metabolism studies to date on benorylate have only partially assessed the potential for interactions at the metabolic level.[5]

The fact that some structures important to prodrug structure, such as the ethinyl group, can act as suicide inhibitors of drug-metabolizing enzymes also needs to be evaluated [refer to structures (**IX**)–(**XII**)]. If prodrugs that contain such structures are used in conjunction with therapeutic agents which have narrow therapeutic windows, life-threatening drug interactions could result.

In conclusion, subtle changes brought about by prodrug structure can markedly influence the metabolism of the prodrug itself, the products of its metabolism, and the metabolism of other drugs. Therefore, metabolism studies of prodrugs need to be carefully designed and performed.

[23] L. R. Pohl and R. V. Branchflower, this series, Vol. 77, p. 43.

[27] Formation of Prodrugs of Amines, Amides, Ureides, and Imides

By HANS BUNDGAARD

A major problem in the general application of the prodrug concept to solve drug delivery problems is the limited possibilities available for making bioreversible derivatives of many drugs. It is relatively simple to design prodrugs of substances possessing hydroxy or carboxyl groups which are readily esterifiable, but for a large number of drugs no apparently readily derivatizable functional groups or entities are present in the molecules. Therefore, to expand the application of the prodrug concept, new types of chemical approaches have to be developed. In our laboratories studies have been performed within this area of prodrug research, focusing on the development of potentially useful prodrug types for amines, amides, imides, and various other NH-acidic compounds. This chapter is largely based on these studies.

N-Mannich Bases

N-Mannich bases are formed by reacting an NH-acidic compound with formaldehyde or, in rare cases, other aldehydes—and a primary or secondary aliphatic or aromatic amine. The process can be considered as an N-aminomethylation or N-amidomethylation (in the case of the NH-acidic compound being an amide).

$$R-CONH_2 + CH_2O + R_1R_2NH \rightleftharpoons R-CONH-CH_2-NR_1R_2 + H_2O$$

Preparation of N-Mannich Bases

N-Mannich bases of carboxamides, thioamides, and various other NH-acidic compounds have been known for a long time. The chemistry of their formation has been the subject of two reviews.[1,2] Generally, N-Mannich bases are simply prepared by heating formaldehyde, the amine (or the amine hydrochloride), and the amide-type component in water, ethanol, methanol, or dioxane according to standard literature procedures.[1-6]

Hydrolysis of N-Mannich Bases

Kinetics. The kinetics of decomposition of a great number of N-Mannich bases in aqueous solution has been the subject of several studies.[7-12] At constant pH and temperature, the decomposition rates of the N-Man-

[1] H. Hellmann and G. Opitz, "α-Aminoalkylierung." Verlag Chemie, Weinheim, West Germany, 1960.
[2] M. Tramontini, *Synthesis* p. 703 (1973).
[3] A. Einhorn, *Justus Liebigs Ann. Chem.* **343**, 207 (1905).
[4] Y. Watase, Y. Terao, and M. Sekiya, *Chem. Pharm. Bull.* **21**, 2775 (1973).
[5] M. B. Winstead, K. V. Anthony, L. L. Thomas, R. Strachan, and H. J. Richwine, *J. Chem. Eng. Data* **7**, 414 (1962).
[6] H. Hellmann, *Angew. Chem.* **69**, 471 (1957).
[7] H. Bundgaard and M. Johansen, *J. Pharm. Sci.* **69**, 44 (1980).
[8] H. Bundgaard and M. Johansen, *Arch. Pharm. Chem., Sci. Ed.* **8**, 29 (1980).
[9] H. Bundgaard and M. Johansen, *Int. J. Pharm.* **8**, 183 (1981).
[10] M. Johansen and H. Bundgaard, *Int. J. Pharm.* **7**, 119 (1980).
[11] H. Bundgaard and M. Johansen, *Int. J. Pharm.* **9**, 7 (1981).
[12] M. Johansen and H. Bundgaard, *Arch. Pharm. Chem., Sci. Ed.* **10**, 111 (1982).

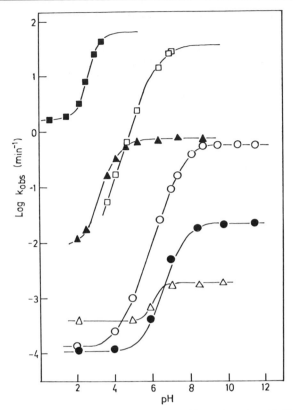

FIG. 1. The pH–rate profiles for the decomposition of various N-Mannich bases in aqueous solution at 37°. (■) N-(Morpholinomethyl)-p-toluenesulfonamide; (□) N-(piperidinomethyl)trichloroacetamide; (▲) N-(morpholinomethyl)trichloroacetamide; (○) N-(diethylaminomethyl)benzamide; (●) N-(isobutylaminomethyl)benzamide; (△) N-(benzylaminomethyl)benzamide. From ref. 8.

nich bases followed strict first-order kinetics and all reactions went to completion. No general acid–base catalysis by the buffers used was apparent. The pH–rate profiles for most compounds have a sigmoidal shape as seen in Fig. 1. These pH dependences of the observed apparent first-order rate constant, k_{obs}, could be accounted for by assuming spontaneous decomposition of the free Mannich bases (B) and their conjugate acids (BH$^+$); the expression for k_{obs} is

$$k_{obs} = \frac{k_1 K_a}{a_H + K_a} + \frac{k_2 a_H}{a_H + K_a} \tag{1}$$

where K_a is the apparent ionization constant of the protonated N-Mannich bases, a_H is the hydrogen ion activity, and k_1 and k_2 are the apparent first-order rate constants for the spontaneous degradation of B and BH$^+$, respectively.

Reaction Mechanism. The reaction mechanism proposed[7,8] for the decomposition involves as rate-determining step a unimolecular N–C bond cleavage with formation of an amide (or imide) anion and an immonium cation. In subsequent fast steps, a solvent molecule transfers a proton to the anion and a hydroxide ion to the immonium ion, giving methylolamine, which rapidly dissociates to formaldehyde and amine. Loudon *et al.*[13] have independently proposed a similar mechanism for N-Mannich bases derived from isopropylaldehyde.

Structural Effects on Decomposition Rate. The structural effects on the decomposition rate of N-Mannich bases derived from carboxamides, thioamides, sulfonamides or imides, and aliphatic or aromatic amines involve steric effects and basicity of the amine component and acidity of the amide-type component.[8,9,12] These factors are most pronounced with respect to the rate constant k_1 and, accordingly, to the decomposition rate in weakly acidic to basic aqueous solutions. The rates of the hydrolysis of unprotonated Mannich bases are accelerated strongly by (1) increasing steric effects within the amine substituent, (2) increasing basicity of the amine component, and (3) increasing acidity of the parent amide-type compound.

For some N-Mannich bases of benzamide and various amines, the rate constant k_1 can be expressed by Eq. (2)[8]:

$$\log k_1 = 2.30\nu - 3.50 \qquad (k_1 \text{ in min}^{-1}; 37°) \qquad (2)$$

where ν is Charton's steric substituent parameter for alkylamino groups.[14] The marked influence of the steric effect on k_1 can be exemplified by comparing the k_1 values for the benzamide Mannich bases of diethylamine (0.52 min^{-1}) and ethylamine (0.0084 min^{-1}).

For amines with the same steric properties but differing in basicity, the rate constants k_1 for the decomposition of the respective N-Mannich bases were shown to increase almost 10-fold with an increase of unity of the pK_a of the amines. Thus, for various *N*-(arylaminomethyl)succinimide derivatives (**I**) the following relationship was derived[9]:

$$\log k_1 = 0.93 \text{p}K_a - 4.81 \qquad (k_1 \text{ in min}^{-1}; 37°) \qquad (3)$$

I

[13] G. M. Loudon, M. R. Almond, and J. N. Jacob, *J. Am. Chem. Soc.* **103**, 4508 (1981).
[14] M. Charton, *J. Org. Chem.* **43**, 3995 (1978).

The structural effect of the amide-type component in the Mannich bases on the decomposition rate was delineated from rate data obtained for several Mannich bases with either piperidine or morpholine.[8] The reactivity was shown to increase strongly with increasing acidity of the parent amide-type compound. For the Mannich bases with piperidine, the following relationship was derived:

$$\log k_1 = -1.42 \mathrm{p}K_a + 19.3 \qquad (k_1 \text{ in min}^{-1}; 37°) \qquad (4)$$

For morpholine derivatives, Eq. (5) was obtained:

$$\log k_1 = -1.15 \mathrm{p}K_a + 13.9 \qquad (k_1 \text{ in min}^{-1}; 37°) \qquad (5)$$

Equations (4) and (5), in which $\mathrm{p}K_a$ refers to the ionization constant for the parent amide-type compounds (at 20–25°), cover both aromatic and aliphatic carboxamides as well as a thioamide and a sulphonamide. N-Mannich bases of urea, thiourea and N-acylthiourea derivatives were found to deviate from these relationships showing a *greater* reactivity than expected on the basis of their $\mathrm{p}K_a$ values. The N-Mannich base (**II**)

$$\langle\bigcirc\rangle\text{—CH—CH—CH}_3$$
$$\quad\ \ \overset{|}{\text{OH}}\ \ \overset{|}{\underset{\underset{\text{CH}_3}{|}}{\text{N}}}\text{—CH}_2\text{—NHCO—}\langle\bigcirc\rangle$$
$$\text{II}$$

formed between (−)-ephedrine and benzamide is also more reactive than predicted, the half-life of hydrolysis at pH 7.4 and 37° being 2.2 min.[12] A positive deviation was further observed with N-Mannich bases of salicylamide.[10]

Some representative rate data for the decomposition of various N-Mannich bases are given in Table I. The breakdown of the N-Mannich bases does not rely on enzymatic catalysis, and identical decomposition rates were observed in solutions with or without addition of human plasma.[15]

Application of N-Mannich Bases as Potential Prodrugs

By appropriate selection of the amine component, it should be feasible to obtain prodrugs of a given amide-type drug with varying degree of *in vivo* lability. Besides, other physicochemical properties such as aqueous solubility, dissolution rate, and lipophilicity can be modified for the parent compounds.[16]

[15] M. Johansen and H. Bundgaard, *Arch. Pharm. Chem., Sci. Ed.* **9**, 40 (1981).
[16] M. Johansen and H. Bundgaard, *Arch. Pharm. Chem., Sci. Ed.* **8**, 141 (1980).

TABLE I
Observed Rate Data for the Decomposition of Various N-Mannich Bases in Aqueous Solution at 37°[a]

Compound	k_1 (min^{-1})	$t_{1/2}$ (min)[b]
N-(Piperidinomethyl)benzamide	0.051	47
N-(Piperidinomethyl)-4-nitrobenzamide	0.17	8.0
N-(Piperidinomethyl)actamide	0.0055	400
N-(Piperidinomethyl)dichloroacetamide	2.48	0.4
N-(Piperidinomethyl)trichloroacetamide	35	0.02
N-(Piperidinomethyl)nicotinamide	0.17	8.0
N-(Piperidinomethyl)thiobenzamide	13	0.06
N-(Morpholinomethyl)benzamide	0.0005	1400
N-(Morpholinomethyl)thiobenzamide	0.52	1.3
N-(Morpholinomethyl)-p-toluenesulfonamide	60	0.01
N-(Phenethylaminomethyl)benzamide[c]	0.0048	205
N-(Phenylpropanolaminomethyl)benzamide	0.0031	225
N-(Methylaminomethyl)benzamide	0.0026	600
N-(Ethylaminomethyl)benzamide	0.0084	190
N-(Diethylaminomethyl)benzamide	0.52	4.0
N-(Dimethylaminomethyl)benzamide	0.032	58
N-(Benzylaminomethyl)benzamide	0.0020	380
N-(Morpholinomethyl)-N'-acetylthiourea	0.91	0.8
N-(Piperidinomethyl)-N'-methylurea	—	5.0
N-(Methylaminomethyl)salicylamide[d]	—	28
N-(Piperidinomethyl)salicylamide[d]	—	14
N-(Morpholinomethyl)salicylamide[d]	—	41
N-(α-Alaninomethyl)salicylamide[d]	—	17
N-(Anilinomethyl)succinimide[e]	0.36	1.9
N-(p-Toluidinomethyl)succinimide[e]	0.76	0.9

[a] From ref. 8, if not otherwise indicated.
[b] At pH 7.40.
[c] From ref. 12.
[d] From ref. 10.
[e] From ref. 9.

TABLE II
Ionization Constants and Rate Data for the Decomposition of Various N-Mannich Bases of Carbamazepine in Aqueous Solution at 37°[a]

Compound	pK_a	k_1 (min^{-1})	k_2 (min^{-1})	$t_{1/2}$[b] (min)
N-(Diethylaminomethyl)carbamazepine	7.95	0.17	0.0017	19
N-(Dipropylaminomethyl)carbamazepine	7.75	0.27	0.0030	7
N-(Piperidinomethyl)carbamazepine	7.90	0.015	0.0011	165

[a] From ref. 18.
[b] At pH 7.40.

Transformation of an amide into an N-Mannich base introduces a readily ionizable amino moiety which may allow the preparation of derivatives with greatly increased water solubilities at slightly acidic pH values where, as a matter of fortune, the stability may be quite high. This has been shown for various Mannich bases using benzamide as a model compound.[16] Whereas N-Mannich bases prepared from secondary amines showed very high solubilities in salt form, the Mannich bases derived from primary amines did not show increased solubility even as salts. This different behavior was attributed to the occurrence of intramolecular hydrogen bonding in the latter derivatives (**III**).

The concept of N-Mannich base formation of NH-acidic compounds to yield more soluble prodrugs has already been utilized in the case of rolitetracycline (**IV**). This highly water-soluble N-Mannich base of tetracycline and pyrrolidine is decomposed quantitatively to tetracycline in neutral aqueous solution, the half-life being 40 min at pH 7.4 and 37°.[17] Similarly, various N-Mannich bases of carbamazepine (**V**) have been developed as water-soluble prodrugs for parenteral administration[18] (Table II).

In addition, the concept may be useful for improving the dissolution behavior of poorly soluble drugs in an effort to improve oral bioavailability. Thus, N-Mannich bases of various NH-acidic compounds (e.g., phthalimide, chlorzoxazone, phenytoin, barbital, *p*-toluenesulfonamide, acetazolamide, chlorothiazide, and allopurinol) with morpholine or piperidine as the amine component were found to possess markedly

[17] B. Vej-Hansen and H. Bundgaard, *Arch. Pharm. Chem., Sci. Ed.* **7**, 65 (1979).
[18] H. Bundgaard, M. Johansen, V. Stella, and M. Cortese, *Int. J. Pharm.* **10**, 181 (1982).

TABLE III
EFFECT OF N-MANNICH BASE FORMATION OF AMINES ON pK_a FOR
THE CORRESPONDING PROTONATED SPECIES

Amine	pK_a value of amine	pK_a value of N-(aminomethyl)-benzamide[a]	pK_a value of N-(aminomethyl)-succinimide[b]
Piperidine	11.1	7.8	
Methylamine	10.7	7.5	
Dimethylamine	10.7	7.6	
Ethylamine	10.7	7.5	
Diethylamine	10.9	7.7	
Isobutylamine	10.7	7.5	
Cyclohexylamine	10.7	7.6	
Benzylamine	9.3	6.4	
Morpholine	8.3	5.6	
Ephedrine[c]	9.7	6.3	
p-Toluidine	5.1		1.6
Aniline	4.6		1.1
Procaine	2.4		1.3
Benzocaine	2.4		1.3

[a] From ref. 8.
[b] From ref. 9.
[c] From ref. 12.

greater (up to a factor of 2000) intrinsic dissolution rates in 0.1 M hydrochloric acid in comparison with the parent compounds.[19,20] Once dissolved, the N-Mannich bases are cleaved very rapidly, with the quantitative release of the parent compounds.

Besides being considered as a possible approach of derivatizing amide-type compounds, N-Mannich base formation can also be thought of as a means of forming prodrugs of primary and secondary amines, in which case the amide-type component would act as a transport group. As can be seen from Table III, N-Mannich base formation lowers the pK_a values of the conjugate acids of amines by up to about 3 units. Therefore, a potentially useful purpose for transforming amino compounds into N-Mannich base transport forms would be to increase the lipophilicity of the parent amines at physiological pH values by depressing their protonation, resulting in enhanced biomembrane-passage properties. Another potential objective for transient derivatizing amino groups in drugs may be to obtain protection against first-pass metabolism, e.g., N-acetylation of primary aromatic amino groups.

[19] H. Bundgaard and M. Johansen, Int. J. Pharm. 7, 129 (1980).
[20] H. Bundgaard and M. Johansen, Acta Pharm. Suec. 18, 129 (1981).

However, the selection of biologically acceptable amide-type transport groups affording an appropriate cleavage rate of a Mannich base of a given amine at pH 7.4 is restricted. In a search for generally useful candidates it was observed[10] that N-Mannich bases of salicylamide and different aliphatic amines (**VI**) including amino acids showed an unexpectedly

$$CONH-CH_2-NR_1R_2$$

(salicylamide structure with OH)

$$\underline{VI}$$

high cleavage rate at neutral pH (cf. Table I), thus suggesting the utility of salicylamide. For aromatic amines, more acidic amide-type transport groups such as succinimide or hydantoins have been suggested.[9]

N-Hydroxmethyl and *N*-Acyloxymethyl Derivatives

When an NH-acidic compound is allowed to react with formaldehyde in the absence of a primary or secondary amine, *N*-hydroxymethylation occurs:

$$R-CONH_2 + CH_2O \rightleftharpoons R-CONH-CH_2OH$$

The kinetics of decomposition of a large number of *N*-hydroxymethylated amides, imides, carbamates, and hydantoins in aqueous solution has been studied.[21,22] It was found that the decomposition exhibited a first-order dependence on hydroxide ion concentration up to pH about 12 and that the rates increased sharply with increasing acidity of the parent compound (Fig. 2). The following linear correlation was found between $\log k_1$ (where k_1 is the apparent hydroxide ion catalytic rate constant) and pK_a for the above-mentioned group of compounds:

$$\log k_1 = -0.77 pK_a + 14.4 \qquad (k_1 \text{ in } M^{-1}; 37°) \qquad (6)$$

This relationship may be useful for the prediction of the reactivity of an *N*-hydroxymethyl derivative solely from a knowledge of the pK_a of the parent compound. Thus, it can be predicted that the requirement for a half-life of the decomposition reaction of less than 1 hr at pH 7.4 and 37° is that the parent NH-acidic compound possesses a pK_a value of less than 13.1.

The *N*-hydroxymethyl derivatives are more water soluble than the parent compounds,[21,22] thus suggesting a potential area of application of *N*-hydroxymethyl prodrugs. However, the most important aspect of

[21] M. Johansen and H. Bundgaard, *Arch. Pharm. Chem., Sci. Ed.* **7**, 175 (1979).
[22] H. Bundgaard and M. Johansen, *Int. J. Pharm.* **5**, 67 (1980).

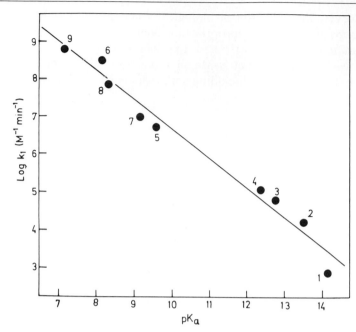

FIG. 2. Plot of the logarithm of the second-order rate constants, k_1, for decomposition of various N-hydroxymethyl derivatives against pK_a of the parent compounds. 1, Chloroacetamide; 2, dichloroacetamide; 3, thiobenzamide; 4, trichloroacetamide; 5, succinimide; 6, 5-chloro-2-benzoxazolinone(chlorozoxazone); 7, 5,5-dimethylhydantoin; 8, phenytoin; 9, nitrofurantoin. From ref. 22.

N-hydroxymethylation resides in the fact that the hydroxy group introduced by this process is readily amenable to bioreversible derivatization, e.g., by esterification to produce water-soluble or lipophilic ester derivatives.[23–26] The N-acyloxymethyl derivative obtained by esterification is stable *in vitro* but is susceptible to enzymatic hydrolysis *in vivo* yielding the unstable N-hydroxymethyl derivative (cf. Fig. 3). Recently, this N-acyloxymethylation prodrug approach has proved useful in enhancing the dermal delivery of various drugs such as theophylline,[27] 6-mercaptopurine,[28] and 5-fluorouracil[29] (Fig. 3). The increased dermal absorption

[23] N. Bodor, *Drugs Future* **6,** 165 (1981).
[24] N. Bodor, *in* "Optimization of Drug Delivery" (H. Bundgaard, A. B. Hansen, and H. Kofod, eds.), p. 156. Munksgaard, Copenhagen, 1982.
[25] I. H. Pitman, *Med. Res. Rev.* **1,** 189 (1981).
[26] M. Johansen and H. Bundgaard, *Arch. Pharm. Chem., Sci. Ed.* **9,** 43 (1981).
[27] K. B. Sloan and N. Bodor, *Int. J. Pharm.* **12,** 299 (1982).
[28] K. B. Sloan, M. Hashida, J. Alexander, N. Bodor, and T. Higuchi, *J. Pharm. Sci.* **72,** 372 (1983).
[29] B. Møllgaard, A. Hoelgaard, and H. Bundgaard, *Int. J. Pharm.* **12,** 153 (1982).

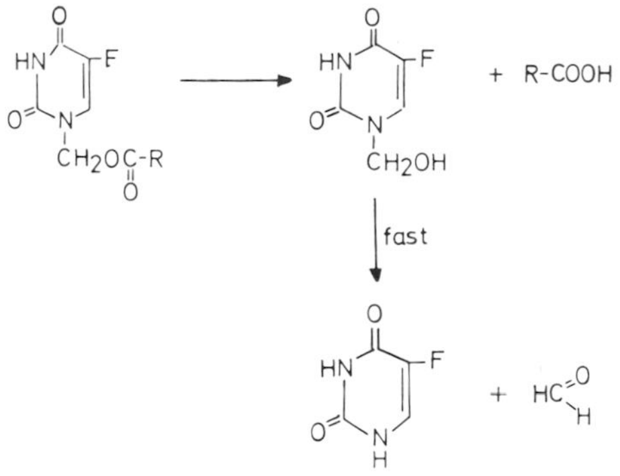

FIG. 3. Improved dermal delivery of 5-fluorouracil via N-acyloxymethyl prodrug derivatives. The compounds are cleaved by enzymes in the skin to N-hydroxymethyl-5-fluorouracil, which spontaneously is decomposed to the parent drug.[29]

observed with the prodrug derivatives is a result of the improved physicochemical properties of the prodrugs, primarily lipophilicity and water solubility along with the susceptibility of the prodrugs to undergo enzymatic cleavage in the skin.

N-Hydroxymethyl derivatives of amines are more unstable than those of amides or imides. For amines with pK_a values >2, the half-lives of hydrolysis of such derivatives at neutral pH are less than a few minutes.[30] N-Acyloxymethylation of primary and secondary amines is not useful because of the extreme lability of such derivatives, but for tertiary amines chemically stable compounds (quarternary ammonium salts) are obtained. Because of a high susceptibility to undergo enzymatic hydrolysis (see below), these compounds are useful as prodrugs for tertiary or N-heterocyclic amines.[24,31]

$$R\text{–}COO\text{–}CHR_1\text{–}N^+\text{—}, \; X^- \rightarrow R\text{–}COOH + R_1CHO + N\text{—}, \; HX$$

N-Acylated Derivatives

N-Acylation of amines to give amide prodrugs has been used only to a limited extent due to the relative stability of amides *in vivo*. However, certain amides formed with amino acids may be so susceptible to undergo enzymatic cleavage that they are useful as prodrugs. Thus, γ-glutamyl derivatives of dopamine (Fig. 4), L-dopa, and sulfamethoxazole are readily hydrolyzed by γ-glutamyltransferase *in vivo* and have been promoted

---

[30] W. R. Abrams and R. G. Kallen, *J. Am. Chem. Soc.* **98,** 7777 (1976).
[31] N. Bodor, U.S. Patent 4,160,099 (1979).

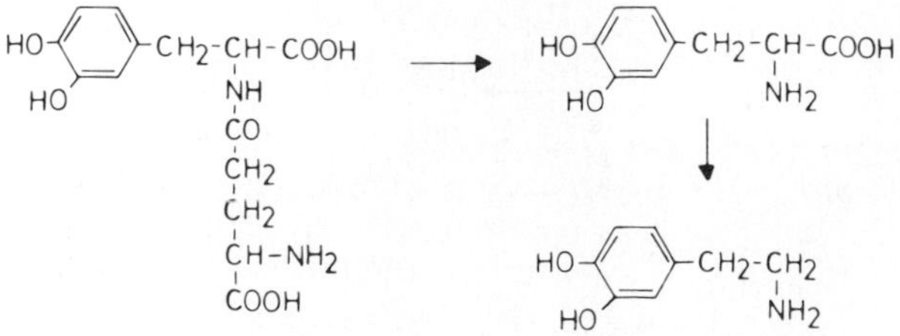

$$HO-\langle\ \rangle-CH_2-\underset{\underset{\underset{\underset{\underset{\underset{COOH}{|}}{CH-NH_2}}{|}}{CH_2}}{\overset{\overset{\overset{\overset{NH}{|}}{CO}}{|}}{CH-COOH}} \longrightarrow HO-\langle\ \rangle-CH_2-\underset{NH_2}{CH-COOH}$$

$$HO-\langle\ \rangle-CH_2-CH_2 \atop HO \qquad NH_2$$

FIG. 4. Selective generation of dopamine in the kidney by sequential action of γ-glutamyltransferase and aromatic L-amino acid decarboxylase on γ-glutamyl-dopa.

as kidney-specific prodrugs because of their preferential bioactivation in the kidney.[32–34] Similarly, various other amides or dipeptides of L-dopa have been shown to be useful as prodrugs.[35] Other enzymatically labile amides include various amino acid derivatives (**VII**) of benzocaine (**VIII**).[36] These compounds are highly water soluble and are cleaved rapidly in the presence of human serum.

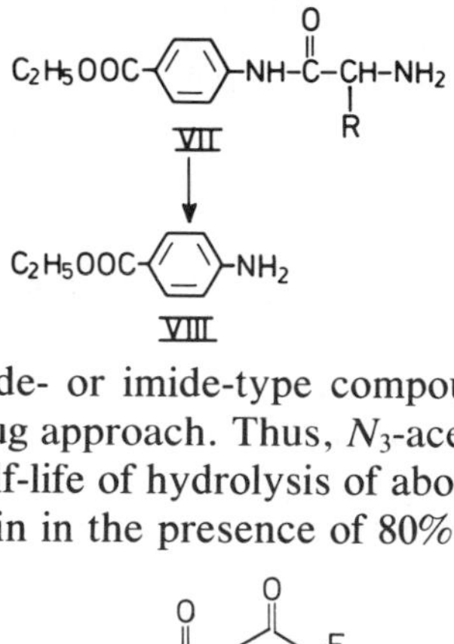

$$C_2H_5OOC-\langle\ \rangle-NH-\overset{\overset{O}{\|}}{C}-\underset{R}{CH}-NH_2$$

VII

$$\downarrow$$

$$C_2H_5OOC-\langle\ \rangle-NH_2$$

VIII

N-Acylation of amide- or imide-type compounds may also in some cases be a useful prodrug approach. Thus, N_3-acetyl-5-fluorouracil (**IX**)[37] was found to have a half-life of hydrolysis of about 40 min at pH 7.4 and 37° and only about 4 min in the presence of 80% human plasma.[38]

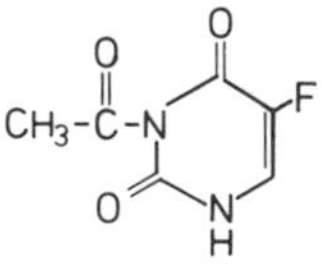

$$CH_3-\overset{\overset{O}{\|}}{C}-N \overset{O}{\underset{O}{\biggl\langle}}\overset{F}{\underset{\underset{H}{N}}{\biggr\rangle}}$$

IX

[32] S. Wilk, H. Mizoguchi, and M. Orlowski, *J. Pharmacol. Exp. Ther.* **206,** 227 (1978).
[33] J. J. Kyncl, F. N. Minard, and P. H. Jones, *Adv. Biosci.* **20,** 369 (1979).
[34] M. Orlowski, H. Mizoguchi, and S. Wilk, *J. Pharmacol. Exp. Ther.* **212,** 167 (1979).
[35] N. Bodor, K. B. Sloan, and T. Higuchi, *J. Med. Chem.* **20,** 1435 (1977).
[36] Z. Slojkowska, H. J. Krasuska, and J. Pachecka, *Xenobiotica* **12,** 359 (1982).
[37] T. Kametani, K. Kigasawa, M. Hiiragi, K. Wakisaka, S. Haga, Y. Nagamatsu, H. Sugi, K. Fukawa, O. Irino, T. Yamamoto, N. Nishimura, and A. Taguchi, *J. Med. Chem.* **23,** 1324 (1980).
[38] A. Buur and H. Bundgaard, *Int. J. Pharm.* **21,** 349 (1984).

Other Prodrug Derivatives

Enamine derivatives of amines with various β-dicarbonyl compounds (such as ethyl acetoacetate and acetylacetone) may be generally useful prodrugs of amines.[39–41] At physiological pH and 25°, enamine derivatives (**X**) of various amino acids with ethyl acetoacetate show half-lives of hydrolysis of 3–30 min.[41]

$$CH_3-C=CH-C-OC_2H_5$$
NH   O
R

X   **XI**   **XII**

Oxazolidines (**XI**) have been proposed as prodrug candidates for β-amino alcohols (**XII**).[42,43] They readily undergo hydrolysis to yield the parent β-amino alcohol (e.g., ephedrine or β-blockers) and aldehydes or ketones, and by varying the carbonyl moiety it is possible to control the rate of formation of a given β-amino alcohol. Thus, the following half-lives of hydrolysis for various (−)-ephedrine oxazolidines were found at pH 7.4 and 37°[42]: 5 min (benzaldehyde), 5 sec (salicylaldehyde), 4 sec (formaldehyde), 17 sec (propionaldehyde), 30 min (pivaldehyde), 4 min (acetone), and 6 min (cyclohexanone). The oxazolidines are much weaker bases than the parent amino alcohol; and, therefore, they are more lipophilic than the parent at physiological pH[42].

Various *ring-opened derivatives* of cyclic drugs have been proposed as potentially useful prodrugs with either increased water or lipid solubility. These include esters of malonuric acids as prodrugs for barbituric acids,[44,45] esters of hydantoic acids as prodrugs for hydantoins,[46] glutaramic acid esters for glutarimides and succinamic acid esters for succinimides.[47] These esters all undergo a base-catalyzed cyclization and by appropriate selection of the alcohol portion of the esters, it is possible to control and modify the physicochemical properties such as rate of prodrug conversion, aqueous solubility, and lipophilicity.

[39] K. Dixon and J. V. Greenhill, *J. Chem. Soc., Perkin Trans. 2* p. 164 (1974).
[40] N. P. Jensen, J. J. Friedman, H. Kropp, and F. M. Kahan, *J. Med. Chem.* **23**, 6 (1980).
[41] T. Murakami, N. Yata, H. Tamauchi, J. Nakai, M. Yamazaki, and A. Kamada, *Chem. Pharm. Bull.* **29**, 1998 (1981).
[42] H. Bundgaard and M. Johansen, *Int. J. Pharm.* **10**, 165 (1982).
[43] M. Johansen and H. Bundgaard, *J. Pharm. Sci.* **72**, 1294 (1983).
[44] H. Bundgaard, A. B. Hansen, and C. Larsen, *Int. J. Pharm.* **3**, 341 (1979).
[45] H. Bundgaard, E. Falch, and C. Larsen, *Int. J. Pharm.* **6**, 19 (1980).
[46] V. Stella and T. Higuchi, *J. Pharm. Sci.* **62**, 962 (1973).
[47] H. Bundgaard and C. Larsen, *Acta Pharm. Suec.* **16**, 309 (1979).

[28] Design of Prodrugs for Improved Gastrointestinal Absorption by Intestinal Enzyme Targeting

By David Fleisher, Barbra H. Stewart, and Gordon L. Amidon

Intake of nutritional substances utilizes enzyme systems which are relatively specific in coupling the processes of digestion and nutrient transport in the gastrointestinal tract (GI tract).[1] In addition, prevention of entry via the oral route of potentially harmful chemical inputs (including some drugs) is carried out by a less specific group of metabolic enzyme systems.[2] It is essential, therefore, that drug delivery strategies for oral dosage formulation take into account both drug properties and biological system parameters. Frequently these strategies require modification of a drug molecule through chemical derivatization to produce a prodrug with desirable delivery properties which will be reconverted in the body to the parent drug.[3]

Delivery problems that suggest a drug chemical modification for oral administration include drug-induced damage to gastrointestinal tissue, poor aqueous stability,[4] extensive first-pass metabolism,[5] poor aqueous solubility,[6] limited water-to-membrane partition coefficient,[7] and a high degree of ionization over the gastrointestinal pH range.[8] The first three problems are usually the result of a particular functional group on the drug molecule which can be derivatized to protect the digestive system from the drug moiety or to protect the drug from enzymatic or nonenzymatic action in the GI tract.

Poor aqueous solubility of a drug may dictate a slow and erratic dissolution profile from a solid oral dosage form[9] and a limitation on the drug concentration gradient driving force from the intestinal lumen to absorp-

---

[1] A. M. Ugolev, *Biomembranes* **4A,** 285 (1974).

[2] P. A. Routledge and D. G. Shand, *Annu. Rev. Pharmacol. Toxicol.* **19,** 447 (1979).

[3] V. J. Stella, *in* "Prodrugs as Novel Drug Delivery Systems" (T. Higuchi and V. Stella, eds.), p. 1. Am. Chem. Soc., Washington, D.C., 1975.

[4] P. K. Banerjee and G. L. Amidon, *J. Pharm. Sci.* **70,** 1299 (1981).

[5] Y. Garceau, I. Davis, and J. Hasegawa, *J. Pharm. Sci.* **67,** 1360 (1978).

[6] Y. Yamaoha, R. D. Roberts, and V. J. Stella, *J. Pharm. Sci.* **72,** 400 (1983).

[7] L. W. Dittert, H. C. Caldwell, H. J. Adams, G. M. Irwin, and J. V. Swintosky, *J. Pharm. Sci.* **57,** 774 (1968).

[8] G. L. Amidon, *in* "Techniques of Solubilization of Drugs" (S. H. Yalkowsky, ed.), p. 183. Dekker, New York, 1981.

[9] J. T. Carstensen, "Pharmaceutics of Solids and Solid Dosage Forms," p. 63. Wiley, New York, 1972.

METHODS IN ENZYMOLOGY, VOL. 112

tion sites at the intestinal wall. Many potentially useful therapeutic agents have molecular structures containing large hydrophobic portions limiting their aqueous solubility. The presence in such molecules of particular functional groups (hydroxyl, carboxyl, amine) suggests chemical derivatization to increase drug polarity and resultant aqueous solubility. While such a modification may improve aqueous drug transport, an increase in polarity will generally result in a decreased water-to-membrane partitioning and may limit transport through intestinal membranes. In addition, derivatization to a small polar group on a large hydrophobic molecule may impart undesirable self-association properties with respect to drug transport.[10,11] This limitation may occur at the high prodrug concentrations generated when large doses are required to achieve sufficient absorption and resultant blood levels.

Many therapeutic agents are weak organic acids or bases; this property and the steep pH gradient which exists from stomach to lower small intestine promote large variations in the fraction of un-ionized drug presented to gastrointestinal absorption sites. This is a significant consideration since it is generally accepted that for passively absorbed drugs, too large to pass through aqueous pores, the un-ionized form is the predominant species absorbed.[12]

Further considerations in designing prodrugs for oral drug delivery include the following items.

1. Choice of derivatizing group to produce the desired modification in the parent molecule.

2. Changed or lost intrinsic pharmacological activity of the prodrug, necessitating its reconversion to the parent drug.

3. Site of reconversion with respect to reconversion extent, kinetics, and mechanism, and proximity to the site of absorption or pharmacological action.

4. The possibility of the release of toxic by-products during reconversion.

Commonly cited prodrug forms are simple organic esters or amides of the parent drug. Such prodrugs can be targeted for regions close to absorption sites or sites of action where physiological conditions are such that nonenzymatic prodrug hydrolysis is a favorable situation. In addition, these prodrugs can be targeted for regions containing high levels of nonspecific esterases. Such strategies have been utilized for drug transport in ophthal-

[10] D. Attwood and J. Gibson, *J. Pharm. Pharmacol.* **30,** 176 (1978).
[11] G. L. Flynn and D. J. Lamb, *J. Pharm. Sci.* **59,** 1433 (1970).
[12] D. Winne, *J. Pharmacokinet. Biopharm.* **5,** 53 (1977).

mic tissue (epinephrine delivery for glaucoma[13]) and skin (vidarabrine use as an antiviral agent[14]) and to overcome problems for gastrointestinal absorption (ampicillin,[15] carbenicillin,[16] and aspirin[17]). The derivatizing group can be selected to increase or decrease drug polarity, thus modifying aqueous or membrane transport properties as well as masking or protecting problematic functional groups. The most significant limitation on use of general ester prodrugs is the lack of specificity with respect to their sites of reconversion.

The body's handling of nutritional substances suggests that the use of a nutrient moiety as a derivatizing group to modify drug physicochemical properties which limit gastrointestinal drug absorption might also permit more specific targeting for enzymes involved in the terminal phases of digestion. Such enzymes include those responsible for the digestion of carbohydrates, fats, protein, and mineral-containing nutrients and the intestinal transport of sugars, lipids and fatty acids, peptides and amino acids, and various electrolyte species. These prodrugs have the additional advantage of producing nontoxic nutrient by-products upon cleavage. For example, prodrugs of aspirin have been made using a 2-deoxyglucose moiety[18] to modify the drug's gastric irritation problems. However, this prodrug was not targeted for specific enzymes but was designed to cleave rapidly in solution over a pH range of 3 to 9. Fatty acid esters of acetaminophen have been suggested as prodrugs when coadministered with pancreatic lipase and calcium. Variation in chain length of the fatty acid derivatizing groups was shown to control the rate of hydrolysis which had resultant impact on the duration of action of the parent drug.[19] Amino acid esters of drugs have been suggested as prodrugs for the purpose of modifying drug solubility, dissolution rate, gastric irritation, and chemical stability.[20,21] The targeting of amino acid prodrugs for enzymes in the small intestine was first proposed by Amidon et al.[22] The amino acids, as a nutrient-derivatizing species, offer a wide range of polarity and thus provide a broad spectrum for drug physical property modification. The po-

[13] A. Hussain and J. E. Truelove, *J. Pharm. Sci.* **65**, 1510 (1976).
[14] C. D. Yu, N. A. Gordon, J. L. Fox, W. I. Higuchi, and N. F. H. Ho, *J. Pharm. Sci.* **69**, 775 (1980).
[15] H. Shindo, K. Fukuda, K. Kawai, and K. Tanaka, *J. Pharmacobio. Dyn.* **1**, 310 (1978).
[16] A. Tsuji, E. Miyamoto, T. Terasaki, and T. Yamana, *J. Pharm. Sci.* **71**, 403 (1982).
[17] T. Loftsson, J. J. Kaminski, and N. Bodor, *J. Pharm. Sci.* **70**, 743 (1981).
[18] J. E. Truelove, A. A. Hussain, and H. B. Kostenbauder, *J. Pharm. Sci.* **69**, 231 (1980).
[19] C. T. Bauguess, J. M. Fincher, F. Sadik, and C. W. Hartman, *J. Pharm. Sci.* **64**, 1489 (1975).
[20] I. M. Kovach, I. H. Pitman, and T. Higuchi, *J. Pharm. Sci.* **64**, 1070 (1975).
[21] E. Aron, B. Delbarre, and J. C. Besnard, *Therapie* **31**, 247 (1976).
[22] G. L. Amidon, G. D. Leesman, and R. L. Elliott, *J. Pharm. Sci.* **69**, 1363 (1980).

tential target reconversion sites include pancreatic enzymes present in the intestinal lumen as well as enzymes associated with the brush border or cytosol of the enterocyte.

Gastrointestinal Enzymes

In discussing the location of intestinal enzymes as prodrug reconversion sites, we will view the gastrointestinal tract as a cylindrical tube. The ability of a drug or prodrug to get to sites of absorption along the tube wall is a function of its GI residence time (how long it takes the drug to travel in an axial direction down the tube) and its radial transport time (how long it takes a drug molecule to move from a point in the tube lumen through the tube wall).

In additional to the steep pH gradient that exists in the axial direction down the GI tract, the distribution, activity, and specificity of enzymes involved in digestion and nutrient transport are very much a function of axial position. For specific enzyme systems, the axial distribution pattern is also a function of species, age, and nutritional state.

Prodrug or drug transport in the radial direction occurs from the lumen, through the mucin layer and glycocalyx at the cell surface, the brush border apical membrane of the enterocyte, the cytosol and basal cell membrane, and the capillary or lymphatic endothelium (Fig. 1). The radial enzyme distribution pattern in the intestinal lumen, glycocalyx, brush border membrane, and enterocyte cytosol is geared to couple sequential digestion and transport in the process of nutrient absorption. Since the radial regions consist of a series of aqueous and membrane components, the site for prodrug reconversion is critical for successful drug transport and subsequent absorption from the small intestine.

The target choice for prodrug reconversion is dictated by drug physical property transport limitations and enzyme distributions, activities, and specificities. From a pharmacokinetic standpoint, the rate of absorption is roughly dependent on radial considerations while the extent of absorption is a function of axial parameters.

Luminal Enzymes

Prodrug reconversion in the intestinal lumen can be utilized as an oral drug delivery strategy when a drug has poor stability at acidic pH or causes gastric irritation. In this regard, a prodrug can be made that is stable in the pH range of the stomach and undergoes hydrolysis to the parent drug at intestinal pH. Alternatively, prodrugs can be targeted for the pancreatic enzymes in the intestinal lumen. These enzymes include

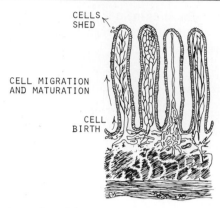

CELLS SHED

CELL MIGRATION AND MATURATION

CELL BIRTH

FIG. 1. Diagrammatic section of the wall of the human small intestine. Representative villi show separately (from left to right) venous drainage, arterial blood supply, lymphatic drainage, and nerve supply.

pancreatic amylase, pancreatic lipase, elastase, trypsin, α-chymotrypsin, and carboxypeptidase A.[4,22–24]

Prodrugs made to increase drug aqueous solubility can be targeted for luminal enzymes in order to gain an advantage with respect to the dissolution rate from a solid oral dosage form. However, luminal reconversion after dissolution regenerates the limitations on aqueous transport of the poorly soluble parent drug. In such cases, it is advisable to target these prodrugs for enzymes associated with the intestinal brush border membrane. These enzymes include adsorbed pancreatic enzymes and enzymes synthesized in the enterocytes which are active at the cell brush border upon maturation at the villus tip.

The binding of pancreatic trypsin and chymotrypsin to intestinal mucosa without loss of activity has been demonstrated and is thought to involve a lipoprotein component.[25,26] These enzymes are endopeptidases and hydrolyze peptide bonds to reduce oligopeptides to smaller peptides and amino acids.[27] Because of their weak association with the intestinal wall, the pancreatic enzymes can be sloughed off by the movement of chyme down the intestine.

[23] P. K. Banerjee and G. L. Amidon, *J. Pharm. Sci.* **70,** 1304 (1981).
[24] P. K. Banerjee and G. L. Amidon, *J. Pharm. Sci.* **70,** 1307 (1981).
[25] D. M. Goldberg, R. Campbell, and A. D. Roy, *Biochim. Biophys. Acta* **167,** 613 (1968).
[26] M. T. Davies, *Analyst* **84,** 248 (1959).
[27] R. L. Crane, *in* "Intestinal Absorption and Malabsorption" (T. Z. Csáky, ed.). Raven Press, New York, 1975.

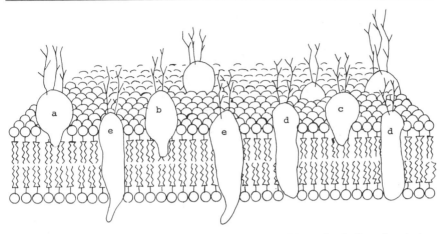

FIG. 2. Membrane-bound enzymes: a,b,c, disaccharidase; d, alkaline phosphatase; e, aminopeptidase.

Mucosal Cell Enzymes

Those enzymes synthesized in the enterocyte and associated with the apical brush border membrane include the disaccharidases, aminopeptidases, some carboxypeptidases, and alkaline phosphatase. These enzymes are more firmly associated with the brush border membrane than the adsorbed pancreatic enzymes.[28] Figure 2 shows a schematic of their extent of membrane association. The disaccharidases penetrate the first lipid layer of the membrane and can be released from membrane vesicles by the action of papain. They are subject to attack by the luminal pancreatic proteases, and as a result the time span of disaccharidase–membrane association is only a few hours. Alkaline phosphatase extends the depth of the bilayer while the aminopeptidases penetrate the bilayer and protrude into the microvillus core. Extraction of aminopeptidase and alkaline phosphatase from membrane vesicles requires the application of a detergent to rupture bonds between lipid molecules. These enzymes are glycoproteins in which the protein component of the enzyme penetrates the membrane lipid matrix while the carbohydrate portion extends out into the lumen, thereby forming part of the outer membrane glycocalyx.[28]

Targeting of prodrugs for these enzymes permits prodrug aqueous transport advantages to be maintained up to the intestinal wall. Reconversion at this radial position releases the more membrane-permeable parent

[28] F. Moog, *Sci. Am.* **245,** 154 (1981).

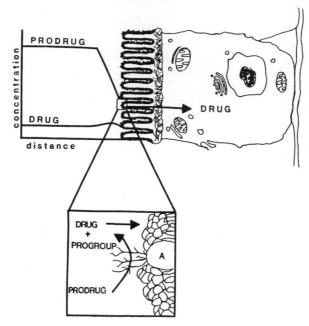

FIG. 3. Proposed mechanism for improving oral absorption of water-insoluble compounds.

drug at a point adjacent to the enterocyte membrane (Fig. 3). A special transport advantage for phosphate and amino acid prodrugs may be gained if enzyme binding and catalytic sites are located so as to permit a favorable positioning of the drug molecule for membrane permeation. Similar transport advantages have been demonstrated in physical systems in which a surface reaction is coupled to diffusion between two phases.[29]

If the intact prodrug can pass intestinal membranes, additional radial reconversion sites become available in the enterocyte cytosol or in the bloodstream. The fact that some of the cytosolic aminopeptidases possess specificities and activities different from those of brush border aminopeptidases and pancreatic proteases provides a potential refinement for this prodrug delivery approach.[30,31] A review on intestinal enzyme specificities and activities for prodrug targeting is available and will not be dealt with extensively in this chapter.[32] With respect to targeting prodrugs for

[29] D. R. Olander, *AIChE J* **6**, 233 (1960).

[30] Y. S. Kim, *in* "Peptide Transport and Hydrolysis" (K. Elliott and M. O'Connor, eds.), p. 151. Elsevier, Excerpta Medica, Amsterdam, 1977.

[31] S. Aurrichio, L. Greco, B. DeVizia, and V. Buonocore, *Gastroenterology* **75**, 1073 (1978).

[32] G. L. Amidon, R. S. Pearlman, and G. D. Leesman, *in* "Design of Biopharmaceutical Properties through Prodrugs and Analogs" (E. B. Roche, ed.), p. 281. Am. Pharm. Assoc., Washington, D.C., 1977.

aminopeptidases and alkaline phosphatase, a brief discussion on their intestinal distribution and specificities is in order.

Alkaline Phosphatase

Alkaline phosphatase activity is not evenly distributed in the axial direction of the vertebrate gastrointestinal tract. In many species, though not in all, enzymatic activity begins abruptly at the pylorus and diminishes gradually from the duodenum to the ileum. This is found in the dog, mouse, and adult rat (Fig. 4). In rats, for example, the ratio of duodenal to jejunal to ileal alkaline phosphatase (per mg protein) is 16:4:1.[33] In humans, high ileal activity has been reported[34,35] (Fig. 4). In several species, including mouse and rat, axial activity distribution varies with growth and development.[36,37] For neonates of these species, alkaline phosphatase activity is more evenly distributed along the small intestinal length. Fasting, fat ingestion, vitamin D deficiency, and dietary zinc have been shown to correlate with increases in intestinal alkaline phosphatase activity while calcium, sucrose, and alcohol intake have been correlated with reduced activities.[38,39] The amino acid L-phenylalanine has been shown to be an "organ-specific" inhibitor of intestinal alkaline phosphatase in mouse, rat, and human.[40]

There is also considerable evidence that intestinal alkaline phosphatases exist as several isoenzymes.[41] In this regard, it has been shown that the ileal enzyme is less easily precipitated than is the duodenal enzyme by mixed antiphosphatase, that L-phenylalanine inhibition is more marked for duodenal than for jejunal enzyme, that heat stability is greatest for duodenal alkaline phosphatase, and that the phenylphosphate to β-glycerophosphate activity ratio is reversed from duodenum (high ratio) to ileum. It is believed this enzyme functions in the small intestine for absorption of phosphate from dietary phosphates, but much uncertainty remains concerning its physiological function.

[33] R. B. McComb, G. N. Bowers, Jr., and S. Posen, in "Alkaline Phosphatase," p. 83. Plenum, New York, 1979.

[34] I. Dawson and J. Pryse-Davies, *Gastroenterology* **44,** 745 (1963).

[35] M. V. Srihantaiah, E. V. Chandrasekaran, and A. N. Radkakrishan, *Indian J. Biochem.* **4,** 9 (1967).

[36] F. Moog and H. S. Glazier, *Comp. Biochem. Physiol. A* **42A,** 321 (1972).

[37] A. Subramoniam and C. V. Ramakrishnan, *J. Exp. Zool.* **214,** 317 (1980).

[38] G. P. Young, S. Friedman, S. T. Yedlin, and D. H. Alpers, *Am. J. Physiol.* **241,** G461 (1981).

[39] S. Miura, H. Asakura, M. Miyairi, T. Morishita, H. Nagata, and M. Tsuchiya, *Digestion* **23,** 224 (1982).

[40] N. K. Ghosh and W. H. Fishman, *Arch. Biochem. Biophys.* **126,** 700 (1968).

[41] D. Chappelet-Tordo, M. Fosset, M. Iwatsubo, C. Gache, and M. Lazdunski, *Biochemistry* **13,** 1788 (1974).

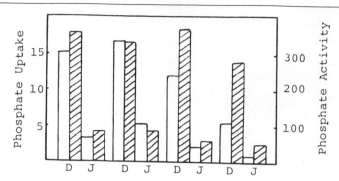

FIG. 4. Phosphate absorption and phosphatase activity in duodenum and jejunum. The first pair of columns in each set of four is for duodenum from a single mouse, the second pair for jejunum from the same mouse. Clear columns, μg P taken into the tissue per mg tissue protein (left ordinate); striped columns, μg P released into the medium per mg tissue protein (right ordinate).

Mucosal Peptidase

Isolation and identification of carboxy- and aminopeptidases from enterocyte brush border and cytosol have shown different, though somewhat overlapping, specificities. It appears that these exopeptidases function at the surface or interior of the absorptive cell to further reduce small oligopeptides released by pancreatic protease action. Their radial distributions appear to be geared to couple small peptide and amino acid intestinal transport. Tetrapeptide activity is associated exclusively with the brush border membrane, tripeptide activity is distributed about equally between membrane and cytosol, while enzymes that cleave dipeptides are more prevalent in the cytosol than in the brush border[31] (Table I). Axial distributions appear to be fairly even over the entire small intestinal length and less subject to species, age, and nutritional input parameters than is alkaline phosphatase. However, this may be attributed to the fact that isoenzymes and specific peptidases have not been well characterized with respect to their association with particular axial regions.[42]

In spite of the fact that limited information is available on the mechanistic details of digestion and transport at the surface and interior of the absorptive cell, the location and specificities of the associated enzymes present a potential target for prodrug reconversion to improve the absorption of some drugs. Potential limitations on this prodrug approach include the following points.

[42] Y. S. Kim, E. J. Brophy, and J. A. Nicholson, *J. Biol. Chem.* **251,** 3206 (1976).

TABLE I

COMPARISON OF THE PROPERTIES OF INTESTINAL CYTOSOL
AND BRUSH BORDER PEPTIDASES

Properties	Cytosol	Brush border
Substrates (percentage of total cellular activity)		
Dipeptides	80–95	5–12
Tripeptides	30–90	10–60
Tetra- and higher peptides	Nil	90
Zymogram pattern	(a) Multiple bands	(a) Two major bands
	(b) Heterogeneity of band pattern	(b) Homogeneity of band pattern
	(c) Non-organ-specific pattern	(c) Organ-specific pattern
Proline-containing dipeptides	Hydrolyzed	Not readily hydrolyzed
Chain length of substrate	2, 3, (4)	2, 3, 4, 5, 6, 7, 8, —
End-terminal specificity	Aminopeptidase	Aminopeptidase
Terminal stability	Heat labile	Relatively heat stable
Effect of p-hydroxymercuri-benzoate	Inhibited	Not inhibited

1. Drastic pH variation along the GI tract poses problems for ensuring prodrug stability in the GI lumen.

2. The natural bacterial flora of the intestine release enzymes which might interfere with more specific targeting.

3. The brush border, exposed to the movement of chyme and to the digestive enzymes in the lumen, is gradually sloughed off, thus releasing brush border and cytosolic enzymes into the lumen and thus reducing the advantage of targeting.

Demonstration that this prodrug approach overcomes limitations on oral absorption in intact biological systems requires the assessment of various pharmacokinetic parameters from drug blood and urine levels versus time data. Such levels are a function of drug distribution and elimination as well as oral input through absorption.[43] To isolate the contribution of the absorption process, experimentally isolated subsystem studies are required both to evaluate mechanistic information and as a screening tool to assess the potential for this approach in whole animal systems.

[43] M. Gibaldi and D. Perrier, in "Pharmacokinetics." Dekker, New York, 1975.

Methodology

In Vitro Experiments

In vitro investigation of drug and prodrug absorption is a useful screening technique prerequisite to whole animal studies which may involve considerably more time and expense. Attention to general details in tissue handling applies regardless of the *in vitro* method chosen. Highest regard must be given to tissue integrity during the experiment, and this may be achieved by careful treatment of the tissue during the experiment, use of physiological buffers such as Krebs-Henseleit,[44] and adequate oxygenation of tissue and buffers. Histological preparations and the monitoring of water flux under experimental conditions are useful indicators of tissue integrity. Morphological changes have been evaluated as a function of time, temperature, and method of animal sacrifice.[45]

The literature is abundant with *in vitro* methods, possibly the earliest being the everted sac technique to monitor mucosal to serosal solute transport.[46,47] A major criticism of this technique is the lengthy incubation times used (at least 1 hr generally) and possible deterioration of the tissue integrity. The tissue accumulation method (TAM) is a widely used technique subject to both positive and negative criticisms.[48,49] A commonly used variation of the TAM is the everted intestinal ring preparation.[50,51] In this procedure, the excised intestinal segment is floated in an oxygenated, physiological buffer and everted with a rod. Transverse sections are made with a razor to produce the rings. The rings are randomized and incubated with the solute of interest in a shaking water bath usually maintained at 37°. The uptake process is quenched at fixed time periods; the tissue is rinsed and weighed; and the levels of drug are analyzed in the tissue. Intracellular water concentrations have been obtained by subtraction of extracellular water (using a nonabsorbable marker) from total tissue water (obtained by total desiccation of the tissue).[50] Results are usually reported as uptake per gram of tissue dry weight. This method enjoys advantages of simplicity and short incubation times such that tissue integrity is main-

[44] H. A. Krebs and K. Henseleit, *Hoppe-Seyler's Z. Physiol. Chem.* **210**, 33 (1932).
[45] R. R. Levine, W. F. McNary, P. J. Kornguth, and R. LeBlanc, *Eur. J. Pharmacol.* **9**, 211 (1970).
[46] T. H. Wilson and G. Wiseman, *J. Physiol. (London)* **123**, 116 (1954).
[47] S. A. Kaplan and S. Cotler, *J. Pharm. Sci.* **61**, 1361 (1972).
[48] F. Alvarado, *in* "Intestinal Ion Transport" (J. W. L. Robinson, ed.), p. 117. MTP Press, Ltd., Lancaster, England, 1976.
[49] G. B. Munck, *J. Membr. Biol.* **53**, 45 (1980).
[50] B. Cheng, F. Navab, M. T. Lis, T. N. Miller, and D. M. Matthews, *Clin. Sci.* **40**, 247 (1971).
[51] J. W. L. Robinson and G. VanMelle, *J. Physiol. (London)* **323**, 569 (1982).

tained; and it is advantageous statistically since many replicates may be performed from each animal.

A large area of work with inorganic ion flux and the Na^+-dependence of active glucose and amino acid absorption has resulted in the development of a number of *in vitro* unidirectional flux systems which can manipulate the electrical potential driving force as well as the chemical potential for solute absorption. The intestine is categorized as leaky, low-resistance epithelium with a transmural (mucosal to serosal) potential difference of 4–5 mV.[52] Accumulated evidence indicates the existence of parallel pathways for the transport of low-molecular-weight solutes such as the amino acids, carbohydrates, and inorganic ions: a relatively high-resistance, transcellular route and a low-resistance, paracellular pathway[53] where standing-gradient osmotic flow may provide a mechanism for water–solute transport coupling.[54] The Ussing apparatus[55,56] in its many modifications has provided the classic tool for studying the impact of electrical potential on unidirectional flux. The apparatus monitors changes in the transmural potential (Ψ_{ms}) as a result of rheogenic solute transport as well as being capable of "short-circuiting" Ψ_{ms} to zero by voltage clamping and examining potential-dependent transport. A model has been developed which separates the transmural flux into potential-dependent (paracellular) and potential-independent (transcellular) components.[57] While exception has been made to its universal applicability,[58] the model has been used in pharmaceutical research to assess the impact of theophylline on salicylate transport[59] and may be of particular utility in the study of absorption of charged drug species. Other *in vitro* methods such as isolated intestinal cell suspensions[60,61] and intestinal and renal brush border vesicles[62] have examined potential-dependent uptake processes through inhibition of biochemical systems with specific chemical blocking agents. These preparations are discussed below.

Some investigators have reduced the level of system complexity by carrying out uptake studies on isolated intestinal epithelial cells. One method of cell isolation involves removal of the small intestine, and perfu-

[52] R. C. Rose, D. L. Nahrwold, and M. J. Koch, *Am. J. Physiol.* **232**, E5 (1977).
[53] R. A. Frizzell and S. G. Schultz, *J. Gen. Physiol.* **59**, 318 (1972).
[54] J. M. Diamond and W. H. Bossert, *J. Gen. Physiol.* **50**, 2061 (1967).
[55] H. H. Ussing and K. Zerhan, *Acta Physiol. Scand.* **23**, 110 (1951).
[56] V. Koefoed-Johnsen and H. H. Ussing, *Acta Physiol. Scand.* **42**, 298 (1958).
[57] S. G. Schultz and R. Zalusky, *J. Gen. Physiol.* **47**, 567 (1964).
[58] B. G. Munck and S. N. Rasmussen, *J. Physiol.* (*London*) **291**, 291 (1979).
[59] G. Barnett, S. Hui, and L. Z. Benet, *Biochim. Biophys. Acta* **507**, 517 (1978).
[60] G. A. Kimmich and J. Randles, *Biochim. Biophys. Acta* **596**, 439 (1980).
[61] F. V. Sepulveda, K. A. Burton, and P. D. Brown, *J. Cell. Physiol.* **111**, 303 (1982).
[62] S. Hilden and B. Sacktor, *Am. J. Physiol.* **242**, F340 (1982).

sion, ligation, and incubation of the excised tissue with two successive salt solutions, the latter containing dithiothreitol (DTT), a disulfide reducing agent.[63] The duration of incubation with the DTT solution is a factor in the heterogeneity of the cell population since the intestinal villus contains a gradient of enterocytes at different stages of development in enzyme activity and functional specialization from the germinal crypts of the villus to the mature state at the villus tip.[30] Also interspersed among the enterocytes are the mucus-secreting goblet cells. Consequently, successive incubations produce serial cell fractions from the villus tip region to the midvillus to the crypt regions. Following the isolation process, the cell suspension is counted on a hemocytometer and the viability percentage determined with a trypan blue exclusion test.[64] The different cell fractions are usually characterized by alkaline phosphatase or sucrase activity, glycoprotein formation, and [^{3}H]thymidine incorporation.[65] Uptake results may be expressed per milligram protein[66] or by DNA content.[67]

The enterocyte is a highly polarized cell with respect to functional and ultrastructural specialization (see Fig. 3). For a number of years, investigators have attempted to further simplify the system and its transport processes by preparing membrane vesicles. The vesicles can be further broken down into brush border versus basolateral vesicles which can be differentially characterized.[68–71] An excellent review on the use of membrane vesicles to study epithelial transport processes has been published.[72]

In Situ Experiments

Rat intestinal perfusion systems combine defined hydrodynamic flow conditions with control of drug input to measure intrinsic intestinal wall

[63] M. M. Weiser, J. Biol. Chem. 248, 2536 (1973).

[64] M. Absher, in "Tissue Culture: Methods and Applications" (P. Kruse, Jr. and M. K. Patterson, eds.), p. 394. Academic Press, New York, 1973.

[65] F. Hartmann, R. Owen, and D. M. Bissell, Am. J. Physiol. 242, G147 (1982).

[66] O. H. Lowry, N. J. Rosebrough, A. L. Farr, and R. J. Randall, J. Biol. Chem. 193, 265 (1951).

[67] B. Fiszer-Szafarz, D. Szafarz, and A. G. deMurillo, Anal. Biochem. 110, 165 (1981).

[68] C. E. Stirling, J. Cell Biol. 53, 704 (1972).

[69] M. Fujita, H. Ohta, K. Kawai, H. Matsui, and M. Nakao, Biochim. Biophys. Acta 274, 336 (1972).

[70] M. Kessler, O. Acuto, C. Storelli, H. Murer, M. Muller, and G. Semenza, Biochim. Biophys. Acta 506, 136 (1978).

[71] V. Scalera, C. Storelli, C. Storelli-Joss, W. Haase, and H. Murer, Biochem. J. 186, 177 (1980).

[72] H. Murer and R. Kinne, J. Membr. Biol. 55, 81 (1980).

permeabilities from experimental (drug output) results.[22,73–76] With this method, a length of small intestine is exteriorized, and proximal and distal cannula are inserted. The proximal cannula is placed distal to the entry of pancreatic and bile secretions to prevent interference with the brush border enzyme systems. In this manner, the mesenteric vascular system remains intact and the continuity of the intestine is maintained but for the subsystem of the segment isolated between the cannula. The proximal cannula leads to a perfusion pump set to a determined flow rate, and the distal cannula leads to a collection tube. The intestine is perfused with buffer until the perfusate is clear (30 min). Wall permeabilities for each compound are calculated from the ratio of the solution concentration exiting the intestine to the concentration entering the intestine (C_m/C_o) under steady-state conditions (45–60 min). At low flow rates, water transport may become significant and must be monitored by including a nonabsorbable marker, such as ^{14}C-labeled PEG-4000, in the perfusing solution to correct C_m values. Water flux may be minimized by preparing solutions which are isosmotic with intestinal fluid for perfusion.[77]

Ligated intestinal loops are frequently used in *in situ* absorption studies. A series of ligations down the small intestine permits a statistical advantage in that more than one compound may be studied and compared in one animal, or that the absorption of one compound may be studied as a function of position in the intestine.[78] A problem with this technique, which requires careful experimental control, is water absorption during the course of the study. A further complication in determination of absorption rate constants is that the system is non-steady state; and, moreover, significant depletion of the solution or suspension may occur. As much as 85% of the administered solution volume has been lost in the course of 1 hr.[79] Aside from these considerations, a great deal of work has been done utilizing the ligated loop method, notably an extensive study comparing the intestinal absorption of ampicillin and pivampicillin. This study employed autoradiography to map carboxylesterase B distribution and noted species differences between rat, dog, and monkey.[15]

[73] R. L. Elliott, G. L. Amidon, and E. N. Lightfoot, *J. Theor. Biol.* **87,** 757 (1980).

[74] I. Komiya, J. Y. Park, A. Kamani, N. F. H. Ho, and W. I. Higuchi, *Int. J. Pharm.* **4,** 249 (1980).

[75] P. F. Ni, N. F. H. Ho, J. L. Fox, H. Leuenberger, and W. I. Higuchi, *Int. J. Pharm.* **5,** 33 (1980).

[76] G. L. Amidon, *in* "Animal Models for Oral Drug Delivery in Man" (W. Crouthamel and A. C. Sarapu, eds.), p. 1. Am. Pharm. Assoc., Washington, D.C., 1983.

[77] D. Fleisher and G. L. Amidon, *Int. J. Pharm.* (submitted for publication).

[78] B. H. Stewart, Master's Thesis, University of Wisconsin, Madison (1982).

[79] K. Pfleger, *Int. Encycl. Pharmacol. Therap., Sect. 39B* Vol. 2, p. 801 (1975).

Whether improved wall flux or permeability for intestinal absorption will be reflected in the drug's potential for getting to its site of action must be assessed in whole animal studies.[76] Drug plasma levels and urine levels are the most accessible data for pharmacological response correlations. Improvements in rate and extent of absorption may be damped out by whole animal elimination and distribution factors dictating pharmacokinetic studies as the final experimental stage for assessing an oral drug delivery strategy. The key pharmacokinetic parameters for evaluating rate and extent of absorption include peak plasma concentration, time of peak plasma concentration, elimination and absorption rate constants, and area under body fluid concentration versus time curves.[43]

Experimentally isolated subsystems are geared to obtain information on the absorption process at the intestinal wall. The *in vitro* experiments and *in situ* ligated loops are analyzed to get some measure of intestinal uptake which can be translated to a rough assessment of flux at the intestinal wall. These experiments may provide a useful screening tool to assess the potential for improved intestinal absorption. The isolated *in situ* perfusion experiments provide further experimental detail by controlling hydrodynamics, thus permitting the separation of aqueous (luminal) resistance from membrane (wall) resistance to drug transport. This results in a resolution of the flux into wall concentration and permeability components, thereby allowing more accurate isolation of wall processes.[73] The utility of the various techniques is illustrated with examples in the Results section.

Results

Experimental work to evaluate the potential for a nutrient prodrug approach to overcome problems in oral drug delivery has been carried out with several compounds at various levels of biological system complexity. Targeting for both intestinal pancreatic enzymes as well as brush border enzymes has been investigated with derivatives of drug carboxyl, amino, and hydroxyl functionalities.

Amino acid derivatization of a problematic drug carboxyl group was studied by Banerjee.[4,23,24,80] In these studies, the phenylalanine ethyl ester of aspirin was targeted for sequential reconversion by the pancreatic enzymes α-chymotrypsin and carboxypeptidase A. Masking of the carboxyl group, the offending moiety in aspirin's gastric irritation properties, has been done previously with several derivatizing groups.[17,21] The choice of phenylalanine as a derivatizing group was motivated not only by targeting

[80] P. Banerjee, Ph.D. Thesis, University of Wisconsin, Madison (1979).

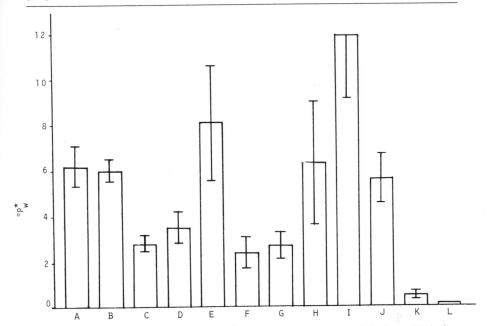

FIG. 5. Intestinal wall permeability ($°P_w^*$) of (A) L-lysine-*p*-nitroanilide alone; (B) lysine; (C) lysine methyl ester; (D) arginine methyl ester; (E) alanine methyl ester; (F) L-arginine-β-naphthylamide; (G) L-prolylglycine; (H) L-glycylproline; (I) L-alanine-*p*-nitroanilide; (J) L-glycine-*p*-nitroanilide; (K) N_α-benzoylarginine-*p*-nitroanilide; (L) N_α-succinylphenylalanine-*p*-nitroanilide.

for small intestinal pancreatic enzymes but also to increase the shelf life stability of aspirin in a suspension formulation. Shelf life studies demonstrated a 400-fold increase in stability for an aspirin phenylalanine ethyl ester suspension over an aspirin suspension of 300 mg/5 ml at equivalent pH. Regeneration of aspirin in solution by the sequential actions of purified α-chymotrypsin (an endopeptidase with preference for aromatic amino acid residues) and carboxypeptidase A (an exopeptidase requiring a free L-carboxyl terminus) was shown *in vitro* to be relatively rapid.

Studies geared to elucidate enzyme specificity requirements for nutrient prodrug targeting were done for compounds in which an amine functionality was derivatized. Glycine, alanine, lysine, and arginine derivatives of *p*-nitroaniline and β-naphthylamine were used in rat intestinal perfusion experiments for the purpose of evaluating substrate and inhibitor structural requirements for enzyme reconversion. While these compounds have no potential therapeutic value, they were chosen because of their relatively good aqueous stability and for ease of analysis. The results (Fig. 5) indicate that for these prodrugs a free α-amino group in the

L-configuration[81] is required to achieve good intestinal wall permeabilities for substrates. These results imply that an aminopeptidase enzyme is involved in this prodrug reconversion. Further experimental work[82] demonstrated specificity for the amino acid side chain as well.

Prodrug derivatives of drugs with hydroxyl functionalities have been the major focus of attempts to couple improved aqueous solubility in the intestinal lumen with reconversion target sites chosen to regain membrane transport advantages of the parent drug. The formation of soluble derivatives of sparingly soluble drug compounds is a recognized technique to increase dissolution of the solid dosage form and/or increase luminal drug concentration levels.[83] Amidon *et al.*[22] first combined this approach with enzyme targeting using estrone and its phenolic ester, estrone lysinate. This derivatization resulted in a solubility increase from 3×10^{-6} M to 0.3 M. *In vitro* experiments with purified trypsin demonstrated reconversion to the parent drug in the presence of this pancreatic protease.[84] *In situ* rat intestinal perfusions were performed in an intestinal segment free of pancreatic enzymes, thereby isolating the intestinal wall enzymes. Dimensionless wall permeabilities of 6.3 and 34 were calculated for the drug and derivative, respectively.[22] Figure 6 illustrates the correlation between absorption rate and initial perfusing concentration for both compounds. The theoretical limits imposed by the respective solubility concentrations serve to emphasize the potential of the derivative for increasing absorption rate which, in this case, is up to five orders of magnitude.

Further studies were undertaken with the normal alcohols, decanol and hexadecanol, and the aliphatic esters, decyl lysinate and hexadecyl lysinate. High water-to-membrane partitioning and poor aqueous solubilities [C_s (decanol), 0.2 mM; C_s (hexadecanol), 0.2 μM[85]] make these compounds good candidates for drug models. The derivatives, with solubilities ≥ 20 mM, were targeted for reconversion by L-aminopeptidase at the intestinal brush border. Intestinal perfusions conducted with decanol and decyl lysinate resulted in infinite wall permeabilities for the alcohol and comparable, though somewhat lower, values for the derivative. Additional experiments utilizing the *in situ* ligated loop method indicated 10–15% higher loss of the lysinate ester from the intestinal lumen.[78]

To conduct uptake experiments above the solubility of the parent alcohols, the four compounds were studied in an *in vitro* intestinal ring

[81] G. L. Amidon, M. Chang, R. Allen, and D. Fleisher, *J. Pharm. Sci.* **71,** 1138 (1982).
[82] G. L. Amidon, M. Lee, and H. Lee, *J. Pharm. Sci.* **72,** 943 (1983).
[83] G. L. Amidon, in "Techniques of Solubilization of Drugs" (S. H. Yalkowsky, ed.), p. 183. Dekker, New York, 1980.
[84] G. L. Leesman, Ph.D. Thesis, University of Wisconsin, Madison (1980).
[85] G. L. Amidon, S. H. Yalkowsky, and S. Leung, *J. Pharm. Sci.* **63,** 1858 (1974).

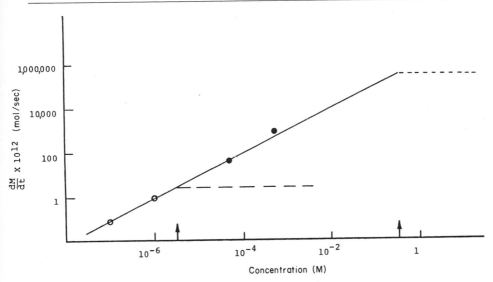

FIG. 6. Amount of estrone/lysine estrone ester transported as a function of initial concentration. ($Q = 0.00825$ ml/sec; length = 6.2 cm).[84] (○), Estrone; (——), solubility of estrone; (●) lysine estrone ester; (–––) solubility of lysine estrone ester.

system.[86] Figure 7 depicts uptake rate as a function of concentration for decanol and decyl lysinate over the range 50 to 2000 μM. Below the solubility of decanol, the slopes of both curves are comparable, providing agreement with the membrane permeabilities obtained from intestinal perfusion experiments. Differences in uptake increase rapidly with concentration as the parent alcohol remains at the plateau dictated by its lower solubility and the lysinate ester maintains a linear increase in uptake over the concentration range studied. Figure 8 summarizes the logarithmic uptake of the four compounds at selected concentrations. Values for hexadecanol and decanol are at the respective solubility concentration or slightly below (200 mM, 150 μM, respectively). The increased solubilities of the lysinate esters demonstrate the impact of a greater concentration gradient driving force on uptake at concentrations at least 10× in excess of the solubilities of the parent alcohols.

Recent work attempting to correlate wall flux and permeability advantages for nutrient prodrugs in isolated subsystems with whole animal bioavailability was carried out with soluble derivatives of the aliphatic alcohol, hydrocortisone. Commercially available soluble derivatives used parenterally, the succinate and phosphate esters, were compared with the

[86] G. L. Amidon and B. H. Stewart, *Int J. Pharm.* (submitted).

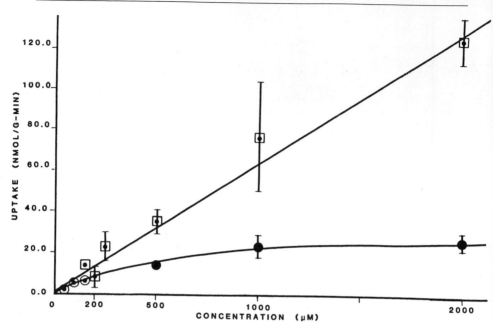

Fig. 7. The intestinal ring uptake of decanol (○) and decyl lysinate (□) as a function of concentration is shown. Values above the solubility concentration of decanol are indicated by (●). Each point represents the mean value of 2–9 intestinal rings from 2–3 rats ± SEM.

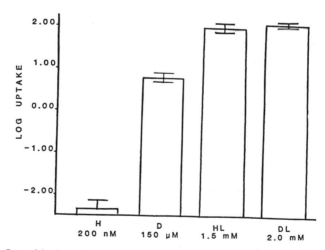

Fig. 8. Logarithmic uptake rates by intestinal rings at selected concentrations. The uptakes are shown at the solubility of hexadecanol (H), slightly below the solubility of decanol (D), and at or below the solubilities of the lysinate esters (HL, DL).

TABLE II
SUMMARY OF WHOLE ANIMAL RESULTS

Hydrocortisone	Rate intestinal perfusion wall permeabilities	Dog plasma levels		
		t_{max} (hr)	C_{max} (μg/ml)	AUC (μg-hr/ml)
Free alcohol	2.9	1.50	0.7	1.4
Succinate	0.3	1.00	0.2	0.4
Phosphate	∞	0.75	0.8	0.9
Lysinate	3.0	0.75	0.6	1.2

soluble lysinate derivative and the parent compound. The lysinate and succinate esters have poor aqueous stability whereas the phosphate derivative combines high aqueous solubility with very good aqueous stability.

Rat jejunal perfusion experiments indicated low permeabilities for the ionized succinate ester, equivalent permeabilities for the lysinate ester and parent compound, and very high permeabilities for the ionized phosphate derivative (see Table II). Exiting perfusate solutions of the stable phosphate derivative contained significant free hydrocortisone levels, a finding implicating the involvement of intestinal alkaline phosphatase. Appearance of free hydrocortisone levels in exiting solutions from hydrocortisone succinate perfusion were consistent with its aqueous stability profile. Similar levels with hydrocortisone lysinate perfusions could not be adequately accounted for by the prodrug's aqueous instability and implicated aminopeptidase reconversion, since the pancreatic duct is bypassed in these experiments.

Bioavailability experiments in beagle dogs were done with oral administration of equivalent molar doses of all four compounds in which endogenous hydrocortisone plasma levels were suppressed with prior administration of dexamethasone. Plasma levels for the succinate ester were low, a finding consistent with permeability results in the isolated intestinal experiments (Fig. 9). Hydrocortisone free alcohol plasma levels indicated that the parent drug was absorbed down the entire length of the small intestine and that the oral dose was not high enough to demonstrate the solubility limitations on the parent drug in intestinal fluids.[87] Plasma levels from hydrocortisone lysinate administration resulted in early peak times consistent with aminopeptidase reconversion at the intestinal wall. The observation of sustained plasma levels may be a function of the prodrug's instability as well as enzyme reconversion. The phosphate derivative,

[87] K. M. Thakker, *J. Pharm. Sci.* **72**, 577 (1983).

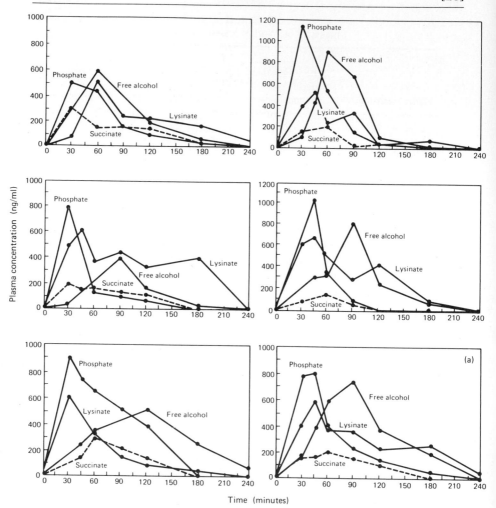

FIG. 9. Hydrocortisone free alcohol plasma levels in nanograms per milliliter after oral administration of 0.15 mmol equivalent hydrocortisone free alcohol in capsule form to beagle dogs.

which had shown interesting potential with its high wall permeability in rat upper small intestine, showed high early peak plasma concentration which dropped off rapidly, thus limiting bioavailability for this prodrug. The fact that intestinal alkaline phosphatase distribution in the dog is predominantly in the upper small intestine and that the ionized prodrug should not be well absorbed accounts for the data in both experimental situations.[77]

These results indicate that prodrugs targeted for intestinal peptidases as well as those targeted for intestinal alkaline phosphatase show potential for improving oral drug delivery. For the stable phosphate derivatives, axial enzyme distribution in different species should determine the relative success for these prodrugs. For those prodrugs targeted for intestinal peptidases, prodrug instability and enzyme specificity with respect to targeting to gain transport advantages for oral administration are key considerations for the utility of this approach.

Acknowledgments

The authors acknowledge the support of NIH Grant No. 1RO1GM3169 for much of the work presented in the Results section and thank Iris Templin for her expert assistance in preparation of the chapter.

[29] Targeting of Drugs to the Brain

By NICHOLAS BODOR

Introduction

The overall membrane transport properties are very important for a drug, as these properties govern the absorption, distribution, and elimination of its intact form, and affect binding, affinity, and other important characteristics. It is of even greater importance to find methods which enable delivery of drugs specifically to a particular organ, or site. This requires more than simply optimizing overall membrane transport characteristics. Among the various possible ways to achieve site-specific or organ-specific delivery, the chemical delivery system (CDS) is the most flexible[1]—and it offers possibilities for specific delivery not only to the brain, or eye[2] but to other organs and sites. Properly designed, a CDS or any other drug delivery system should concentrate the desired active agent at its site of action and reduce its concentration in other locations. The main result of this manipulation is not only an increase in the efficacy of the drug entity but also a decrease in its toxicity. A site-specific delivery designed for the central nervous system (CNS) would be especially useful. It is well known that many of the pharmacologically active and

[1] N. Bodor and H. H. Farag, *J. Med. Chem.* **25**, 313 (1983).
[2] N. Bodor and G. Visor, *Exp. Eye Res.* **38**, 621 (1984).

important agents, including endogenous neurotransmitters, cannot be transported from the blood into the brain because of a set of specialized barriers present at the blood–brain interface. This barrier system is generally called the blood–brain barrier (BBB) and is composed of a set of anatomical and enzymatic components.

The existence of a barrier which separates the general circulation from the central nervous system was first postulated by Ehrlich at the end of the nineteenth century.[3-5] The actual barrier was later described as acting against complexes formed between the dyes and plasma proteins with which these compounds are extensively bound. However, it was discovered that many small compounds are also unable to pass this barrier.

The morphological basis of the BBB was for long a rather controversial issue. Most recently, the anatomical basis for the BBB was identified as the endothelial lining of the cerebral capillaries, and most importantly, not the perivascular cells.

There are several ultrastructural differences between systemic capillaries and cerebral capillaries which explain the difference in their permeabilities. The main difference is in the manner in which endothelial cells in cerebral capillaries are joined. Cerebral junctions are characterized as tight or closed junctions which gird the cell circumferentially, forming zona occludens and providing an absolute barrier. Structurally, the junctions consist of aligned intramembranous ridges and grooves which are in close apposition.[6]

Systemic capillaries lack this closed junction. Morphologically, this can be traced to a lack of continuity in the intercellular appositions. Materials pass easily between these leaky cells, while in the brain the sealing of the intercellular fissures severely restricts this nonspecific transport. Since intercellular transport is impossible, only intracellular or transcellular transport remains. Lipophilic compounds can readily pass through these phospholipoidal membranes, but hydrophilic compounds and high-molecular-weight substances are excluded.

A second main difference between cerebral and systemic capillaries is the paucity of vesicles and vesicular transport in the CNS.[7] A third difference is the lack of fenestrae in the cerebral capillaries.

Cerebral vessels have a number of perivascular accessory structures which appear to be involved in the function of the BBB. Astrocytic foot processes may be involved in the regulation of the amino acid flux. Phago-

[3] W. M. Pardridge, J. D. Conner, and I. L. Crawford, *CRC Crit. Rev. Toxicol.* **3,** 159 (1975).
[4] E. Levin, *Exp. Eye. Res.* **25,** Suppl., 191 (1977).
[5] B. Van Deurs, *Int. Rev. Cytol.* **65,** 117 (1980).
[6] M. W. Brightman and T. S. Reese, *J. Cell Biol.* **40,** 648 (1969).
[7] M. W. Brightman, *Exp. Eye Res.* **25,** Suppl., 1 (1977).

cytic pericytes which are present abluminally may play a similar role. It is possible that the basement membrane of endothelial capillaries acts as a mass filter, preventing large molecules from penetrating it.

In addition to these structural features, the BBB maintains a number of enzymes which appear to augment barrier function.[4,8,9] Optimal neuronal control requires a careful balancing between neurotransmitter release, metabolism, and uptake, as it is of vital importance to restrict the entry of blood-borne neurotransmitters into the CNS. This is why high concentrations of such enzymes as catechol O-methyltransferase (COMT), monoamine oxidase (MAO), γ-aminobutyric acid aminotransferase (GABA-T), and aromatic-L-amino-acid decarboxylase (DOPA-decarboxylase) are found in the BBB. The enzymatic BBB may also play a role in the exclusion of some lipophilic compounds which otherwise might passively diffuse through the barrier.

The transport of nutrients is brought about by a number of carriers which are situated in the endothelial cells and which are assumed to be proteinaceous materials. They are equilibrative, i.e., nonenergy dependent and bidirectional in nature, and can be saturated.[10,11] The net movement of compounds is always along a concentration gradient, and since nutrients are readily utilized as soon as they pass into the brain, this gradient is in the direction of the brain.

The BBB, therefore, consists of a relatively impermeable membrane superimposed on which are mechanisms for allowing the entrance of essential nutrients and the exit of metabolic wastes. If a compound is to gain access to brain parenchyma, it may do so via several routes. If the molecule has affinity for one of the carriers previously described, it may diffuse across the BBB by association with this carrier. A compound with high intrinsic lipophilicity can diffuse passively through the phospholipid cell membrane matrix.

These two avenues, namely, passive diffusion and carrier mediation, represent the major components for influx. Other minor mechanisms may also allow the entry of substrates to the CNS.[7,12] Retrograde axoplasmic transport has been observed in such areas as the nucleus ambiguous and the abducens nucleus.

There are several areas in the brain which lack a BBB.[13,14] These

[8] J. E. Hardebo and B. Nilsson, *Acta Physiol. Stand.* **107**, 153 (1979).
[9] J. E. Bardebo and C. Owman, *Ann. Neurol.* **8**, 1 (1980).
[10] W. M. Pardridge, *Diabetologia* **20**, 246 (1981).
[11] W. M. Pardridge and W. H. Oldendor, *J. Neurochem.* **28**, 5 (1977).
[12] E. Westergaard, *Adv. Neurol.* **28**, 55 (1980).
[13] C. W. Wilson and B. B. Brodie, *J. Pharmacol. Exp. Ther.* **133**, 332 (1961).
[14] G. D. Pappas, *J. Neurol. Sci.* **10**, 241 (1970).

include locations near the ventricles such as the area postrema, the subfornical organ, the median eminence of the neurohypophysis, the organum vasculosum of the lamina terminalis, and the choroid plexus. Collectively, these areas are termed the circumventricular organ. In addition, the pineal gland also lacks a BBB. These areas constitute a small fraction of the total surface area of the BBB and may allow a limited nonspecific flux.

Cerebral spinal fluid (CSF) is produced at the choroid plexus and drains from the ventricles through the foramina of Magendie and Luschka into the subarachnoid space of the brain.[15] Cerebral spinal fluid, along with any dissolved materials, leaves the subarachnoid space via the arachnoid villi, which protrude into a venous sinus. The arachnoid villi act as a one-way valve and prevent backflow.[16] This loss of CSF provides a slow mechanism for nonspecific efflux of compounds from the CNS. The mechanism rids the brain of polar compounds such as metabolic wastes at a fairly constant rate regardless of the molecular size. Therefore, while lipophilicity is very important for influx to the brain, the efflux of a compound is only partially dependent on this parameter.[17–19] The BBB excludes a number of pharmacologically active agents and, as such, treatment of many cerebral diseases is severely limited. To increase the effectiveness of drugs which are active against these diseases, the specific transit time of a desired drug in the brain must be increased. This should increase the therapeutic index of an agent since not only is the concentration and/or the residence time of the agent increased in the vicinity of the bioreceptor, but of equal importance, the peripheral concentration of the drug is reduced, thereby decreasing any associated toxicity.

Unfortunately, there are very few methods for circumventing the BBB, and these are of limited usefulness. The direct administration of drugs in the CNS, i.e., an intrathecal injection (i.t.), has been used to deliver various pharmacologically active agents. While i.t. injections are of potential value, they are notorious for their high incidence of deleterious reactions. These side effects can arise from a number of different causes. Polar compounds injected into the CSF are restricted to the CSF, and the distribution of a drug administered i.t. is uneven and incomplete in the CNS. Also, since the rate of distribution is related to the rate of CSF movement, it is often slow.[17] Furthermore, since the ventricular

[15] T. H. Maren, in "Medical Physiology" (V. B. Mountcastle, ed.), p. 1218. Mosby, St. Louis, Missouri, 1980.
[16] R. Welch and V. Friedman, Brain 83, 454 (1960).
[17] L. S. Schanker, Antimicrob. Agents Chemother. p. 1044 (1905).
[18] L. D. Prockop, L. S. Schanker, and B. B. Brodie, Science 134, 1424 (1961).
[19] H. Davson, J. Physiol. (London) 255, 1 (1976).

volume of the CSF is small, increases in intracerebral pressure can occur with repeated injections. Improper needle or catheter placement can result in seizure, encephalitis, tuberculosis, meningitis, arachnoiditis, necrotizing encephalopathy, or cellulitis. The administered compounds themselves can produce toxic reactions.

Although macromolecular carriers[20–22] proved to be of limited use, a general method which can possibly be applied to the delivery of otherwise unusable drugs to the brain is the prodrug approach.[23–27]

A major advance in the area of drug delivery to the brain came about as a result of work with the polar quaternary salt, N-methylpyridinium-2-carbaldoxime chloride (2-PAM), which is the drug choice in the treatment of organophosphate poisonings. However, 2-PAM as a quarternary pyridinium salt will not pass the BBB and is, therefore, ineffective in treating central intoxication. Bodor et al.[28] approached this problem by transiently removing the positive charge of 2-PAM (1) by chemically reducing the quarternary compound to its corresponding dihydropyridine derivative. This reduction produced a tertiary amine which is far more lipophilic than 2-PAM (Pro-2-PAM; 2). N-Substituted dihydropyridines are known to be relatively unstable and are rapidly oxidized to the parent quarternary compound. This redox system is the basis of the important coenzyme system, NADH $\rightleftharpoons$ NAD.

As a result of this, it was found that *in vivo* pro-2-PAM was dramatically more effective in reactivating phosphorylated enzymes than 2-PAM (Table I).

[20] G. Gregoriadis, *Nature (London)* **265,** 407 (1977).
[21] G. Gregoriadis, ed., "Drug Carriers in Biology and Medicine." Academic Press, New York, 1979.
[22] R. L. Juliano, "Drug Delivery Systems." Oxford Univ. Press, London and New York, 1980.
[23] A. A. Sinkula and S. H. Yalkowski, *J. Pharm. Sci.* **64,** 181 (1975).
[24] N. Bodor, in "Design of Biopharmaceutical Properties through Prodrugs and Analogs," (E. B. Roche, ed.), p. 98. Am. Pharm. Assoc., Washington, D.C., 1977.
[25] N. Bodor, *Drugs Future* **6,** 165 (1981).
[26] N. Bodor, in "Strategy in Drug Research" (J. A. Buisman, ed.), p. 137. Elsevier, Amsterdam, 1981.
[27] V. Stella, in "Prodrugs as Novel Drugs Delivery Systems" (T. Higuchi and V. Stella, eds.), p. 1. Am. Chem. Soc., Washington, D.C., 1975.
[28] N. Bodor, E. Shek, and T. Higuchi, *Science* **190,** 155 (1975).

TABLE I
ACTIVITY AND REACTIVATION OF ACETYLCHOLINESTERASE[a]

Drug	Dose (mg/kg)	Activity ($\times 10^6$)	SE	Activity (%)	Reactivation (%)
2-PAM	30	1.053	0.076	10.27	0.00
2-PAM	40	1.683	0.041	16.42	5.93
2-PAM	50	2.246	0.187	21.92	12.12
Pro-2-PAM	20	2.781	0.163	27.14	18.00
Pro-2-PAM	30	3.111	0.063	30.36	21.63
Pro-2-PAM	40	4.871	0.355	47.53	40.95
Pro-2-PAM	50	7.429	0.183	72.49	69.04

[a] In mouse brain pretreated with DFP and treated minutes later with either 2-PAM (**1**) or Pro-2-PAM (**2a**) i.v. Taken from ref. 29.

The loss of the delivered compound from the brain is not as well characterized as its influx. As was discussed earlier, the brain possesses a number of specialized systems for eliminating compounds from the CNS. These include specific molecular pumps which actively remove compounds at the choroid plexus and the simple bulk flow rate of CSF. The efflux, therefore, of a compound from the brain is not precisely proportional to lipophilicity, and 2-PAM is an example of this. The quaternary salt **1** formed *in situ* is rapidly lost from all locations including the brain.[30]

The rapid loss of this quaternary salt from the CNS was attributed to an active process. It was assumed, however, that not all quaternary salts formed in the CNS are good substrates for the active transport system. An appropriately chosen pharmacologically active agent (i.e., one which contains a pyridinium moiety) could be reduced to its corresponding dihydropyridine derivative.[31] After systemic administration, this lipophilic species will penetrate the BBB as well as into the peripheral tissues. After oxidation to the quaternary salt in all body compartments, it can rapidly be eliminated by renal or biliary mechanisms (k_{out2}). In the CNS, however, the salt, because of its charge, would be retained ($k_{out1} \ll k_{out2}$). This yields significant concentration of the compound in the brain relative to the periphery and should increase the efficacy of the compound in the brain while reducing its systemic toxicity (Scheme 1).

The developed scheme was successfully applied to berberine (**3**). When dihydroberberine (**4**) or its hydrochloride were administered i.v., the concentration of berberine found in the brain was large.[32] The ber-

[29] E. Shek, T. Higuchi, and N. Bodor, *J. Med. Chem.*, **19**, 108, 113 (1976).
[30] N. Bodor, R. G. Roller, and S. S. Selk, *J. Pharm. Sci.* **67**, 685 (1978).
[31] N. Bodor, H. Farag, and M. E. Brewster, *Science* **214**, 1370 (1981).
[32] N. Bodor and M. E. Brewster, *Eur. J. Med. Chem.* **18**, 235 (1983).

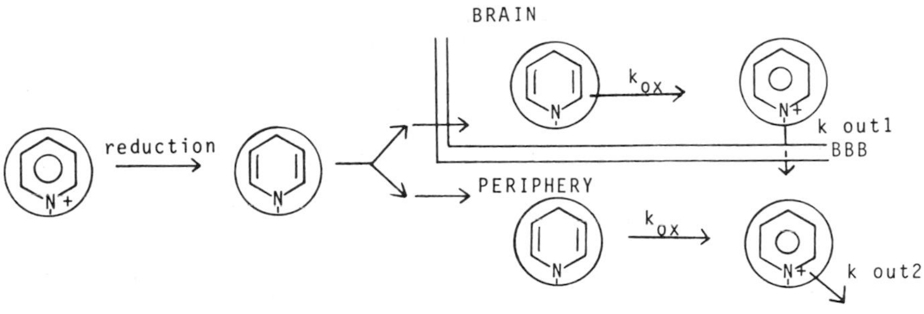

BRAIN

SCHEME 1

berine delivered to the brain in this way was slowly eliminated ($t_{1/2} \cong$ 11 hr). This is somewhat slower than the bulk flow of CSF, and this difference is attributable to the more complete distribution of berberine in the CNS afforded by its prodrug. The systemically formed (**3**) was eliminated via the kidney.

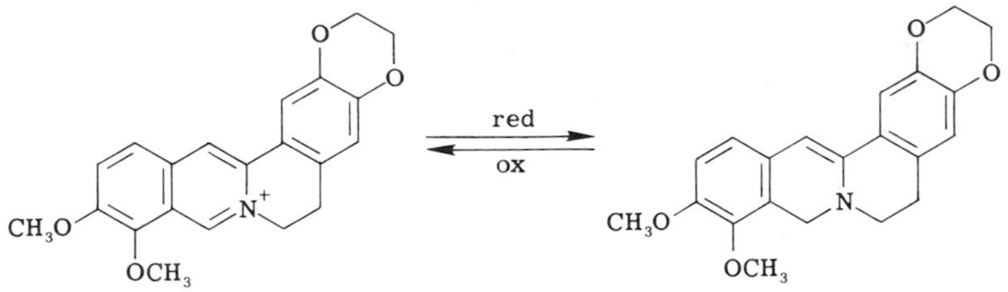

        3   **4**

The brain specificity of the dihydroberberine $\rightleftharpoons$ berberine system could further be enhanced by a slow infusion because of the differences in the bidirectional flux rates and the oxidation rates in the brain and blood[33] (Table II). For example, the concentration of berberine in the brain is higher than in any other tissue analyzed at 45 min.

On the basis of these results—namely, that large quaternary salts formed *in situ* in the brain are lost slowly while smaller quaternary salts are lost rapidly—a carrier-mediated drug delivery scheme was proposed. According to this, a pharmacologically active agent whose ability to pass the BBB is low is chemically coupled to a pyridinium carrier (for example, an N-substituted nicotinic acid or nicotinamide). After coupling, the drug–carrier complex would be reduced, thus yielding the dihydropyridine. This reduced complex would then be systemically administered. It

[33] M. E. Brewster and N. Bodor, *J. Parenter. Sci. Technol.* **37**, 159 (1983).

TABLE II
BERBERINE CONCENTRATIONS AFTER
SLOW INFUSION OF DIHYDROBERBERINE

Organ	Concentration (μg/g) after infusion	Concentration (μg/g) iv. bolus	Δ (μg/g)
30 min			
Brain	135.95 $\pm$ 13	91.8 $\pm$ 20	+44
Kidney	185.5 $\pm$ 26	351.8 $\pm$ 54	-166
Lung	71.4 $\pm$ 10	210.2 $\times$ 14	-139
Liver	101.2 $\pm$ 23	67.8 $\pm$ 11	+33
45 min			
Brain	162.2 $\pm$ 8	88	+84
Kidney	121.4 $\pm$ 19	315	-194
Lung	62.8 $\pm$ 6	165	-102
Liver	79.4 $\pm$ 10	52	+27

would cross the BBB because of its enhanced membrane permeability and also would be distributed elsewhere in the body. In all locations, oxidation (k_{ox}) would occur. The resultant positively charged drug–carrier complex is rapidly eliminated from the periphery by renal and/or biliary processes (k_{out2}) while, in the brain, the compound is retained because of its size and charge. The cleavage of the drug ($k_{cleavage}$) from the oxidized carrier will also occur ubiquitously. If the rate of this cleavage is more rapid than the rate of efflux (i.e., $k_{out1} \ll k_{cleavage}$) of the complex from the brain, a sustained release of the drug could be obtained. This concept is shown in Scheme 2. However, when $k_{out2} \gg k_{cleavage}$, the periphery will be void of the active D, thereby resulting in significant reduction in toxicity.

The dihydropyridine drug complex is not, therefore, a prodrug but rather a pro-pro-. . .-drug or better stated, a chemical delivery system (CDS). It is becoming increasingly apparent that simple prodrugs cannot, in many cases, solve complex drug delivery problems.

This carrier-mediated delivery scheme was successfully applied to phenylethylamine (5), dopamine[34] (6), tryptophan[35] (7),

     5   **6**   **7**

and other related neurotransmitters.

[34] N. Bodor and H. H. Farag, *J. Med. Chem.* **26**, 528 (1983).
[35] N. Bodor and T. Nakamura, unpublished results.

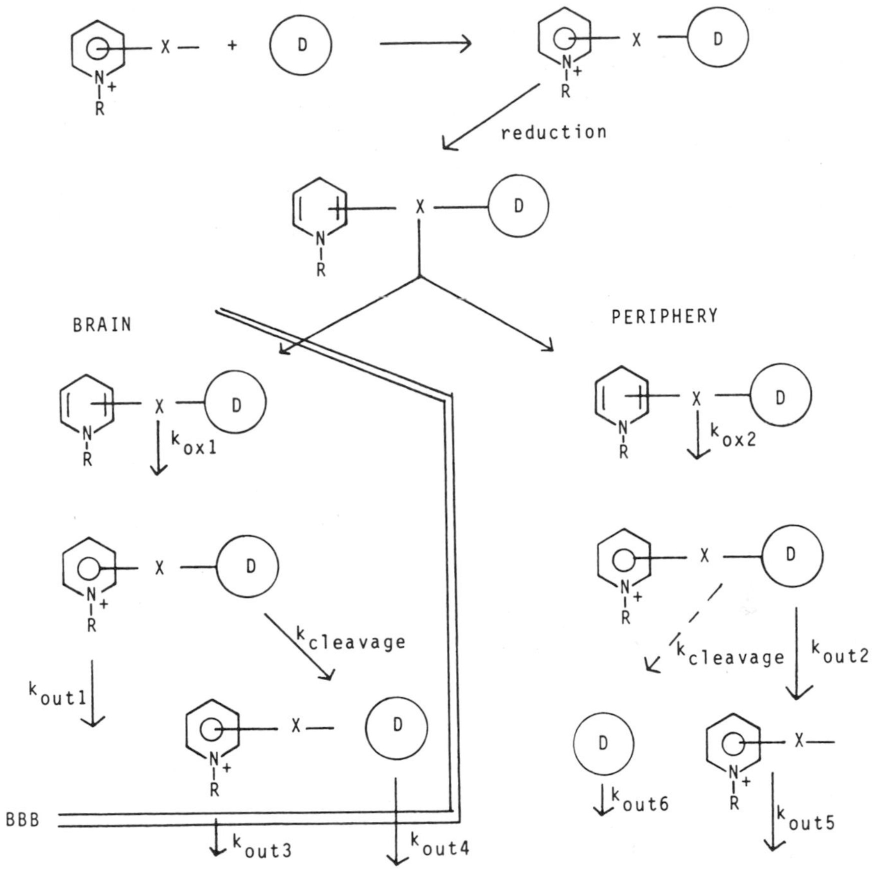

SCHEME 2

For example, Fig. 1 shows that when the dihydropyridine derivative (**8**) of (**5**) is administered systemically, the concentration of the quaternary derivative (**9**) slowly rises in the brain, reaching a maximum at 80 min, after which the concentration slowly declines.

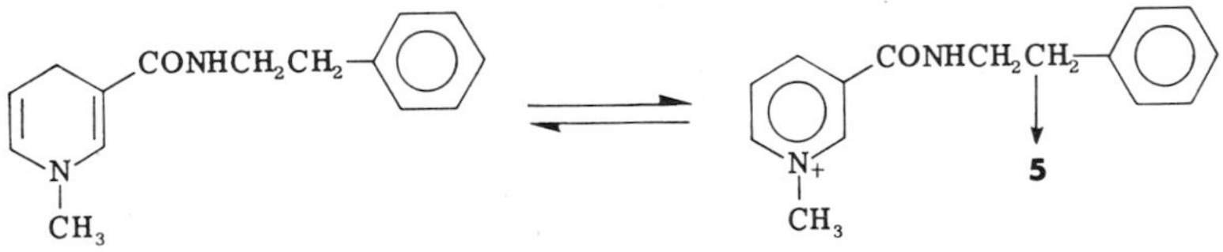

The demonstration of "locking in" of a compound to the brain is unique to this scheme and this technique.

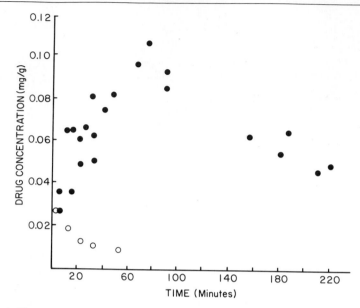

FIG. 1. The concentration of 1-methyl-3-(*N*-β-phenethyl)carbamoylpyridinium salt in the brain (●) and the blood (○) of rats after systemic administration of 125 mg/kg of the corresponding dihydro derivative.

The chemical delivery system was also applied to the very important class of compounds, the steroids. There are a number of clinical situations in which the delivery of sex hormones to the brain is desirable. These include dysfunctions such as impotency and centrally controlled contraception.

For example, when the dihydrocarrier–testosterone complex is administered systemically,[36] the quaternary compound is found (as shown in Fig. 2) in the brain. It is "locked in,"[6] thereby providing a sustained release form for the testosterone. Conversely, the quaternary salt, after systemic injection, was rapidly lost from the peripheral circulation (half-life of efflux of 54 min).

Synthetic Examples

N-Nicotinoyldopamine (10). To a pyridine solution containing 11.7 g (0.05 mol) of dopamine hydrobromide and 6.15 g (0.05 mol) of nicotinic acid at 0° was added 10.3 g (0.05 mol) of dicyclohexylcarbodiimide (DCC). The reaction mixture was stirred at room temperature for 24 hr, and the

[36] N. Bodor and H. H. Farag, *J. Pharm. Sci.* **73**, 385 (1984).

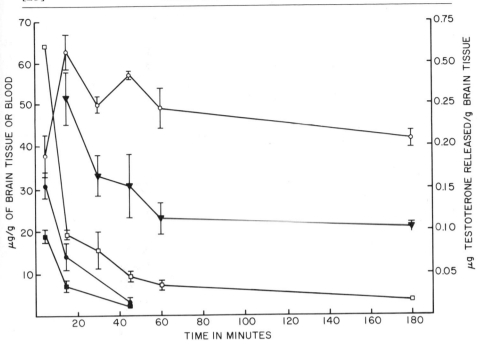

Fig. 2. Concentration of testosterone 17-nicotinate–*N*-methyl cation in the blood (□) and brain (○) of rats and the concentration of released testosterone (▼) following an administration of 28.2 mg/kg of the testosterone-CDS. Also given is the concentration of testosterone in the brain (■) and in the blood (●) following the administration of testosterone itself. The concentration of the released testosterone (▼) is indicated by the right abscissa; all other concentrations are indicated by the left.

formed dicyclohexylurea was removed by filtration. The pyridine was removed *in vacuo*, and the residue was crystallized from water at 0°. The product was isolated by filtration and dried over phosphorous pentoxide. Recrystallization from 2-propanol gave 0.9 g (0.035 mol, 70%) of *N*-nicotinoyldopamine, m.p. 159–162°.

3-(N-{β-[3,4-Bis(pivalyloxy)phenyl]ethyl}carbamoyl)pyridine (11). To a suspension of 5.16 g (0.02 mol) of finely powdered nicotinoyldopamine in 100 ml of chloroform was added 7.23 g (0.06 mol) of trimethylacetyl chloride while stirring. The mixture was refluxed for 6 hr and then filtered. The filtrate was washed with water free of chloride ions, washed once with a 5% solution of NaHCO₃, and then with water. The chloroform was evaporated, and the residue was chromatographed by using a silica gel G column and 2% methanol in chloroform as the eluant. The first fraction was collected and evaporated, and the residue was crystallized from ether–petroleum ether: yield 6.2 g (73%) of a white crystalline solid; m.p. 112–114°.

1-Methyl-3-(N-{β-[3,4-bis(pivalyloxy)phenyl]carbamoyl})pyridinium Iodide (12). To a solution of 5.0 g (11.7 mmol) of compound (11) in 20 ml of acetone was added 3.3 g (23.4 mmol) of methyl iodide; the mixture was refluxed while stirring for 6 hr and then cooled. An orange crystalline solid separated; this solid was filtered, washed with ether, and crystallized from acetone–ether: yield 5.6 g (85%).

17β-[(1,4-Dihydro-1-methyl-3-pyridinylcarbonyl)oxy]androst-4-en-3-one (14). To an ice-cold solution of 1.1 g (2 mmol) of testosterone nicotinate–N-methyl iodide (13) in 150 ml of deaerated 10% aqueous methanol was added 0.67 g (8 mmol) of sodium bicarbonate and 1.37 g (8 mmol) of sodium dithionite. The mixture was stirred for 20 min at room temperature, and the separated pale-yellow material was removed by filtration, washed with water, and dried over P_2O_5 under vacuum to yield 0.82 g (98%) of (14), m.p. 172–175°.

Analytical Methods

Dopamine Delivery. A high-pressure liquid chromatography (HPLC) method was developed for the studies of the degradation of a dihydropyridine derivative. The chromatographic analysis was performed on a component system consisting of Waters Associates Model 6000A solvent delivery system, Model U6K injection, and Model 440 dual-channel absorbance detector operated at 254 and 280 nm. A 30 cm × 3.9-mm (internal diameter) reverse-phase Bondapak C_{18} column (Waters Associates), operated at ambient temperature, was used for all separations. The mobile phase used for the separation of the dihydropyridine derivative, its degradation products, and oxidation products consisted of a 0.005 M solution of 1-heptanesulfonic acid sodium salt (PIC B-7 Eastman Kodak) in CH_3CN–0.01 M aqueous dibasic ammonium phosphate (2.5 : 1), at a flow rate of 2.0 ml/min.

Testosterone Delivery. An HPLC method was developed for the studies of the degradation of the quaternary (13) and dihydropyridine derivatives (14) of testosterone using the system described earlier. The absorbance detector was operated at 254 nm. A 15 cm × 4.6-mm i.d., 5-μm particle size, Ultrasphere reverse-phase C_{18} column, operated at ambient temperature was used for all separations. The mobile phase used for the separation of the dihydropyridine derivative products and the oxidation products consisted of a 0.002 M solution of 1-heptanesulfonic acid sodium salt in CH_3CN–0.01 M aqueous Na_2HPO_4 (7 : 3). At a flow rate of 2.0 ml/ min, (13) has a retention time of 12 min and (14) has a retention time of 5 min. For the analysis of testosterone in the *in vivo* brain delivery studies, the solvent consisted of a 0.002 *M* solution of 1-heptanesulfonic acid sodium salt in CH_2CN–0.01 *M* aqueous Na_2HPO_4 (1 : 1). At a flow rate of 2.0 ml/min, testosterone has a retention time of 3.3 min and (**13**) has a retention time of 36.5 min (very broad peak).

Determination of the Enzymatic Hydrolytic Cleavage and Rate of Oxidation of the Delivery System (*15*) for Dopamine

In Human Plasma. The freshly collected plasma used was obtained at the Civitan Regional Blood Center, Inc. (Gainesville, FL) and contained about 80% plasma diluted with anticoagulant citrate–phosphate–dextrose solution U.S.P. The plasma was stored in a refrigerator and used the next day. One hundred microliters of freshly prepared 0.61 *M* solution of (**15**) in methanol was added to 20 ml of plasma, previously equilibrated to 37° in a water bath, and mixed thoroughly to result in an initial concentration of 3.05×10^{-3} mol/liter. One-milliliter samples of plasma were withdrawn from the test medium, added immediately to 5 ml of ice-cold acetonitrile, shaken vigorously, and placed in a freezer. When all samples had been collected, they were centrifuged, and the supernatant fluids were filtered through Whatman No. 1 filter paper and analyzed by HPLC.

In Rat Brain Homogenate. The brain homogenate was prepared by the following method. Five Sprague–Dawley rats were killed by decapitation. The brains were removed, weighed (total weight 98.5 g), and homogenized in 49.3 ml of aqueous 0.11 *M* phosphate buffer, pH 7.4. The homogenate was centrifuged, and supernatant fluid was used for the test. One hundred microliters of a 0.18 *M* solution of (**15**) was mixed with 10 ml of homogenate, previously equilibrated to 37° in a water bath, to result in an initial concentration of 1.8×10^{-3} mol/liter. Samples of 1.0 ml were withdrawn every 10 min from the test medium, added immediately to 5 ml of ice-cold acetonitrile, and placed in a freezer. When samples had been collected, they were centrifuged. Each supernatant was filtered through two Whatman No. 1 filter papers and analyzed by HPLC.

In Rat Liver Homogenate. The liver homogenate was prepared by the following method. Three Sprague–Dawley rats were killed by decapitation. The livers were removed, weighed, and homogenized in a tissue homogenizer in 0.11 *M* aqueous phosphate buffer, pH 7.4, to make 20% liver homogenate. The homogenate was centrifuged, and the supernatant was used for the test. One hundred microliters of a 0.1 *M* solution of (**15**) in methanol was mixed with 20 ml of the homogenate, previously equilibrated to 37° in a water bath, to result in an initial concentration of 9×10^{-4} mol/liter. Samples of 1.0 ml were withdrawn every 5 min from the test medium, added immediately to 5 ml of ice-cold acetonitrile, shaken vigorously, and placed in a freezer. When all samples had been collected, they were centrifuged, and each supernatant was filtered through Whatman No. 1 filter paper and analyzed by HPLC.

Determination of the Concentration of the Locked-in Delivery Form of Dopamine in Brain and Blood of Rats after Parental Administration of (15)

Male Sprague–Dawley rats (average weight, 150 g) were used. The rats were anesthetized with an intramuscular injection of Inovar, and the jugular was exposed. Compound (15) was injected intrajugularly in the form of solution in dimethyl sulfoxide at a dose of 50 mg/kg and at a rate of 24 μl/min by using a calibrated infusion pump. After appropriate time periods, 1 ml of blood was withdrawn from the heart and dropped immediately into a tared tube containing 3 ml of acetonitrile, which was afterward weighed to determine the weight of the blood taken. The animals were perfused with 20 ml of saline solution and decapitated, and the brain was removed. The weighed brain was homogenized with 0.5 ml of distilled water; 3 ml of acetonitrile was added; the mixture was rehomogenized thoroughly, centrifuged, and filtered; and the filtrate was analyzed for the compounds by using the HPLC method. The tubes containing the blood were shaken vigorously, centrifuged, decanted, and also analyzed using the HPLC method. Quantitation was done by using a recovery standard curve obtained by introducing a known amount of the compound in either brain homogenate or blood and then treated in the same manner.

Testosterone Studies

Table III shows the results of kinetic studies in biological fluids.

Determination on in Vitro Rates of Oxidation of (14)

In Biological Media. One hundred microliters of a freshly prepared 0.024 M solution of (14) in dimethyl sulfoxide was added to 10 ml of plasma, previously equilibrated to 37° in a water bath, and mixed thoroughly to result in an initial concentration of 2.4×10^{-4} mol/liter. One-milliliter samples of plasma were withdrawn every 20 min from the test medium, added immediately to 5 ml of ice-cold acetonitrile, shaken vigorously, and placed in a freezer. When all samples had been collected, they were centrifuged and the supernatants were filtered through nitrocellulose membrane filters (0.45-μm pore size) and analyzed by HPLC, following the appearance of (13).

TABLE III

KINETICS OF *in Vitro* OXIDATION OF THE
DIHYDROPYRIDINE ESTER (**14**) TO THE QUATERNARY DERIVATIVE
(**15**) IN BIOLOGICAL FLUIDS

Medium	k (s^{-1})	$t_{1/2}$ (min)	r
80% Plasma	8.12×10^{-5}	142	0.959
20% Brain homogenate	1.72×10^{-4}	67	0.997
Whole blood	1.74×10^{-4}	66	0.997

In Rat Brain Homogenate. Five female Sprague–Dawley rats were decapitated, and the brains were removed, pooled, weighed (total weight 9.2 g), and homogenized in 36.8 ml of aqueous 0.11 M phosphate buffer, pH 7.4. One hundred microliters of 0.024 M solution of (**14**) in dimethyl sulfoxide was mixed with 20 ml of the homogenate, previously equilibrated to 37° in a water bath, to result in an initial concentration of 2.4 × 10^{-4} mol/liter. Samples of 1.0 ml were withdrawn every 10 min from the test medium, added immediately to 5 ml of ice-cold acetonitrile, shaken vigorously, and placed in a freezer. When all samples had been collected, they were centrifuged and the supernatants were filtered through nitrocellulose membrane filters (0.45-μm pore size) and analyzed by HPLC.

In Vitro Determination of the Site-Specific Conversion of (13) to Testosterone

A fresh brain homogenate was prepared as previously described. One hundred microliters of a 0.017 M solution of the quaternary compound (**13**) in methanol was mixed with 10 ml of the brain homogenate, previously equilibrated to 37°, to result in an initial concentration of 1.7 × 10^{-4} M. Samples of 1.0 ml were withdrawn every 20 min from the test medium, added immediately to 5 ml of ice-cold acetonitrile, and placed in a freezer. When all samples had been collected, they were centrifuged, and the supernatant was filtered through a nitrocellulose membrane filter (0.45-μm pore size) and analyzed for the quaternary compound (**13**).

Conclusions

The inability of many potentially useful agents to cross the BBB has limited the treatment of cerebral diseases. The basis of this impermeability is the peculiar way in which cerebral capillaries are fused. This tight joining prevents all but the most lipophilic compounds from passing into the brain parenchyma. Attempts to circumvent this barrier have included direct injection of therapeutic agents into the CSF, but the technique is replete with dangerous side effects. Another method by which a drug can gain access into the brain is by temporarily making the compound more lipophilic. Unfortunately, when this is done indiscriminately, all tissues are exposed to higher levels of the drug. The approach which seems to be the most useful, and demonstrates the most potential, is the dihydropyridine $\rightleftharpoons$ pyridinium redox system. There are two major areas in this technique. The first involves dealing with molecules which contain a pyridinium partial structure (2-PAM, berberine), whereas the second involves using a pyridine carrier to which a drug can be attached (phenethylamine, dopamine, testosterone). Both approaches are based on the ability to transiently convert highly polar molecules (pyridinium salts) to nonpolar species (dihydropyridines). By performing this conversion, the ability of a molecular to pass biologically important membranes can be altered. This method is superior to simple ester- or amide-type prodrugs in that it delivers compounds specifically to the brain, while maintaining a lower peripheral concentration. This increases the efficacy of the agent centrally, while reducing any associated peripheral toxicity.

Acknowledgments

Financial support from the National Institutes of Health (Grant GM 27167), from Otsuka Pharmaceutical Co., and from Pharmtec, Inc. is gratefully acknowledged.

Section IV

Polymer Systems

[30] Controlled Release and Magnetically Modulated Release Systems for Macromolecules

By ROBERT LANGER, LARRY BROWN, and ELAZER EDELMAN

Over the past decade, a large number of polymeric delivery systems have been developed with the capability of controlling the release rate of a drug over a long period of time (e.g., anywhere from 12 hr to 5 years).[1,2] Almost all of these systems have been limited by the small size (MW <500) of the drugs that could be released as a result of the impermeability of high molecular weight drugs through biocompatible polymers.[3] However, in 1976, several biocompatible polymer systems were developed using ethylene–vinyl acetate copolymer and other polymers with the capability of releasing macromolecules up to 2,000,000 MW.[4] These polymer systems were used subsequently for insulin[5] or interferon[6] delivery systems, as a depot for single-step immunizations,[7] as a source of chemotactic gradients,[8] as a means of releasing histochemically useful polypeptides,[9,10] and as a delivery system for numerous growth factors[11-20] and inhibitors.[21-25]

[1] R. Langer, *Drug Ther.* **13,** 217 (1983).

[2] R. Langer, *Tech. Rev.* **83,** 26 (1981).

[3] R. W. Baker and H. K. Lonsdale, *Adv. Exp. Biol. Med.* **47,** 15 (1974).

[4] R. Langer and J. Folkman, *Nature (London)* **263,** 797 (1976).

[5] H. Creque, R. Langer, and J. Folkman, *Diabetes* **29,** 37 (1980).

[6] R. Langer, D. S. T. Hsieh, L. Brown, and W. Rhine, *in* "Better Therapy with Existing Drugs: New Uses and Delivery Systems" (A. Bearn, ed.), p. 179. Merck & Co., Biomedical Information Corporation, New York, 1981.

[7] I. Preis and R. Langer, *J. Immunol. Methods* **28,** 193 (1979).

[8] R. S. Langer, M. Fefferman, P. Gryska, and K. Bergman, *Can. J. Microbiol.* **26,** 274 (1980).

[9] M. Moskowitz, M. Mayberg, and R. Langer, *Brain Res.* **212,** 460 (1981).

[10] M. Mayberg, R. Langer, N. Zervas, and M. Moskowitz, *Science* **213,** 228 (1981).

[11] K. Falterman, D. Ausprunk, and M. Klein, *Surg. Forum* **27,** 358 (1976).

[12] P. Polverini, R. Cotran, M. Gimbrone, and E. Unanue, *Nature (London)* **269,** 804 (1977).

[13] A. M. Schor, S. L. Schor, and S. Kumar, *Int. J. Cancer* **24,** 225 (1979).

[14] R. McAuslan and G. A. Gole, *Trans. Ophthalmol. Soc.* **100,** 354 (1980).

[15] M. E. Plishkin, S. M. Ginsberg, and N. Carp, *Transplantation* **29,** 255 (1980).

[16] D. Ausprunk, K. Falterman, and J. Folkman, *Lab. Invest.* **38,** 284 (1978).

[17] B. M. Glaser, P. A. D'Amore, R. G. Michels, A. Patz, and A. Fenselau, *J. Cell Biol.* **84,** 298 (1980).

[18] B. M. Glaser, P. A. D'Amore, R. G. Michels, S. L. Brunson, A. H. Fenselau, T. Rice, and A. Patz, *Ophthalmology (Rochester, Minn.)* **87,** 440 (1980).

[19] B. M. Glaser, P. A. D'Amore, G. A. Lutty, A. H. Fenselau, R. G. Michels, and A. Patz, *Trans. Ophthalmol. Soc. U.K.* **100,** 369 (1980).

This report concerns the methodology of formulating these polymeric systems. It includes three main sections: methods of preparing polymeric delivery systems, methods of regulating the release kinetics of these systems, and a discussion of the advantages and limitations of the systems as well as potential directions for future research in this area.

Methods of Preparing Polymeric Delivery Systems

Materials

Ethylene–vinyl acetate copolymer (40% by weight vinyl acetate, Elvax 40P) and poly(vinyl alcohol) (Elvanol), in powder or pellet form, were obtained from Dupont Corporation, Wilmington, DE. PolyHEMA (Hydron-S or Hydron-N, a polymer of hydroxyethyl methacrylate) was supplied by Hydron Laboratories, New Brunswick, NJ.

As obtained from the manufacturer, poly(vinyl alcohol) or ethylene–vinyl acetate copolymer may cause a mild inflammatory response. This can be eliminated for ethylene–vinyl acetate, or reduced for poly(vinyl alcohol), by washing with solvents to extract impurities. A suggested procedure is 10 washes in distilled water, followed by 40 washes in 95% ethanol and 10 washes in reagent-quality absolute ethanol. The washes should be conducted at 37°, with the polymer constituting no more than 10% (w/v) of the mixture. Mild agitation is helpful; washes are usually conducted in large sterile, roller bottles with gentle shaking. Each wash should take 3 hr or more. After the final wash, the alcohol should be decanted and the polymer poured into a sterile glass petri dish. The remaining alcohol can be removed by mild vacuum drying or by placing the beads in a loosely covered petri dish at 37°.

The water wash extracts certain polymer surface impurities. (The amounts of these impurities vary depending on the batch received from the manufacturer.) The alcohol wash sterilizes the polymers and removes the antioxidant butylhydroxytoluene (BHT). Since BHT absorbs ultraviolet light, the alcohol can be analyzed spectrophotometrically at 230 nm to monitor the effectiveness of the washing procedure. The alcohol's ab-

[20] D. Gospodarowicz, H. Bialceki, and T. K. Thakral, *Exp. Eye Res.* **28**, 501 (1979).

[21] R. Langer, H. Brem, K. Falterman, M. Klein, and J. Folkman, *Science* **193**, 70 (1976).

[22] A. Lee and R. Langer, *Science* **221**, 1185 (1983).

[23] J. Folkman, R. Langer, R. Linhardt, C. Haudenschild, and S. Taylor, *Science* **221**, 719 (1983).

[24] S. Taylor and J. Folkman, *Nature (London)* **297**, 307 (1982).

[25] S. Brem, I. Preis, R. Langer, H. Brem, J. Folkman, and A. Patz, *Am. J. Ophthalmol.* **84**, 323 (1977).

sorbance after the first wash is between 1 and 2 absorbance units, depending on the initial size of the polymer particles, and this absorbance decreases with each successive wash. After the final wash, absorbance at 230 nm should be less than 0.03 absorbance units.

It should be noted that more efficient techniques for washing, such as the use of a Soxhlet extractor, could be employed. The alcohol washes do not affect the release properties of the polymer. Even without alcohol washing, the polymer induces only mild inflammatory responses (again varying from batch to batch).

Proteins were obtained from Sigma Chemical Co., St. Louis, MO, unless otherwise specified. Dichloromethane was purchased from Fisher Scientific, Fairlawn, NJ. The sieves used were American Standard Sieves, Nos. 40, 60, and 200, from Dual Manufacturing Co., Chicago, IL.

A number of different variations of fabricating polymeric delivery systems are described in the following sections. Some variations are superior for particular applications and therefore all are included for completeness.

Methods for Producing Systems to Release Macromolecules

Method 1. This method was the original procedure for fabricating polymeric systems.[4,26] Polymer casting solutions were made by dissolving polymers in appropriate solvents: Hydron in absolute alcohol at 37° overnight; ethylene–vinyl acetate copolymer in dichloromethane at 37° for several hours; and poly(vinyl alcohol) by dissolving in distilled water in an autoclave for 1 hr.

The slow-release pellets were made by first adding a specified amount of powdered chemical (in any of the three polymer systems) or solution [if desired for the Hydron or poly(vinyl alcohol) systems] into a test tube. A specific volume of polymer casting solution was then added and mixed with the chemicals, using a spatula. [In many studies, we found that 5 mg of powdered macromolecule mixed with 100–200 μl of 10% (w/v) ethylene–vinyl acetate copolymer solution were useful quantities.] The mixtures were then pipetted (or, if the viscosity was extremely high, the mixture was placed on the edge of the spatula) into an appropriate mold. The molds were dried under vacuum overnight, causing the solvent to evaporate with the macromolecule trapped within the polymer matrix.

The dried polymer matrix was then hydrated by adding a drop of lactated Ringer's solution and gently removed from the mold. The entire procedure can be performed simply, quickly, and at room temperature.

[26] R. S. Langer and J. Folkman, *in* "Polymeric Delivery Systems, Midland Macromolecular Monograph" (R. J. Kostelnik, ed.), p. 175. Gordon & Breach, New York, 1978.

To make sterile pellets, the macromolecule should be passed through a Millipore filter and lyophilized, the glassware should be autoclaved, and all operations should be performed under a sterile hood.

The apparatus used to dry the polymers consisted of a laboratory desiccator attached to wall suction. The desiccator's inlet contained a 25-mm diameter, 0.45-μm Millipore filter, which allowed a constant exchange of sterile air at slightly less than atmospheric pressure. [Drying under vacuum only accelerates the rate of solvent evaporation; polymer pellets displaying the same release kinetics can be obtained without this apparatus. In particular, the ethylene–vinyl acetate copolymer systems dry rapidly at atmospheric pressure because of the high volatility of the solvent (dichloromethane).]

Molds were used to cast the polymer solution. The molds used in these studies for Hydron and poly(vinyl alcohol) preparations were plastic Co-star Histo-Plates with conical wells 4 mm in diameter and 2.5 mm deep. The dichloromethane in the ethylene–vinyl acetate copolymer solution dissolved the plastic, therefore, molds for the ethylene–vinyl acetate copolymer were made using 4-mm-thick plate glass. A No. 761, 1/8 in.-diameter, carbide-tipped glass drill was used to bore holes 1.5 mm deep, resulting in conical molds 2 mm in diameter.

To retard the diffusion of the macromolecules from the polymer matrix, the pellets were sometimes coated with pure polymer solution. Coating was accomplished by one of two procedures. In the first procedure, the polymer–macromolecule pellet was made as described above. After drying, it was removed from its mold and placed in a glass petri dish. A 50-μl drop of polymer casting solution was pipetted directly over the pellet. The pellet was allowed to sit in this puddle of pure polymer solution for approximately 20 sec. Pure polymer adhered to the pellet's surface, thereby coating it. The pellet was removed from the puddle with a forceps and placed in a second dish to be dried.

In the second coating procedure, pure polymer casting solution was placed in the mold and allowed to dry under vacuum; the solvent was evaporated overnight. The polymer–macromolecule mixture was placed over this layer and dried under vacuum. Finally, a top layer of pure polymer solution was added and vacuum dried, resulting in a three-layer pellet with the macromolecules in the middle layer.

Method 2. In this method, polymer pellets were made as described in method 1, but without molds. Solutions or mixtures of polymer plus macromolecules were cast as 5- to 50-μl drops directly onto a glass surface, such as a microscope slide or petri dish, and dried as in method 1. When extremely high ratios of macromolecule to polymer were used, the mixtures were so viscous that they could not easily be pipetted. In this case, a spatula was used to lift the mixture out of the test tube and onto the glass surface.

The dried pellets could be used directly or they could be cut with a scissors and forceps into desired sizes. Coated polymer pellets were made using the first coating procedure discussed in method 1. Method 2 does not allow control of shape, but we have generally found that with a little practice it can be performed with greater facility and equally good kinetic results.

Method 3. One of the limitations of methods 1 and 2, particularly with ethylene–vinyl acetate copolymer, was the mediocre reproducibility of the release kinetics (see Kinetic Studies). This difficulty was compounded when larger volume polymeric delivery systems were fabricated. This was because significant drug settling and redistribution occurred during casting and drying. At room temperature, the insoluble macromolecular drugs migrated vertically in the polymer solvent, and there was also visible lateral motion caused by currents (possibly thermal) in the mixture. We therefore developed a low-temperature casting and drying procedure to minimize drug movement during matrix formation.[27] The reproducibility of release kinetics of matrices prepared using this low-temperature method was markedly improved.

In this low-temperature method, ethylene–vinyl acetate copolymer (40% vinyl acetate by weight) was dissolved in dichloromethane to give a 10% solution (w/v). If desired, protein or other macromolecular powder was sieved to give particles of specified sizes. A weighed amount of powder from a single size range was added to 15 ml of the polymer solution in a glass vial (Wheaton Scientific, Millville, NJ), and the mixture was agitated (Vortex Genie set at speed 10, Scientific Industries, Bohemia, NY) for at least 10 sec to yield a uniform suspension. This mixture was poured quickly from corner to corner into a leveled square glass mold (7 × 7 × 0.5 cm), which had been previously cooled by placing it on a flat slab of dry ice for 5 min. During precooling, the mold was covered with a Kimwipe and a glass plate to prevent excess frost formation on the mold. After the mixture was poured, the mold remained on the dry ice for 10 min, during which time the mixture froze. The mold was covered with a Kimwipe for the last 7 min of this stage. The frozen slab was easily pried loose with a cold spatula, transferred onto a wire screen (0.5 mm spacing, common steel mesh), and kept at −20° for 2 days in a freezer under a chemical hood. The slab was then dried for 2 more days at room temperature in a desiccator kept under a mild, house-line vacuum. (Drying for up to 60 days results in slabs with the same release kinetics as slabs dried for

[27] W. Rhine, D. S. T. Hsieh, and R. Langer, *J. Pharm. Sci.* **69**, 265 (1980).

2 days.) Drying caused the slabs to shrink in area to approximately 5 × 5 cm. The central 3 × 3 cm square was excised with a scalpel (No. 10 blade, Bard-Parker, Rutherford, NJ) and straight edge, and further divided into nine 1 × 1 cm squares.

If the dried polymer squares made using method 3 were to be coated, each square was attached to a 30-gauge hypodermic needle and was dipped into a vial containing 15 ml of polymer solution of a desired weight-volume percentage (usually 10%) of ethylene–vinyl acetate copolymer in dichloromethane. After 1 min in solution, the square was removed from solution and held to dry at room temperature for 2 min. The matrix was dried for an additional 24 hr at room temperature under mild vacuum.

Method 4. In some cases it may be preferable to incorporate aqueous protein or macromolecule solutions directly into the matrix rather than lyophilizing the molecules to obtain powders. Unfortunately, the most desirable release kinetics were obtained with ethylene–vinyl acetate copolymer (see Kinetic Studies), which has dichloromethane as a solvent. We reasoned, however, that we could make an emulsion of the aqueous macromolecule solution in the organic polymer solution and use the low-temperature procedure to trap that emulsion. Upon drying, both water and dichloromethane would be removed and the macromolecule would be trapped within the dried polymer matrix.[28]

In this method, the macromolecule was dissolved in distilled water and the resultant solution was added to a 10% (w/v) solution of the polymer in dichloromethane. This mixture was agitated for 60 sec on a Vortex Genie mixer at speed 10 to yield a uniform emulsion; this emulsion was quickly poured into a leveled glass mold that had been previously cooled by placing it on dry ice for 10 min. The slab froze within 10 min and was then removed from its mold by first encircling the inside wall of the mold with a cold spatula tip and then prying the slab loose. The slab was transferred onto a cold wire screen as in method 3 and kept in a −20° freezer for 2 days. It was then removed from the freezer and placed in a lyophilizer at a vacuum of less than 60 mtorr until no weight change was observed.

Method 5. In many cases, it would be desirable not to have to use a solvent at all. This is particularly true if the molecule could be easily denatured. The following procedure[29,30] takes advantage of the low glass transition temperature (−36.5°) of ethylene–vinyl acetate copolymer and

[28] R. Langer, this series, Vol. 73, p. 57.

[29] J. Cohen, R. Siegel, and R. Langer, *J. Pharm. Sci.* **73,** 1034 (1984).

[30] R. A. Siegel, J. M. Cohen, L. Brown, and R. Langer, *in* "Recent Advances in Drug Delivery System" (J. M. Anderson and S. W. Kim, eds.), 315. Plenum, New York (1984).

its ability to flow under moderate temperature and high pressure to form an intact matrix. Ethylene–vinyl acetate copolymer was converted into a powder by one of two methods. The first method involved the dissolution of 3 g ethylene–vinyl copolymer in dichloromethane to form a 5% (w/v) solution. The solution was extruded dropwise, using a 50-cm³ syringe that was fitted with a 20-gauge hypodermic needle into a 250-ml beaker containing 100 ml of liquid nitrogen. From this time on, all instruments that came into contact with the frozen polymer solution were cooled with liquid nitrogen and, wherever possible, precooled in a freezer to minimize the quantity of liquid nitrogen necessary for cooling.

The frozen droplets were ground for 5 min with a mortar and pestle. The powder was then spread evenly over three 8 × 8 in. glass sheets that had been cooled to $-10°$. The glass sheets were returned to a $-10°$ freezer for 2 hr. At the end of that time most of the solvent had evaporated, leaving a stringy powder. This powder was removed from the glass sheet with a razor blade, bathed in a 100-ml pyrex beaker with 30 ml of liquid nitrogen, and then ground to a fine powder with a mortar and pestle as before. This powder was placed under vacuum for 2 hr. The bulk powder was sieved to specific size ranges using a stack of graduated sieves. Polymer powder prepared by this method will be referred to as "powder type I."

Powder type I was produced under two aging conditions. In one set of experiments, the powder was dried under vacuum for 2 hr at room temperature. This will be referred to as "fresh" powder type I. In the other set of experiments, the powder was dried under vacuum for 2 hr and then left in a covered petri dish for 1 week at room temperature. This will be referred to as "aged" powder type I.

In the second method of powder preparation, 20 g of ethylene–vinyl acetate copolymer beads were cooled in 40 ml liquid nitrogen and placed in an electric mill (Micromill; Technilab Instruments). The mill was set for 90-sec grinding intervals. Between grindings, the polymer beads were cooled with 20-ml portions of liquid nitrogen. During the grinding process, cold nitrogen vapor was circulated around the sample chamber through the chamber's cooling ducts. The powder collects around the outer edges of the sample chamber and can be extracted with a spatula after the second grinding, and after every successive grinding. After the eighth grinding, approximately 4 g of frozen pellets were added to restore the original volume. This process was repeated until sufficient powder was collected to prepare the samples. The ground polymer powder was then sieved to specific size ranges using a stack of graduated sieves in an automatic sieve shaker at $-40°$. Polymer powder prepared by this second method will be denoted "powder type II."

To formulate the controlled release system, macromolecular drug powder was sieved to a 90- to 180-μm particle size range. Then, macromolecule and polymer powders, totaling 1.0 gm, were placed in a plastic weighing boat which was then transferred to a Pyrex baking dish containing liquid nitrogen at a depth of 1 cm. The powders are mixed in the weighing boat for 5 min with a spatula that was chilled with liquid nitrogen. After mixing, the powders were poured into a 1 in.^{2} Carver press piston mold. The mold had been chilled in a $-10°$ freezer for 1 hr and then chilled with 20 ml liquid nitrogen immediately before the powder mixture was poured in. After the mixed powder was poured into the piston, the piston mold assembly was warmed to 37° in an oven for 1 hr and then placed in a Carver hydraulic press. The pressure on the mold was increased over a 90-sec interval from 0 Pa to the maximum pressure desired. After 30 min, the pressure was released, leaving a cohesive, heterogeneous slab.

The slab was removed from the mold with the aid of a scalpel and forceps. A small amount (less than 2%) of the fused polymer mixture was extruded into the space between the moving piston and the jacket of the mold. The scalpel was used to trim any of the polymer mixture that may have leaked. The slab was then gently peeled from the mold with the forceps.

Method 6. One limitation of all the systems discussed thus far is that the release rates are not constant but decrease with time. This is because release occurs by diffusion through pores in the matrix[31] and the diffusion distance increases as time increases. To provide constant release, a shape is required so that the area of available drug increases as time increases in such a way as to compensate for the increased diffusion distance. We have demonstrated that a hemispheric system laminated in all places except for a cavity in the center face (Fig. 1) is such a system.[32,33] The molds used for fabricating hemisphere-shaped systems for macromolecules were made using the bottoms of glass test tubes and had hemispheric bottoms (13 mm diameter $\times$ 11 mm height). To prevent these molds from falling over, an embedding platform was made. This platform was constructed by pouring 15 ml of molten paraffin into a bacterial petri dish and then placing empty glass molds into the paraffin so that they touched the bottom. After allowing the paraffin to harden, the molds were removed and

[31] R. Langer, W. Rhine, D. S. T. Hsieh, and R. Bawa, *in* "Controlled Release of Bioactive Materials" (R. W. Baker, ed.), p. 83. Academic Press, New York, 1980.
[32] W. Rhine, V. Sukhatme, D. Hsieh, and R. Langer, *in* "Controlled Release of Bioactive Materials" (R. W. Baker, eds.), p. 177. Academic Press, New York, 1980.
[33] D. Hsieh, W. Rhine, and R. Langer, *J. Pharm. Sci.* **72,** 17 (1982).

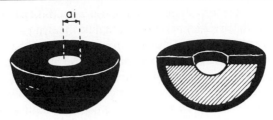

FIG. 1. Schematic diagram of hemispheric system to achieve constant release rates. Black represents impermeable coating. Hatched lines represent the polymer–drug matrix.

the indentation left behind in the paraffin could be used to subsequently support the molds.

To make the hemisphere polymer systems, earlier methods of preparing polymer–macromolecule slabs were adapted. First, empty glass molds were positioned in the indentations in the paraffin-embedding platform. Then, 6 ml of 20% ethylene–vinyl acetate copolymer solution in dichloromethane and a desired amount of drug [600 mg yields a 33% (w/w) loaded matrix] were mixed in a glass vial at room temperature and vortexed for 1 min to yield a uniform suspension. Into each hemispheric glass mold, 0.8 ml of the protein–polymer dispersion was pipetted. The petri dish containing these samples was then transferred onto a block of dry ice for 10 min. The samples gelled within 1 min. The petri dish containing the samples was then dried first at $-20°$ for 2 days and then at $20°$ for another 2 days, as discussed above.

The hemisphere pellets were coated twice with 15% ethylene–vinyl acetate (EVA) copolymer solution (containing no macromolecules) to form an impermeable barrier. The coating procedure is as follows. (1) The tip of a cylindrical metal stick (1.8 mm diameter $\times$ 30 mm length) was inserted into the center of the flat surface of each hemisphere pellet to a depth of 3 mm. (2) The hemisphere was placed on a slab of dry ice for 10 min to freeze the matrix. (3) Again using the metal stick as a handle, the cooled hemisphere was immersed into 15% ethylene–vinyl acetate copolymer solution at $20°$ for 10 sec, removed, and then placed immediately on dry ice for 10 min. (4) The hemispheres were then put in the freezer ($-20°$) for 2 days followed by further drying at $20°$ for 2 days in a desiccator under a house-line vacuum (600 mtorr) to remove residual solvent. (5) Finally, to create the exposed cavity in the face of the hemisphere pellet, the metal stick was removed by gently encircling the polymer surface immediately surrounding the stick with a scalpel blade.

Since the method described above necessitates the fabrication of individual matrices, the interpolymer matrix reproducibility is low. The most important feature of this geometry is that all of the drug diffusion from the device occurs through the aperture alone. Thus, thick-coated disks with an aperture can be used as a matrix in place of the hemispheric shape in an effort to improve the system's reproducibility. In this case, method 3 was adapted by preparing a 15% (w/v) EVA copolymer dissolved in dichloromethane. Protein powder was added to the polymer solution in a glass vial and the mixture was vortexed to give a suspension of the drug in the polymer solution. The suspension was poured into a $5 \times 5 \times 2$-cm glass mold and dried as previously described. The 15% (w/v) EVA copolymer in dichloromethane resulted in a thicker slab than with the 10% (w/v) EVA copolymer solution. After drying, the resulting slab was 3 mm thick. A No. 3 cork borer was used to excise a round matrix 5 mm in diameter and 3 mm thick. The matrix was coated by the following procedure. A 30-gauge needle was inserted into the flat surface of the matrix. The matrix was frozen in liquid nitrogen as described in method 6 to prevent the dissolution of the matrix when dipped in the EVA copolymer solution for coating. The frozen matrix was then dipped into 15% (w/v) EVA copolymer in dichloromethane. The coating was allowed to dry for 2 days at $-20°$ and for 2 days at room temperature under vacuum. The coating was approximately 150 μm thick.

The aperture was then opened by removing the 30-gauge needle. The coated matrix was frozen in liquid nitrogen for 3–5 sec. The aperture was widened using a 0.58-mm-diameter drill bit on a Dremel drill press. The depth of the hole was 1.5 mm from the surface of the coating.

The sintering technique (method 5) can also be used to fabricate thicker disk matrices. The sintering technique results in a matrix which has greater porosity than those made with the solvent casting technique. The increased porosity allows for greater release rates and more efficient utilization of marginally soluble macromolecules such as insulin. After the fabrication of the 3-mm-thick slabs via sintering, individual disks can then be punched out with a No. 3 cork borer and coated as described for the solvent cast matrix. The aperture can be opened by the same procedure described above.

Method 7. Microspheres using ethylene–vinyl acetate can also be prepared.[34] Two milliliters of 10% ethylene–vinyl acetate polymer solution was added to a weighed amount of macromolecule powder in a 5-ml glass vial and the mixture was vortexed to give a uniform suspension. The mixture was quickly drawn into a disposable 3-ml syringe via a 0.5 in. long 16-gauge needle that had been trimmed to a blunt end with a standard

[34] L. R. Brown, M. Sefton, and R. Langer, *J. Pharm. Sci.* **73**, 1859–1861 (1984).

machine shop grinder wheel. After vortexing again for a few seconds, the suspension was carefully extruded through the same 16-gauge needle drop by drop into 20 ml of cold, unstirred absolute ethanol in a 50-ml beaker immersed in a dry ice–ethanol bath ($-78°$). The mixture gelled virtually immediately on contact with the cold ethanol into the near spherical shape of the droplets. The hard, gelled spheres sank to the bottom of the beaker. The mixture was extruded as quickly as possible drop by drop to minimize dichloromethane evaporation and plugging of the needle. It was also found that extrusion was facilitated by conducting it outside a fume hood to prevent dichloromethane evaporation.

After 5–10 min, the beaker containing the microspheres was removed from the dry ice bath and allowed to warm to room temperature. The microspheres turned white as the solvent was extracted into the ethanol. After at least 1 hr, the liquid was replaced with approximately 10 ml of fresh ethanol and set aside overnight in a 20-ml glass vial. The ethanol was decanted and the microspheres were dried for 5 hr in a desiccator under a mild house-line vacuum. The duration of the various process steps were not considered critical as long as solvent and ethanol removal was largely complete. For example, removal of the microspheres from the cold ethanol immediately after extrusion caused the microspheres to agglomerate since insufficient dichloromethane had been removed for the microspheres to remain gelled at room temperature.

Beads prepared from 10% polymer in dichloromethane were almost perfectly spherical immediately after formation in cold ethanol. As the beads warmed and the dichloromethane removed, the polymer became softer and deformable. The beads flattened slightly and became more ellipsoidal in shape; this was accentuated for the beads with low loadings of macromolecule (i.e., <20% w/w macromolecule). The softness of the beads with lower macromolecule loadings was noted particularly when they were dry; it was difficult to handle the beads with forceps or a spatula without further changing their shape. Since these lower loading beads were also tacky (in the "dry" state), they stuck to the walls of glass or plastic containers, thus further enhancing handling difficulties. Scanning electron micrographs of beads containing 50% (w/w) bovine serum albumin are shown in Fig. 2, illustrating the sphericity of these beads.

Five percent polymer in dichloromethane, while less viscous and therefore easier to use in making microspheres with high drug loadings, resulted in beads with poorly defined shapes. Beads with 50% drug loading were tear drop shaped while smaller loadings (<20%) resulted in large aggregates of irregularly shaped beads. It appeared that the low ratio of polymer to solvent in these beads resulted in a slower cold ethanol solidification process than when beads were made from a 10% polymer solution.

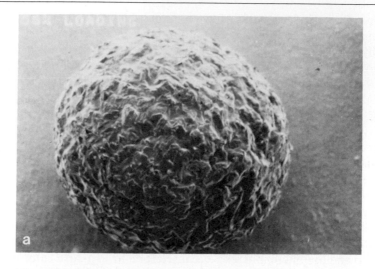

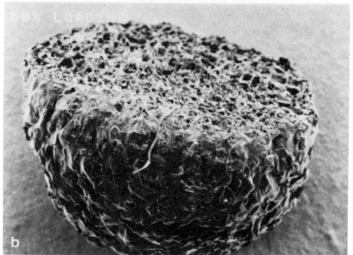

FIG. 2. Scanning electron micrograph of controlled-release microsphere. (a) Microsphere; (b) microsphere sliced in half.

Consequently, the beads were deformable and changed shape while the newly formed beads sank to the bottom of the cold ethanol. The beads with 50% loading were less deformable so that only a small tail was produced while the beads with lower loadings were still tacky in the cold ethanol solution and stuck to the previously formed beads. The most desirable shape was obtained with the 10% solution.

Only to a limited extent can smaller microspheres be prepared effectively by using a smaller gauge needle or by using a lower polymer concentration. Smaller beads may possibly be prepared more readily by a vibratory prilling process[35] or by blowing droplets off the end of a needle by a coaxial air stream.[36] In the latter case, it would be necessary to prevent the volatile dichloromethane from evaporating before the droplets have left the needle.

Method 8. In the preceding systems, milligram quantities of macromolecules were required for polymer pellet formulation, even for pellets as small as 1 mm^3, to ensure the release of physiologically significant amounts during the experimental period. The requirement for milligram quantities is a consequence of the mechanism by which macromolecules are released from polymeric matrices. During the formulation process, pores are formed in the macromolecule–polymer matrix as a result of the presence of particulate macromolecule.[31] When small amounts of macromolecules are present, few pores are formed. On the other hand, when large amounts are present, pores are numerous and interconnect to form tortuous channels that run through the matrix to the surface through which macromolecules can be released. Because of the requirement for relatively large amounts of macromolecules, these systems have been less suitable for the delivery of growth factors or other substances which are obtainable only in microgram or nanogram amounts. This problem can be solved by lyophilizing milligram amounts of albumin mixed with the test macromolecule[37,38] and then using any of methods 1 through 7. Albumin is a useful agent because of its high aqueous solubility, its inertness, and its availability from a variety of different species.

Methods of Producing Magnetic Polymeric Systems

Method 9. The procedure[39,40] for preparing magnetic, sustained-release, polymer systems was modified from our methods of preparing controlled-release systems for macromolecules. Polymer casting solution was made by dissolving ethylene–vinyl acetate copolymer in dichloromethane to achieve a 10% (w/v) solution. One-half gram of powdered drug was mixed with 10 ml casting solution. The suspension was poured quickly

[35] W. E. Yates and N. B. Åkesson, *Proc. Int. Conf. Liq. Atom. Spray Sys., 1st, 19* p. 181 (1978).

[36] S. Wagaosa, H. Matsui, N. Tokuoka, and G. T. Sato, *Proc. Int. Conf. Liq. Atomization Spray Syst., 1st, 1978* p. 29 (1979).

[37] J. Murray, L. Brown, M. Klagsbrun, and R. Langer, *In Vitro* **19**, 743 (1983).

[38] J. Murray and R. Langer, *Cancer Drug Delivery* **1**, 119 (1984).

[39] R. Langer, W. Rhine, D. S. T. Hsieh, and J. Folkman, *J. Membr. Sci.* **7**, 333 (1980).

[40] D. S. T. Hsieh, R. Langer, and J. Folkman, *Proc. Natl. Acad. Sci. U.S.A.* **78**, 1863 (1981).

onto a leveled glass mold ($7 \times 7 \times 0.5$ cm) which had been previously cooled by placing it on dry ice for 5 min. The mold remained on the dry ice throughout the procedure. Immediately following the pouring of the polymer–drug mixture, magnetic steel beads (79.17% iron, 17% chromium, 1% carbon, 1% manganese, 1% silicone, 0.75% molybdenum, 0.4% phosphorus, and 0.4% sulfur, type 440C; 1.4 mm diameter; from Ultraspherics, Inc., Marie, MI) were placed onto the mixture using a loading device.

The loading device was made of one bacterial culture petri dish (Falcon 1001, Oxnard, CA) with the bottom sitting inside the inverted lid. Both the bottom and the lid have an identical arrangement of 263 holes (1.8 mm diameter) with 3-mm spacing. While the plates were shifted with respect to each other such that the upper and lower holes were offset, the upper holes were filled with magnetic beads. The plates with magnetic beads were positioned over the polymer slab in the mold. When the plates were shifted back so that all the holes were aligned, the magnetic beads dropped onto the polymer in a uniform array.

Thirty seconds after the magnetic beads were added, a top layer of polymer–drug mixture identical to the bottom layer was cast over the beads. After the entire mixture had solidified (approximately 10 min), the slab was transferred to a $-20°$ freezer for 48 hr.

The use of low-temperature casting and drying prevents migration of the drug powder. The use of a three-step procedure to embed the beads between two layers of partially fluid polymer–drug mixture provides precise vertical positioning of the beads while the device used to place the beads between polymer layers provides uniformity of orientation and position.

Method 10. To achieve near constant baseline release rates, a hemispheric magnetic system was formulated.[41] The molds used for fabricating hemispheric magnetic devices and the embedding platform for device formulation are those described in method 6.

The procedure for preparing the hemispheric magnetic system consists of four stages: (1) casting, (2) drying, (3) coating, and (4) opening a cavity.

CASTING. A layer of protein–polymer mixture was cast in a glass mold with a hemispheric bottom (13 mm o.d. $\times$ 11 mm i.d. $\times$ 11 mm height) placed on a block of dry ice. The protein loading was one-third of the mixture (w/w). A single hollow cylindrical Samarium Cobalt permanent magnet (3 mm o.d. $\times$ 1.5 mm i.d. $\times$ 1.5 mm height, obtained from

---

[41] D. S. T. Hsieh and R. Langer, *in* "Controlled Release of Bioactive Materials" (Z. Mansdorff and T. J. Roseman, eds.), p. 121. Dekker, New York, 1983.

Permag Northeast Co., Billerica, MA) was then placed in the center after 1 min. Another layer of protein–polymer mixture identical to the first layer was immediately cast onto it. As in method 9, control of the casting time is important to prevent the magnet from settling out and to assure that the two layers of protein–polymer mixture will adhere tightly.

DRYING. The hemispheric pellet was dried in a freezer ($-20°$) for 2 days followed by drying in a desiccator ($20°$) under a house-line vacuum (600 mtorr) for 2 days. The finished pellet was approximately 6 mm in height and 8 mm in diameter. Its composition was approximately 51.5 mg protein, 103.0 mg polymer, and 71.6 mg magnet.

COATING. This is done as described in method 6.

OPENING A CAVITY. This is done as described in method 6.

Magnetic Triggering System

To prevent collisions against the glass walls of the scintillation vials in *in vitro* studies, a poly(vinyl chloride) rod (World Plastics, Waltham, MA) was passed through the cap of the vial and the sample attached to its end with paraffin (Paraplast, Fischer Scientific). The paraffin was placed in a glass beaker and melted atop a lab hot plate at $60°$. Two drops were then applied to the end of the rod with a glass pipet, and after 15 sec the medium had cooled somewhat and the samples were pressed into it.

The application of an alternating magnetic field to "trigger" these samples was accomplished with a number of different devices designed to alternatively apply and withdraw a permanent magnet for the samples. Originally a commercial speed-controlled rocker was modified (Minarik Electric Co., Los Angeles, CA) by placing a permanent magnet bar (Crucore Magnet Bar, No. RE80108, Permag Northeast Co.) on one end of the rocker and a balanced weight on the other end. The frequency of the motion was 18 cycles/min. Thus, an oscillating magnetic field ranging from background levels (0.5 Gs) to approximately 1000 Gs (gauss) was created for triggering.

However, electric and mechanical noise, the limitations on oscillation frequency, and the inability to expose more than a few samples at a time to the same field dictated modifications. In a new device, vials were securely held in holes drilled through a 1/4-in. Plexiglas plate below which a second plate rotated as it was driven by direct motor drive. All samples were exposed in the same fashion to the field for magnets with surface field strength of 2.5 kGs mounted to their plate. A motor (hp $+70$, 8 lb. torque, 24 rpm) rotated this bottom disk at speeds regulated by a Variac electronic control. This system could accommodate 24 test vials.

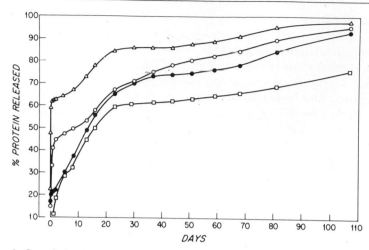

FIG. 3. Cumulative percentage released of proteins from coated ethylene–vinyl acetate copolymer pellets prepared using method 1. (△) Lysozyme; (○) soybean trypsin inhibitor; (●) alkaline phosphatase; (□) catalase.

Kinetic Studies

To examine release kinetics of pellets made using method 1, we incorporated different proteins into polymers and tested for the amount of protein released as a function of time. In one study,[4] the release of soybean trypsin inhibitor (MW 21,000) from coated Hydron, poly(vinyl alcohol), and ethylene–vinyl acetate copolymer pellets was examined. Casting solutions were 12% Hydron, 10% poly(vinyl alcohol), and 10% ethylene–vinyl acetate copolymer. These concentrations minimized initial release rates from each of the individual polymer systems and still permitted viscosities low enough so the solutions were easy to work with.

Protein diffused out of Hydron relatively rapidly, somewhat more slowly from poly(vinyl alcohol), and least rapidly from ethylene–vinyl acetate copolymer. A "burst" effect, in which a large percentage of protein was released during the first few hours of incubation, was evident in each of the slow-release systems.

In Fig. 3, release profiles from coated ethylene–vinyl acetate copolymer pellets for four different proteins ranging in molecular weight from 14,400 (lysozyme) to 250,000 (catalase) are shown. In every case, release continued for over 100 days. Additional studies verified that over 80% of the protein being released was in biochemically active form.[4] Kinetic profiles of pellets made using methods 1 and 2 are very similar.

Release kinetics from matrices prepared using method 3 were ex-

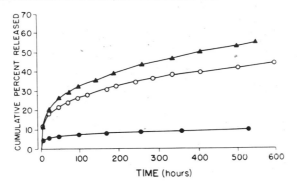

FIG. 4. Effect of protein particle size on the cumulative release of bovine serum albumin. The loading (protein content) is 25% by weight. Method 3 was used. (▲) 250–425 μm; (○) 75–250 μm; (●) <75 μm.

tremely reproducible. Three slabs were prepared with bovine serum albumin (BSA), each was cut into 9 1 × 1 cm squares, and all 27 squares were measured for release for more than 24 days. The mean daily release rates and their standard deviations were monitored. Standard deviations of release rates were generally within 10% of the respective means. Standard deviations of release rates from matrices prepared at room temperature were only within 75%.

Both drug particle size and loading significantly affect on release kinetics. Loading is defined as the percentage of matrix weight that is protein. Figure 4 shows an example of the effect of drug particle size on release rates. This graph is representative, and detailed studies have been conducted demonstrating the importance of both particle size and loading on release kinetics.[27] The use of larger particles or increases in drug loading increased release rates. Coating matrices prepared by method 3 also significantly affected drug release rates. Release rates decreased with increases in coating solution concentrations.

The most critical factor controlling release rates in matrices prepared by method 4 was the ratio of water to dichloromethane used during the casting procedure. In one set of experiments, either 1.0, 1.2, 1.4, or 1.6 ml of 5% BSA in water was added to 9 ml of ethylene–vinyl acetate copolymer solution in a 20-ml glass scintillation vial. Each was poured into a glass mold 4 × 4 × 0.5 cm. Complete drying for this volume took 20 days (drying time also affects release kinetics, and slabs dried for 2 rather than 20 days released BSA at a rate about five times more slowly); however, drying time could be considerably reduced using either smaller-volume preparations or higher vacuum drying procedures. Squares 1 × 1 cm were cut from the center of the dried slab and analyzed for release kinetics. As

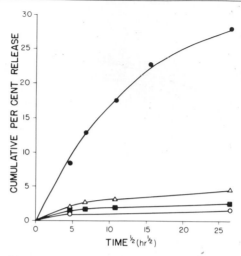

FIG. 5. Release of bovine serum albumin from matrices prepared by method 4 with differing volumes of aqueous protein. The changing parameter is the initial percentage of aqueous volume. Each point represents the mean of 16 samples. (●) 15.1%; (△) 13.5%; (■) 11.8%; (○) 10%.

shown in Fig. 5, greater quantities of water in the casting solution increased release rates. Reproducibility using method 4 is good, the release rates from different slabs consistently being within 10% of one another.

Release kinetics were investigated for slabs cast at 10.5 MPa with polymer powder granules according to method 5.[29] Both the powder type and aging condition affect the rate of drug release. Specifically, powder type II matrices release drug rapidly. Aged powder type I matrices release drug at an intermediate rate, and fresh powder type I matrices release drug at the slowest rate. It was found, in general, that formation pressure and polymer powder size distribution have small effects on release kinetics using method 5.

Additionally, release kinetics of albumin from the hemisphere-shaped systems prepared using method 6 were studied. Each time point represented the mean of eight samples. A linear relationship between cumulative percentage release and time was observed for 60 days.[33]

Typical release kinetics from ethylene–vinyl acetate copolymer microspheres (method 7) are shown in Fig. 6A as a plot of cumulative release against time. Figure 6B replots the cumulative percentage release data against $t^{1/2}/r$ where r is the average radius for the spheres, thereby normalizing the data for the variability in sphere size.

The presence of a carrier protein such as albumin (method 8) facilitates the release of other macromolecules from polymer squares (Fig. 7).

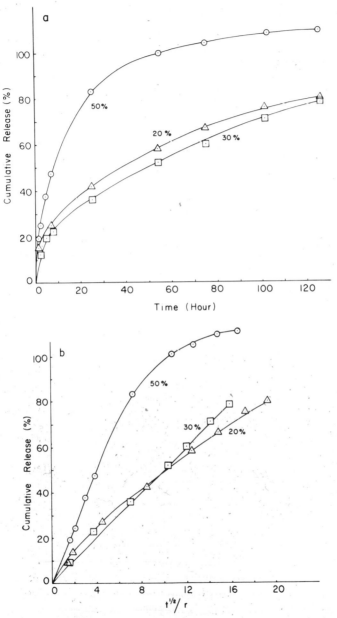

FIG. 6. Cumulative release of albumin from ethylene–vinyl acetate microspheres made using method 7. Values are the mean of two measurements; release value were within ±10% of the mean or better. (a) Release vs time; (b) release vs $t^{1/2}/r$ ($hr^{1/2}/mm$) to account for bead size variation with loading.

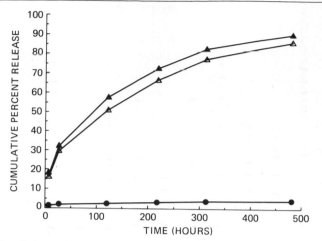

Fig. 7. Cumulative percentage macromolecule released as a function of time (mean of 6 samples in each case). Method 8 was used. (▲), Inulin released from polymer square containing 39% by weight bovine serum albumin (BSA); (△), BSA released from the same polymer square; (●), inulin released from polymer square containing no BSA.

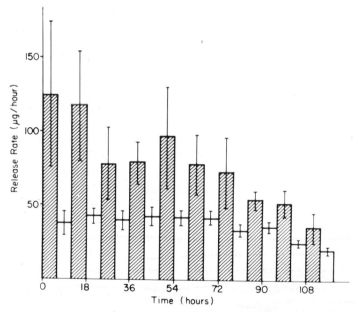

Fig. 8. Modulated sustained release of bovine serum albumin (BSA) from polymeric vehicles by magnetism. The systems were made via method 9. Each histogram represents the average release rate of bovine serum albumin from eight polymer squares (1 cm × 1 cm). There was a 72-hr prerelease period in this experiment. (▨) Moving magnet; (□) no magnet.

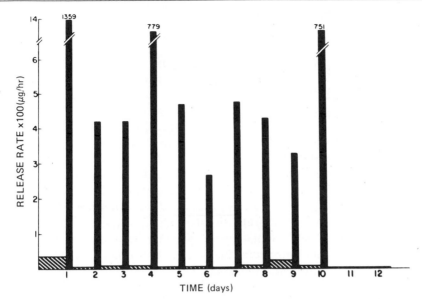

FIG. 9. Magnetic modulation of bovine serum albumin released from a hemisphere pellet. Before triggering, the pellets were preleased for 1 day. The pellets were triggered for 5 hr, followed by nontriggering for 19 hr. This cycle of triggering and nontriggering was repeated daily. The systems were made via method 10. (■) Quantity of bovine serum albumin (BSA) released during magnetic modulation period; (▨) quantity of BSA released during nontriggering period.

In an experiment using inulin as a model macromolecule, when no albumin was present, only 5% or about 5 μg of the incorporated inulin was released by the end of a 3-week experimental period. In contrast, when albumin was present as 39% of the polymer square's weight, 90% of the incorporated inulin (31.5 μg) was released during the same time period. Furthermore, the percentage of incorporated inulin that was released at each time point was not significantly different from the percentage of incorporated BSA that was released at the same time point.[38] Similar results have been achieved using albumin as a carrier to enable the release of picogram quantities of epidermal growth factor for up to 3 weeks.[37]

When exposed to the oscillating magnetic field, magnetic polymer slabs (method 9) released up to 100% more drug than when the magnetic field was discontinued (Fig. 8). For example, the first 6-hr exposure period showed an average release rate of 125 μg/hr compared to 40 μg/hr in the following 6 hr of no exposure. The differential decreased with increasing time but was still significant at the end of the experiment (37 μg/hr versus 25 μg/hr). Numerous controls were conducted and in no case was magnetic modulation observed, including a control in which a constant 1000-Gs magnetic field was maintained through the entire triggering interval.[40]

The hemispheric systems (method 10) were also exposed to the magnetic field. Bovine serum albumin was used as a model compound to demonstrate the feasibility of magnetic modulation with this design. The results showed that the average baseline release rate was 14.8 μg/hr, and the average burst of modulated release was 424.1 μg/hr (Fig. 9). This represents a 29-fold increase.[41] The systems could deliver drugs for 1 month and the burst of release could be repeatedly modulated when desired. Other simulated test drugs such as pepsin and insulin also showed modulated release when exposed to the magnetic field. Control hemisphere pellets which were not exposed to the magnetic field delivered drugs at a constant baseline release rate without significant fluctuation.

Both magnetic and nonmagnetic polymer systems appear to be biocompatible. When implanted into the rabbit cornea or subcutaneous tissue, no inflammation was observed.[40,42,43] In addition, identically formulated ethylene–vinyl acetate copolymer slabs containing test macromolecules displayed identical *in vitro* and *in vivo* release rates.[43]

Discussion

The basis for the slow release of macromolecules through these polymers appears to be diffusion through a series of interconnecting macrochannels in the polymer matrix. These channels do not normally exist within the polymer but are caused by the incorporation of solid or liquid in the matrix during the casting procedure. Thus, factors that increase the size of these channels or provide simpler pathways (lower tortuosity) for diffusion out of the matrix would be expected to increase release rates. This would explain the effect of particle size and loading in increasing release rates in matrices prepared by method 3 and the effect of increased aqueous volume in accelerating release rates of method 4 matrices. For the same reason, coating the matrix would reduce surface access of the pores and therefore decrease release rates. The choice of polymers is also important, since hydrophobic polymers, such as ethylene–vinyl acetate copolymer, release macromolecules more slowly than hydrogels, such as Hydron or poly(vinyl alcohol). Another factor important in determining release rates is the macromolecule's aqueous solubility. The ability to have high solution concentration within the matrix channels provides a high driving force for diffusion and hence faster release rates. The molec-

[42] R. Langer, H. Brem, and D. Tapper, *J. Biomed. Mater. Res.* **15**, 267 (1981).
[43] L. Brown, C. Wei, and R. Langer, *J. Pharm. Sci.* **72**, 1181 (1983).

ular weight of the macromolecule itself is of only minor importance in determining release rates. The advantage of having different methods available for producing controlled-release systems for macromolecules enables one to achieve desired characteristics, e.g., constant release, no exposure to solvent, convenience of formulations, or high reproducibility.

While the current polymer preparations successfully release macromolecules for prolonged periods of time without significant tissue damage, recent studies have suggested that alternative methods of controlled release may also be useful. Particularly promising in this regard is the prospect of biodegradable polymers (for reviews, refs. see 44, 45), which release drugs by erosion rather than diffusion and are absorbed by the host animal after release has occurred. A current focus in our laboratory is the synthesis and use of polyanhydrides to release such molecules.[46]

The mechanism of magnetic modulation is not known. One might speculate that the beads cause alternating compression and expansion of the matrix pores, thereby squeezing more drug out in the presence of the magnetic field. The major problem encountered with the magnetic systems thus far is irreproducibility, with respect to differences in release rates both among different pellets within a given triggering period and among the same pellets within different triggering periods. These irreproducibilities are particularly significant in the case of the hemisphere. Possible causes leading to these irreproducibilities may result from the techniques used to open the cavity on the flat surface of the hemispheric magnetic pellets and from the inconsistencies in manipulations conducted during exposure to the alternating magnetic field. Computer-aided drilling may provide a more reproducible method of opening the cavity, while an electromagnet may allow for better control over the applied magnetic field.

Modulated delivery systems controlled by external means may ultimately improve the release pattern of certain drugs such as insulin. For example, studies have shown that excellent control of blood glucose can be achieved using pumps that infuse insulin at a constant rate and are capable of increasing the rate before meals. We have recently shown that a pellet of ethylene–vinyl acetate copolymer containing insulin, without magnetic particles, will bring about normoglycemia in diabetic rats for nearly 4 months after a single implant.[47] These pellets, though at a much earlier stage of development than pumps, contain 1000-fold greater quan-

[44] R. Langer and N. Peppas, *Biomaterials* **2**, 195 (1981).
[45] R. Langer and N. Peppas, *J. Macromol. Sci.* **23**, 61 (1983).
[46] H. Rosen, J. Chang, G. Wnek, R. Linhardt, and R. Langer, *Biomaterials* **4**, 131 (1982).
[47] L. Brown, C. Munoz, L. Siemer, E. Edelman, J. Kost, and R. Langer, *Diabetes* **32** Suppl. 1, 35a (1983).

tities of insulin per unit volume (at 50% loading by weight a 10-cm^3 pellet contains a 10-year supply of insulin for a human) and contain no moving parts. One might speculate that with further development, insulin pellets might be implanted subcutaneously (e.g., wrist area) and triggered to release more drug by a watchlike device worn on the wrist. Modulated release systems triggered by magnetism or other means may also be useful for delivering hormones for birth control and other therapies.

[31] Poly(ortho ester) Biodegradable Polymer Systems

By J. HELLER and K. J. HIMMELSTEIN

The importance of controlled systemic drug delivery from bioerodible implants is well established, and several bioerodible drug delivery systems are under active development.[1] Our approach to developing such systems is to prepare hydrolytically labile polymers into which a drug can be physically dispersed under relatively mild conditions and in which release of the drug is controlled by the hydrolytic erosion of the matrix. We are concentrating on devices in which the hydrolytic erosion process is confined to the surface of a solid device. In such devices, and for a constant rate of polymer hydrolysis, rate of drug release is directly proportional to drug loading, and lifetime of the device is directly proportional to the physical dimensions of the device. However, because rate of drug release is directly proportional to the total surface area of the device, as the physical dimensions of the device decrease because of the erosion process, rate of drug release will also decrease. The decrease in the rate of drug release is predictable and can be calculated from the following relationship[2]

$$M_t/M_\infty = 1 - [1 - k_0t/C_0a]^n$$

where M_t is drug released at time t; M_∞ is drug released at device exhaustion; k_0 is an erosion constant; C_0 is the uniform initial drug concentration in the matrix; a is the radius of sphere or cylinder, or half-thickness of the slab; and n is 1 for a slab, 2 for a cylinder, and 3 for a sphere.

There are two possible approaches to the achievement of controlled surface erosion. In one approach, the interior of the matrix is stabilized so

---

[1] J. Heller, *CRC Crit. Rev. Ther. Drug Carrier Syst.* **1**, 39 (1984).

[2] H. B. Hopfenberg, in "Controlled Release Polymeric Formulations" (D. R. Paul and F. W. Harris, eds.), Chapter 3. Am. Chem. Soc., Washington, D.C., 1976.

that only the outer layers can erode. The other approach is to develop a highly hydrophobic polymer that contains linkages in the polymer backbone whose rates of hydrolysis are pH dependent and to incorporate into the polymer excipients that, in contact with water, produce a pH that induces the desired polymer hydrolysis rate. Because the polymer is highly hydrophobic, only the excipient in the surface layers is exposed to water and polymer hydrolysis occurs only in the surface layers.

One such polymer system is the poly(ortho esters), and this chapter describes our work on polymer synthesis and use of various excipients to achieve controlled drug release of various therapeutic agents physically dispersed in these polymers.

Polymer Synthesis

Poly(ortho esters) were first disclosed in a series of patents assigned to the ALZA Corporation.[3-6] They were prepared by the following transesterification reaction:

$$\text{EtO} \quad \text{OEt} + \text{HO—R—OH} \rightarrow \left[\begin{array}{c} O \quad O\text{—R} \\ O \end{array} \right]_n + \text{EtOH}$$

To achieve reasonable molecular weights, the reaction mixture was heated to 100–115° for 1.5–2 hr and then further heated at 180° and 0.01 torr for 24 hr to shift the equilibrium of the transesterification reaction to the right by removal of the evolved ethanol. Different R groups can be used, and the resultant polymer family was initially known as Chronomer and currently is known as Alzamer.

More recently, another family of poly(ortho esters) not related to the Alzamer system has been developed at SRI International. These polymers are prepared by adding diols to diketene acetals,[7] which is shown schematically as follows:

$$\underset{\substack{| \\ CH_2=C—O—R'—O—C=CH_2}}{\overset{OR \qquad\qquad OR}{}} + \text{HO—R''—OH} \rightarrow \left[O—\underset{\substack{| \\ CH_3}}{\overset{\substack{OR \\ |}}{C}}—O—R'—O—\underset{\substack{| \\ CH_3}}{\overset{\substack{OR \\ |}}{C}}—O—R'' \right]_n$$

[3] N. S. Choi and J. Heller, U.S. Patent 4,093,709 (1978).
[4] N. S. Choi and J. Heller, U.S. Patent 4,131,648 (1978).
[5] N. S. Choi and J. Heller, U.S. Patent 4,138,344 (1979).
[6] N. S. Choi and J. Heller, U.S. Patent 4,180,646 (1979).
[7] J. Heller, D. W. H. Penhale, and R. F. Helwing, *J. Polym. Sci., Polym. Lett. Ed.* **18**, 619 (1980).

Unlike the transesterification reaction, which requires long reaction times at high temperatures and vacuum, the addition reaction is conducted at atmospheric pressure and without external heating. It procccds either spontaneously or can be catalyzed by traces of acid. The reaction is exothermic and proceeds to completion virtually instantaneously. Because no small molecule by-products are evolved, dense, cross-linked matrices can be produced by using varying proportions of monomers having a functionality greater than two.

Because of difficulties arising from a transesterification reaction between a poly(ortho ester) and unreacted diol leading to cross-linking, it is necessary to use a cyclic diketene acetal; and the only cyclic diketene acetal described in the literature is 3,9-bis(methylene-2,4,8,10-tetraoxaspiro[5,5]undecane) prepared by a dehydrochlorination reaction as follows[8,9]:

(I)

Although the dehydrochlorination reaction proceeds readily, the diketene acetal (I) is difficult to handle because it contains two electron donor groups on a methylenic double bond and is thus highly susceptible to cationic polymerization. Furthermore, special precautions must be observed during the acid-catalyzed polymer formation with diols to prevent a competing cationic polymerization of the diketene acetal, leading to cross-linking. Even though the purification of thc diketene acetal is a serious problem, poly(ortho esters) have been successfully synthesized using iodine in pyridine as the catalyst system.[7]

A more satisfactory monomer is one in which the cationic polymerization is sterically inhibited by replacing a hydrogen with a methyl group. Such a monomer is readily prepared by a base-catalyzed rearrangement of the commercially available precursor:

(II)

Monomer (II), 3,9-bis(ethylidene-2,4,8,10-tetraoxaspiro[5,5]undecane), is relatively stable and can be easily purified; poly(ortho esters) made from a

[8] B. G. Yasnitskii, S. A. Sarkisyants, and E. G. Ivanyuk, *Zh. Obshch. Khim.* **34,** 1940 (1964).

[9] F. V. Zalar, *Macromolecules* **5,** 539 (1972).

variety of diols can be easily prepared using acidic catalysts, without problems due to competing cationic polymerizations.

Polymers formed by the addition of diols to monomer (I) or (II) have the following structure:

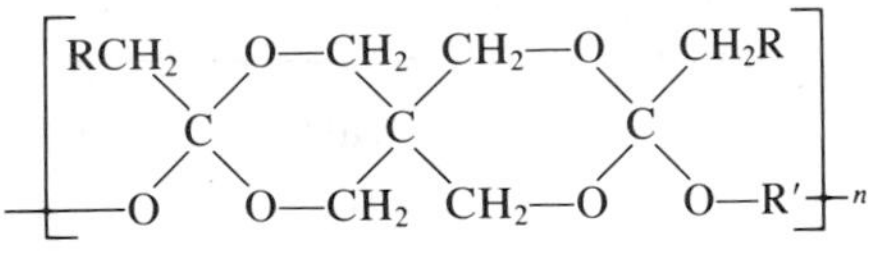

$$\left[\begin{array}{ccc} RCH_2 & O—CH_2\quad CH_2—O & CH_2R \\ \diagdown C\diagup & \diagdown C\diagup & \diagdown C\diagup \\ O & O—CH_2\quad CH_2—O & O—R' \end{array}\right]_n$$

where R = H for monomer (I) and R = CH_3 for monomer (II). The structure of the polymers has been verified by ^{13}C NMR spectroscopy,[7] and the spectrum of a polymer formed from monomer (I) and 1,6-hexanediol is shown in Fig. 1.

Polymer Hydrolysis

As expected, the polymer first hydrolyzes to the mono- and diesters [acetates for polymer based on monomer (I) and propionates for polymer based on monomer (II)] of pentaerythritol and the diol HO—R'—OH. These mono- and diesters then hydrolyze further to acetic or propionic acids and the corresponding alcohols.

Drug Release Studies

Use of Basic or Neutral Water-Soluble Excipients

Because poly(ortho esters) are stable in base and become progressively more labile as the pH of the surrounding medium decreases, devices containing dispersed drug and the basic salt sodium carbonate were expected to hydrolyze and concomitantly release their incorporated drug. The drug release was expected to occur only at the outer surface of the device, where the basic salt is neutralized by the external buffer.[10–13]

In such experiments, disk-shaped devices containing dispersed 10 wt% norethindrone and 10 wt% Na_2CO_3 exhibited excellent zero-order kinetics for long periods, and, as shown in Fig. 2, release was linear for 240 days, at which time the experiment was discontinued. However, at

[10] J. Heller, D. W. H. Penhale, R. F. Helwing, B. K. Fritzinger, and R. W. Baker, *AIChE Symp. Ser.* **77**, No. 206, 28 (1981).

[11] J. Heller, D. W. H. Penhale, R. F. Helwing, and B. K. Fritzinger, *Polym. Eng. Sci.* **21**, 727 (1981).

[12] J. Heller, D. W. H. Penhale, R. F. Helwing, and B. K. Fritzinger, *in* "Controlled Release Delivery Systems" (T. J. Roseman and S. Z. Mansdorf, eds.), p. 91. Dekker, New York, 1983.

[13] J. Heller, D. W. H. Penhale, B. K. Fritzinger, J. E. Rose, and R. F. Helwing, *Contracept. Delivery Syst.* **4**, 43 (1983).

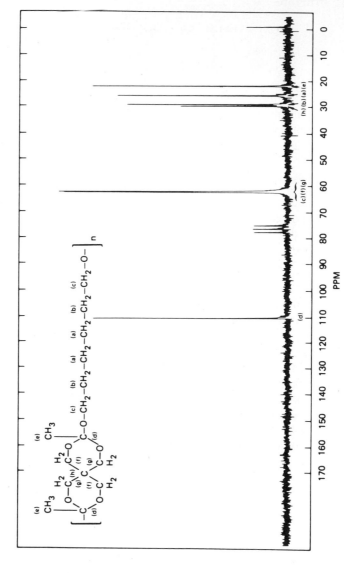

Fig. 1. 25.5 MHz ^{13}C NMR spectrum of 3,9-bis(methylene-2,4,8,10-tetraoxaspiro[5,5]undecane)/1,6-hexanediol polymer in CDCl$_3$ at room temperature. (From Heller *et al.*[7])

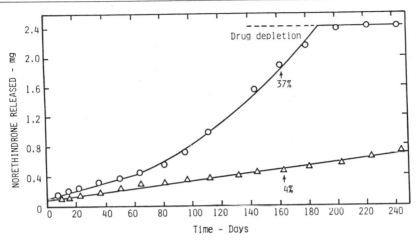

FIG. 2. Norethindrone release from 3,9-bis(methylene)-2,4,8,10-tetraoxaspiro[5,5]undecane/1,6-hexanediol poly(ortho ester), 6.3 mm-diameter disks at pH 7.4 and 37°. (○) 10 wt% norethindrone, 10 wt% Na_2SO_4; disk thickness, 0.6 mm; total drug content, 2.4 mg. (△) 10 wt% norethindrone; 10 wt% Na_2SO_3; disk thickness, 1.2 mm; total drug content, 4.0 mg. Numbers below arrows indicate total weight loss. (From Heller *et al.*[13])

day 160, only 4% weight loss was measured. Because at that point about 13% of the drug had been released, the rate of drug release was clearly not controlled by polymer erosion and was instead controlled by an osmotic imbibing of water driven by the incorporated water-soluble Na_2CO_3.[14] The linearity of drug release was initially attributed to a uniform movement of a swelling front and release of the drug from the swollen layer.[15] However, it now appears more likely that the movement of a swelling front is not the dominant release mechanism but that instead the kinetics of release are principally determined by the rate of solubilization of the highly insoluble steroid. Because poly(ortho esters) are stable in base, the polymer swells without erosion.

Even though these studies showed that no erosion took place, when all Na_2CO_3 has been leached out of the device, erosion of the matrix should ultimately take place. Therefore, such devices could be used as implants where drug release is determined by controlled release from the swollen matrix and where the drug- and excipient-depleted polymer will ultimately bioerode. Consequently, an *in vivo* study was performed in which rod-shaped devices were implanted in rabbits and the drug plasma level was measured by radioimmunoassay. The results of that study are shown

[14] R. F. Fedors, *Polymer* **21**, 207 (1980).
[15] H. B. Hopfenberg and K. C. Hsu, *Polym. Eng. Sci.* **18**, 1186 (1978).

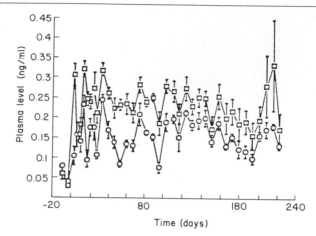

Fig. 3. Daily rabbit blood plasma levels of levonorgestrel from a 3,9-bis(ethylidene 2,4,8,10-tetraoxaspiro[5,5]undecane)/(70/30) *trans*-cyclohexanedimethanol/1,6-hexanediol polymer containing 30 wt% drug and 10 wt% Na_2CO_3; rods, 2 × 10 mm. (○) 1 device per rabbit; (□) 2 devices per rabbit.

in Fig. 3. The study was discontinued before drug depletion occurred, and no weight-loss data of the devices were determined. Despite the scatter, which was partially due to the radioimmunoassay method, reasonably constant blood levels were maintained for about 8 months.[1,16]

Several studies in which the basic Na_2CO_3 salts were replaced with the neutral salts Na_2SO_4 or NaCl were also performed, and the results of drug release studies of a device containing 10 wt% norethindrone and 10 wt% Na_2SO_4 are shown in Fig. 2. These results are significantly different from those obtained with Na_2CO_3 and have been rationalized as the combined effect of polymer erosion and osmotically driven swelling.[12,13]

Use of Acidic Excipients

Even though reasonably linear drug release kinetics were achieved with devices that contain incorporated Na_2CO_3, the use of this compound is undesirable for several reasons. Even though the polymer should bioerode after the depletion of the additive, this methodology relies on polymer swelling and drug diffusion and does not satisfy the original design criteria of drug release control by polymer erosion. Furthermore, because Na_2CO_3 is a highly basic salt, a localized high pH might lead to

[16] J. Heller, D. W. H. Penhale, B. K. Fritzinger, and S. Y. Ng, *in* "Long Active Contraceptive Delivery Systems" (G. I. Zatuchni, A. Goldsmith, J. D. Shelton, and J. Sciarra, eds.), p. 113. Harper, New York, 1984.

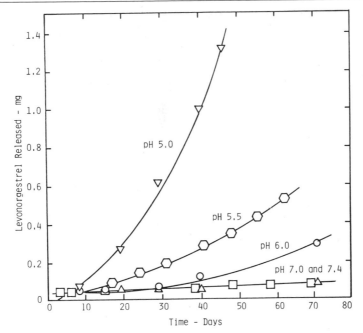

FIG. 4. Levonorgestrel release from a 3,9-bis(ethylidene-2,4,8,10-tetraoxaspiro[5,5]un-decane)/1,6-hexanediol poly(ortho ester) containing 30 wt% drug at various external buffer pH values and 37°. (From Heller *et al.*[13])

local tissue irritation. Use of osmotically active neutral salts such as Na_2SO_4 or NaCl also relies on polymer swelling and drug diffusion and could also lead to tissue irritation because of localized osmolarity problems.

For these reasons, a better approach is to use excipients that can lower the pH at the polymer–water interface relative to that in the interior of the matrix. The sensitivity of the polymer to an acidic environment was ascertained by measuring the erosion of the polymer as the rate of release of incorporated levonorgestrel. The results of these studies are shown in Fig. 4. These data show that at the physiological pH of 7.4 and 37° polymer erosion is very slow but that a relatively modest lowering of the external pH results in a significant increase in the polymer erosion rate.[13,16]

Use of Calcium Lactate

The release of levonorgestrel catalyzed by varying amounts of the low water solubility, slightly acidic salt calcium lactate is shown in Fig. 5.[16]

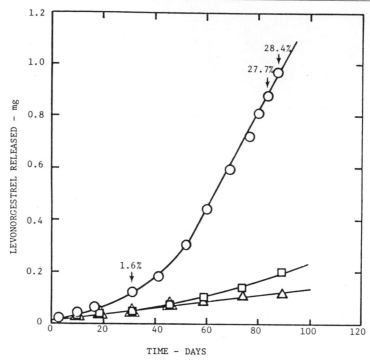

FIG. 5. Cumulative release of levonorgestrel from 3,9-bis(ethylidene-2,4,8,10-tetraoxa-spiro[5,5]undecane)/(70/30) *trans*-cyclohexanedimethanol/1,6-hexanediol polymer containing 30 wt% levonorgestrel and varying amounts of calcium lactate. Total drug, 8.5 mg; rods, 1.6 × 15 mm; pH 7.4 and 37°. Numbers above arrows indicate total weight loss. (○) 2 wt% Ca-lactate; (□) 0.5 wt% Ca-lactate; (△) no Ca-lactate.

Thus, the addition of 0.5 wt% calcium lactate has a very slight accelerating effect, but the addition of 2 wt% has a significant effect. As can be seen from the weight-loss data shown above the arrows in Fig. 5, polymer erosion significantly leads drug release.

Cumulative and daily *in vitro* release of levonorgestrel shown in Figs. 6 and 7 indicate that levonorgestrel is released at a reasonably constant rate.[17] However, a comparison of total weight loss indicated by the numbers above the arrows and percentage drug released again indicates that erosion significantly leads drug release. These results indicate that release by polymer erosion occurs faster than the rate at which the highly insoluble levonorgestrel can solubilize, and release kinetics become controlled by the maximum rate at which levonorgestrel can solubilize.

[17] J. Heller and B. K. Fritzinger, *J. Controlled Release* (in press).

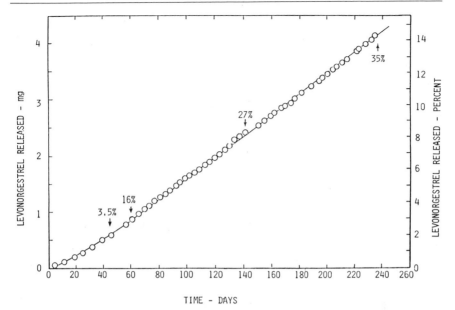

FIG. 6. Cumulative release of levonorgestrel from a 3,9-bis(ethylidene-2,4,8,10-tetraoxaspiro[5,5]undecane)/(60/40) *trans*-cyclohexanedimethanol 1,6-hexanediol polymer containing 30 wt% levonorgestrel and 2 wt% calcium lactate. Total drug, 29.0 mg; rods, 2.4 × 20 mm; pH 7.4 and 37°. Numbers above arrows indicate total weight loss.

Figure 8 shows levonorgestrel blood plasma levels measured by radioimmunoassay of rabbits with implanted cylindrical devices identical to those used in the *in vitro* study. Again, reasonably constant blood levels for many months have been achieved.[16,17]

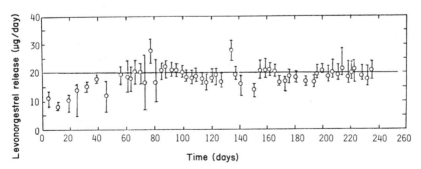

FIG. 7. Daily release of levonorgestrel from a 3,9-bis(ethylidene-2,4,8,10-tetraoxaspiro[5,5]undecane)/(60/40) *trans*-cyclohexanedimethanol/1,6-hexanediol polymer containing 30 wt% levonorgestrel and 2 wt% calcium lactate. Total drug, 29.0 mg; rods, 2.4 × 20 mm; pH 7.4 and 37°.

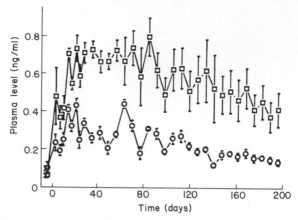

FIG. 8. Daily rabbit blood plasma levels of levonorgestrel from a 3,9-bis(ethylidene 2,4,8,10-tetraoxaspiro[5,5]undecane)/(60/40) *trans*-cyclohexanedimethanol/1,6-hexanediol polymer containing 30 wt% levonorgestrel and 2 wt% Ca-lactate. Rods 2.4 × 20 mm. (○) 1 device per rabbit; (□) 3 devices per rabbit.

Use of Anhydrides

The catalytic action of acid anhydrides depends on reaction with water to yield a diacid, which then catalyzes hydrolysis of the matrix. Because the poly(ortho esters) used in this study are highly hydrophobic, only the

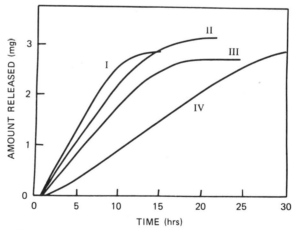

FIG. 9. Cumulative release of timolol maleate from a 7/3 blend of 3,9-bis(ethylidene 2,4,8,10-tetraoxaspiro[5,5]undecane)/1,6-hexanediol and 3,9-bis(ethylidene 2,4,8,10-tetra-oxaspiro[5,5]undecane)*trans*-cyclohexanedimethanol polymers containing 2 wt% timolol maleate and varying amounts of maleic anhydride. Disks about 0.8 × 10 mm at pH 7.4 and 37°. Maleic anhydride concentrations: I, 0.5%; II, 0.4%; III, 0.3%; IV, 0.2%.

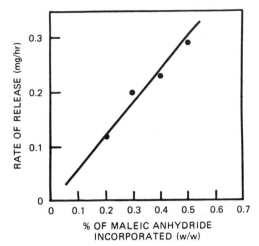

FIG. 10. Release rate of timolol maleate as a function of maleic anhydride concentration. Data taken from Fig. 9.

anhydride in the surface layers is converted to the diacid, so that the catalyzed erosion is essentially confined to a layer at the surface of the device. Additionally, because the erosion is taking place at only the surface, the acid concentration at the erosion site does not vary appreciably with time. Therefore, the system is not autocatalytic and erosion does not accelerate except during the lag-phase at the onset of delivery.

Polymer erosion, and hence rate of drug release, can be controlled by the amount of incorporated anhydride, and Fig. 9 shows release of timolol maleate from thin polymer disks for differing amounts of maleic anhydride. The terminal plateaus in Fig. 9 correspond to greater than 99% release of the contained drug. The variable plateau levels are due to slight variations in device thickness. As shown in Fig. 10, the steady-state zero-order rate of release of timolol maleate increases linearly with maleic anhydride concentration.

Figure 11 shows rate of release of a marker dye physically incorporated into the polymer and appearance of one of the polymer degradation products, pentaerythritol dipropionate.[18] Because appearance of both compounds is closely coupled, the release mechanism is polymer erosion and not diffusion.

A further verification of a surface erosion process can be found in Fig. 12, which shows duration of drug release as a function of device thick-

[18] C. Shih, T. Higuchi, and K. J. Himmelstein, *Biomateials (Guildford, Engl.)* **5,** 237 (1984).

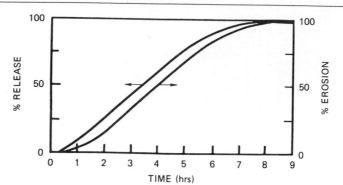

FIG 11. Polymer erosion and methylene blue released from a 3,9-bis(ethylidene 2,4,8,10-tetraoxaspiro[5,5]undecane)/ethylene glycol polymer containing 0.1 wt% methylene blue and 4 wt% phthalic anhydride. Disks 0.8 × 10 mm; pH 7.4 and 37°. Drug release measured spectrophotometrically, erosion measured by HPLC analysis of pentaerythritol dipropionate degradation product.

ness.[19] As predicted by a surface erosion process, duration of release is directly proportional to device thickness.

Because rate of hydrolysis of poly(ortho esters) increases as the pH of the surrounding medium decreases, rate of polymer erosion should be sensitive to the pK_a of the incorporated anhydride. Figure 13 demonstrates that dependence on acid strength and amount.[19] Over a given range, the release rate is proportional to the amount of anhydride incorporated. Above a particular concentration, increasing amounts of anhydride do not further accelerate the release rate, probably because diffusion of water into the reaction zone becomes rate limiting.

Other factors that affect changes in the release rate are molecular weight and glass transition temperature.[20] In general, as the molecular weight increases, the concomitant zero-order release rate decreases, and as the glass transition temperatures falls below the temperature at which erosion is occurring, release rate increases dramatically.

Conclusions

Poly(ortho ester) bioerodible polymers are suitable materials for the systemic or topical administration of a wide variety of therapeutic agents, and varying the nature and amounts of excipients physically incorporated

[19] R. V. Sparer, C. Shih, C. Gruen, and K. J. Himmelstein, *Proc. Int. Symp. Controlled Release, 10th,* p. 37 (1983).

[20] R. V. Sparer, C. Shih, C. Ringeisen, and K. J. Himmelstein, *J. Controlled Release* **1,** 23 (1984).

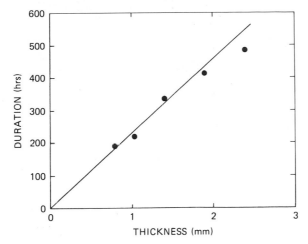

FIG. 12. Effect of disk thickness on duration of methylene blue release from a 3,9-bis(ethylidene 2,4,8,10-tetraoxaspiro[5,5]undecane)/(50/50) *trans*-cyclohexanedimethanol/ 1,6-hexanediol polymer containing 4 wt% methylene blue and 0.2 wt% poly(sebasic anhydride). Disks, 10 mm diameter of varying thickness; pH 7.4 and 37°.

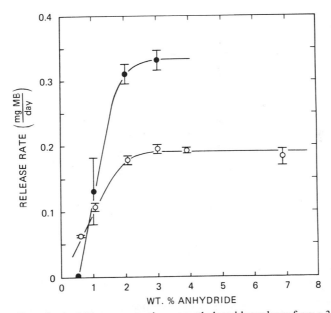

FIG. 13. Effect of anhydride concentration on methylene blue release from a 3,9-bis(ethylidene 2,4,8,10-tetraoxaspiro[5,5]undecane)/(35/65) *trans*-cyclohexanedimethanol/1,6-hexanediol polymer containing 0.2 wt% methylene blue and varying amounts anhydrides. Disks, 0.8 × 10 mm; pH 7.4 and 37°.

into the polymer will vary the erosion rates from a few hours to many months. With this range of control, delivery systems for short-term applications such as oral 12- to 24-hr tablets, intermediate ophthalmic products lasting 1 week, or implants lasting as long as 1 year can be produced.

Although no extensive toxicological studies have as yet been performed, *in vivo* release rate experiments performed over many months have not shown any adverse tissue reactions beyond the expected encapsulation.

Acknowledgment

A portion of the work reported here was supported by Contraceptive Development Branch of the National Institute of Child Health and Human Development under Contract 1-HD-7-2826. Work with acid anhydrides was performed at the Interx Corporation, a division of Merck, Sharp and Dome.

[32] Poly(lactic/glycolic acid) Biodegradable Drug–Polymer Matrix Systems

By JUDITH P. KITCHELL and DONALD L. WISE

Introduction

Poly(lactic/glycolic acid) Polymers

Poly(lactic acid), poly(glycolic acid), and the poly(lactic/glycolic acid) polymers (PLGA) have been used for many years as surgical sutures because they are biodegradable, biocompatible, and physically strong. These same features make PLGA desirable for use in other medical applications. Over the past decade, PLGA systems have been developed for delivery of active compounds such as proteins and pharmaceutical products at controlled release rates.

The physical characteristics of PLGA include strength, hydrophobicity, and pliability. The polymer is water insoluble; however, it is reduced by hydrolysis to the water-soluble monomers: lactic acid and glycolic acid. The structures of these acids are shown in Fig. 1. PLGA is miscible with a wide variety of active compounds, and solid formulations of polymer with active compound may be prepared. In an aqueous environment, the solid drug–polymer formulations slowly release active compound and

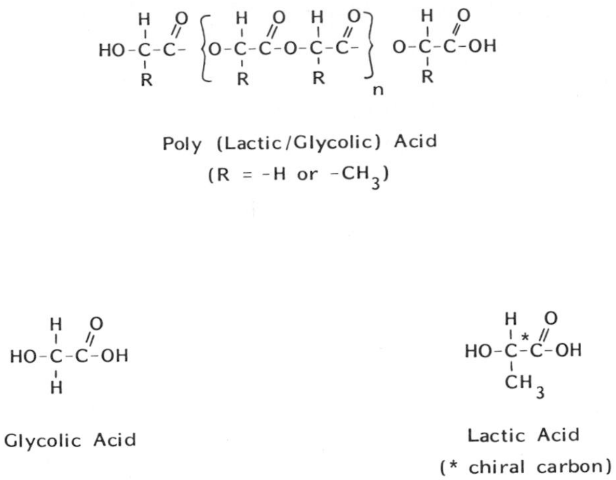

$$\text{HO}-\overset{\overset{\text{H}}{|}}{\underset{\underset{\text{R}}{|}}{\text{C}}}-\overset{\overset{\text{O}}{\|}}{\text{C}}-\left\{\text{O}-\overset{\overset{\text{H}}{|}}{\underset{\underset{\text{R}}{|}}{\text{C}}}-\overset{\overset{\text{O}}{\|}}{\text{C}}-\text{O}-\overset{\overset{\text{H}}{|}}{\underset{\underset{\text{R}}{|}}{\text{C}}}-\overset{\overset{\text{O}}{\|}}{\text{C}}-\right\}_n \text{O}-\overset{\overset{\text{H}}{|}}{\underset{\underset{\text{R}}{|}}{\text{C}}}-\overset{\overset{\text{O}}{\|}}{\text{C}}-\text{OH}$$

Poly (Lactic/Glycolic) Acid

(R = -H or -CH$_3$)

$$\text{HO}-\overset{\overset{\text{H}}{|}}{\underset{\underset{\text{H}}{|}}{\text{C}}}-\overset{\overset{\text{O}}{\|}}{\text{C}}-\text{OH}$$

Glycolic Acid

$$\text{HO}-\overset{\overset{\text{H}}{|}}{\underset{\underset{\text{CH}_3}{|}}{\text{C}}}\overset{*}{-}\overset{\overset{\text{O}}{\|}}{\text{C}}-\text{OH}$$

Lactic Acid

(* chiral carbon)

FIG. 1. The chemical structures of poly(lactic/glycolic acid), glycolic acid, and lactic acid.

lactic and glycolic acids. Wise *et al.* have recently reviewed the medical uses of PLGA.[1]

Controlled Release of Biologically Active Compounds from PLGA

Many drug–PLGA systems for long-term maintenance of therapeutic drug levels have been developed. Systems have been prepared for delivery of a variety of active compounds with durations of delivery varying from days to years. Drug–PLGA systems have been prepared in several forms, e.g., drug encapsulation, drug–polymer matrix, and encapsulated matrix. Only matrix-type systems will be discussed in this chapter. Drug delivery devices are characterized, not only by chemical composition, but by physical features such as size, shape, density, and porosity. The latter properties are results of manufacturing procedures.

PLGA controlled-release systems may be introduced orally or subcutaneously. The predominant application has been subcutaneous injectable systems in the form of suspended powders, thin rods, or small beads. Such devices have been made for release of fertility-regulating drugs, narcotic antagonists, antimalarials, and medication for prevention of alcohol abuse.[2-9] Bolus devices, placed in the bovine rumen for prevention of

[1] D. L. Wise, T. D. Fellmann, J. E. Sanderson, and R. L. Wentworth, *in* "Drug Carriers in Biology and Medicine" (G. Gregoriadis, ed.), pp. 237–270. Academic Press, New York, 1979.

[2] J. P. Kitchell, S. C. Crooker, D. L. Wise, and L. J. D. Zaneveld, *in* "Long-acting

infection or increase in feed efficiency, have also been developed. These systems have target lifetimes of 1 month to 1 year. While the implantable and bolus polymer matrix systems can achieve deliveries of extended duration, oral systems are limited to a functional period of 12–24 hr, which is the retention time of material in the gastrointestinal tract. In addition, the lifetime of the device may be shortened by acid- or base-catalyzed hydrolysis of polymer or by mechanical agitation.

The mechanism of drug release from PLGA compressed matrices is a combination of diffusion and erosion. As drug on the surface of the device diffuses away, exposed polymer hydrolyzes and a greater matrix surface area is exposed. Drug may also diffuse through the polymer to the surface. The loss of drug is more rapid than the loss of polymer, and the device passes through a very porous stage before it is totally eroded. Figure 2 illustrates the change in appearance of a drug–polymer matrix which occurs *in vivo*. The device was a 1.6-mm bead of 70% drug, 30% PLGA and was designed to deliver naltrexone, a narcotic antagonist, for 3–4 weeks. The cross section of the material as manufactured (Fig. 2a) illustrates a tightly packed and homogeneous interior; however, after 1 week as a subcutaneous implant (Fig. 2b), the bead cross section reveals many empty areas where material has been lost by erosion.[10]

System Design

Polymer Composition

PLGA polymers may be prepared in any molar ratio of lactic to glycolic acids; the chosen proportion is an important factor in the system design. The *in vivo* polymer degradation rate depends strongly on the mole ratio of the monomers; polymers prepared in a 50 : 50 proportion are hydrolyzed much faster than those with a higher proportion of either

Contraceptive Delivery Systems" (G. I. Zatuchni, ed.), pp. 164–168. Harper & Row, Philadelphia, Pennsylvania, 1984.

[3] L. C. Anderson, D. L. Wise, and J. F. Howes, *Contraception* 13, 375 (1976).

[4] G. E. Benagiano, D. Schmitt, D. L. Wise, and M. Goodman, Jr., *J. Polym. Sci., Polym. Symp.* 66, 129 (1979).

[5] J. D. Gresser, D. L. Wise, L. R. Beck, and J. F. Howes, *Contraception* 97, 253 (1978).

[6] A. D. Schwope, D. L. Wise, and J. F. Howes, *NIDA Res. Monogr.* 4, 13 (1976).

[7] A. C. Sharon and D. L. Wise, *NIDA Res. Monogr.* 28, 194 (1981).

[8] D. L. Wise, J. D. Gresser, and G. J. McCormick, *J. Pharm. Pharmacol.* 31, 294 (1979).

[9] D. L. Wise, J. Rosenkrantz, J. B. Gregory, and H. J. Esber, *J. Pharm. Pharmacol.* 32, 399 (1980).

[10] D. L. Wise, "Development of Drug Delivery Systems for Use in the Treatment of Narcotic Addiction." Reports on Contract No. HSM-42-73-267 with NIDA. Dynatech R/D Company, Cambridge, Massachusetts, 1973–1978.

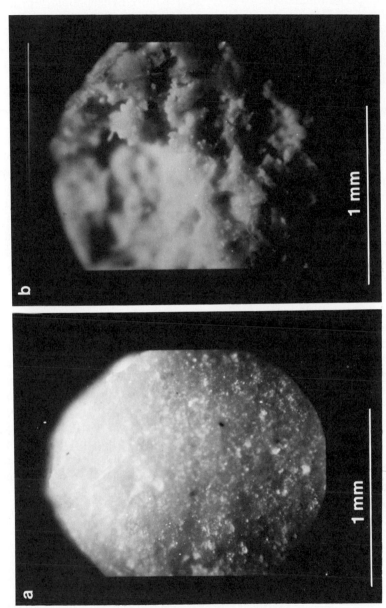

FIG. 2. The appearance of PLGA–naltrexone beads (a) before subcutaneous implantation in monkeys, and (b) after 1 week *in vivo*. The 1.6-mm (diameter) beads were sliced with a blade to show the interior matrix. The initial drug loading was 70%.

monomer.[11] In drug delivery systems, lactic acid is more frequently chosen as the predominant species because it is more hydrophobic. Further, lactic acid is optically active and the option for use of optically pure or racemic monomer in the preparation of PLGA offers an additional measure of system design control. Use of either the L(−) or D(+) form is generally preferred because polymers prepared with the racemic starting material exhibit accelerated hydrolysis rates.[12]

Molecular Weight and Dispersity

Both the molecular weight and the molecular weight distribution of the polymer affect the lifetime of the device.[13] The dependence of release rate on polymer molecular weight was demonstrated by Wise *et al.* in work on the sustained release of sulfadiazine from poly(lactic acid). A decrease in release rate occurred as polymer molecular weight increased up to 100,000; at larger sizes no difference was observed.[14]

The PLGA polymers may be prepared in a specified molecular weight range. The crude preparation mixture contains polymer chains in a wide range of sizes. This range is narrowed by fractionation, an important step in the manufacturing process. Lowering the polymer dispersity from 2.0 to 1.4 produced a PLGA system which had a reduced initial level of release and a longer period of constant release. In this experiment, two groups of rats were given injections of a matrix-containing labeled drug. One group received preparation containing low-dispersity polymer; the other group received unfractionated polymer. Each rat was given 500 mg of composite containing 100 mg of drug. The group receiving unfractionated polymer had a maximum initial release rate of 5.0 mg per day, and the rate fell below 1.0 mg per day by day 27. The group that received fractionated polymer had a maximum early release rate of 2.7 mg per day, and the rate remained between 1.0 and 2.1 mg/day from day 7 to day 44.[15,16] The elimination profile of estradiol-containing polymer rods implanted in rats is shown in Fig. 3. Drug release is linear over the time course of the study.

[11] R. A. Miller, J. M. Brady, and D. E. Cutright, *J. Biomed. Mater. Res.* **11**, 711 (1977).

[12] R. K. Kulkarni, E. G. Moore, A. F. Hegyeli, and F. Leonard, *J. Biomed. Mater. Res.* **5**, 169 (1971).

[13] D. L. Wise, J. B. Gregory, P. M. Newberne, L. C. Bartholow, and J. B. Stanburg, *Midl. Macromol. Monogr.* **5**, 121–138 (1977).

[14] D. L. Wise, G. V. McCormick, G. P. Willet, and L. C. Anderson, *J. Pharm. Pharmacol.* **30**, 686 (1978).

[15] J. D. Gresser and D. L. Wise, "Development and Testing of a Sustained Release System for Delivery of Pyrimethamine." Dynatech Rep. No. 2190 to World Health Organization. Dynatech R/D Company, Cambridge, Massachusetts, 1982.

[16] J. D. Gresser, J. F. Howes, D. F. Worth, and D. L. Wise, *in* "Biopolymeric Controlled Release Systems" (D. L. Wise, ed.). CRC Press, Boca Raton, Florida (1984).

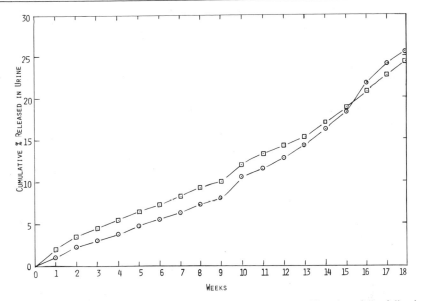

FIG. 3. Weekly cumulative percentage of ¹⁴C materials excreted in urine of rats following implantation of 75L+/25G, 56K, 50% estradiol rods; dose of estradiol, 6000 μg.

Drug Loading

The single most important factor in the design of the PLGA controlled-release system is the percentage of drug loading. This factor is based on the solubility of the drug. The table shows the approximate lifetimes for

APPROXIMATE SYSTEM DURATION AS A FUNCTION OF DRUG SOLUBILITY

Drug	Solubility (μg/ml)	Approximate system duration (days)
Quinazoline derivative WR-158122	0.02	1600
Levonorgestrel	1.0	750
Acedapsone	3.0	600
Norethisterone	10.0	50
Pyrimethamine	100.0	80
Naltrexone pamoate	800.0	70
Sulfadiazine	2000.0	40
Naltrexone (free base)	4000.0	30

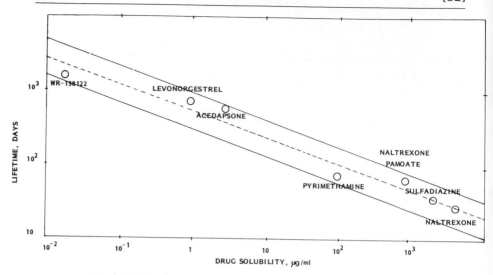

FIG. 4. PLGA/drug system lifetimes vs drug solubility in water.

drug–PLGA systems with drug solubilities that vary from 0.02 μg/ml (antimalarial quinazoline derivative, WR-158122) to 4000 μg/ml (naltrexone free base). System durations vary from ~30 days for the latter to ~1600 days for the former. These durations are determined by extrapolation from cumulative release profiles for each drug, usually to about 80% exhaustion. For short-term systems, a lifetime is defined as the time over which release of drug is sufficient for therapeutic efficacy.

In Fig. 4, the approximate lifetimes for these sustained-release systems are shown as a function of drug solubility in water (pH 7). Dose form and size varied from drug to drug. Levonorgestrel was implanted as 1/32-in. cylindrical rods, and the naltrexones and sulfadiazine as 1.5-mm beads. The other compounds were delivered as injectable powders. Doses ranged from 40 to 700 mg; over this range, duration for the specific drug should vary by no more than a factor of 1.7. This range is indicated by the dashed lines parallel to the regression line. The empirical relationship obtained from a linear regression analysis of the data in the table is

$$\log T = -0.355 \log S + 2.7625$$

or approximately,

$$T = (5.8 \times 10^2)S^{-0.4}$$

where T is system duration (days) and S is solubility (μg/ml).

The release rate of a specific drug from a PLGA matrix is selected by manipulation of the percentage of drug content in the drug–polymer composite. Such a dependence is clearly demonstrated by release of the narcotic antagonist, naltrexone, from PLGA matrix beads. Four different dry loadings, ranging from 50 to 80%, were prepared as 1/16 in.-diameter rods with 75L/25G PLGA. *In vitro* release studies of these rods showed that release rates increased with loading. The time taken to achieve 80% release varied from 8 to 45 days.[6]

PLGA systems release very soluble active macromolecules, such as proteins, quite rapidly. Long-term delivery of proteins (e.g., antigens for single-dose immunization systems, or peptide hormones for fertility, growth, or neurological control) will require PLGA matrix systems with very low drug content. Release rate profiles for several loadings of bovine serum albumin from a poly(lactic acid) polymer are shown in Fig. 5.

Dose Form

The gross physical features of the dosage, size, and shape are less significant in determining the release rate. The choice of form is made primarily on the specific delivery requirements. Oral systems may be coated beads of powders. Bolus devices are often monolithic cylinders of 3/4 in. diameter. Subcutaneous or intramuscular implantation may be made with a rod, bead, or suspended powder. Retrieval of implanted devices is sometimes desirable, and this is most easily accomplished with rods; retrieval is nearly impossible with powders or beads. In clinical testing of the narcotic antagonist release system mentioned above, tiny beads were used as controlled-release vehicles; the subjects could not easily remove the devices themselves; however, the beads could be retrieved surgically if necessary.

System Manufacture

Polymer Synthesis

Discussion. Low molecular weight PLGA may be prepared by the direct polyesterification of lactic and/or glycolic acids. Condensation of α-hydroxy acids occurs as water is removed by boiling or by azeotropic distillation with an aromatic hydrocarbon solvent. At reaction temperatures below 120°, an acid catalyst increases the reaction rate; but above this temperature, water removal is generally the rate-limiting step; little benefit is gained by using a catalyst. A practical molecular weight limit, ~10,000, exists for polymers prepared by this direct route.

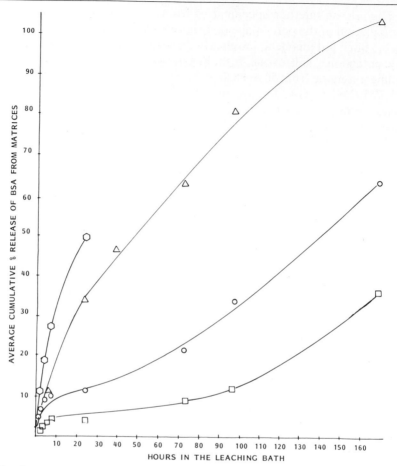

FIG. 5. *In vitro* release patterns of bovine serum albumin from several poly(lactic acid) matrix devices.

For most applications, it is preferrable to prepare the polymer or copolymer using the cyclic dimers as the starting material. This method is advantageous because higher molecular weight polymers may be obtained without carrying the dehydration to such a high degree of completion. Many kinds of catalysts have been used to initiate polymerization. These include compounds of cadmium, tin, lead, titanium, zinc, antimony, and a variety of amines. For intermediate molecular weights (~10,000–40,000), acid-catalyzed bulk polymerizations are convenient and effective; the molecular weight may be determined by the catalyst concentration. PLGA polymer preparation has recently been reviewed.[1]

Method. Purified lactide and glycolide are bulk polymerized at 130° in an evacuated and sealed bulb, with *p*-toluenesulfonic acid as a catalyst. Water exclusion is essential for reaction completion. The polymerization takes 7–10 days; at completion, the polymer solution is a solid golden mass.

Polymer Purification

Discussion. The crude polymer is a glassy solid containing residual starting material and catalyst.

Method. The crude polymer is dissolved in THF to make a 15% (w/v) solution. This solution is precipitated in excess distilled water, with stirring. As precipitation occurs, the polymer is collected as cocoons on glass rods. The white polymer cocoons are dried under vacuum. Polymer is analyzed for residual catalyst, for percentage of optical activity (if optically active lactide was used), and for molecular weight size and distribution by gel permeation chromatography.

Polymer fractionation is achieved by slowly adding an excess of isopropanol, with constant stirring, to a 1% solution of polymer in dichloromethane. Fractionated polymer is isolated as fibrous cocoons in a yield of less than 50%.

Matrix Prepration

Discussion. A matrix is formed by casting a solution of polymer and drug onto a glass plate, then removing the solvent by evaporation.

Method. PLGA polymer is placed in dichloromethane in a jar with a tight cover and a polyethylene film liner. The container is rolled until all the polymer is in solution. After pressure filtration through a 5.0-μm Telfon Millipore filter, the solution is placed in a clear jar containing the appropriate proportion of drug, about one-third volume of 1/4-in. glass beads and several 1/2-in. porcelain rolls. The jar is placed on a ball mill and rolled overnight. The suspension is then cast directly onto a clean, level piece of plate glass and spread to a thickness of 0.625 mm with a Boston–Bradley adjustable blade. The solvent is evaporated in a fume hood, and the film is peeled from the plate and vacuum desiccated. The matrix is analyzed for percentage of drug content and for residual solvent.

Compression

Discussion. A hydraulic press is used for compression. The compression may be combined with extrusion into a dose form such as a rod or bead. The process of compression generally employs gradual heating and

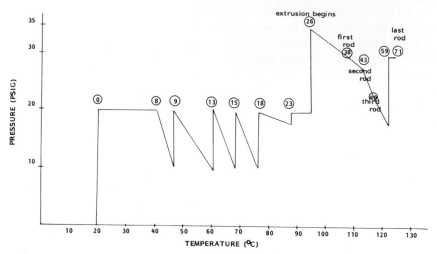

FIG. 6. The extrusion of levonorgestrel–PLGA rods, a plot of pressure vs temperature. The cumulative time (min) is shown inside circles at each data point.

maintenance of pressure at about 20 psig. The softening point of the polymer being used may be determined experimentally, typically ranging from 80 to 120°. The melting point of the drug and the ratio of drug to polymer alter this softening point. Proteins lower the temperature of compression, generally to about 40°. High-melting compounds may raise the softening temperature to over 150°.

Method. In the extrusion of rods of 50% levonorgestrel and 50% PLGA, pressure was applied and the temperature raised as shown in Fig. 6. The material compressed considerably between 40 and 80° and then extruded between 100 and 120°. The total time for compression and extrusion was 71 min.

Sterilization

Discussion. Sterilization is the removal and/or destruction of all microorganisms on and in a given object or space. No distinction is made between the different kinds of microorganisms that are destroyed (i.e., pathogenic or nonpathogenic, vegetative or sporous). Sterilization also includes the inactivation of viruses.

The most dependable, economical, and universally standard method used to destroy all forms of microbial life is steam sterilization; this process involves heating the object with 121° steam for at least 15 min. Unfortunately, this method cannot be used with the poly(lactic/glycolic acid) implant system for two reasons. First, the polymers soften at approximately 65° and begin to flow at 100°. This softening would cause a deformation of the matrix form. Second, the penetrating, high-pressure steam would initiate hydrolysis of the copolymer.

Dry heat is generally regarded as less efficient than moist heat for sterilization. Consequently, longer periods of exposure and higher temperatures are required. Such an environment would be destructive to both polymer and drug.

Gases, such as ethylene oxide, propylene oxide, and formaldehyde, are often utilized for sterilization when other methods, e.g., heat, cannot be used. These gases are not recommended for sterilization of materials that are to be implanted in the body, however, because the residual vapor has been found to be mutagenic, damaging the cytoplasm and nuclei of rapidly growing cells.

There are two types of radiation sterilization: neutral particles (γ rays) and charged particles (high-energy electrons). Both methods are considered more reliable than gaseous sterilization where applicable, but they can have unwanted effects such as gas evolution, cross-linking, degradation, and double bond formation. Polyesters, such as poly(lactic acid) and poly(glycolic acid), have been found to be more resistant than other polymers to the aforementioned side effects. Therefore, radiation sterilization is considered to be the most promising technique available to produce sterile drug–polymer matrices for clinical uses.

Method. The recommended sterilization system for PLGA matrix products is γ radiation. The cobalt-60 process, at a 2.5 Mrad dosage, is effective.

Conclusion

The drug–PLGA matrix delivery systems previously described are diverse. Many types of drug compounds have been incorporated in PLGA matrices; those mentioned included steroid hormones, quinazoline antimalarials, and proteins. These systems have applications in subcutaneous, rumenal, and oral delivery. The duration of the drug release is primarily limited by the retention time of the device at the site; short-term delivery is easily accomplished in any route of application; however, subcutaneous systems are required for lifetimes greater than 1 year.

The rate and duration of release are controlled by system design, and the system design parameters are selected by the choice of manufacturing processes. Polymer preparation and purification control the composition, size, and dispersity; matrix preparation is a function of the drug loading; and the shaping into the dosage forms determines the density, size, and shape.

The two largest areas of growth for PLGA matrix delivery applications are expected to be in protein hormone and enzyme therapy and in single-dose immunizations. PLGA will be important in these areas because proteinaceous drugs will require protection until released from aqueous environment. Other applications of PLGA matrix systems are *in vitro* systems such as nutrient delivery to cell growth media and pesticide delivery. Another novel application may be plant hormone release.

[33] Transdermal Dosage Forms

By Jane E. Shaw

Transdermal pharmaceutical products—whether ointments, matrix formulations, or reservoir systems—provide the considerable advantages of a noninvasive parenteral route for drug therapy.

Rate-controlled transdermal dosage forms can provide, in addition, precise regulation of drug concentrations in plasma and thus a high degree of safety and selectivity of action for some drugs. Another corollary of rate-controlled drug delivery is its capability of lengthening dosage intervals to an unprecedented degree. Because of the opportunities that such dosage forms open to advances in pharmaceutical development, they will constitute the main focus of this chapter.

Transdermal dosage forms provide a safer and more convenient method of parenteral therapy than, for example, intravenous infusions; yet they retain many advantages associated with drug therapy via the parenteral route. These stem mainly from avoidance of the gastrointestinal tract variables (acidity, motility, enzymatic activity, food intake, and transit time) that frequently make absorption from the gut unpredictable. A first pass through the liver prior to reaching the systemic circulation is also avoided, minimizing drug degradation by that organ. Thus, transdermal systemic drug input can be more reliable than oral dosing, for some drugs, and smaller daily doses of drug may be efficacious.

The reliability of transdermal drug administration is dependent on elimination of any variables that can make permeation of drugs through skin itself unpredictable. This topic will be addressed in some detail later.

No single transdermal drug delivery system has yet realized all the potential advantages of this route. The decision to develop a drug in a transdermal dosage form, however, may justifiably be based on only one

METHODS IN ENZYMOLOGY, VOL. 112

or two significant benefits, such as effective utilization of a drug with a short half-life or elimination of an important side effect.

In fact, many drugs are neither appropriate nor suitable for transdermal administration. Drugs may have excellent efficacy, with only minor side effects, and convenient regimens when taken in ordinary dosage forms; for these agents a rationale is lacking to justify transdermal administration. Many other drugs fail to meet one or more of the principal criteria that make transdermal dosing possible: namely, topical safety and appropriate physicochemical properties permitting transmission of efficacious doses through a feasible area of intact skin. Nevertheless, important drugs in virtually every therapeutic category have properties that make their transdermal delivery practical and therapeutically advantageous.

The Course of Transdermal Development

Only in the last decade or so has any major effort been made to utilize the limited permeability of skin for systemic drug administration. Prior to that, drugs were applied topically for localized effects only; any other actions constituted unwanted side effects.

Systemic drug administration by the transdermal route first became a part of modern therapeutics in the form of drug-releasing ointments or creams. Preparations of nitroglycerin, etofenamate, and 17β-estradiol, for example, were found efficacious to varying degrees.

Systemic drug absorption from ointments or creams was unpredictable, however, because patients inevitably did not apply the ointments to the skin in a reproducible manner. Variations in area of ointment applications could lead to an undershoot or overshoot of drug input to the circulation. Differing thicknesses of application unpredictably affected duration of drug input, which in most cases was only a few hours. Additional inconveniences were the messiness of applications and the need to cover them to prevent evaporation of drug and stained clothing. Nevertheless, ointment formulations demonstrated that systemic therapy via the transdermal route could be efficacious. For this route to become more practical and reliable, it was necessary that dosage forms provide predictable drug input, with simplified application procedures and regimens.

In the 1970s attention was initially paid to the development of such dosage forms. Transdermal dosage forms of defined surface area are now available that deliver drug to the surface of intact skin for periods up to 1 week after application. These dosage forms provide a preprogrammed rate and duration of drug delivery; the rate is such that the dosage form and not the barrier properties of skin predominantly control or limit systemic drug input. Thus, many of the nuisance features and unreliability of ointment applications for systemic therapy are avoided.

Requirements for Developing Rate-Controlled Transdermal Forms

The design of rate-controlled transdermal dosage forms raised issues that could not be resolved merely by recourse to the literature. Scheuplein[1-3] had already laid the basis for understanding of the permeability of skin to solutes in terms of their solubility in oil and water. The main resistance to drug permeation was identified as residing in the stratum corneum. Little was known, however, of many other factors affecting skin permeability—and data on the permeation properties of individual drugs were virtually nonexistent.

The complexity of developing the first rate-controlled transdermal drug delivery systems is best explained by reviewing the multiplicity of factors requiring consideration. These include drug properties, patient characteristics, and dosage form parameters requiring definition for each drug of interest. Table I lists certain of these factors.

The highly variable factors shown in columns 1 and 2 of the table affect those in column 3 to an extent that makes each rate-controlled transdermal system unique. That is, each drug requires a dosage form specifically designed to fit its unique combination of properties. An off-the-shelf approach is not feasible; one cannot simply substitute one drug for another in a given system.

The consideration of first importance is topical safety; extensive testing must be done to eliminate agents that induce severe localized irritation, contact sensitivity, or phototoxicity in a significant proportion of patients. Another primary consideration is ascertaining whether the drug of interest (1) is potent enough and (2) has the appropriate physicochemical properties to permeate skin in therapeutic amounts over an area of acceptable size.

The factors determining the suitability of a drug for transdermal administration are shown in Table I (column 1). Moreover, *in vitro* tests are now available, using cadaver skin mounted as a membrane in a diffusion chamber, to follow the permeation and immobilization of drugs in human skin. The results obtained have been predictive of transdermal permeation *in vivo*.[4]

[1] R. J. Scheuplein, *J. Invest. Dermatol.* **45,** 334 (1965).
[2] R. J. Scheuplein, *J. Invest. Dermatol.* **48,** 79 (1967).
[3] R. J. Scheuplein, "Molecular Structure and Diffusional Processes across Intact Epidermis," Edgewood Lab. Contract Rep. No. 18. Edgewood Arsenal, Edgewood, Maryland, 1967.
[4] S. K. Chandrasekaran, W. Bayne, and J. E. Shaw, *J. Pharm. Sci.* **67,** 1370 (1978).

TABLE I
CONSIDERATIONS IN THE DESIGN OF A RATE-CONTROLLED TRANSDERMAL
DOSAGE FORM FOR ANY DRUG

Drug properties	Patient characteristics	Dosage form design parameters
Potency	Skin permeability to drug of	Range of therapeutic concentrations
Therapeutic index	interest: differences	in plasma
Effective parenteral	among individuals and	Delivery rate/pattern required
dose	body sites	to achieve desired concentra-
Half-life	Condition of skin: intact,	tions
Solubility in oil	inflamed, sunburned, etc.	Constant or varying delivery
and water	Allergies to components of	rate(s)?
Molecular weight	system, including drug	Degree of concentration control
and size	Any other contraindicating	needed
Extent of binding	factors: age (e.g., prema-	Optimum dosage intervals
in skin	turity), systemic disease,	Dosage form components
	skin pathology, etc.	Nonallergenic materials
		Suitable form of drug (base,
		salt, etc.)
		Appropriate vehicle
		Type of rate controller
		Flux enhancer

Subsequently, for drugs whose actions correlate with their concentrations in plasma, studies are made to identify the plasma profile that will elicit and maintain a full therapeutic effect while minimizing side effects. These data provide the basis for programming the pattern of drug release from the dosage form that will assure the appropriate plasma concentrations over time.

The first transdermal therapeutic system to be studied clinically delivers the drug scopolamine at a rate that prevents motion-induced nausea; its efficacy exceeds that of the leading oral antimotion sickness preparation. However, the system avoids, in all but a small minority of subjects, other parasympatholytic effects of the drug. The system functions for 3 days. Because development of this system required solutions to common problems of transdermal therapy, it is appropriate to describe its components and functionality in some detail.

Design of Rate-Controlled Scopolamine System

Scopolamine as the free base has the required potency and permeation characteristics for transdermal delivery. Its therapeutic index is narrow, however, a property which dictated that its concentration in plasma had to be precisely controlled. That precision posed challenges during development of the dosage form because *in vitro* studies revealed that about a 5-fold difference in skin permeability to the drug existed among individuals with the most and least permeable skin.[5] Precise dosing thus required some means of negating the effect of differences in skin permeability on systemic drug input. Clinical rate-ranging studies showed that antiemetic prophylaxis required a scopolamine release rate to the systemic circulation of approximately 5 μg/hr.[5] *In vitro* studies showed that scopolamine flux through skin obtained from the postauricular area was variable but averaged 20 μg/cm^2 hr; that rate was substantially higher than the flux through skin obtained from most other regions of the body. This finding made it feasible to develop a transdermal dosage form—for placement in the postauricular area—that would exert the major degree of control over the rate of drug delivery to the systemic circulation. Thus, a transdermal dosage form, with a rate of drug release to the circulation of only 2 μg/cm^2 hr, was developed. To deliver 5 μg/hr of scopolamine (the defined therapeutic rate), the area of the system had to be 2.5 cm^2, a size suitable for postauricular application. Because the rate of scopolamine delivery is below the rate at which the drug can permeate the least permeable postauricular skin, the system rather than the skin controls systemic drug input. That control negates individual differences in skin permeability.

Before steady-state drug release from a transdermal dosage form can produce steady-state input to the systemic circulation, immobilization sites for drug in skin must be saturated.[5] For drugs that are significantly absorbed by skin, saturation may be long delayed: scopolamine is such a drug. Therefore, to hasten saturation of drug immobilization sites, the scopolamine transdermal dosage form incorporates a priming dose of about 140 μg of drug, which is delivered at an asymptotically declining rate over the 6 hr immediately following application to the skin[5]; drug input then stabilizes at 5 μg/hr for the remainder of the product's functional lifetime.

The system has the appearance of a small circular adhesive film. It consists of the following four functional layers: (1) an adhesive face containing the priming dose; (2) a rate-controlling microporous polypropylene membrane; (3) a drug reservoir of scopolamine; and (4) an impermeable backing layer. After application to skin, the drug diffuses through microscopic channels in the membrane in the direction of the concentration gradient. Drug release is governed by modifications of the Fick diffusion equation; the energy source is the difference in the drug's chemical potential between the reservoir and the system's exterior. Virtual constancy of rate of drug delivery, after release of the priming dose, is as-

[5] J. Shaw and J. Urquhart, *Trends Pharmacol. Sci.* **1**, 208 (1980).

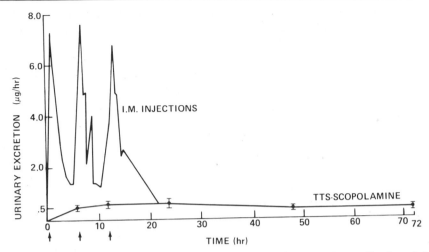

FIG. 1. Mean urinary excretion rate of scopolamine free base after application of the transdermal therapeutic system for scopolamine (TTS-scopolamine) and intramuscular injections of scopolamine hydrobromide (200 μg). Arrows indicate times of injections; vertical bars at point represent SE.

sured as long as drug is present in excess in the reservoir (i.e., a saturated solution exists in the vehicle).

Figure 1 demonstrates the consistent rate of scopolamine permeation through skin, as reflected by rate of urinary excretion of unchanged scopolamine. Data are shown for subjects wearing a transdermal scopolamine system for 72 hr and for subjects receiving intramuscular administration of the drug (200 μg scopolamine hydrobromide). The drug delivery pattern from the transdermal scopolamine system has been demonstrated to be bioequivalent to a 72-hr constant i.v. infusion of scopolamine at a rate of 5 μg/hr.[5,6]

Rate-Controlled, Transdermal 17β-Estradiol Administration

Transdermal delivery of 17β-estradiol provides a highly pertinent example of the advantages provided by that route vs the oral route of drug administration.[7] Both estradiol and conjugated estrogens, when given by mouth, are substantially metabolized on the first pass through the liver to

[6] L. G. Schmitt, J. E. Shaw, P. F. Carpenter, and S. K. Chandrasekaran, *Clin. Pharmacol. Ther.* **29**, 282 (1981).
[7] L. Schenkel, J. Balestra, L. Schmitt, and J. Shaw, *In* "Rate Control in Drug Therapy, Abstracts of the Second International Conference on Drug Absorption," p. 41. 1983.

estrone or conjugates while simultaneously modifying hepatic function. The estradiol/estrone (E2/E1) ratio is typically only 0.1–0.3—a range much closer to the postmenopausal range of 0.3–0.8 than to the 0.8–2 range prevailing in premenopausal women.

Thus, the aim of developing a rate-controlled transdermal dosage form for 17β-estradiol was to deliver this ovarian hormone in a pattern that would restore its premenopausal concentrations and metabolic pattern in postmenopausal women. Other estrogenic products either are not physiological hormones or are not administered in a physiological pattern.

Continuous, rate-controlled transdermal administration offered the prospect of achieving appropriate estradiol plasma levels with much lower daily doses of estradiol than are required orally. Moreover, appropriate E2/E1 ratios could be maintained because avoidance of the first pass through the liver would prevent immediate estrogen metabolism to estrone.

Rate-controlled transdermal systems for administering 17β-estradiol through intact skin utilize a multilaminate design of the type already described; they utilize a membrane of ethylene–vinyl acetate copolymer to regulate release of the hormone. The systems are programmed to deliver *in vivo* 0.025, 0.05, or 0.1 mg estradiol daily. Thus, estrogen replacement can be titrated to patients' individual needs. (The systems are identical in composition per unit area and hence titration can be achieved by modifying surface area.)

A study[8] in postmenopausal women—with pretreatment estradiol and estrone levels in plasma of 7.4 and 32.3 pg/ml, respectively—showed that the transdermal systems produced therapeutic levels of 17β-estradiol in less than 4 hr, which persisted for the entire application period. The E2/E1 ratio was elevated to—and maintained within—the premenopausal range. Within 24 hr of removal of the systems, serum concentrations of 17β-estradiol and estrone returned to untreated levels; daily urinary output of estradiol conjugates approached baseline values within 3 days. In contrast, orally administered 17β-estradiol (2 mg/day) produced mean peak levels of estradiol and estrone of 133 and 709 pg/ml, respectively, 8 hr after administration. Table II lists the increase in serum concentrations of estradiol and estrone following estradiol administration by the two routes.

A clinical study[9] compared the transdermal estradiol system with placebo in women experiencing severe and frequent hot flushes (>10/day). The frequency of hot flushes as measured by thermography was signifi-

[8] ALZA Corporation, data on file.
[9] L. R. Laufer, J. L. DeFazio, J. K. H. Lu, D. R. Meldrum, P. Eggena, M. P. Sambhi, J. M. Hershman, and H. L. Judd, *Am. J. Obstet. Gynecol.* **146,** 533 (1983).

TABLE II

17β-ESTRADIOL (E2) AND ESTRONE (E1) SERUM
CONCENTRATION INCREMENTS AFTER 3 DAYS OF E2
ADMINISTRATION

17β-Estradiol route and dose	Mean increase above baseline (pg/ml)	
	17β-Estradiol	Estrone
Oral		
2 mg/day	59	302
Transdermal		
0.025 mg/day	16	0.3
0.05 mg/day	32	9
0.1 mg/day	67	27

cantly reduced with transdermal administration in this double-blind, parallel, placebo-controlled study, following 21 days of treatment. At baseline the subjects experienced a mean of 0.76 hot flushes per hour, and following treatment that number decreased to 0.25 per hour; hot flushes were totally eliminated in 3 of the 10 subjects. In placebo-treated patients, the frequency of hot flushes remained unchanged. The estradiol levels in plasma increased during drug treatment from 7 to 72 pg/ml, and estrone increased from 16 to 37 pg/ml. Pre- and posttreatment E2/E1 ratios were 0.44 and 1.95, respectively.

With use of transdermal estradiol, no significant changes have been seen to date in the circulating concentration of renin substrate or other hepatic proteins (sex hormone-binding globulin capacity, thyroxine-binding globulin, corticosteroid-binding globulin). Absence of such increases indicates that transdermal estradiol does not effect the changes in hepatic function known to occur when conjugated or synthetic estrogens are administered orally.[9]

In a separate series of skin safety studies, transdermal administration of estradiol showed no potential for inducing phototoxicity, photocontact allergy, or contact sensitization. Only mild erythema was observed when systems were applied repeatedly to the same skin site. (In routine use, each fresh system is to be placed on a different site.)

In summary, rate-controlled transdermal estradiol administration appears to offer several advantages over the oral route:

Maintenance of controlled, therapeutically effective, physiologic levels of estrogen with low daily doses of 17β-estradiol

Maintenance of physiologic E2/E1 ratios

Rapid clearance of estradiol and its metabolites after removal of transdermal systems, providing a ready pause in estrogen therapy when cycling is desirable

A convenient therapeutic regimen

Rate-Controlled Transdermal Clonidine System

The transdermal clonidine system (Catapres-TTS, registered trademark of ALZA Corporation) represents state-of-the-art in regimen optimization via the transdermal route. It provides a dosage interval of unprecedented length for routine outpatient therapy. After one application, Catapres-TTS delivers clonidine for 7 days, reducing dosage frequency (which is b.i.d. orally) by an order of magnitude. That length of time may represent the biological limit on duration of drug delivery by the transdermal route, since the skin sloughs a layer daily and factors such as adhesion become unsatisfactory after a week.

Regimen optimization and compliance are especially important in the therapy of hypertension. When hypotensive medications give rise to more symptoms than the underlying hypertension, patients are not highly motivated to comply with the prescribed regimen. Thus, one objective for a clonidine transdermal system was to reduce the peaks and valleys associated with oral drug administration in the hope that side effects of the drug (dry mouth, drowsiness) would be minimized.

Clonidine is effective in oral doses of 1 mg/day or less and hence its potency made it an ideal candidate for prolonged, rate-controlled, transdermal administration. The system (Catapres-TTS) has the same four-layer structure as the scopolamine system, though the individual components are not the same.

In vitro release of clonidine from Catapres-TTS into water under infinite sink conditions is illustrated in Fig. 2. Initially, release of drug is rapid, a finding reflecting absorption of a priming dose incorporated into the contact adhesive. As the clonidine content of the adhesive falls below saturation, drug is released from the drug reservoir at a rate predetermined by the properties of the rate-controlling membrane. From 24 hr through the remainder of the lifetime of the system (168 hr), drug release rate is constant.

Figure 3 shows permeation of clonidine from a Catapres-TTS placed in contact with the stratum corneum surface of human cadaver thigh skin. An initial lag time reflects absorption of drug from the adhesive by skin immobilization sites; subsequently drug permeation reaches a steady

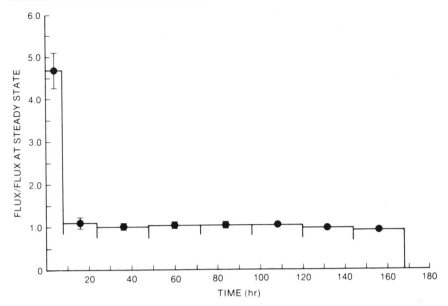

FIG. 2. Mean ratio of flux to flux at steady state with Catapres-TTS. Measurements were made in water (32°). $N = 10$; error bars represent SD.

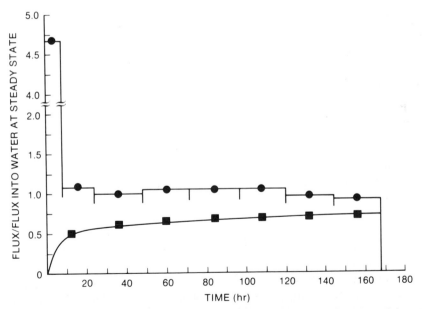

FIG. 3. *In vitro* drug release into water (32°) and transdermal drug flux through human epidermis for Catapres-TTS: (●) into water; (■) through human epidermis.

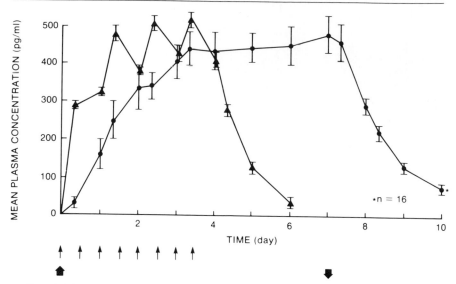

FIG. 4. Mean plasma concentrations ($n = 17$) of clonidine following repeated doses of oral Catapres (▲) or application of Catapres-TTS (●). ↑, Catapres dose; ↕ , application/removal of Catapres-TTS. Error bars represent SE.

value that approximates 75% of the steady-state release rate into an infinite sink of water.

In a biovailability study, the plasma levels of clonidine associated with transdermal and oral administration of the drug were compared. The data indicated that continuous rate-controlled administration of clonidine via the transdermal route could minimize the daily peaks and valleys in drug plasma concentration known to be associated with oral drug therapy.

Thus, during 7-day wearing of Catapres-TTS on the upper arm by 17 normal volunteers, drug plasma levels rose over a 48-hr period to a steady value that was maintained throughout the 7 days of wearing. Following system removal, plasma levels of drug declined over the following 48 hr (Fig. 4).

In contrast, following oral administration of 0.1 mg Catapres b.i.d., plasma levels rose to a maximum 2 hr following dosing, and then declined over the subsequent time between dosage intervals. This pattern in the plasma profile of clonidine was repeated after each oral dosage (Fig. 4).

If, as one transdermal system is removed, a second system is placed on a fresh skin site, constancy of drug plasma levels is maintained. Average clonidine plasma concentration—measured at steady state (48–168 hr) during wearing of different dosages (increasing sizes) of the Catapres-

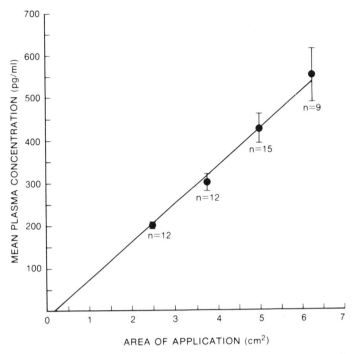

FIG. 5. Effect of area of application on steady-state (120, 144, 168 hr) mean drug plasma concentration of clonidine for Catapres-TTS. $y = 92.30x - 22.5$; $r = .992$; error bars represent SE.

TTS—is directly proportional to the area of the dosage form applied (Fig. 5).

In clinical studies, patients—whose blood pressure control was adequate while they were receiving a diuretic and clonidine orally—continued taking their diuretic but used transdermal clonidine instead of oral clonidine. Patients were titrated—according to therapeutic response—to the area of the transdermal dosage form that controlled their blood pressure when applied once weekly. In these patients, Catapres-TTS lowered blood pressure as effectively as the oral drug while minimizing unwanted pharmacological effects.[10,11] Thus, rate-controlled transdermal clonidine administration offers the advantages, over the oral route, of providing convenient dosage regimen and minimum oscillations in plasma levels of clonidine.

[10] W. J. Mrozcek, M. Ulrych, and S. Yoder, *Clin. Pharmacol. Ther.* **31**, 252 (1982).
[11] R. C. Michael, A. K. Jain, J. R. Ryan, and F. G. McMahon, *Clin. Pharmacol. Ther.* **33**, 229 (1983).

Transdermal Nitroglycerin Systems

Nitroglycerin exemplifies a drug whose therapeutic applications have been extended by its incorporation into a transdermal dosage form. It is now being widely used for prophylaxis of angina; previously it was employed mainly for treatment of anginal attacks as they occurred. Ointment forms of the drug were previously utilized for prophylaxis of angina but had the disadvantages already described.

Nitroglycerin has a wide therapeutic index, a finding indicating that fluctuations in its levels in plasma can be tolerated. Therefore, a monolithic transdermal system without a rate-controlling membrane was initially fabricated, to ascertain whether it could deliver nitroglycerin at predictable rates to the systemic circulation. Data showed that dose dumping from monolithic systems would be a risk for certain patients[12]; for this reason a rate-controlling membrane was incorporated into the Transderm-Nitro, registered trademark of ALZA Corporation, system. Its clinical testing indicates that it provides a significant reduction in the number of anginal attacks for many patients.

Other types of nitroglycerin systems do not incorporate either an enclosed reservoir or a rate-controlling membrane. Nitro-Dur (Key) contains the drug in a polymer matrix that is held on the skin by a microporous tape. The manufacturer has stated that the absorption of nitroglycerin from transdermal systems is limited only by the properties of the skin.[13] Nitrodisc (Searle) also employs a matrix system; drug is dispersed throughout a silicone polymer in microscopic compartments. The polymer is said to release drug at a known rate; it is not clear whether this rate was selected so as to limit transmission through highly permeable skin.

Conclusions

Developments discussed in this chapter indicate the following points with respect to the future of transdermal systemic therapy:

This mode of treatment seems destined to persist and expand.

Its expansion will require the identification of drugs suitable for dosing by this noninvasive parenteral route.

The principal limitations in finding suitable drugs relate to the need to get therapeutically effective amounts of drug through an acceptably small

[12] J. E. Shaw, "Development of the Transdermal Therapeutic System," Presented at the Seminar on Transdermal Therapeutic System: A Major Advance in Angina Prophylaxis, in conjunction with the 16th Am. Soc. Hosp. Pharm. Midyear Clin. Meet., 1981.

[13] Anonymous, *Therapaeia* February issue, 6 passim (1983).

skin patch and the potential of many drugs to cause irritation and sensitization.

Thus, judicious selection of drug candidates is the key to the successful development of transdermal therapy. Its extension will become increasingly feasible as means are found to increase the flux of drugs through skin and to formulate drugs so as to minimize their irritancy potential.

[34] Microsealed Drug Delivery Systems: Fabrications and Performance

By Yie W. Chien

Much interest has recently been generated in the pharmaceutical and biomedical fields[1,2] in the development of controlled-release drug delivery systems for achieving one or more of the following objectives: (1) to maximize the bioavailability of a therapeutic agent in a target tissue; (2) to minimize the adverse side effects of a therapeutic agent or its metabolites; (3) to optimize the onset, rate, and duration of drug delivery; (4) to maintain the steady-state plasma drug level within a therapeutic range for as long as required for an effective treatment; (5) to improve the patient compliance to a therapeutic regimen.

The Microsealed Drug Delivery (MDD) system was recently developed to achieve the same objectives.[3-5] It is a drug delivery system fabricated from a biocompatible elastomer by microdispersion of the drug reservoir, as immobilized microscopic spheres, in the cross-linked polymer matrix (Fig. 1). The result is that the release of drugs is controlled at a programmed rate, as described by Eq. (1), for a prolonged time period[6]:

[1] 1982 Industrial Pharmaceutical R & D Symposium on "Transdermal Controlled Release Medication," sponsored by Rutgers University, College of Pharmacy, Piscataway, New Jersey, 1982: *Drug Dev. Ind. Pharm.* **9**(4), 497 (1983).

[2] 1983 Industrial Pharmaceutical R & D Symposium on "Oral Controlled Drug Administrations," sponsored by Rutgers University, College of Pharmacy, Piscataway, New Jersey, 1983: *Drug Dev. Ind. Pharm.* **9**(7), 1077 (1983).

[3] Y. W. Chien and H. J. Lambert, U.S. Patent 3,946,106 (1976).

[4] Y. W. Chien and H. J. Lambert, U.S. Patent 3,992,518 (1976).

[5] Y. W. Chien and H. J. Lambert, U.S. Patent 4,053,580 (1977).

[6] Y. W. Chien, "Novel Drug Delivery Systems: Fundamentals, Developmental Concepts and Biomedical Assessments," Chapter 9. Dekker, New York, 1982.

$$\frac{Q}{t} = \frac{D_p D_s \gamma' K}{D_p \delta_d + D_s \delta_p \gamma' K} \left[\beta S_p - \frac{D_l S_l (1 - \beta)}{\delta_l} \left(\frac{1}{K_l} + \frac{1}{K_p} \right) \right] \tag{1}$$

where Q/t is the rate of drug release from a unit surface area of MDD system; K, the partition coefficient for the interfacial partitioning of drug molecules from polymer coating membrane toward the elution solution; K_l, the partition coefficient for the interfacial partitioning of drug molecules from the microscopic liquid compartment toward the polymer matrix; K_p, the partition coefficient for the interfacial partitioning of drug molecules from the polymer matrix toward the polymer coating membrane; D_l, the diffusivity of the drug molecules in the microscopic liquid

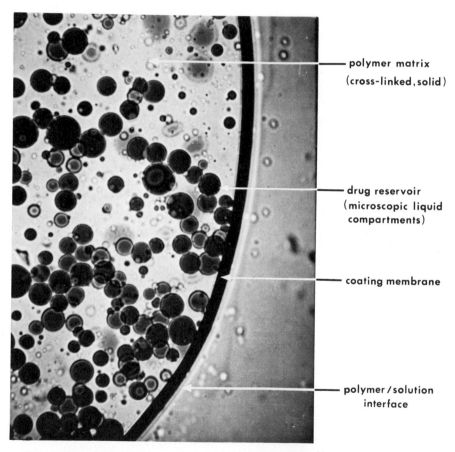

polymer matrix
(cross-linked, solid)

drug reservoir
(microscopic liquid
compartments)

coating membrane

polymer/solution
interface

FIG. 1. Photomicroscopic view of a cross section of the microsealed drug delivery system, in which numerous microscopic liquid compartments of drug reservoir are dispersed as discrete, immobilized, unleachable spheres in the cross-linked polymer matrix. ×625.

compartment; D_p, the diffusivity of the drug molecules in the polymer coating membrane; D_s, the diffusivity of the drug molecules in the elution solution; S_l, the solubility of the drug molecules in the microscopic liquid compartment; S_p, the solubility of the drug molecules in the polymer matrix; δ_l, the thickness of the liquid layer around the drug particles; δ_p, the thickness of the polymer coating membrane around the polymer matrix; δ_d, the thickness of the hydrodynamic diffusion layer surrounding the polymer coating membranes; β, the ratio of drug concentration at the inner edge of the interfacial barrier over drug solubility in the polymer matrix; γ', α'/β' in which α' is the ratio of drug concentration in the bulk of elution solution over drug solubility in the elution solution, and β' is the ratio of drug concentration at the outer edge of the polymer coating membrane over drug solubility in polymer coating membrane.

Equation (1) suggests that the controlled release of drugs from the MDD system is zero-order as defined by the linear Q vs t relationship.

Methods of Fabrication

Several physical shapes and sizes of controlled-release drug delivery devices can be fabricated from this MDD system for various biomedical applications. For example, a bandage-type transdermal therapeutic system, called Nitrodisc system (Searle Pharmaceuticals, Inc.), was developed from the MDD system for transdermal controlled administration of nitroglycerin for 24-hr continuous medication in patients suffering from angina pectoris.[7,8] On the other hand, for subcutaneous controlled drug administration, a cylinder-shaped subdermal implant was fabricated also from the MDD system for easy subcutaneous implantation and the controlled release of deoxycorticosterone acetate in rats at zero-order rates for a duration of up to 129 days for cardiovascular pharmacology studies.[9] These are just two examples of the many controlled-release drug delivery devices that can be fabricated from the patented MDD system.

Transdermal MDD Device

For the transdermal controlled administration of systemically effective drugs through an intact skin, a bandage-type transdermal MDD device was first developed to release an androgen, testosterone, at a constant rate (Fig. 2) as defined by Eq. (1).

Topical application of this testosterone-releasing transdermal patch over the skin in the umbilical area of monkeys was observed to achieve a

[7] A. Karim, *Drug Dev. Ind. Pharm.* **9**, 671 (1983).
[8] D. R. Sanvordeker, J. G. Cooney, and R. C. Wester, U.S. Patent 4,336,243 (1982).
[9] Y. W. Chien, L. F. Rozek, and H. J. Lambert, *J. Pharm. Sci.* **67**, 214 (1978).

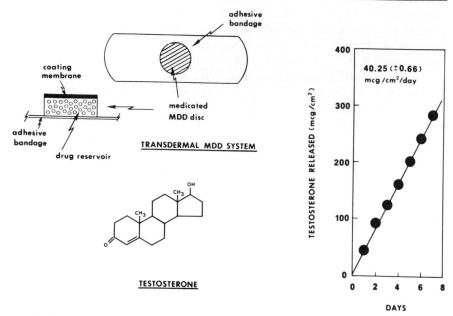

Fig. 2. Development of a bandage-type transdermal MDD device for the transdermal controlled administration of testosterone at constant rate for once-a-month male fertility control.

relatively constant plasma level of testosterone for a duration of up to 31 days.[10] The encouraging results thus paved the foundation for the recent development and marketing of Nitrodisc system (Fig. 3) for the 24-hr continuous transdermal controlled administration of nitroglycerin in patients suffering from angina pectoris.[7,8,11] The Nitrodisc system is fabricated by the following procedure.

1. Thoroughly mix 11 parts of nitroglycerin–lactose triturate (10% w/w) with 7 parts of aqueous polyethylene glycol 400 solution (10% v/v) to form a uniform paste. A drug dispersion compartment is produced.

2. Add the drug dispersion compartment to 57 parts of silicone (medical grade) elastomer (MDX-4-4210). Thoroughly mix the combination, using a high-torque mixer, to form a homogeneous dispersion. A drug–polymer microdispersion is produced.

3. With continuing agitation, 13 parts of isopropyl palmitate, 6 parts of mineral oil, and 6 parts of curing agent are added, in sequence, into the drug–polymer microdispersion to form a pre-MDD formulation.

[10] Y. W. Chien, *J. Pharm. Sci.* **73**, 1064 (1984).
[11] Y. W. Chien, *Drug Dev. Ind. Pharm.* **9**, 497 (1983).

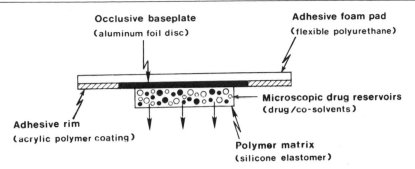

FIG. 3. Cross-sectional view of a unit of Nitrodisc system, a transdermal therapeutic system for transdermal controlled administration of nitroglycerin at controlled rate for one-a-day medication of anginal pectoris.

4. Expose the well-mixed pre-MDD formulation to a vacuum of 71 cm, with continuing agitation, for at least 30 min or until all the entrapped air is removed.

5. Pour an adequate amount of the deaerated pre-MDD formulation into disk-type molds.

6. Polymerize the pre-MDD formulation in the molds by heating at 60° in an air-circulating oven (or a vacuum oven) for 2 hr. MDD disks are thus formed.

7. Remove the MDD disks from the molds and glue them, individually, onto the occlusive baseplate at the center of the adhesive foam pad. After covered with surlyn laminated foil, a unit of Nitrodisc system is produced (Fig. 3).

In vitro skin permeation studies in hairless mouse[12] indicated that a constant rate of skin permeation of nitroglycerin, as expected from Eq. (1), was achieved. It maintained a constant plasma level of nitroglycerin in humans up to 32 hr (Fig. 4).

Subdermal MDD Implants

Several kinds of cylinder-shaped subdermal implants have been developed from the MDD technology for easy implantation in subcutaneous tissues and for the subcutaneous controlled administration of a number of systemically effective drugs. Examples include desoxycorticosterone acetate-releasing subdermal implants for producing a long-acting (100-day)

[12] Y. W. Chien, P. R. Keshary, Y. C. Huang, and P. P. Sarpotdar, *J. Pharm. Sci.* **72**, 968 (1983).

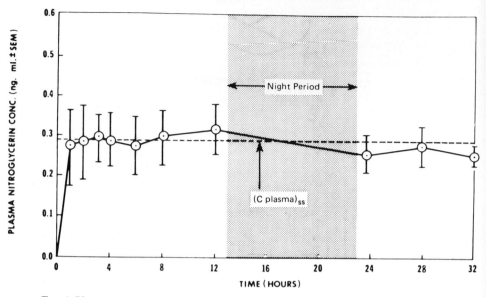

FIG. 4. Plasma concentration profile of nitroglycerin in 12 healthy human volunteers after 32-hr topical application of Nitrodisc system, one unit (16 cm²) for each volunteer.[7]

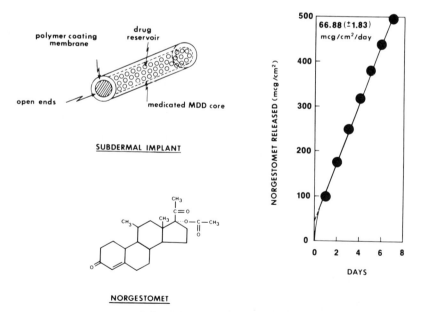

FIG. 5. Development of a cylinder-shaped subdermal implant for the 16-day subcutaneous controlled administration of norgestomet at constant rate for estrous synchronization in domestic animals.

hypertensive rat for cardiovascular pharmacology studies[9] and the norgestomet-releasing subdermal implants for synchronization of estrus in domestic animals.[13,14] The general procedure used for the fabrication of subdermal MDD implants is outlined below.

1. Thoroughly mix 4 parts of finely milled norgestomet crystals with 6 parts of aqueous polyethylene glycol 400 solution (40% v/v) to form a uniform paste. A drug dispersion compartment is produced.

2. Add the drug dispersion compartment to 82.5 parts of silastic (medical grade) elastomer 382. Thoroughly mix the combination, using a high-torque mixer, to form a homogeneous dispersion. A drug–polymer microdispersion is produced.

3. With continuing agitation, 7.5 parts of Dow Corning medical fluid 360 and 15 drops of catalyst M (for every 100 g of mixture) are added, in sequence, into the drug–polymer microdispersion to form a pre-MDD formulation.

4. Expose the well-mixed pre-MDD formulation to a vacuum of 71 cm, with continuous agitation, for 5 min or until all the entrapped air is released.

5. Deliver the deaerated pre-MDD formulation into polymeric tubings, e.g., silastic (medical grade) tubings, with a controlled membrane thickness, via a specially designed multihead extrusion pump.

6. Polymerize the deaerated pre-MDD formulation in the polymeric tubings by heating at 60° in an air-circulating oven for 1 hr. MDD implants are thus formed.

7. Cut the MDD implants into sections of 2–4 cm in length. The subdermal MDD implants are thus produced (Fig. 5).

As illustrated in the *in vitro* results (Fig. 5), the *in vivo* studies[9,13,14] also demonstrated that the rate of release of drugs from the MDD implants in the subcutaneous tissues is constant (Fig. 6) as expected from Eq. (1). The open ends on the subdermal implants did not affect the mechanism of subcutaneous drug release and the clinical efficacy of the subcutaneous controlled estrous synchronization.[14]

Vaginal MDD Ring

A donut-shaped vaginal device can be fabricated from the MDD system for the intravaginal controlled administration of an orally active con-

[13] Y. W. Chien, "Novel Drug Delivery Systems: Fundamentals, Developmental Concepts and Biomedical Assessments," Chapter 8. Dekker, New York, 1982.
[14] Y. W. Chien, *Int. Conf. Drug Absorption: Rate Control Drug Ther.*, 2nd, 1983, Abstracts, p. 18 (1983).

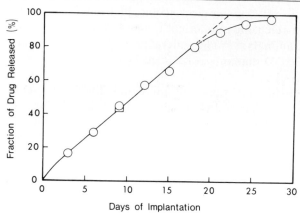

Fig. 6. Subcutaneous controlled release of norgestomet from the MDD-subdermal implants in 20 United States cows is at constant rate for up to 18 days of implantation and 80% of the loading dose.

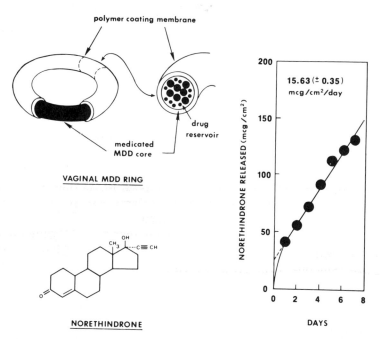

Fig. 7. Development of a donut-shaped vaginal MDD device for 3-week intravaginal controlled administration of norethindrone at controlled rate for cyclic contraception in humans.

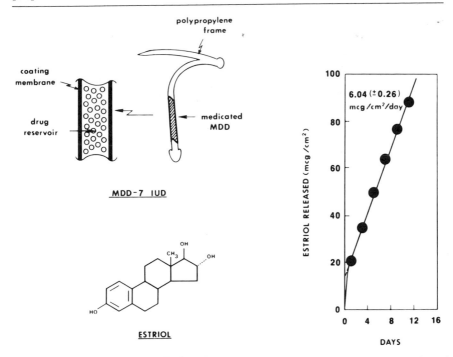

FIG. 8. Development of a 7-shaped IUD for the intrauterine controlled administration of estriol at controlled rate for yearly contraception in humans.

traceptive progestin, such as norethindrone, or of a combination of one progestin with one estrogen, such as the combination of ethynodiol diacetate and mestranol to simulate the composition in the oral contraceptive pills.

The procedure for the fabrication of vaginal MDD rings is similar to that outlined earlier for subdermal MDD implants, except that in the final stage (step 7) the MDD implants are cut into sections of 16 cm in length. The ends of each section of MDD implant are then bound together to form a donut-shaped vaginal ring by applying a thin layer of Silastic Medical Adhesive (Silicone Type A) and then inserting both ends into a small section (2 cm in length) of silastic (medical grade) tubing (which has been swollen reversibly in diameter by an organic solvent) (Fig. 7).

MDD-7 IUD

A 7-shaped intrauterine device (IUD) can be fabricated from the MDD system for intrauterine controlled administration of an estrogen, such as estriol, or a progestin, such as progesterone, to achieve a localized contraceptive action.

The procedure for the fabrication of MDD-7 IUD is similar to that outlined earlier for subdermal MDD implants, except that in the final stage (step 7) the MDD implants are cut into sections of 2 cm in length. The implant is then bound to a unit of 7-shaped polypropylene frame by gluing the ends of each MDD implant to the vertical stem of the plastic frame (Fig. 8).

[35] Osmotic Delivery Systems for Research

By JOHN. W. FARA

Over the past 10 years the concept of rate-controlled drug delivery has become an integral part of pharmacological research. During this time new insights into drug action and interaction have emerged, largely through the introduction and use of drug delivery systems which enable the investigator to choose the rate and duration at which particular agents are given. Two of these delivery devices for experimental research are based on osmotic technology: one is the ALZET (registered trademark of ALZA Corporation) osmotic pump for studies in animals[2-3]; the other is the OSMET (trademark of ALZA Corporation) module for administration of drugs in clinical research.[2,3] This chapter will focus on the functionality of these osmotic devices and on the impact their use is having on pharmacological research.

ALZET Osmotic Pumps

Developed in the mid-1970s, these osmotic pumps have nearly replaced the clumsy alternative of connecting animals to heavy infusion pumps (Fig. 1). Because they are implantable and of relatively small size, they provide for the unattended administration of solutions and suspensions of drugs and other bioactive agents to animals as small as mice and as large as dogs, sheep, and baboons. Pumps available at this time have delivery rates and durations of 1 or 10 μl/hr for 7 days, 0.5 or 5 μl/hr for 14 days, or 2.5 μl/hr for 28 days.

[1] F. Theeuwes and S. I. Yum, *Ann. Biomed. Eng.* **4**, 343 (1976).
[2] B. Eckenhoff, F. Theeuwes, and J. Urquhart, *Pharm. Technol.* **5**, 35 (1981).
[3] B. Eckenhoff and S. I. Yum, *Biomaterials (Guildford, Engl.)* **2**, 89 (1981).

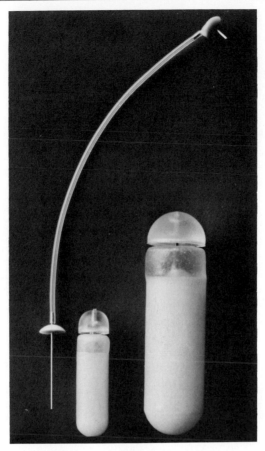

FIG. 1. The ALZET osmotic pump.

Internally, each pump consists of the following components (Fig. 2): an inert, impermeable, flexible drug reservoir open to the exterior via a single orifice; a thin sleeve of osmotic agent surrounding the reservoir; and, a semipermeable membrane surrounding the sleeve of the osmotic agent. The researcher fills the reservoir through the orifice with the solution or suspension of interest, and a flow moderator is inserted into the orifice. This restricts diffusion of the agent from the exit portal, thereby assuring osmotic control of delivery. After the delivery system is filled and implanted, water from surrounding tissues begins to move osmotically through the membrane into the osmotic sleeve. The permeability of the membrane controls the rate at which water moves into the osmotic sleeve; the rigidity of the membrane causes the swelling of the osmotic

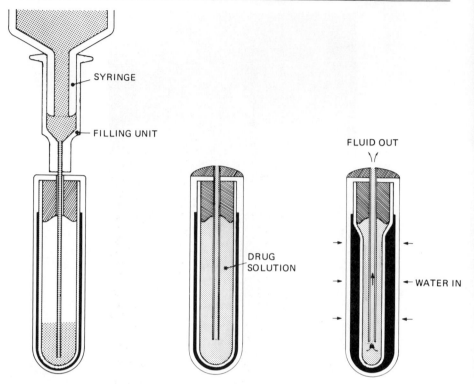

Fig. 2. Cross section of the osmotic pump: (left) being filled using a syringe and filling unit; (middle) filled with solution and with flow-moderating cap in place; and (right) in operation, with water entering the membrane by osmosis and solution being pumped out.

sleeve to displace the liquid drug formulation within the reservoir, out through the portal, continuously and at a controlled rate.

In addition, a transparent overcap can be removed from the flow moderator to permit a catheter to be attached to carry the drug solution to areas remote from the site of implantation, such as one of the cerebral ventricles, a specific brain center, and the lumen of the stomach or intestine. With attached catheters, the pumps have also been used to deliver drugs intravenously or intraarterially.

The osmotic delivery mechanism is capable of generating the forces necessary to pump even quite viscous solutions through the smallest plastic catheters or metal cannulae. Also, the rate specifications of the osmotic pumps include both temperature and osmolarity correction factors, to allow each researcher to use the pump accurately in diverse applica-

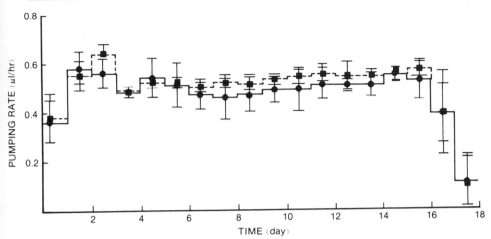

FIG. 3. *In vivo* and *in vitro* pumping rates of an osmotic pump designed to deliver 0.5 μl/hr of agent for 14 days.

tions in both warm- and cold-blooded animals, in seawater and in the laboratory. The continuous-delivery performance characteristics of one model of the osmotic pump are shown in Fig. 3.

These osmotic systems are designed to deliver at a constant rate by incorporating into the sleeve compartment a defined mass of osmotic driving agent that suffices to maintain constant osmotic activity in the compartment throughout the functional lifetime of the system.

The constant or zero-order mass delivery rate from the system is represented by Eq. (1).

$$K_0 = (dV/dt)C_d \tag{1}$$

Here, C_d represents the concentration of drug formulated by the researcher and dV/dt the volume imbibition rate of water into the osmotic sleeve compartment.

With π_o the osmotic pressure of the osmotic agent, and π_e the osmotic pressure of the environment, the volume delivery rate from the system can be written as

$$(dV/dt) = P_o(\text{area}/h)(\pi_o - \pi_e) \tag{2}$$

where area and h are the membrane area and thickness, respectively, and P_o is the osmotic permeability coefficient of the membrane. Design characteristics of the osmotic pump are described in detail by Theeuwes and Yum.[1]

AGENTS DELIVERED BY OSMOTIC PUMPS IN RECENTLY
REPORTED EXPERIMENTS

Peptides
ACTH	Insulin
Angiotensin	LH
Bombesin	LHRH
Bungarotoxin snake venom	LRHR analogs, agon/antag
Calcitonin	α-MSH
Cholecystokinin	Muramyl dipeptide
Dermorphin	Neurotensin
Endorphins	Parathyroid hormone
Enkephalins	Pentagastrin
Erythropoietin	Pituitary extract
FSH	Prolactin
Gastrin	Somatomedins
Glucagon	Teprotide
Growth hormones	Tetragastrin
IGF	Vasopressin

Steroids
Aldosterone	Estriol
Catecholestrogens	Estrone
Corticosterone	Methoxyestrone
Dexamethasone	Progesterone
DOCA	Spironolactone
Estradiol	Testosterone

Other substances
Anesthetics	GI motility modulators
Antibacterials	Heavy metals
Anticancer agents	Immunologic agents
Anticoagulants	Indicator substances
Antiepileptics	Metabolites
Antihypertensives	Neurotransmitters
Anti-Parkinson agents	Nerve growth factors
Antivirals	Nucleosides and nucleotides
Carbonic anhydrase inhibitors	Prostaglandins
Carcinoma antibodies	Radioisotopes
Catecholamines	Renin–angiotensin and inhibitors
Chelators	Thyroid/thyroid-related hormones
Cholinergics	Vitamins and minerals
CNS-acting agents	Miscellaneous test vehicles and solvents
Enzymes	

Pump Applications

Use of this osmotic technology in animal research has provided convenience as well as the logic for selecting the appropriate drug delivery
pattern, eliminating the need for painstaking empirical investigations. The

systems have, moreover, allowed the development of new approaches to the evaluation of potent hormones, peptides, and other agents.

Review of over 950 scientific publications in which ALZET pumps have been used reveal several general areas of application. More than 150 agents have been delivered by this means (see table) into various animal species including mice, rabbits, rats, dogs, monkeys, baboons, and sheep.[4] In addition, novel approaches have been described in cold-blooded animals including goldfish, trout, hermit crabs, and snakes (personal communications).

Delivery of Short Half-Life Agents

The ALZET osmotic pumps have been extensively used in research to deliver short half-life agents, notably peptides. Because most of these compounds have a half-life of minutes, their bolus delivery via syringe for example will produce only a short period of tissue exposure to them. Thus, one must make observations at precisely the right moment to recognize certain of their actions. In such cases rate-specified administration—but not necessarily constant-rate administration—is the proper delivery mode, for the peptide action may be maintained after delivery has ceased.

On the other hand, effects of some agents may fade during constant-rate administration because receptors become refractory to the stimulus. In this case, the logical step is not to retreat back to single-dose, bolus administration, but rather to advance to rate-varying patterns of administration. This will be further discussed shortly.

Delivery Systems as Artificial Organs

One method employed in research in endocrinology and other biomedical disciplines is to remove an organ or part of an organ by surgery or other means and then observe the resulting effect on a discrete system or the whole animal. Replacement of substances normally produced by that organ or tissue is then used to explore the mechanisms of hormone release and hormone action. This approach has been applied to the adrenal glands, pancreas, parathyroid, pituitary, thyroid, and indeed to most, if not all, other endocrine structures.

As an artificial pancreas, osmotic delivery systems have been implanted to infuse insulin to rats with experimentally (streptozotocin) induced diabetes.[5,6] Since the rat does not eat meals but rather nibbles day

[4] J. Urquhart, J. W. Fara, and K. L. Willis, *Annu. Rev. Pharmacol. Toxicol.* **24,** 199 (1984).

[5] S. I. Yum, S. A. Tillson, and F. Theeuwes, *Abstr., Short Commun. Poster Presentations, 5th Int. Congr. Endrincol.* p. 366 (1976).

[6] D. G. Patel, *Proc. Soc. Exp. Bio. Med.* **172,** 74 (1983).

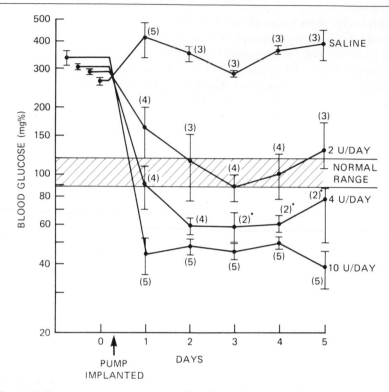

Fig. 4. Delivery rate-response curves of insulin administered at 2, 4, and 10 U per day in streptozotocin-diabetic rats (200 g).

and night, large pulses of insulin release, typical of meal-eaters, do not ordinarily occur in this species. Thus, continuous infusion is a means of assaying the total daily need for insulin. In the example illustrated in Fig. 4, osmotic pumps were filled with different concentrations of unmodified crystalline bovine insulin in isotonic saline and implanted subcutaneously to infuse doses of 2, 4, and 10 U per 200-g rat per day.[5] Isotonic-saline-filled pumps were implanted subcutaneously in other (control) rats with induced diabetes. The temporal dose–response curve (Fig. 4) showed that a dose of 2 U/day was sufficient to return the streptozotocin-diabetic rat to normoglycemic levels of 90–120 mg%.

In another study of insulin replacement regimens, Patel[6] was able to maintain a streptozotocin-diabetic colony over 60–80 days by the repeated replacement of 2 week-duration osmotic pumps.

Although rate control is built into these osmotic delivery systems, it is not yet clear which drugs or bioactive agents are best given by a *constant* rate. For some drugs, a *pattern* of input at different rates may be optimal. Additionally, the true physiological effect of some hormones may be observable only when the agent is given by an on–off or phasic–tonic administration.

For example, a common temporal pattern of endogenous signals is the circadian rhythm, which has recently been mimicked for melatonin by an adaptation of the osmotic pump.[7] In this study, Lynch and colleagues connected each osmotic pump to a coil of fine-gauge polyethylene tubing containing an alternating sequence of vehicle–drug solution–vehicle–drug solution; the lumen volume of the coiled tubing equaled the reservoir volume of the pump. The pump and attached coil were then implanted in rats subcutaneously. Constant inflow of fluid from the pump into the coil displaced the alternating sequence of vehicle–drug solution, thereby causing delivery of melatonin in an on–off time pattern programmed over 6 days. The agent was subsequently recovered from urine in a pattern equilvalent to its subcutaneous delivery.

This is one of the first published accounts of the use of ALZET osmotic pumps to achieve a temporally patterned delivery. It stimulates one to consider to what extent, with properly designed experiments, other circadian patterns can be mimicked or uncovered.

Minimizing Stress

Biomedical researchers are aware of the biochemical and hence physiological changes that stress causes. Generally, researchers deal with this by administering only vehicle to a control group of animals, according to the same regimen that they use to administer drug to the other animals. This procedure does not eliminate stress; it only reveals the effects of drugs or other administered agents in stressed animals. Therefore, a large portion of the endocrine and pharmacological literature probably deals with stress as a unintended, unmeasurable variable. The implantable osmotic delivery systems help to solve this problem of experimental design; their use can lead to very different results from those seen when stress is present.[4] This statement applies particularly to behavioral studies where chronic administration without the stress of frequent injections is essential.

Exploring Schedule Dependency of Drug Action

As mentioned earlier, although rate control is built into these new osmotic delivery systems, it is not yet clear which drugs or bioactive

[7] H. J. Lynch, R. W. Rivest, and R. J. Wurtman, *Neuroendocrinology* **31,** 106 (1980).

agents are best given by a constant rate. For some drugs, a pattern of input may be optimal. For others, a steady-state level may be most efficacious. Thus, it is becoming increasingly common in biomedical research to test drugs at very early stages of development for their regimen-dependent actions, that is, administering them continuously to achieve steady-state drug levels and comparing the efficacy and toxicity to that obtained by administering them in a multiple-injection or pulsed regimen. Indeed, by comparing drug regimens, it has been shown that one can quite readily sort out effects of the agent associated with peak and trough levels in the blood and those associated with constant plasma concentrations.[8,9]

In general, such studies of regimen dependence or schedule dependence of drug action have not been a standard part of assessing drug action. Not only were such studies difficult and expensive to conduct, but they offered little prospect of having any practical importance. However, the introduction of these osmotic drug delivery systems made it increasingly possible to design drug regimens and delivery patterns at all stages of drug research. Two examples of these will be briefly discussed: one addresses toxicity testing, and the other the delivery of a cancer chemotherapeutic agent.

Nau and co-workers[10,11] recently reported that the embryo toxicity of valproic acid (VPA) in mice is markedly dependent on its regimen of administration. The half-life of this widely used antiepileptic drug is only 0.8 hr in the mouse compared with 8–16 hr in humans. Nau and co-workers administered it in the same total dose to mice on days 7–15 of gestation in two different regimens: (1) by injection once daily and (2) by continuous infusion from implantable pumps. With the continuous infusion, drug levels were maintained within the concentrations used for therapy. The once-daily injections causes valproic acid concentrations in plasma to peak and decline quickly (Fig. 5); for long periods between injections, the drug was undetectable. In humans the peaks are only one-tenth as high, and the drug is not totally eliminated from the plasma between doses. Thus, extrapolating the results of mouse toxicity data to humans can give rise to two types of errors: the high peak levels in mice may lead to overestimating the human toxicity of VPA; or conversely, the long periods of no detectable drug in mice may lead to underestimating its human toxicity. With the continuous infusion of VPA, Nau found that a higher total dose was required to produce embryo toxicity than with a

[8] J. Urquhart, *Drugs* **23,** 207 (1982).
[9] J. Fara and J. Urquhart, *Trends Pharmacol. Sci.* **5,** 21 (1984).
[10] H. Nau, R. Zierer, H. Spielmann, D. Neubert, and C. Gansau, *Life Sci.* **29,** 2803 (1981).
[11] H. Nau and H. Spielmann, *Lancet* **1,** 763 (1983).

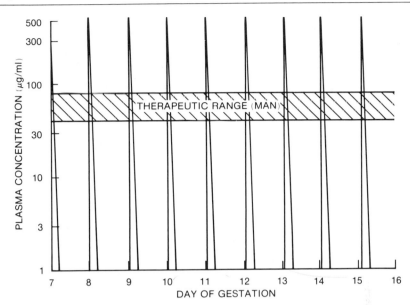

FIG. 5. Concentrations of valproic acid in mouse plasma versus time following subcutaneous administration of the drug (400 mg/kg) once daily between days 7 and 15 of gestation. The shaded area indicates the human plasma concentrations observed during pregnancy, which are lower than those observed in nonpregnant adult epileptics. (From ref. 10.)

single daily administration, as shown by exencephaly, resorptions, and fetal weights (Fig. 6).

Thus, the results of the once-daily injection regimen, considered alone, give an incomplete view of the toxicity of valproate. However, when the injection data are considered together with the data from the constant-rate regimen, the regimen dependence of the drug's toxicity becomes evident.

Extrapolating this situation to other drugs and other toxicology test protocols, one wonders the extent to which drug toxicity is sometimes underestimated, because of the very low trough concentrations that may occur with daily or even twice-daily injection regimens in small animals.

These same considerations apply to efficacy studies. The second example of a regimen-dependent expression of drug action is the work of Sikic and colleagues.[12] These investigators had previously observed regimen-dependent effects associated with continuous infusion of the anticancer drug bleomycin to mice with lung carcinoma. Identical doses were

[12] B. Sikic, J. M. Collins, E. G. Mimnaugh, and T. E. Gram, *Cancer Treat. Rep.* **62,** 2011 (1978).

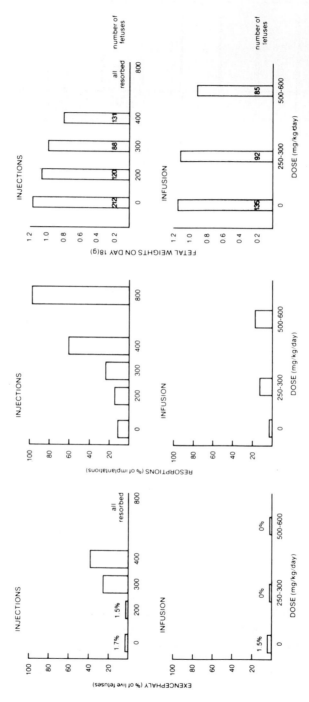

FIG. 6. Incidence of exencephaly, resorption, and fetal weight loss in mice following administration of valproic acid to dams between days 7 and 15 of gestation. Upper graphs: Multiple injections (once daily). Lower graphs: Constant rate infusion via implanted ALZET osmotic pumps. (From ref. 10.)

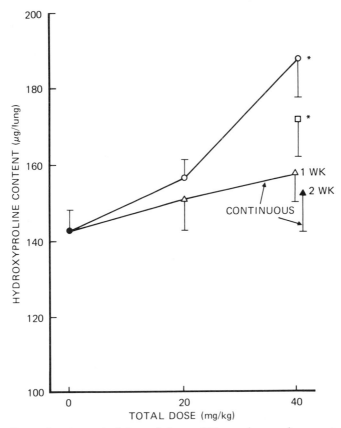

FIG. 7. Effects of various schedules and doses of bleomycin on pulmonary toxicity in nontumored animals, as measured by lung hydroxyproline content 10 weeks after treatment. (From ref. 12.)

administered in three different 7-day regimens (injections twice daily, injections twice weekly, and continuous subcutaneous infusions). With drug infusion, results differed in two ways from those obtained with the other regimens: (1) the drug's pulmonary toxicity (Fig. 7) was less (e.g., fewer deaths, less pulmonary fibrosis); and (2) tumor size was smaller (Fig. 8). These results suggest that bleomycin should be administered clinically by continuous infusion instead of by injections. This enhanced efficacy observed with continuous infusion of bleomycin was confirmed by Peng and colleagues,[13] and subsequent clinical studies also appear to

[13] Y. M. Peng, D. S. Alberts, H. S. Chen, N. Mason, and T. E. Moon, *Br. J. Cancer* **41,** 644 (1980).

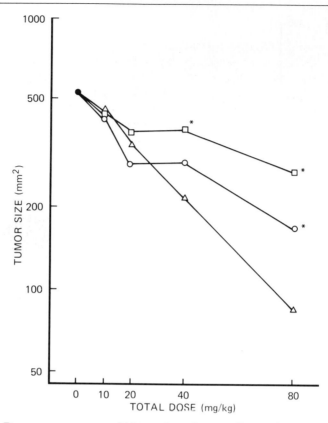

FIG. 8. Dose–response curve of bleomycin antitumor effect against Lewis lung carcinoma comparing three schedules of administration. These measurements, made on day 15 after osmotic pump implantation, are representative of differences that existed throughout the course of tumor growth. (From ref. 12.)

confirm the prediction that the human use of bleomycin may be made both safer and more efficacious by use of a constant-rate infusion regimen.[14,15]

These examples are merely illustrative of the growing number of studies underway to test the advantages and disadvantages of constant-rate infusion vs. intermittent injections in the administration of bioactive agents. Rate- and duration-specified drug delivery in the osmotic pumps is playing a key role in these studies.

[14] G. Coonley, D. Vugrin, C. La Monte, and M. J. Lacher, *Proc. Am. Assoc. Cancer Res.* **22,** 369 (1981).
[15] K. R. Cooper and W. K. Hong, *Cancer Treat. Rep.* **65,** 419 (1981).

OSMET Pumps

The osmotic delivery system technology also has been adapted to provide orally or rectally administered drug delivery systems used in clinical pharmacology. These drug delivery modules have the same internal volume, cross section, and operational principles as described above, but they deliver their content over shorter durations. The modules are supplied empty and filled by the investigator with 0.2 ml of test agent solution or suspension, which they deliver continuously to the gastrointestinal tract at a near-constant rate of 8, 15, or 25 μl/hr, respectively, for 24, 12, or 8 hr. The 2-ml capacity systems for rectal or vaginal administration deliver 60 μl/hr for 30 hr or 120 μl/hr for 15 hr. Their use permits assessments of the effects of constant-rate delivery on the agent's pharmacological, pharmacokinetic, and pharmacodynamic attributes, e.g., biological actions, bioavailability, and time course of plasma concentrations. The modules also provide what may be the simplest means of defining the gastrointestinal absorption window of a drug. This follows from measured discrepancies between the extent of drug absorbed vs extent delivered and the time course of absorption rate vs that of delivery rate. Since the modules simulate the action of rate-controlled solid oral dosage forms, they provide specifications of drug delivery rate and duration that are basic to efficient pharmaceutical development of such forms.

De Leede and co-workers[16,17] have utilized the modules in human subjects to study rectal administration of antipyrine and the choline salt of theophylline over 98 and 72 hr, respectively. For both agents, use of OSMET systems sequentially by each subject resulted in prolonged maintenance of virtually constant plasma levels of drug and good agreement of *in vitro* and *in vivo* functionality of the dosage form.

These OSMET modules have also been used in clinical investigative studies[18] to deliver antiinflammatories, antihypertensives, vitamins, and various types of receptor-blocking agents.

Conclusions

Practical methods for the rate-controlled administration of drugs and other bioactive agents are bringing new capabilities to basic biomedical experimentation and clinical research. The ALZET osmotic pumps and

[16] L. G. J. De Leede, A. G. De Boer, and D. D. Breimer, *Biopharm. Drug Dispos.* **2,** 131 (1981).

[17] L. G. J. De Leede, A. G. De Boer, S. L. Van Velzen, and D. D. Breimer, *J. Pharmacokinet. Biopharm.* **10,** 525 (1982).

[18] C. G. Wilson, J. G. Hardy, and S. S. Davis, *Pharm. J.* **231,** 334 (1983).

the OSMET drug delivery module differ from conventional sustained-release, depot, or slow-release products by their ability to precisely specify rate and duration of drug release *in vivo* on the basis of simple *in vitro* tests.

Until the multipurpose, rate-controlled ALZET implants became available for animal use, rate-controlled drug administration in preclinical studies was complicated by the need for animal restraint and special equipment. Therefore, researchers in general seem to have regarded the elucidation of an agent's steady-state actions—based on extended-duration, rate-controlled administration—as largely of academic interest. Thus, no one yet knows the scope of the opportunities that await use of rate- and duration-controlled delivery in research. Investigators are only beginning to explore the potential of such delivery in the numerous fields encompassing biomedical experimentation. Little use of ALZET and OSMET pumps has been reported from outside the fields of biology and medicine, although doubtless applications await the innovative investigator.

[36] Enzymatically Controlled Drug Release Systems

By B. D. RATNER and T. A. HORBETT

Introduction

The methods used to deliver drugs to the body have evolved through a number of stages. The simplest methods involve a relatively large bolus of drug introduced by any of a number of means (e.g., oral, intravenous) into the body. On the average, the drug is administered at the appropriate dosage, but at certain time points, an inappropriate level of drug can be observed (see Fig. 1A). A more recent innovation has utilized various barrier membranes or erosion schemes to deliver drug over an extended period of time at a constant level to the body (Fig. 1B). Of course, the drug is delivered at this level whether the body needs it or not. The future of therapeutically effective drug delivery systems may lie with devices which will sense the needs of the body and deliver a drug at appropriate levels in response to the physiological or pathological indicators produced by the body (Fig. 1C). Such a device would act much like an organ in the body (e.g., the pancreas).

Enzymes are well suited for incorporation into such bioresponsive drug delivery devices. The high specificity and high catalytic activity of

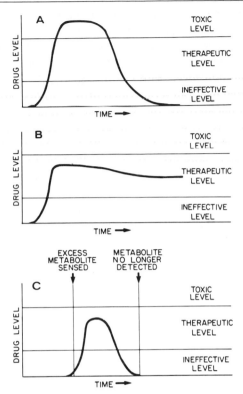

FIG. 1. Physiological drug level versus time for three generations of drug delivery systems. (A) Direct administration (via injection or oral ingestion) of a bolus of drug. (B) Zero-order release controlled by an erosion or barrier membrane system. (C) Drug release in response to the physiological needs of the body.

enzymes permit properly designed drug delivery devices to respond only to a metabolite of interest in a sensitive and rapid manner. This chapter will review the use of enzymes in drug delivery devices and will describe two methods which can be used to fabricate such devices.

Descriptions of Enzyme-Controlled Drug Release Systems

Very few studies of enzyme-controlled polymeric drug delivery systems have been carried out to date and none are as yet close to clinical application. However, the concept of using enzymes to control drug delivery is sufficiently appealing that a number of systems have been conceived. Many problems remain to be overcome before the promise of this approach can be fulfilled. Specific areas in which further development will be required for all the systems described in this chapter include increasing response speed, increasing sensitivity to small changes in the concentration of the molecule being detected, ensuring reproducibility and reversibility, inhibiting activity loss, and improving biocompatibility.

The use of immobilized enzymes in conjunction with electrochemical sensors has been studied extensively because of the ease with which an electrical signal can be coupled to a mechanically activated drug delivery system (e.g., the various glucose-controlled insulin infusion systems employing immobilized glucose oxidase in the electrochemical glucose sensor). In this article, however, we consider only systems in which the enzyme is intended to directly control drug delivery by affecting the properties of a polymer.

A model system for enzyme control of drug delivery was developed by Heller and Trescony[1] and constitutes the earliest known experimental test of this concept. The system was made by encapsulating a hydrolytically unstable polymer (n-hexyl half-ester of a methyl vinyl ether–maleic anhydride copolymer) inside a glutaraldehyde cross-linked albumin–urease "hydrogel." The encapsulated polymer contained hydrocortisone, whose release kinetics were studied. The hydrolysis and solubility of the anhydride side groups in this type of polymer had previously been shown to be extremely sensitive to pH and dependent on the size of the alkyl group in the copolymer ester.[2] The encapsulation with an enzyme coating capable of generating basic product was therefore capable of altering the dissolution rate of the anhydride-containing polymer and its dissolved hydrocortisone in response to variations in the concentration of the enzymes' substrate (urea, in this case). Because many enzymes produce acidic or basic products and because enzymes are sensitive and specific catalysts, this approach offered the possibility of tailoring the device to the requirement for controlling the delivery of a variety of drugs in response to changes in the concentration of a wide variety of metabolites. Heller and Trescony chose to demonstrate the approach with urease and hydrocortisone in a model system with no specific therapeutic application in mind. They found that hydrocortisone release rates were enhanced severalfold when urea was added to the media, that the enhancement depended somewhat on the urea concentration, and that the rate of release decreased when the polymer was removed from the urea environment. Repetitive transfers between urea-free and urea-containing solutions resulted in consistent decreases or increases in hydrocortisone release rates. The increase in rates of release upon transfer to urea-containing solutions oc-

[1] J. Heller and P. V. Trescony, *J. Pharm. Sci.* **68,** 919 (1979).
[2] J. Heller, R. W. Baker, R. M. Gale, and J. O. Rodin, *J. Appl. Polym. Sci.* **22,** 1991 (1978).

curred over an approximately 5-min period while the decrease in rate upon transfer to urea-free solutions was slower (typically 15 min) and exhibited exponential decay. Deposition of the hydrolyzed polymer on the surface of the albumin–urease hydrogel was noted, apparently as a result of loss of solubility as the hydrolyzed polymer entered the lower pH region of the bulk phase.

This model system thus illustrates the fundamental feasibility of enzyme-controlled drug release polymers since all the essential features of such systems were observed, i.e., specificity to a selected metabolite and metabolite-controlled variations in drug delivery. These studies also suggest certain difficulties in putting such systems into practice, e.g., the limited range of sensitivity to urea and insolubilization of the eroded polymer. They did not address many other relevant issues such as the toxicity, biocompatibility, and actual usefulness of a system which might be encapsulated by relatively impermeable inflammatory tissue when implanted. Nonetheless, this pioneering study suggested the feasibility of the development of enzyme-controlled drug delivery from polymers.

The development of noneroding polymers capable of reversible swelling in response to a specific metabolite has been pursued in our laboratories.[3–5] Glucose-sensitive membranes which could be used for glucose detection or for directly controlling insulin permeation rates have been the specific focus of this work, but the approach is general. Thus, by immobilizing an appropriate enzyme in an ionizable, swellable membrane, changes in the swelling of the membrane in response to changes in the concentration of the metabolite will result as long as the enzyme-catalyzed turnover of the metabolite results in changes in the local pH in the domain of the membrane. The need for acid or basic product and a highly pH-sensitive polymer in close proximity to the site of enzyme catalysis is conceptually identical to the system described in Heller and Trescony's work, but the use of polymers capable of undergoing swelling rather than erosion is a key difference. Thus, for example, questions about the toxicity of the erosion products and the ability to use such polymers as biosensitive transport rate-limiting barriers between a drug reservoir and the body environment provide additional flexibility to enzyme-controlled drug delivery from polymers.

The glucose-sensitive membranes we have studied employ glucose

[3] J. Kost, T. A. Horbett, B. D. Ratner, and M. Singh, *J. Biomed. Mater. Res.* (in press).
[4] T. A. Horbett, B. D. Ratner, J. Kost, and M. Singh, p. 209 *in* "Recent Advances in Drug Delivery Systems" (S. W. Kim and J. M. Anderson, eds.). Plenum, New York, 1984.
[5] T. A. Horbett, J. Kost, and B. D. Ratner, *in* "Polymers as Biomaterials" (S. Shalaby, A. S. Hoffman, T. A. Horbett, and B. D. Ratner, eds.). Plenum, New York, 1985 (in press).

oxidase entrapped in an amine-containing hydrogel of relatively high water content. Hydrogels of this type undergo relatively large changes in swelling in response to pH changes. Glucose oxidase (GO)-catalyzed turnover of glucose results in gluconic acid formation. Membranes of this type do, in fact, swell and change their permeability in response to variations of glucose in the physiological concentration. The changes in swelling are not directly proportional to the concentration of glucose, however, and appear to reach a maximum below the maximum physiological concentration of glucose. However, substantial effects of membrane formulation on glucose sensitivity have been observed. It appears that optimum properties for such systems will require further investigation to begin to sort out the rather complex and as yet poorly understood relationships between enzyme loading, membrane formulation, and physical configuration of the system (e.g., the effect of membrane shape and thickness may have major effects on the swelling kinetics because of mass transfer changes). Some of the other feasibility criteria for such a system have been briefly examined and do not seem to preclude the eventual workability of this system. The reader is referred to our publications[3–5] for a full description of these systems as space does not permit it here. Thus, for example, the enzyme is relatively stable and its presence in great excess in the membrane renders the latter relatively stable in its response to glucose. Also, the membrane exhibited no toxicity or excessive inflammatory response in the mouse peritoneal and subcutaneous spaces after 4 weeks. On the other hand, insulin permeation rates are much lower than we expected from literature studies of other hydrogels. This is presently under active investigation. As with Heller and Trescony's work, these studies of glucose-sensitive membranes have suggested the fundamental feasibility of the approach but revealed many problems to be overcome in developing enzyme-controlled drug delivery from polymers.

A third approach to enzyme-controlled drug delivery from polymers taken by Ishihara *et al.* uses membranes capable of oxidation–reduction reactions to control insulin permeation rates.[6] Their system consisted of two polymeric layers held in direct contact between the halves of a permeation cell in which insulin transport was studied. One layer was made of polyacrylamide-containing entrapped glucose oxidase while the second layer was a complex redox copolymer made from 3-carbamoyl-1-(*p*-vinylbenzyl)pyridinium chloride (prepared from the reaction of nicotinamide and 4-chloromethylstyrene) and 2-hydroxypropyl methacrylate. The nicotinamide side chain in this copolymer is neutral in the reduced state but

[6] K. Ishihara, M. Kobayashi, and I. Shinohara, *Makromol. Chem., Rapid Commun.* **4,** 327 (1983).

becomes positively charged upon oxidation. Insulin permeation through the membrane was shown to increase from 3.98×10^{-7} cm^2/sec to 5.85×10^{-7} cm^2/sec upon addition of glucose to the buffer reservoir downstream of the insulin reservoir. Infrared spectra on membranes exposed to hydrogen peroxide demonstrated the transformation of the reduced form to the oxidized form. It is therefore clear that H_2O_2 derived from glucose oxidase catalyzed turnover of glucose in the polyacrylamide membrane was able to oxidize and enhance the permeability of the adjacent redox-sensitive copolymer to insulin. The authors attributed the enhancement to increased swelling in the positively charged membrane as a result of water uptake, although swelling measurements were not reported. A fundamental limitation of such a system for controlling insulin delivery is that no mechanism for reversing the oxidation to cause a return of the membrane to its reduced glucose-sensitive state is apparent, i.e., the glucose-induced changes appears irreversible. Ishihara *et al.* allude to the difficulties of this limitation but only suggest that "research is underway" to overcome it.

Experimental

Preparation of a Membrane Which Swells in Response to Glucose

Materials. 2-Hydroxyethyl methacrylate (HEMA) monomer, obtained in a highly purified form from Hydron Laboratories, Inc. New Brunswick, NJ, was used as received. N,N-Dimethylaminoethyl methacrylate (NNDMAEM), obtained from Polysciences, Inc., Warrington, PA., was purified using a spinning band distillation column (57°, 7 mm Hg). Tetraethylene glycol dimethacrylate (TEGDMA) was used as received from Polysciences, Inc. Glucose oxidase type VII from *Aspergillus niger* (125,000 units/g solid) was obtained from Sigma Chemical Co., St. Louis, MO.

The buffers used in the study of this type of enzymatically controlled polymeric system were 0.01 M citrate, 0.01 M phosphate, 0.12 M NaCl, 0.02% sodium azide, pH 7.4 (CPBSz); 0.01 M phosphate, 0.15 M NaCl, pH 7.4 (PBS); and 0.01 M citrate at pH 3 and 4.

Membrane Preparation. Membranes were usually prepared at low temperature, using a radiation-initiated polymerization technique shown in previous work by others to enhance the retention of enzyme activity.[7] HEMA and NNDMAEM were mixed with cross-linking agent and an ethylene glycol–H_2O solvent mixture containing glucose oxidase. The

---

[7] I. Kaetsu, M. Kumakura, and M. Yoshida, *Biotechnol. Bioeng.* **21**, 847 (1979).

MEMBRANE FORMULATIONS[a]

Membranes	HEMA	DMAEM	TEGDMA	Ethylene glycol	H₂O	Glucose oxidase
1	5.0	0.5	0.2	0.5	4.5	105
2	4.5	1.0	0.2	0.5	4.5	105
3	3.5	2.0	0.2	0.5	4.5	105
4	5.0	0.5	0.2	0.5	4.5	10.5
5	5.0	0.5	0.2	0.5	4.5	210

[a] Figures given are volume in milliliters except glucose oxidase, which is given in milligrams.

monomer–solvent–enzyme mixture was poured between two glass plates separated by shims. This "sandwich" assembly was then set on a level shelf in a −70° cold storage box and allowed to freeze. It was then removed, quickly sealed into a polyester bag, and placed in a Dewar flask containing dry ice and acetone. The Dewar flask and glass plates were then irradiated in a ⁶⁰Co source (~10,000 Ci) with a dose of 0.25 Mrad. After irradiation, the plates were set in buffer solution in a refrigerator (4–7°) until they easily separated, whereupon the membrane could be removed. The membranes were then placed in CPBSz buffer solution at room temperature for at least 1 week. The buffer was changed frequently. The membranes were never dehydrated. Typical membrane formulations used in this study are listed in the table.

Swelling Measurements. Membrane disks (12 mm diameter) cut from a larger sheet of membrane were kept in solutions under a pure oxygen atmosphere. The water content and diameter of the disk and the solution pH were measured at various times. Water content was determined with a gravimetric method previously described.[8] Water content was calculated using the following relationship:

$$\%H_2O = \frac{\text{weight wet disk} - \text{weight dry disk}}{\text{weight wet disk}} \times 100$$

The differences between duplicate measurements of water content were always less than 2%.

Enzyme Assay. In the presence of oxygen, glucose oxidase catalyzes the oxidation of glucose into gluconic acid, with the formation of H_2O_2. The rate of formation of gluconic acid was measured with a pH stat (model RTS 822, Radiometer, Copenhagen). The pH was maintained at

[8] B. D. Ratner and I. F. Miller, *J. Polym. Sci., Part A-1* **10**, 2425 (1972).

7.4 $\pm$ 0.02 during the reaction by automatic addition of 0.01 N NaOH. The temperature was held at 37°. The assay was performed as follows: 15 ml of the glucose solution of interest adjusted to pH 7.4 was placed in the pH stat cell. Oxygen was bubbled through the stirred solution. When the temperature reached 37°, a membrane disk (10 mm diameter) was placed in the solution, and the assay was initiated. Each assay was performed for at least 28 min at 100 mg% initial glucose concentration.

Transport Measurements. These glucose-sensitive polymers can be used to control the rate of permeation of compounds through them. Permeation rate will be affected by glucose concentration. The following procedure was used to measure permeation through these membranes. A transport cell with two equal 42-ml compartments separated by the membrane of interest was employed to obtain permeability data. The solution in each chamber was constantly mixed with magnetic stirring bars. Transported insulin was detected colorimetrically or by measuring its radioactivity. The Bio-Rad protein assay was used to detect insulin colorimetrically. A Varian/Cary 219 spectrophotometer was used to measure absorbance. A gamma counter (Searle model 1085) was used to measure ^{125}I-labeled insulin. A beta scintillation counter (Packard) was used to determine [^{14}C]ethylene glycol transport. The quantity $\ln(1 - 2C_2/C_{10})$ was plotted versus time, where C_{10} is the solute concentration in the high-concentration compartment (C_1) at time zero and C_2 is the solute concentration in the other compartment at time t. The concentrations, C_{10} and C_2, were calculated from the radioactivity or absorbance of the appropriate solution. The slope of such a plot is related to permeability (P) by

$$P = (\text{slope})(V)(\xi)/2A$$

where A is the membrane area, V is the volume in each cell compartment, and ξ is the membrane thickness.[9]

The thickness (ξ) of the membrane was obtained by the following procedure. Wet membranes were lightly blotted and then immediately placed between two microscope slides. The thickness of the membrane was obtained by measuring the thickness of the sample assembly using a micrometer and subtracting the thickness of the two glass slides, measured previously.

Results. Typical data showing the swelling response of a number of membranes of the types described here to glucose are shown in Fig. 2. Permeation data as a function of glucose concentration are presented in Fig. 3. The membrane alters its water content and, consequently, its permeability, as the glucose concentration increases. A more complete

[9] J. H. Northrop and M. L. Anson, *J. Gen. Physiol.* **12,** 543 (1928).

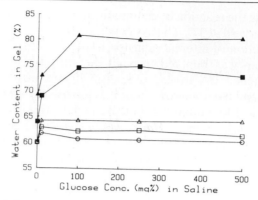

Fig. 2. Effect of glucose concentration on water content of glucose-sensitive membranes swollen in saline solution. (○) 0.1× glucose oxidase; (□) 1× glucose oxidase; (△) 2× glucose oxidase; (■) 2× amine; (▲) 4× amine. Swelling was done in 0.15 M NaCl solutions adjusted to pH 7.4. (From ref. 3.)

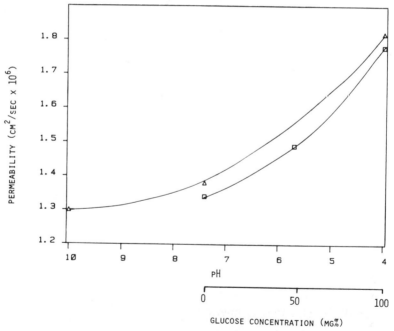

Fig. 3. Effect of pH (△) and glucose concentration (□) on the permeability of ethylene glycol through glucose-sensitive membranes. (From ref. 3.)

description of the swelling and transport properties of these membranes has been presented.[3-5]

Preparation of a Material Which Erodes in Response to Urea

The following experimental section has been excerpted from ref. 1 with permission of the author.

Preparation of the n-Hexyl Half-Ester of a Methyl Vinyl Ether–Maleic Anhydride[10] Copolymer (50:50). A three-necked, 2000-ml, round-bottom flask equipped with a mechanical stirrer, a heating oil bath, a condenser, and a nitrogen inlet and exit was charged with 109.30 g of the copolymer (1.400 equivalent) and 794.82 g of 1-hexanol (7.779 mol). The alcohol, present in an 11:1 molar excess, was sufficient to produce a 20% (w/w) solution of the half-ester product.

After the flask was purged with nitrogen, the reactants were vigorously stirred and heated over 0.5 hr to 145°. The reaction was followed by IR analysis of the carbonyl peaks associated with residual cyclic anhydride in the copolymer and with the formed ester linkage in the half-ester product. The reaction was judged complete after 2.5 hr at 145°. The solution was cooled to room temperature with an ice bath and precipitated in 8 liters of methanol–water (1:1 v/v). The precipitated polymer was dissolved in 2 liters of acetone and precipitated in 8 liters of methanol–water (1:2 v/v). This step was repeated twice. The product was dried in a forced-air oven for 3 days at 50°. A tough, clear-to-slightly-hazy material (148.10 g) was collected in an 82% yield.

Test Specimens

FILM CASTING. The polymer and micronized hydrocortisone[11] were added to 2-ethoxyethyl acetate–isopropylacetone (7:3 w/w) to produce a 9% polymer solution containing 10 parts of dispersed hydrocortisone and 100 parts of resin. A homogeneous solution was obtained by using a jar mill. Films were cast by pouring this solution into level Polytef-lined molds. The molds were then partially covered and, to prevent bubble formation in the film, were dried slowly for 8 days. They were further dried in a forced-air oven at 35° for 1 day and then *in vacuo* for 1 day.

Disks, 9.5 mm in diameter, were cut from the film using a drill press and hole cutter. The disks were weighed after being placed in a vacuum oven for 48 hr at room temperature and then being equilibrated to a

[10] Gantrez AN-169, GAF Corp., New York.
[11] The Upjohn Co., Kalamazoo, Michigan.

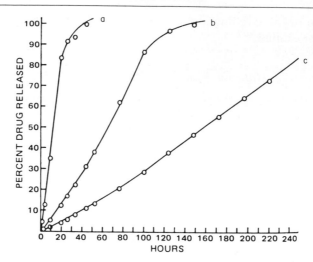

FIG. 4. Hydrocortisone release rate at 35° from a *n*-hexyl half-ester of a copolymer of methyl vinyl ether and maleic anhydride at pH 6.25 in the absence and presence of external urea. (a) 10^{-1} M urea; (b) 10^{-2} M urea; (c) no urea. Used with permission from ref. 1.

constant weight in air. The disks selected for release-rate testing weighed 47.5–52.5 mg and were ~0.75 mm thick.

ENZYME COATING. Successful urease coupling with the polymer matrix was based on the work of Mascini and Guilbault.[12] A small locking forceps was affixed to the edge of each polymer disk so that it could be manipulated without touching the surfaces during the immobilized enzyme coating procedure. A 30% aqueous solution of bovine serum albumin was prepared, and 1 g of urease was added to 10 ml of this solution. After quick stirring until the urease had dissolved, the solution was chilled in an ice bath. Each disk was held horizontally by the attached forceps, and 1 drop of the albumin–urease solution was added to the upper disk face. The disk was quickly rotated, and a drop was added to the opposite face. Similarly, 1 drop of 25% aqueous glutaraldehyde was added to each face. One minute after the glutaraldehyde addition, the coating had gelled sufficiently to allow the disks to be hung vertically.

After standing in air for 15 min, the coated disks were immersed in cold, deionized water from 15 min, in 0.1 M glycine for 15 min, and in pH 5.75 phosphate buffer for 2 hr. Finally, they were immersed in fresh pH 5.75 phosphate buffer for 4 hr.

Release Rate Measurements. The coated disks were heat-sealed in polypropylene mesh bags having 0.5-mm openings. Care was taken to

[12] M. Mascini and G. G. Guilbault, *Anal. Chem.* **49**, 795 (1977).

apply heat only to the edges of the bags and to avoid applying heat to the disks. The mesh bags were attached to 10-gauge stainless steel wires and moved vertically up and down at ~140 cm/min in test tubes, as described previously.[2] Hydrocortisone release was followed spectrophotometrically at 242 nm.

Results. The effect of urea on hydrocortisone release rate is shown in Fig. 4. Increased urea concentration results in a more rapid erosion of the polymer matrix and, consequently, a more rapid liberation of the hydrocortisone.

Conclusion

Enzyme-containing drug delivery systems have tremendous potential because of their ability to respond directly and specifically to metabolites of interest. This controlled response may result in improved drug therapy which mimics the action of healthy glands and organs. Devices in this category of drug delivery systems are very new and their properties have not yet been explored in great detail.

Acknowledgment

Portions of the work described in this chapter have been funded under National Institutes of Health Grant AM 30770.

[37] Membrane Systems: Theoretical Aspects

By K. L. Smith and H. K. Lonsdale

Introduction

Controlled release can be defined as the delivery of an active agent from a device to a target site at a rate and for a duration that are controlled by the device itself. Thus, controlled release of a drug to human patients is determined by the drug-delivery device rather than by external (and variable) factors such as gastrointestinal motility, absorption rates, and patient compliance. Such device-controlled delivery offers several important advantages, which are rapidly becoming widely known and accepted, over conventional therapy. These advantages include (1) maintenance of optimal drug concentrations in body tissues, resulting in greater efficacy of treatment; (2) avoidance of toxic drug "peaks" and insufficient drug "valleys," resulting in greater safety; (3) ability to use new drugs such as those with narrow therapeutic indices or short biological half-lives; (4) greater patient convenience and compliance; and in some cases (5) targeting of drugs to specific tissues.

Most controlled-release technologies fit into one of four major technical categories: membrane-coated reservoirs, monolithic matrices, osmotic systems, or erodible polymers. The first two of these technologies involve diffusion of drug through a polymer membrane and are discussed in some detail in this chapter. A later chapter (see this volume [38]) will describe practical applications of such membrane systems. The latter two technologies involve different mechanisms of drug release and are treated in other chapters of this volume. In addition to these four technologies, other, less common or more exotic techniques have been under investigation in recent years: mechanical devices including pumps, chemically based systems such as prodrugs, and magnetically controlled systems.

Principles of Diffusion in Polymers

The most important type of membrane system for controlled release consists of a nonporous, homogeneous polymer. In some cases, liquid membranes supported by porous polymers or by laminated polymeric membranes are of interest, but drug transport through these modified systems follows the general principles of diffusion through homogeneous polymers. Transport through membranes occurs by a solution–diffusion process in which the solute (drug) dissolves in the membrane and diffuses through the membrane in the direction of lower solute concentration. This process is governed by an expression known as Fick's first law:

$$J = D(dC/dx) \tag{1}$$

where J is the solute flux (usually expressed in units of g/cm^2 sec), D is the diffusivity of the solute in the membrane (cm^2/sec), and dC/dx is the concentration gradient of the solute in the membrane (g/cm^3 cm). In most cases, the diffusivity can be assumed to be constant, and Eq. (1) can be integrated to give the steady-state flux:

$$J = D(\Delta C/\Delta x) = D(\Delta C/l) \tag{2}$$

where l is the thickness of the membrane (cm). Furthermore, if the process involves diffusion from a source of pure or saturated solute (thermodynamic activity of unity) to an infinite sink (negligible drug concentration maintained), the drug concentration within the membrane is the saturation concentration at the reservoir–membrane interface and zero at the sink–membrane interface. The drug concentration profile in a system of

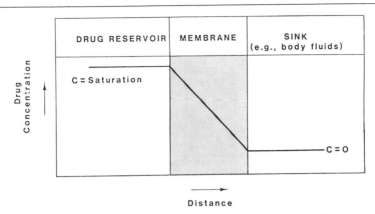

FIG. 1. Drug concentration profile in a membrane-coated reservoir system.

this type is diagrammed in Fig. 1. The concentration difference, ΔC, is then equal to the solubility of the solute in the membrane, S (g/cm³). Consequently, Eq. (2) reduces to

$$J = DS/l \tag{3}$$

More rigorous theoretical treatments of diffusional transport are readily available in the literature.[1-5]

From Eq. (3), it can be seen that the rate of diffusion through a membrane is governed by two solute–polymer properties: diffusivity and solubility. Diffusivity is a kinetic property of the system and can be viewed as the ease with which solute movement takes place in the membrane. Thus, factors that make such movement easier increase the diffusivity, whereas factors that restrain movement decrease the diffusivity. The key parameters that influence diffusivity of solutes in polymers are the size and shape of the solute, the tightness of packing of the polymer chains, and the stiffness of the polymer chains.

Empirical correlations between the diffusivity and solute molecular weight are well established, as shown in Fig. 2 for numerous solutes in

[1] J. Crank, "The Mathematics of Diffusion," 2nd ed. Oxford Univ. Press (Clarendon), London and New York, 1975.

[2] R. W. Baker and H. K. Lonsdale, in "Controlled Release of Biologically Active Agents" (A. C. Tanquary and R. E. Lacey, eds.), p. 15. Plenum, New York, 1975.

[3] R. M. Barrer, "Diffusion in and Through Solids." Cambridge Univ. Press, London and New York, 1955.

[4] J. Crank and G. S. Park, in "Diffusion in Polymers" (J. Crank and G. S. Park, eds.), p. 1. Academic Press, London, 1968.

[5] C. E. Rogers, in "Physics and Chemistry of the Organic Solid State" (D. Fox, M. M. Labes, and A. Weissberger, eds.), Vol. 2, p. 509. Wiley (Interscience), New York, 1963.

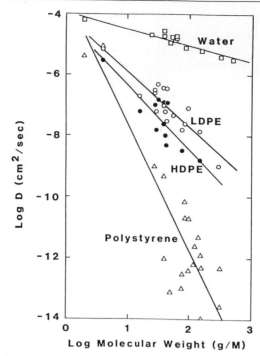

FIG. 2. Dependence of diffusivity on solute molecular weight in water, polystyrene, low-density polyethylene (LDPE), and high-density polyethylene (HDPE). Data at 25° from refs. 2, 5–9.

water and in three polymers.[6-9] In each case, greater solute molecular weight results in lower diffusivity. As evidenced by the scatter of data for each polymer, however, molecular shape and other factors also influence the diffusivity: in general, more spherical molecules tend to exhibit lower diffusivities in a given polymer than do more linear molecules, for example.[6] Theoretical correlations of molecular size and shape with diffusivity have been derived,[10] but generally have limited practical utility.

Most polymers can exist in either a glassy or a rubbery state, depending on temperature. Rubbery polymers exhibit much greater chain flexibil-

[6] H. J. Bixler and O. J. Sweeting, in "The Science and Technology of Polymer Films" (O. J. Sweeting, ed.), Vol. 2, p. 1. Wiley (Interscience), New York, 1971.

[7] F. Grun, Experientia 3, 409 (1947).

[8] V. Stannett, in "Diffusion in Polymers" (J. Crank and G. S. Park, eds.), p. 41. Academic Press, London, 1968.

[9] A. S. Michaels, P. S. L. Wong, R. Prather, and R. M. Gale, AIChE J. 21, 1073 (1975).

[10] A. S. Michaels and H. J. Bixler, Prog. Sep. Purif. 1, 143 (1968).

ity than do glassy polymers and, therefore, also permit higher diffusivities, especially for large solute molecules. This is also apparent from Fig. 2. The lowest diffusivities, and greatest dependence on molecular weight, in this figure are exhibited by polystyrene, a glassy polymer at 25°. The two polyethylene polymers, which are rubbery, yield higher diffusivities and a lesser dependence on molecular weight. Water, which is representative of most liquids in this regard, exhibits very high diffusivities with only a minor dependence on molecular weight, because of the much greater molecular movement that liquids offer. Polymer cross-linking, of course, (not shown in Fig. 2) also reduces the chain flexibility of polymers and results in correspondingly lower diffusivities.

Finally, polymer crystallinity greatly influences solute diffusivity. Crystalline or semicrystalline polymers contain small crystallites, which consist of regions of high molecular order and tight chain packing. It has been well established that penetrant molecules are not soluble in and cannot penetrate polymer crystallites and are constrained to diffuse in the amorphous regions between crystallites.[10] Crystallinity thus reduces diffusivity by causing diffusion to take place through irregular, tortuous pathways. This is evidenced by comparing the data in Fig. 2 for low-density polyethylene (crystallinity ~50%) with the data for high-density polyethylene (crystallinity ~75%). Diffusivities in the more crystalline polymer are lower than those in the less crystalline polymer, again with the greatest differences seen with the largest solutes.

The solubility of a drug in a polymer is an equilibrium property of the system, rather than a kinetic property, and is dependent primarily on chemical factors rather than physical ones. The solubility is governed largely by the heat of mixing of the solute and polymer, which has been related to the Hildebrand solubility parameters of the solute and polymer.[11] The solubility parameter is dependent on the extent of hydrogen bonding, the polarity, and nonpolar dispersion forces and has been extensively studied and tabulated.[12-14] The solubility of a solute in a polymer is generally inversely proportional to the difference between the solute solubility parameter and the polymer solubility parameter. Thus, solubilities are greatest for solutes and polymers with nearly identical solubility pa-

[11] J. Hildebrand and R. Scott, "Solubility of Non-Elecrolytes," 3rd ed. Van Nostrand-Reinhold, Princeton, New Jersey, 1949.

[12] C. M. Hansen and A. Beerbower, in Kirk-Othmer "Encyclopedia of Chemical Technology" (A. Standen, H. F. Mark, J. J. McKetta, Jr., and D. F. Othmer, eds.), 2nd ed., Supplement Volume, p. 889. Wiley, New York, 1971.

[13] H. Burrell, in "Polymer Handbook" (J. Brandup and E. H. Immergut, eds.), 2nd ed., p. IV-337. Wiley, New York, 1975.

[14] P. A. Small, J. Appl. Chem. 3, 71 (1953).

rameters, or, more simply, with similar chemical functionality. The crystallinity of a polymer also affects the drug solubility. As mentioned earlier, solutes are not soluble in polymer crystallites, and their solubilities in semicrystalline polymers are therefore reduced in direct proportion to the crystallinity.

The consequences for drug diffusion through polymers as a result of these effects on diffusivity and solubility can be summarized by noting the following factors, which decrease the drug flux: (1) higher molecular weight of the drug; (2) glassy polymers; (3) polymers with greater crystallinity; (4) polymers with a greater degree of cross-linking; and (5) polymers and drugs with dissimilar solubility parameters.

Delivery Rates and Mechanisms

The optimal rate of drug delivery from controlled-release devices is usually a constant rate. However, other release kinetics are commonly seen and are useful. The release rate is often described in terms of its dependence on the quantity of drug remaining in the device. When the release is at a constant rate, it is termed zero-order release, due to its dependence "to the zero power" (i.e., independence) of the amount of remaining drug. First-order release occurs at a rate that is directly proportional to the quantity of remaining drug and thus decreases exponentially with time. It is typical of conventional, non-controlled-release systems as well as some controlled-release systems. An additional common type of release is referred to as "$t^{-1/2}$ kinetics," because the release rate decreases proportionally to the square root of time. A comparison of these release kinetics for hypothetical systems is presented in Fig. 3. As shown, virtually all of the drug is released at a constant rate of 10 mg/hr from the zero-order device, and the duration was 10 hr. As explained in the next section, the brief non-zero-order initial release from this device is due to a "burst effect," and there is a similar brief tailing-off period of non-zero-order release near drug depletion. With $t^{-1/2}$ kinetics, approximately 85% of the drug is released at rates between 5 and 15 mg/hr, and the duration is nearly as long as for the zero-order case. With first-order kinetics, only about 5% of the drug is released at rates between 5 and 15 mg/hr, and the duration is much shorter. Depending on drug efficacy, toxicity, pharmacokinetics, and cost considerations, either zero-order release or $t^{-1/2}$ release is generally preferable for a given controlled-release system.

Membrane-Coated Reservoirs

Membrane-coated reservoirs consist of a reservoir of active ingredient (drug) enclosed by a polymeric membrane. Drug release occurs by diffu-

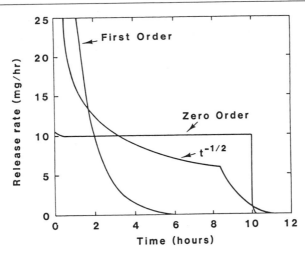

Fig. 3. Release kinetics of drug delivery from controlled-release systems. Assumptions: 100 mg drug loading; D, 1 cm²/sec; S, 1 g/cm³.

sion from the reservoir through the membrane to the external medium. If the thermodynamic activity of the drug within the reservoir remains constant and if drug transport to and from the membrane surfaces is rapid, then the concentration profile shown in Fig. 1 will be established (steady state), the drug release rate will be controlled by diffusion through the membrane, and it will be constant, as predicted by Eqs. (1)–(3). The most useful of such reservoir systems is the case where saturated drug is present within the reservoir for the bulk of the useful lifetime of the system, so that the thermodynamic activity is unity, and the release rate is zero order. For such a case, the drug release rate, dM_t/dt, is simply the steady-state flux multiplied by the surface area, or, for slab or film geometry,

$$(dM_t/dt) = AJ = ADS/l \qquad (4)$$

where M_t is the mass of drug released at time t, A is the releasing surface area of the device, and D, S, and l are as defined earlier. Corresponding expressions for cylindrical (e.g., hollow tubes or fibers) and spherical (e.g., microcapsules) geometries are as follows[1,2]:

Cylinder

$$dM_t/dt = 2\pi h DS/\ln(r_o/r_i) \qquad (5)$$

Sphere

$$dM_t/dt = 4\pi DS[r_o r_i/(r_o - r_i)] \qquad (6)$$

where h is the length of the cylinder and r_o and r_i are the outside and inside radii, respectively, of the cylinder or sphere. It should be noted that edge effects (such as open ends of a cylinder or uncoated edges of a slab or laminate) have not been considered; the result of these effects is a non-zero-order rate of drug release.

There are two interesting and common non-steady-state delivery regimes (where the drug concentration profile within the membrane is changing): the initial release period of a device, and the period after saturated drug no longer remains in the reservoir. During the initial period of release, the release rate can be either higher or lower than the steady-state rate, depending on the storage history of the device. A freshly prepared device will exhibit an initially low release rate while the drug concentration within the membrane increases from zero to its steady-state profile (see Fig. 1). This delay in reaching the steady-state release rate is called the "time-lag effect" and is a kinetic phenomenon. It is thus dependent only on the drug diffusivity in the membrane and on the membrane thickness. A useful indication of how rapidly steady-state release is achieved is an expression called the time lag, τ:

$$\tau = l^2/6D \qquad (7)$$

The time lag is an easily measured value and is often used to calculate diffusivities. Alternatively, when the diffusivity is known, the time lag for a given membrane thickness can be calculated. Typical time lags range from a few minutes in the case of systems with high diffusivities ($>10^{-8}$ cm^2/sec) and thin membranes (<100 μm) to several hours or days in the case of systems with low diffusivities ($<10^{-9}$ cm^2/sec) and thick membranes (>100 μm).

If a membrane device is stored for a sufficiently long period of time under conditions where drug release does not occur, then the entire membrane will become saturated with drug, and the initial release rate will be much higher than the steady-state rate. This is known as the "burst effect." The quantity of drug released during the burst effect is dependent on the drug solubility in the membrane and on the membrane volume. The duration of the effect depends on the diffusivity and membrane thickness, as was the case for the time lag. In fact, typical durations of burst effects are in the same range as time lags for equivalent systems. The drug release during the burst effect is approximately first order.

When the drug concentration in the reservoir of a membrane-coated system falls below saturation, then, as drug release continues, the thermodynamic driving force for diffusion through the membrane continuously decreases, and the release rate decreases with time as well. Drug release under these conditions is first order and can be expressed as follows for slab or film geometry[2]:

$$dM_t/dt = (ADS/l) \exp(-ADSt/VC_s) \tag{8}$$

where V is the volume of the reservoir and C_s is the solubility of the drug in the reservoir. The quantity of drug released at first order is a function of the drug solubility in the membrane and in the reservoir as well as the size of the device.

Membrane-coated reservoirs allow zero-order delivery of a drug, and the rate of delivery is easily controlled by modifying the membrane permeability, thickness, and area. The release rate is independent of the initial drug loading in the reservoir. The duration of release is also easily controlled, for it is directly proportional to the initial loading of the device. Potential disadvantages of membrane-coated reservoirs include cost, as the methods of encapsulating the drug are not incidental, and the possibility of "load dumping" if the membrane is broken or otherwise fails.

Monolithic Matrices

A monolithic system consists of a polymeric matrix containing a uniformly dispersed or dissolved drug. Drug release occurs by diffusion through the polymeric matrix to the surface of the device. The principles discussed earlier in this chapter concerning diffusion in polymers also govern drug release from matrix systems. However, in this case, because the drug near the surface is released first and the path length for release of drug from the device interior is greater, the release rate from a monolithic matrix decreases with time. Drug release kinetics depend on whether (1) the drug is initially entirely dissolved in the polymeric matrix, or (2) the solubility limit is exceeded so that it is both dispersed and dissolved in the matrix. Rigorous expressions for the drug release in each of these cases have been derived.[1,2,15–17]

In the case where all of the drug is dissolved in the polymer matrix, the release rate can be approximated for slab geometry by the following two expressions, which are valid for different portions of the release process:

Early time
$$dM_t/dt = 2M_\infty(D/\pi l^2 t)^{1/2} \qquad M_t/M_\infty < 0.6 \tag{9}$$
Late time
$$dM_t/dt = (8DM_\infty/l^2) \exp(-\pi^2 Dt/l^2) \qquad M_t/M_\infty > 0.4 \tag{10}$$

[15] T. Higuchi, *J. Pharm. Sci.* **50**, 874 (1961).
[16] T. Higuchi, *J. Pharm. Sci.* **52**, 1145 (1963).
[17] D. R. Paul and S. K. McSpadden, *J. Membr. Sci.* **1**, 33 (1976).

where M_∞ is the initial mass of drug in the device. As can be seen from these expressions, drug release follows $t^{-1/2}$ kinetics for the first 60% of release and is first order after that. Similar expressions hold for cylindrical and spherical geometry.[2]

In the case where the drug is initially dispersed and dissolved in the matrix, the initial release rate from a slab can be expressed as

$$dM_t/dt = (A/2)[(DS/t)(2C_0 - S)]^{1/2} \tag{11}$$

where C_0 is the initial concentration of dispersed and dissolved drug in the polymer matrix. This expression holds as long as dispersed drug remains in the matrix; following this, the release expressions for dissolved drug [Eqs. (9) and (10)] govern the drug release. Thus, for virtually the entire duration of release, the release rate follows $t^{-1/2}$ kinetics. For very high drug loadings (>40%), drug release still follows the form of Eq. (11), but the release rate is higher than predicted.[18] This is due to the decreased resistance to diffusion afforded by fluid-filled holes created by dissolution and release of drug particles.

Although monolithic matrices do not result in zero-order release, they are relatively inexpensive to fabricate and do provide prolonged and controlled drug release. In addition, significant damage to a monolithic device will not appreciably alter its drug release characteristics, in contrast to membrane-coated reservoir systems.

[18] R. M. Barrer, in "Diffusion in Polymers" (J. Crank and G. S. Park, eds.), p. 165. Academic Press, London, 1968.

[38] Membrane Systems: Practical Applications

By K. L. SMITH

Introduction

Theoretical applications of drug release from membrane systems have been described in [37] of this volume. In this chapter, key representative applications of membrane systems are presented, and their structure, function, and release kinetics are discussed with reference to the theory addressed earlier. The two types of membrane systems considered are membrane-coated reservoirs and monolithic matrices. In each type of system, the drug is released by diffusion through a rate-controlling membrane or matrix. As noted below, applications range from ocular inserts to transdermal patches to oral dosage forms. A comprehensive review of all applications is beyond the scope of this chapter; however, the principles involved in most membrane systems are covered by the examples discussed below.

Membrane-Coated Reservoir Systems

The first major (although noncommercial) application of membrane-coated reservoirs for controlled drug delivery was the use of silicone-rubber-tubing implants for delivery of cardiovascular drugs by Folkman and Long in 1964.[1] The reservoir in these devices was simply powdered drug, and the membrane was the wall of the tubing. It was subsequently discovered that silicone rubber was exceptionally permeable to most hydrophobic drugs, and numerous studies of similar systems[2-6] followed, especially steroid implants for contraception. These studies, both *in vivo* and *in vitro,* clearly demonstrated the long-term release characteristics of these devices, and led to the development of membrane-based steroid-releasing intrauterine devices, also for contraception.

Progestasert Intrauterine Contraceptive System

Although the silicone rubber devices filled with solid drug mentioned above did exhibit long-term drug release, the release rates were not very reproducible and usually decreased significantly with time, in contrast to the constant release rates predicted by theory. Probably the main reason for such performance was lack of uniform or complete contact of the drug with the membrane (inside wall of the tubing), which resulted in a smaller effective membrane surface area than expected. As drug was released from these devices, the membrane contact area decreased, resulting in a decreasing release rate with time. To overcome these problems, Alza Corporation developed a device (Progestasert) that incorporates a reservoir consisting of drug (progesterone) crystals suspended in silicone oil. The use of a liquid reservoir enables uniform contact of the reservoir with the membrane surface and greatly improves reproducibility.[7] In addition,

[1] J. Folkman and D. M. Long, *J. Surg. Res.* 4(3), 139 (1964).
[2] P. J. Dziuk and B. Cook, *Endocrinology* 78, 208 (1966).
[3] F. A. Kincl, G. Benagiano, and I. Angee, *Steroids* 11(5), 673 (1968).
[4] H. B. Croxatto, S. Diaz, R. Vera, M. Etchart, and P. Atria, *Am. J. Obstet. Gynecol.* 105, 1135 (1969).
[5] E. M. Coutinho, C. E. R. Mattos, A. R. S. Sant'Anna, J. A. Filho, M. C. Silva, and H. J. Tatum, *Contraception* 2, 313 (1970).
[6] A. S. Lifchez and A. Scommegna, *Fertil. Steril.* 21(5), 426 (1970).
[7] D. R. Swanson, P. Wong, and B. B. Pharriss, *in* "Conception and Contraception," p. 45. Excerpta Medica, Amsterdam, 1975.

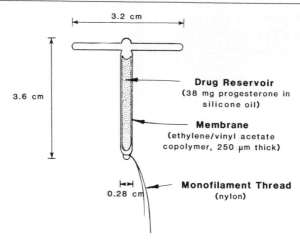

3.2 cm

3.6 cm

Drug Reservoir
(38 mg progesterone in silicone oil)

Membrane
(ethylene/vinyl acetate copolymer, 250 μm thick)

0.28 cm

Monofilament Thread
(nylon)

FIG. 1. Diagram of progestasert intrauterine contraceptive system. From ref. 8a.

since the solubility of progesterone in silicone oil is very low (about 700 ppm),[8] a large excess of solid progesterone is present, which maintains the maximum thermodynamic driving force for diffusion of progesterone through the membrane.

A sketch of the Progestasert system is shown in Fig. 1.[8a] The drug reservoir is contained within the stem of the device, which consists of an extruded tube of ethylene/vinyl acetate (9 wt% vinyl acetate)[9] copolymer. The wall of the tube constitutes the membrane, which is 250 μm thick, and the total membrane surface area is approximately 2 cm². Ethylene/vinyl acetate was selected as the material for the rate-controlling membrane for three main reasons[10]; (1) it has a permeability to progesterone appropriate for obtaining the desired release rate subject to geometrical considerations of the device, (2) it is an easily formed thermoplastic, unlike silicone rubber, and (3) it is sufficiently flexible and biocompatible for placement in the uterus.

Shown in Fig. 2 are release rates of progesterone from the Progestasert system *in vitro* and *in vivo*.[11] As is evident from the figure, there is a small discrepancy between the release rates measured under the different

[8] F. Theeuwes, R. M. Gale, and R. W. Baker, *J. Membr. Sci.* **1**, 3 (1976).

[8a] K. Heilman, "Therapeutic Systems—Pattern-Specific Drug Delivery: Concept and Development," p. 94. Thieme, Stuttgart, 1978.

[9] Based on device dimensions and on permeability measurements from A. S. Michaels, P. S. L. Wong, R. Prather, and R. M. Gale, *AIChE J.* **21**(6), 1073 (1975).

[10] W. P. O'Neill, *in* "Controlled Release Technologies: Methods, Theory, and Applications" (A. F. Kydonieus, ed.), Vol. I, p. 129. CRC Press, Boca Raton, Florida, 1980.

[11] S. K. Chandrasekaran, R. Capozza, and P. S. L. Wong, *J. Membr. Sci.* **3**, 271 (1978).

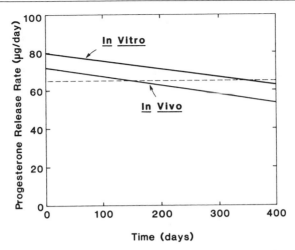

FIG. 2. Progesterone release rates from the progestasert system *in vitro* and *in vivo*. The dashed line indicates the rated release rate. *In vitro* rate, $78 - 0.040t$ μg/day; *in vivo* rate, $72 - 0.048t$ μg/day. (Data from ref. 11.)

conditions: the *in vivo* release is consistently lower than the *in vitro* release. The apparent reason for the lower *in vivo* release rate is that, unlike the *in vitro* conditions, there is an appreciable steady-state progesterone concentration in the uterus (5% of saturation),[12] which lowers the driving force for diffusion of progesterone from the device. The effect of this lower driving force should be an equivalent (5%) lowering of the release rate. In fact, the *in vivo* release rate is approximately 10–15% lower than the *in vitro* rate. The balance of the effect is probably due to better stirring conditions *in vitro* then *in vivo*, which avoids a boundary-layer build-up of drug *in vitro*, but not *in vivo*. Such a boundary layer, which is caused by an insufficient rate of drug removal from the membrane surface, results in a lower driving force for diffusion across the membrane, and a consequent lower release rate.

The second notable feature of the release rate data shown in Fig. 2 is that the release rate is not constant, as predicted by theory, but decreases with time. Over the course of a year, the release rate decreases by about 25% from its initial value (*in vivo* data). This decrease is probably due to a decrease in the volume of reservoir solution, and a consequent decrease in membrane surface area contacted, resulting from the release of drug. That is, initially there is about 38 mg of drug present in a total reservoir capacity of about 110 mg. During the first year, approximately 24 mg of progesterone is released,[10] which reduces the reservoir contents (and, to a

---

[12] J. Urquhart, *in* "Controlled-Release Pharmaceuticals" (J. Urquhart, ed.), p. 6. Am. Pharm. Assoc., Washington, D.C., 1981.

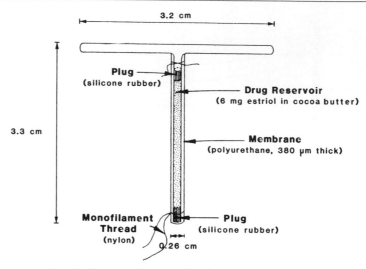

FIG. 3. Diagram of estriol-releasing intrauterine device.

first approximation, the reservoir solution volume) by about 22%, nearly the exact decrease seen in the release rate. Thus, even though the system is a membrane-coated reservoir and a constant driving force is maintained, the release rate is not entirely constant.

Estriol-Releasing Intrauterine Device

A more recently developed steroid-releasing intrauterine device is similar in many ways to the Progestasert but contains estriol rather than progesterone; it was developed by Bend Research.[13] The primary advantage of using estriol is that it is effective at much lower doses than is progesterone, and an estriol device can therefore theoretically be used for several years without replacement. The structure of the estriol device developed by Bend Research is shown in Fig. 3. It consists of a reservoir of estriol dissolved in cocoa butter, surrounded by a membrane of an ether-based polyurethane. The release of estriol is controlled by diffusion through the polyurethane membrane and is shown in Fig. 4. Following a short initial "burst effect," the release rate was nearly constant for the duration of the test. The burst effect is due to saturation of the membrane with estriol during storage and to subsequent rapid release, as was discussed in [37], this volume. In separate measurements, it was determined

[13] R. W. Baker, M. E. Tuttle, H. K. Lonsdale, and J. W. Ayres, *J. Pharm. Sci.* **68**(1), 20 (1979).

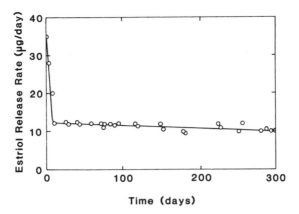

FIG. 4. *In vitro* release rate of estriol from an intrauterine drug delivery device. Release rate, $12.3 - 0.007t$ μg/day. (Data from ref. 13.)

that the solubility of estriol in the polymer was 0.2%; therefore, based on the weight of polymer in the device, approximately 170 μg of estriol would be dissolved in the polymer at saturation.[13] It was determined from the data in Fig. 4 that very nearly that quantity of drug was released during the burst period. As was the case with the Progestasert system, there is a slight decrease in the release rate during the device lifetime, again caused by depletion of the reservoir volume during release.

The above mentioned drug depletion effect was studied as a function of the drug carrier used in the reservoir.[13] With no carrier, i.e., with only dry, powdered drug in the reservoir, the release rate was nearly a linear function of the extent of drug depletion, as shown in Fig. 5. The use of drug carriers in the reservoir substantially improved the constancy of release as a function of drug depletion. With cocoa butter as the carrier— in which estriol has a very low solubility (70 ppm)—the drug release rate was essentially constant until more than 80% of the drug was depleted.

Ocusert Ocular Drug Delivery System

At about the same time that the Progestasert system was being developed, Alza also developed a membrane-based ocular system for the delivery of pilocarpine to treat glaucoma. This system, diagrammed in Fig. 6, is in the form of a thin, elliptical wafer that consists of a rate-controlling membrane on each side of the drug-containing reservoir. The reservoir consists of pilocarpine base in alginic acid as a gelling agent. Each membrane is made of an ethylene–vinyl acetate copolymer and is approximately 100 μm thick.[10] Encircling the reservoir is an ethylene–vinyl ace-

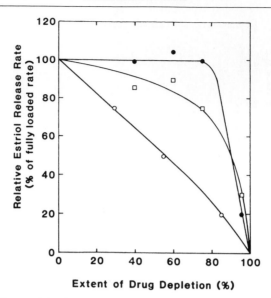

FIG. 5. Relative estriol release rate as a function of drug depletion in the reservoir. (●) Cocoa butter; (□) Triglyceride; (○) powdered drug, no carrier. Data from ref. 13.

tate annulus containing titanium dioxide for visibility. The annulus functions to prevent rapid release of the drug from the device edges. There are two commercial Ocusert systems—one designed to deliver pilocarpine at a rate of 20 µg/day and the other designed to deliver at a rate of 40 µg/day, with the duration of each system being 1 week. The higher-release-rate system was designed with two main differences from the lower-release-rate system: the reservoir is thicker and contains more drug (11 mg vs 5 mg), and the membranes were made more permeable by

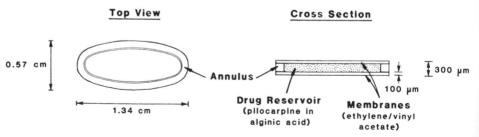

FIG. 6. Diagram of Ocusert pilocarpine delivery system. Dimensions are for 20 µg/day system.

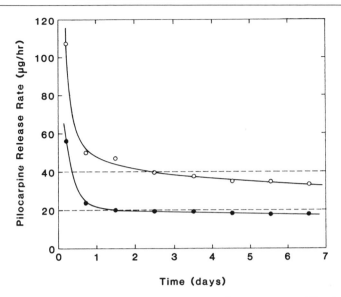

FIG. 7. *In vitro* pilocarpine release rates from the two Ocusert systems. The dashed lines indicate the rated release rates. (Data from ref. 14a.)

the addition of a plasticizer, di(2-ethylhexyl) phthalate. In use, about 90 μg of the plasticizer is released during the 1-week duration.[14]

Based on the structure and composition of the devices, zero-order, or constant, release rates would be predicted from the Ocusert systems. *In vitro* release rate data for these devices are presented in Fig. 7.[14a] As shown, there is an initial high release rate from each system that lasts for a few hours. This is another example of the burst effect mentioned earlier. During storage, pilocarpine diffuses into the membranes until they are saturated (i.e., in equilibrium with the reservoir). When the device is placed in a release environment, this drug is released rapidly, in accordance with the theory for release from a monolithic matrix. As this "excess" drug is released, the drug concentration profile within the membrane approaches that for steady-state release from a membrane-coated reservoir, and the overall release rate approaches a constant value. The total quantity of pilocarpine released in this initial burst is approximately

[14] K. Heilmann, "Therapeutic Systems—Pattern-Specific Drug Delivery: Concept and Development," p. 68. Thieme, Stuttgart, 1978.

[14a] S. I. Yum and R. M. Wright, *in* "Controlled Drug Delivery" (S. D. Bruck, ed.), Vol. 2, p. 71. CRC Press, Boca Raton, Florida, 1983.

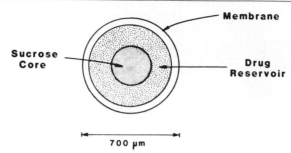

FIG. 8. Diagram of Theo-Dur Sprinkle theophylline pellet.

0.3 mg from the 20 μg/day system, and about twice that for the 40 μg/day system.[15]

As is evident from the release rate data in Fig. 7, the release rate from each device decreases slightly throughout the lifetime of the device. The primary cause for this decrease is the extremely high water solubility of pilocarpine base. Water from the release medium is osmotically imbibed into the device during use, which results in dilution of the drug in the reservoir, and consequent lowering of the driving force for diffusion of pilocarpine through the membrane. Because the ethylene–vinyl acetate membranes are relatively impermeable to water, the rate of dilution is slow and the effect on the release rate is small.

Theo-Dur Sprinkle Oral Pellets

In 1977 Key Pharmaceuticals began marketing a controlled-release tablet containing theophylline for treatment of asthma and bronchitis. This was the first theophylline product that allowed 12-hr dosing and has become widely accepted by physicians and patients. Although this dosage form apparently results in relatively constant serum theophylline levels, it has a complex composite structure consisting of coated pellets embedded in a tablet base; insufficient information concerning this structure has been published to enable analysis. A more recently introduced oral theophylline product, Theo-Dur Sprinkle (also Key Pharmaceuticals), apparently exhibits equivalent bioavailability and has a simpler structure. This product, diagrammed in Fig. 8, consists of a drug layer coated onto sucrose pellets and then overcoated with a polymer layer. The drug layer is the reservoir, and the polymer layer is the rate-controlling membrane. Since the reservoir consists basically of solid drug, the release rate should remain constant, or zero-order, for virtually the entire duration of release.

[15] Y. W. Chien, "Novel Drug Delivery Systems," p. 31. Dekker, New York, 1982.

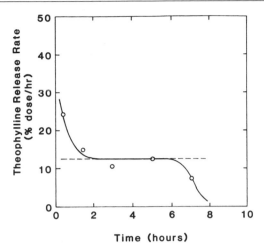

FIG. 9. *In vitro* theophylline release rate from Theo-Dur Sprinkle pellets. The dashed line indicates the rated release rate. (Data from ref. 16.)

The kinetics of release of theophylline from Theo-Dur Sprinkle are shown in Fig. 9.[16] As was the case with systems described above, there is an initial burst effect due to the drug dissolved in the membrane, followed by a relatively constant release rate. In this case, however, the initial burst period lasts for a significant portion of the release duration, and there is also a substantial "tail-off" period as the pellets become depleted of drug. Both of these effects are probably due to a range of membrane thicknesses on the pellets, which results in pellets with different release rates and different durations from the average.[16] That is, pellets with thin coatings exhibit higher release rates and shorter durations than those exhibited by pellets with thicker coatings.

Transderm-Scop Transdermal Delivery System

Over the last 3 or 4 years, transdermal drug delivery has received wide publicity. Although the number and types of drugs that will permeate the skin at useful rates are limited, there are obviously some drugs (nitroglycerin chief among them) that are especially attractive for transdermal delivery. The first commercial, controlled-release transdermal delivery system was developed by Alza Corporation and marketed by Ciba Pharmaceuticals as Transderm-Scop for the delivery of scopolamine to treat motion sickness. Scopolamine was an attractive candidate for transdermal deliv-

[16] M. A. Gonzalez and A. L. Golub, *Drug Dev. Ind. Pharm.* **9**(7), 1379 (1983).

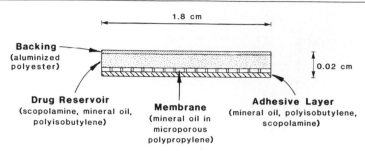

FIG. 10. Diagram of Transderm-Scop transdermal delivery system.

ery because of its serious side effects and obvious difficulties (for a nauseated person) when given orally. Human skin is also reasonably permeable to this drug.[17] The structure of this system is shown in Fig. 10. It is a circular laminated device, consisting of an impermeable backing layer, a drug reservoir, a rate-controlling membrane, and an adhesive layer. The reservoir contains approximately 1.3 mg scopolamine mixed with mineral oil and polyisobutylene. The rate-controlling membrane consists of mineral oil contained within the pores of a microporous polypropylene support, which is, in effect, a liquid membrane through which the scopolamine diffuses. The adhesive layer consists of polyisobutylene and mineral oil, with 200 μg scopolamine added as a loading dose.[15,18] This loading dose is necessary to rapidly saturate drug-binding sites in the skin and thereby minimize the time lag between device administration and onset of drug action.

This system is also an example of a membrane-coated reservoir; however, it has two interesting variations. First, the drug contained in the adhesive layer is released rapidly and at a decreasing rate, following the theory of release from monolithic matrices. This is actually a designed burst effect. Second, the membrane, as mentioned above, is a liquid membrane rather than a polymeric membrane. This results in a much higher permeability to drugs than polymeric membranes exhibit. The combined effects of these variations are that scopolamine is released at an initially rapid rate, quickly reaching a lower (but still relatively rapid) constant rate, as shown in Fig. 11. In this figure, the experimental *in vitro* release rate is compared with the rate calculated from theoretical considerations of a combined membrane and matrix system.[11] The agreement is clearly very good.

[17] J. E. Shaw and S. K. Chandrasekaran, *Drug Metab. Rev.* **8,** 223 (1978).
[18] "Physician's Desk Reference," 38th ed., p. 874. Medical Economics Co., Oradell, New Jersey, 1984.

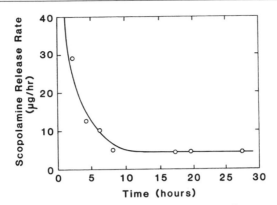

F<small>IG</small>. 11. *In vitro* scopolamine release rate from Transderm-Scop transdermal system. Comparison of theoretical (line) and experimental (data points) release rates. (Data from ref. 11.)

Monolithic Matrices

A mixture of a dispersed or dissolved drug in a polymeric matrix is called a monolithic system. Drug release from such a system is controlled by diffusion of the drug through the polymeric matrix and usually follows $t^{-1/2}$ kinetics, as explained in the previous chapter (see this volume [37]). That is, the drug release rate is not constant but decreases with time in a nonlinear manner. Monolithic matrices for controlled or sustained drug release have been investigated for many years, and several commercial systems have resulted. A few representative examples are discussed in the following sections.

Nitrodisc and Nitro-Dur Transdermal Nitroglycerin Systems

As mentioned in the previous section, the transdermal drug delivery system is a recent development that has received much attention in the last few years. Since 1981, three different commercial transdermal systems have been introduced for delivery of nitroglycerin to treat angina. One of these, Transderm-Nitro (Ciba Pharmaceuticals) is a membrane-coated reservoir type of system and releases nitroglycerin according to the principles discussed in the previous section. The other two nitroglycerin products are monolithic matrices and are discussed in this section. The Nitrodisc system (G. D. Searle and Co.) consists of a nitroglycerin–lactose mixture dispersed as microscopic droplets within a matrix of cross-linked silicone rubber containing isopropyl palmitate or similar solvent.[19] There is a partitioning process between the droplets and the ma-

[19] Y. W. Chien, P. R. Keshary, Y. C. Huang, and P. P. Sarpotdar, *J. Pharm. Sci.* **72**(8), 968 (1983).

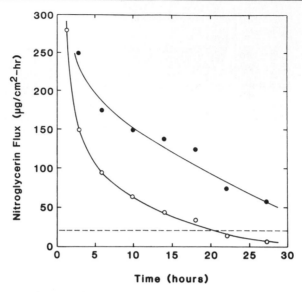

Fig. 12. *In vivo* nitroglycerin flux from Nitro-Dur (●) and Nitrodisc (○) systems compared with average *in vitro* flux through skin (dashed line). (Data from ref. 19.)

trix, but the drug release appears to be controlled by diffusion through the matrix.[19] The Nitro-Dur system (Key Pharmaceuticals) consists of a hydrophilic matrix of poly(vinyl alcohol), poly(vinylpyrrolidone), sodium citrate, glycerin, lactose, and nitroglycerin.[20] Nitroglycerin diffuses rapidly through this gel matrix. One notable difference in the two devices is the total drug loading: Nitro-Dur contains over three times the quantity of drug that Nitrodisc contains. This, combined with the highly liquid nature of the Nitro-Dur matrix, should result in a higher rate of release from this device.

As would be expected from these matrix systems, the *in vitro* release rate of nitroglycerin decreases according to $t^{-1/2}$ kinetics[19] as shown in Fig. 12. Each system releases nitroglycerin for more than 24 hr and generally at rates that are much higher than the rate at which nitroglycerin permeates skin. This implies that much, if not all, of the control of the release rate resides in the skin itself, rather than in the device. (An interesting note is that, even though the release rate from the Transderm-Nitro is at zero-order, it is approximately twice the rate at which the drug permeates average skin, so that much of the release rate control with this

[20] "Physician's Desk Reference," 38th ed., p. 1033. Medical Economics Co., Oradell, New Jersey, 1984.

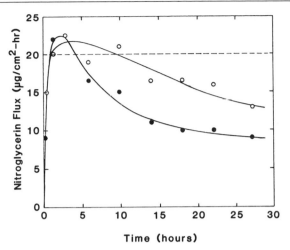

FIG. 13. *In vitro* nitroglycerin flux from Nitro-Dur (●) and Nitrodisc (○) systems through mouse skin, compared with average *in vitro* flux through mouse skin (dashed line). (Data from ref. 19.)

device also resides in the skin.) As expected, nitroglycerin release from the Nitro-Dur system is more rapid than that from the Nitrodisc system. When these devices are applied to the skin, the rate of nitroglycerin delivered is indeed largely controlled by the skin, as shown by the data in Fig. 13 (in these studies, hairless mouse skin was used, which has been shown to have a permeability to nitroglycerin similar to that of human skin).[19] Interestingly, however, there does appear to be some control of the release rate in the devices, as the release rates during the latter portion of the test were significantly lower than the rate through skin. The reason for the apparent lower rate of delivery for the Nitro-Dur system is not clear. It should be noted that previous studies have reported zero-order delivery rates from these systems through skin,[21,22] probably as a result of the rate control effected by the skin.

Synchron Hydrogel Matrix

Over the last few years, there have been a number of sustained-release formulations developed for oral delivery of drugs, including diffusional matrices and slowly dissolving or erodible matrices. One such type of matrix is the Synchron system, developed by Forest Laboratories, which generally consists of a compressed tablet of hydroxypropylmethylcellu-

[21] A. D. Keith, *Drug. Dev. Ind. Pharm.* **9**(4), 605 (1983).
[22] A. Karim, *Drug. Dev. Ind. Pharm.* **9**(4), 671 (1983).

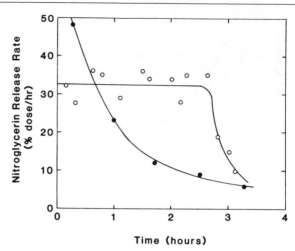

Time (hours)

FIG. 14. *In vivo* release of DTPA (nitroglycerin drug model) from Synchron tablets administered orally (●) and buccally (○). (Data from ref. 25.)

lose mixed with the drug.[23] The polymer is quite hydrophilic and swells when exposed to water. The drug then diffuses through the swollen gel matrix, following the expected $t^{-1/2}$ kinetics. A useful property of the system is that, when hydrated, it adheres to mucosal tissue. Because of this, the system has recently been used in the development of a commercial buccal tablet for systemic delivery of nitroglycerin. The tablet is placed on the buccal mucosa, where it adheres to the tissue and releases nitroglycerin through the mucosa to the systemic circulation. Because mucosal tissue is many times more permeable than skin,[24] the onset of action is much more rapid than with the transdermal nitroglycerin products. The system has also been investigated for use as a sustained-release oral (gastrointestinal) dosage form. In many of these studies, a radiolabeled model compound, such as ^{99m}Tc-diethylenetriaminepentaacetic acid (DTPA), has been used for analysis of the release kinetics of the system.[25]

A comparison of *in vivo* release kinetics of DTPA from buccal and oral Synchron systems is shown in Fig. 14. The release rate from the oral tablet, as expected, follows $t^{-1/2}$ kinetics, showing a rather rapid decrease over a 3-hr period. The release rate from the buccal tablet, however, is quite constant for nearly the entire duration of release. The reason for this

[23] H. Lowey and H. H. Stafford, U.S. Patent 3,870,790 (1975).
[24] W. R. Galey, H. K. Lonsdale, and S. Nacht, *J. Invest. Dermatol.* **67,** 713 (1976).
[25] S. S. Davis, P. B. Daly, J. W. Kennerly, M. Frier, J. G. Hardy, and C. G. Wilson, *Adv. Pharmacother.* **1,** 17 (1982).

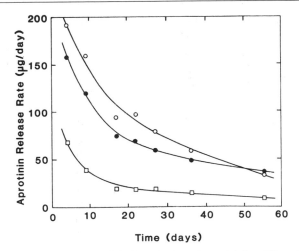

FIG. 15. *In vitro* release of aprotinin from hydrogels containing different drug loadings. Drug loadings: (O) 32 mg; (●) 16 mg; (□) 8 mg. (Data from ref. 26.)

unexpected performance is that the buccal tablet in use is surrounded by buccal tissue, which, although quite permeable, is still less permeable than the hydrogel matrix. The buccal tissue thus acts as a rate-controlling membrane, much as the skin does for the transdermal systems discussed above. A similar tablet given orally is surrounded by gastric fluid, which acts as a sink for the released drug, enabling the $t^{-1/2}$ kinetics to be observed.

Aprotinin-Releasing Hydrogel

A final example of a drug-releasing monolithic matrix is a hydrogel containing a proteinase inhibitor for reduction of the pain and bleeding caused by intrauterine devices. This system, developed by Bend Research, constitutes a hydrogel coating applied to the fundal surface of an intrauterine device, such as a Lippes Loop. The hydrogel consists of a cross-linked matrix of hydroxyethyl methacrylate containing dispersed drug. The drug, aprotinin, is a polypeptide of molecular weight 6500, which would not diffuse at appreciable rates through solid, homogeneous membranes, such as those usually used in membrane-coated reservoir devices. In addition, the desired therapeutic regimen was a high initial dose over the first several days, followed by a lower, sustained dose for a 30- to 60-day period, corresponding to the incidence of undesired side effects associated with intrauterine devices. For both of these reasons, a monolithic matrix is more likely to provide the target release kinetics than a membrane-coated reservoir would be, for example.

The two main parameters that were used to adjust the drug release rate were the extent of cross-linking and the drug loading. Cross-linking acts to decrease the permeability of the hydrogel, primarily by decreasing the diffusivity of the large aprotinin molecule in the matrix. The drug loading also directly affects the drug release rate, with higher loadings resulting in higher release rates [see Eq. (11) in the previous chapter [37], this volume). Aprotinin release rates from this cross-linked hydrogel matrix are presented in Fig. 15.[26] As expected, the release rates all followed $t^{-1/2}$ kinetics, with the highest release rates exhibited by the hydrogels with the highest drug loadings. The system with a drug loading of 16 mg (approximately 2.5 wt%) showed a satisfactory combination of a rapid release of about 3 mg drug in the first 2 weeks, followed by a slowly decreasing release rate in the range of 40 to 90 μg/day over the next 6 weeks.

Final Note

Representative examples of drug-releasing membrane systems have been discussed in some detail to provide a basis for appreciation of the operational principles involved. These principles, which were explained on a theoretical basis in the previous chapter of this volume [37], all focus on the diffusion of solutes in polymers, whether the system is a membrane-coated reservoir or a monolithic matrix. It must be emphasized that the examples discussed were selected for their value in highlighting these principles, and numerous other examples, both commercial and experimental, could have been cited. The reader is referred to the list of references for this chapter as well as those for the previous chapter for a more complete record of the work in this field.

[26] M. E. Tuttle, R. W. Baker, and L. E. Laufer, *J. Membr. Sci.* **7**, 351 (1980).

[39] Implantable Infusion Pumps: Clinical Applications

By PERRY J. BLACKSHEAR

Introduction

The use of totally implantable infusion pumps for the long-term delivery of drugs in human patients has several potential advantages over conventional methods of drug administration. For example, total implantation of the device is largely successful in preventing the entrance of bacteria leading to common complications of external infusion systems, local infection, and even systemic septicemia. Because of this lack of percutaneous access and the fact that the drug infusion set is totally implanted, no hospitalization is required for long-term, continuous drug infusion under appropriate circumstances; such chronic hospitalization is still the norm for long-term infusion of certain drugs such as heparin for serious thromboembolic disease, or antibiotics for chronic, refractory infections such as bacterial endocarditis or osteomyelitis. The patients are afforded complete freedom of movement and can lead a normal life as outpatients if the devices are appropriately implanted and anchored in position.

From a pharmacological point of view, the advantages of basal-rate continuous or modulated continuous drug infusions obtain both with implanted and external infusion devices; that is, the continuous infusion of drugs avoids the problems associated with peaks and valleys of drug concentrations in the plasma, which can lead to toxic effects or the unrestrained problems associated with the underlying disease, respectively. In addition, in the case of drugs such as 5-fluorodeoxyuridine (FUdR), continuous infusion of the drug has been shown to potentiate its effect on certain tumors. In other cases, described in more detail below, the continuous infusion of certain drugs has opposite effects to those achieved with intermittent bolus administration of the same drug.

Finally, an implantable infusion device can provide access for drug delivery to a wide variety and type of specific body sites. Examples of sites in which drugs have been delivered directly using such devices are the intrathecal or subdural spinal spaces, the cerebral ventricles, numerous arteries leading directly to organs at which the drug was directed, the intraperitoneal space, and the intravenous and subcutaneous systems.

The aforementioned advantages are those which apply to a simple, continuous infusion device with a single reservoir and single delivery cannula. However, such devices also provide the potential for much more versatile methods of drug administration. For example, two or more devices can be used to deliver drugs to different body sites simultaneously; a variant of this possibility is to use a single device with two or more drug chambers, connected to the same number of delivery catheters, so that drug can be infused into various sites from a single device. Such multiple-site infusion devices are already being used for the treatment of certain tumors by simultaneous intravenous and intraarterial drug infusion. In addition, two or more different drugs can also be infused simultaneously at the same or different sites using additional devices, additional drug chambers, or certain auxiliary sideports for bolus drug administration.

Combined with these two increases in complexity is the possibility of superimposing various schedules of drug administration, since it is becoming clear that for many different drugs, not only the toxicity but the therapeutic effectiveness of the drug is markedly dependent on schedule of administration as well as route and quantity of drug administered. Thus, the possible permutations available from combinations of several body sites, drugs, and schedules of administration, while still largely theoretical, should eventually allow for a dramatic increase in our ability to provide appropriate drug administration for specific disease states.

Such implantable infusion devices also have certain advantages over the other types of polymeric or liposomal drug delivery systems described elsewhere in these volumes. For example, macroscopic infusion pumps can often use commercially available drugs rather than specifically modified drugs, which is very convenient from a regulatory point of view. In addition, the properties, effectiveness, and toxicities of commercially available drugs, as used in a given application, are often well known, and this information can be utilized in determining the appropriate uses of these drugs in implantable delivery devices. With rare exceptions, commercially available drugs cannot be used in polymeric or liposomal drug delivery systems. A second advantage is that comparatively large implantable infusion pumps can use relatively large volumes of these drugs. This has the advantage that pump refills can be relatively infrequent; in addition, it allows for the convenient use of many dilute drugs which form precipitates when concentrated. Many such drugs are now in common clinical use.[1] Finally, the use of a convenient and radiopaque drug delivery catheter makes vascular or other site-specific access often easier than in the case of polymeric drug delivery systems; in some cases, this is also an advantage over liposomes in drug delivery.

On the other hand, such devices also have several inherent and serious drawbacks when compared to either conventional therapeutic methods or polymeric or liposomal drug delivery systems. For example, the fact that the devices are totally implanted under the skin makes it inconvenient to repair or modify the device once it has been implanted. This difficulty may be obviated to some extent through the use of the newer, transcutaneously programmable devices, where flow rate and rate of bolus administration can be altered through the use of transcutaneous signaling mechanisms, either magnetic or telemetric. Another disadvantage has to do with the fact that implanted foreign bodies can serve as a nidus for infection which is extremely difficult to eradicate once it has taken hold; the

[1] P. J. Blackshear, in "Proceedings of the Second World Conference on Clinical Pharmacology and Therapeutics" (M. M. Reidenberg, and L. Leinberger, eds.). American Society for Pharmacology & Experimental Therapeutics, Bethesda, Maryland, 1984, pp. 459–468.

long experience with pacemaker pocket infections is a well-known example of this problem. This requirement for extreme care and sterility in the refilling of implantable drug delivery devices means, in general, that such refill maneuvers must be performed by personnel trained in aseptic technique and that refills cannot generally be accomplished by the patient and his family, thus requiring inconvenient trips to the doctor's office for refills. Although infection appears to be rare with totally implanted devices compared with external infusion systems, when an infection around the device occurs, it usually necessitates pump removal. Finally, there is always the potential for deleterious pump–drug interactions. We have noted a number of such interactions,[1] the most notorious of which occurs with insulin, which forms insoluble aggregates when incubated with moderate shaking at body temperature, resulting in occlusion of insulin delivery devices.

Available Devices

The small number of totally implantable drug infusion pumps currently available has been described recently.[2,3] In general, all of the available implantable drug infusion devices are titanium canisters designed for minimum tissue reactivity once implanted and have plastic or silicone rubber drug delivery cannulas and rubber refill septa so that the drug reservoirs can be refilled by percutaneous needle injections. However, internally, pumps vary considerably. In one category are peristaltic pumps, which include a pump manufactured by Siemens AG, Erlangen, West Germany.[2,3] This pump has been used in preliminary studies of insulin delivery in diabetic patients[4,5] and is programmable through a hand-held, telemetry-controlled programmer. Another peristaltic device which is also programmable was developed by the University of New Mexico in collaboration with Sandia Laboratories; this has also been implanted in a relatively small number of diabetic patients, generally for intraperitoneal insulin infusion.[6] Somewhat more complicated devices are being

[2] P. J. Blackshear and T. D. Rohde, in "Controlled Drug Delivery" (S. D. Bruck, ed.), Vol. II, pp. 111–147. CRC Press, Boca Raton, Florida, 1983.
[3] Chapter 5, in "Diabetes Treatment with Implantable Insulin Infusion Systems" (K. Irsigler, H. Kritz, and R. Lovett, eds.), p. 29. Urban & Schwarzenberg, Munich, 1983.
[4] K. Irsigler, H. Kritz, G. Hagmuller, M. Franetzki, K. Prestele, H. Thurow, and K. Geisen, *Diabetes* **30**, 1072 (1981).
[5] J. L. Selam, A. Slingeneyer, P. A. Chaptal, M. Franetzki, K. Prestele, and J. Mirouze, in "Diabetes Treatment with Implantable Insulin Infusion Systems" (K. Irsigler, H. Kritz, and R. Lovett, eds.), p. 119. Urban & Schwarzenberg, Munich, 1983.
[6] D. S. Schade, R. P. Eaton, W. S. Edwards, R. C. Doberneck, W. J. Spencer, G. A. Carlson, R. E. Blair, J. T. Love, R. S. Urenda, and J. I. Gaona, Jr., *J. Am. Med. Assoc.* **247**, 1848 (1982).

manufactured by Medtronics, Inc. in Minneapolis and the Applied Physics Laboratory at Johns Hopkins University.[2,3] The Medtronic device has been used in preliminary studies for cancer chemotherapy at the time of this writing; to my knowledge, the Johns Hopkins pump has not yet been implanted in man.

In another category is the Infusaid pump (Infusaid Corp, Norwood, MA) which is a simple device used only for continuous, single-rate drug infusion at the present time, although the model 400 pump has an auxiliary sideport through which bolus injections of drug, contrast media, or radionuclides can be injected. The design and clinical usefulness of this pump have been reviewed[1,2,7,8]; practical aspects of its use in animals and man are reviewed in the following chapter [40]. Briefly, it differs from the other types of infusion pumps currently available in that the power source is provided not by a battery but by the vapor pressure of a fluorocarbon vapor–liquid mixture; this vapor pressure, which remains constant at body temperature, exerts pressure against the infusate contained within a collapsible titanium or stainless steel bellows, thereby extruding the infusate through a series of filters and a flow-regulating capillary tube into a silicone rubber delivery cannula. Reservoir refills are accomplished by puncturing a subcutaneously located rubber refill septum with a special needle; the pressure of the hand-held or mechanically assisted infusate injection simultaneously refills the pump's drug chamber and condenses the volatile vapor power source, thus effectively refilling and recycling the pump in a single operation. Since its invention at the University of Minnesota in 1969, this device has been widely used in both animal and human drug infusion studies, and at the time of the present writing has been implanted in nearly 7000 human patients, with the longest implant being in place for greater than 5 years. Although this is the simplest of the devices currently available, it has the longest and broadest clinical experience; this experience will be described briefly in the following section as an example of the potential clinical usefulness of all devices of this type, both simple and complex.

Current Clinical Applications of the Infusaid Pump

Applications Approved by the Food and Drug Administration (FDA)

The first clinical application of the Infusaid pump was as a means of infusing the anticoagulant agent heparin in patients with refractory throm-

[7] P. J. Blackshear, T. D. Rohde, F. Prosl, and H. Buchwald, *Med. Prog. Technol.* **6,** 149 (1979).

[8] P. J. Blackshear, *Sci. Am.* **241,** 66 (1979).

boembolic disease of the venous system. It had been shown previously that continuous intravenous heparin infusions were a safer means of administering the drug than intermittent heparin injections; in addition, heparin is either inactivated in or not absorbed from the gastrointestinal tract and thus must be given by a parenteral route. Dog studies of more than 3 years duration had indicated that the pump was effective in maintaining constant blood levels of heparin in this species for long periods of time and that the device itself was well tolerated by the animals.[9] In a series of 21 patients with recurrent thromboembolic disease, Buchwald *et al.*[10] demonstrated that continuous intravenous heparin infusions were a safe and effective means of administering this drug in these patients, most of whom had failed conventional therapy with oral anticoagulants. In this trial, the pumps were refilled with heparin at 4- to 8-week intervals; this relatively long refill interval was possible because heparin is available in extremely concentrated solutions. Of the 21 patients treated, one patient appeared to be refractory to heparin as well as oral therapy; several problems with delivery cannula plugging were encountered, most of which could be traced to inappropriate refill technique; and evidence of osteopenia, a known complication of chronic heparin therapy, was noted in one patient. One patient developed the syndrome of heparin-induced thrombocytopenia. Most of the other patients, however, maintained stable and appropriate blood levels of heparin, had no recurrent thromboembolic episodes, and had no significant bleeding episodes. Thus, although a controlled trial of this modality of anticoagulant administration has not yet been performed, it appears that this form of heparin administration is reasonably safe and effective as a treatment for patients who have failed conventional anticoagulant therapy for their venous thromboembolic disease.

Other patients have received continuous intravenous heparin infusions for refractory arterial clotting problems. For example, Chapleau and Robertson[11] were able to demonstrate resolution of a spontaneous cervical carotid artery dissection after several months of continuous heparin infusion using the implanted infusion pump. Other patients with transient ischemic attacks due to atherosclerotic disease in inoperable locations were treated with continuous intravenous heparin infusions by these investigators; transient ischemic attacks appeared to be less frequent than when the patients were receiving conventional oral anticoagulant therapy.

[9] P. J. Blackshear, T. D. Rohde, R. L. Varco, and H. Buchwald, *Surg., Gynecol. Obstet.* **141**, 176 (1975).
[10] H. Buchwald, T. D. Rohde, R. L. Varco, P. D. Schneider, and P. J. Blackshear, *Surgery* **88**, 507 (1980).
[11] L. E. Chapleau and J. T. Robertson, *Neurosurgery* **8**, 83 (1981).

Again, a controlled trial of the effectiveness of this form of anticoagulation in the treatment of refractory transient ischemic attacks has not been performed to date.

By far the most common use of the Infusaid implantable infusion pump is as a means of infusing cancer chemotherapeutic agents into one or more vascular or other body locations. The widest application has been for the treatment of both primary and metastatic cancer of the liver by means of intraarterial infusions of FUdR. This was first shown to be possible with this pump by Buchwald et al.[12]; several more recent and larger series of patients have been published.[13-16] In general, the devices have been well tolerated by the patients, and intraarterial infusion chemotherapy to the liver can be performed readily on outpatients using this totally implanted device. This application exemplifies two of the major advantages of this type of delivery system compared to conventional methods of drug delivery. The first is that site-specific chemotherapy can be delivered, in this case by placing the delivery catheter in the hepatic artery so that the liver receives the full brunt of the cytotoxic agent while the remainder of the systemic circulation is relatively spared. The second advantage is that the totally implanted nature of the device makes the frequency of pump and catheter complications much lower than that seen with conventional infusion techniques. Although complications have been rare and palliation has been excellent in many cases, a controlled trial of this modality of therapy for hepatic metastatic cancer has not been completed; such a trial is now underway under the auspices of the National Cancer Institute.

Other forms of cancer are currently being treated with this device as well. Examples include intracarotid artery infusion chemotherapy for head and neck cancer[17-19] and intraarterial or intraventricular infusion for central nervous system tumors.[20]

[12] H. Buchwald, T. B. Grage, P. P. Vassilopoulos, T. D. Rohde, R. L. Varco, and P. J. Blackshear, *Cancer* **45,** 866 (1980).

[13] W. D. Ensminger, J. E. Niederhuber, S. Dakhil, J. Thrall, and R. H. Wheeler, *Cancer Treat. Rep.* **65,** 393 (1981).

[14] R. M. Barone, J. E. Byfield, P. B. Goldfarb, S. Frankel, C. Ginn, and S. Greer, *Cancer* **50,** 850 (1982).

[15] A. M. Cohen, A. Greenfield, W. C. Wood, A. Waltman, R. Novelline, C. Athanasoulis, and N. J. Schaeffer, *Cancer* **51,** 2013 (1983).

[16] C. M. Balch, M. M. Urist, S. J. Soong, and M. McGregor, *Ann. Surg.* **198,** 567 (1983).

[17] S. R. Baker, R. H. Wheeler, W. D. Ensminger, and J. E. Niederhuber, *Head Neck Surg.* **4,** 118 (1981).

[18] S. R. Baker and R. H. Wheeler, *J. Surg. Oncol.* **21,** 125 (1982).

[19] S. R. Baker, R. H. Wheeler, and B. Medvec, *Arch. Otolaryngol.* **108,** 703 (1982).

[20] T. W. Phillips, W. F. Chandler, G. W. Kindt, W. D. Ensminger, H. S. Greenberg, J. F. Seeger, K. M. Doan, and J. W. Gyves, *Neurosurgery* **11,** 213 (1982).

Finally, relatively recently this device has been used as a means of delivering analgesic opiates such as morphine into the epidural or intrathecal spinal spaces for relief of severe, chronic pain associated with malignancy.[21-24] The devices have been well tolerated by the patients, and pain relief over the relatively short term has been excellent in patients with severe pain of malignant origin. As expected, tolerance or refractoriness to the original drug has limited therapy in some cases and has necessitated huge doses of analgesics in others. Patients with chronic pain of nonmalignant origin have not, in general, responded to this form of drug infusion as favorably as those with malignant pain. Again, this application utilizes the ability of the device to deliver drugs to a specific target area; in this case, analgesia is provided in the body region of interest, while avoiding the unpleasant and sometimes dangerous side effects of systemic morphine administration, including respiratory and central nervous system depression.

Experimental Applications

Several other clinical trials are currently underway to evaluate the potential usefulness of this device as a means of delivery drugs by continuous, parenteral infusion. It has been suggested for many years that a device of this type might be useful as a means of delivering parenteral insulin in patients with diabetes; conventional insulin therapy does not mimic the normal pancreatic response to glycemic and other challanges, and it is widely accepted that this abnormal control of blood glucose leads eventually to the complications characteristic of long-standing diabetes mellitus. Several groups of investigators have used this pump as a means of delivering continuous, intravenous or intraperitoneal insulin in patients with either type II (adult-onset), or type I (juvenile-onset) diabetes.[25-28] In

[21] B. M. Onofrio, T. L. Yaksh, and P. G. Arnold, *Mayo Clin. Proc.* **56**, 516 (1981).
[22] D. W. Coombs, R. L. Saunders, M. S. Gaylor, M. G. Pageau, M. G. Leith, and C. Schaiberger, *Lancet* **2**, 425 (1981).
[23] H. S. Greenberg, J. Taren, W. D. Ensminger, and K. Doan, *J. Neurosurg.* **57**, 360 (1982).
[24] D. W. Coombs, R. L. Saunders, M. S. Gaylor, A. R. Block, T. Colton, R. Harbaugh, M. G. Pageau, and W. Mroz, *J. Am. Med. Assoc.* **250**, 2336 (1983).
[25] H. Buchwald, J. Barbosa, R. L. Varco, R. M. Rupp, R. A. Schwartz, F. J. Goldberg, T. G. Rublein, and P. J. Blackshear, *Lancet* **1**, 1233 (1981).
[26] W. M. Rupp, J. Barbosa, P. J. Blackshear, H. B. McCarthy, T. D. Rohde, F. J. Goldenberg, T. G. Rublein, F. D. Dorman, and H. Buchwald, *N. Engl. J. Med.* **204**, 817 (1982).
[27] P. J. Blackshear, T. D. Rohde, W. M. Rupp, and H. Buchwald, *in* "Artificial Systems for Insulin Delivery" (P. Brunetti, K. G. M. M. Alberti, A. M. Albisser, K. D. Hepp, and M. Massi Benedetti, eds.), p. 131. Raven Press, New York, 1983.
[28] H. Kritz, G. Hagmuller, R. Lovett, and K. Irsigler, *Diabetologia* **25**, 78 (1983).

addition, several studies in diabetic animals have been conducted with the basic pump modified to provide bolus delivery of insulin at appropriate times; this modification is expected to be of great use in clinical models for the treatment of type I diabetes.[29-31] All of these studies have been made possible by modifying conventional insulin (which inevitably precipitates in implanted pumps within 1–2 months) by the addition of high concentrations of glycerol, which prevents this precipitation and subsequent blocking of flow passages.[32] In general, the glycemic control achieved with these basal rate intravenous or intraperitoneal infusions has been excellent in type II patients, without side effects or frequent or severe hypoglycemic episodes; it is too early to tell whether this type of infusion without supplementary insulin will provide reasonable glycemic control in type I patients. Another small group of patients in which this type of infusion is being successfully applied includes those who are resistant to subcutaneous insulin because of subcutaneous inactivation of the hormone; in such cases, intravenous drug delivery is almost the only route by which insulin can be delivered to the patient, and in these cases, insulin infusion by means of implanted pumps appears to be a reasonable therapeutic approach.[33]

Other experimental applications are being pursued in the animal laboratory. For example, for eventual use in clinical states in which insufficient tears are produced to lubricate the cornea, long-term continuous infusion of artificial tears by means of pumps connected to artificial tear ducts may be possible.[34] Another experimental application now being pursued is the use of the pump to deliver several different types of antiarrhythmic agents in animals[35]; such drugs are often best administered parenterally, and their very rapid half-life means that continuous delivery is necessary to maintain steady plasma levels of the drug.

[29] P. J. Blackshear, T. D. Rohde, J. C. Grotting, F. D. Dorman, P. R. Perkins, R. L. Varco, and H. Buchwald, *Diabetes* **28**, 634 (1979).

[30] H. Buchwald, T. D. Rohde, F. D. Dorman, J. G. Skakoon, B. D. Wigness, F. D. Prosl, E. M. Tucker, T. G. Rublein, P. J. Blackshear, and R. L. Varco, *Diabetes Care* **3**, 351 (1980).

[31] M. W. Steffes, H. Buchwald, B. D. Wigness, T. J. Groppoli, W. M. Rupp, T. D. Rohde, P. J. Blackshear, and S. M. Mauer, *Kidney Int.* **21**, 721 (1982).

[32] P. J. Blackshear, T. D. Rohde, J. L. Palmer, B. D. Wigness, W. M. Rupp, and H. Buchwald, *Diabetes Care* **6**, 387 (1983).

[33] I. W. Campbell, H. Kritz, C. Najemnik, G. Hagmueller, and K. Irsigler, *Diabetes Res.* **1**, 83–88 (1984).

[34] P. G. Rehkopf, B. J. Mondino, S. I. Brown, and D. B. Goldberg, *Invest. Ophthalmol. Visual Sci.* **19**, 428 (1982).

[35] J. L. Anderson, E. M. Tucker, S. Pasyk, E. Patterson, A. B. Simon, W. E. Burmeister, B. R. Lucchesi, and B. Pitt, *Am. J. Cardiol.* **49**, 1954 (1982).

Potential Uses of Implantable Infusion Pumps

As the available devices become more sophisticated or are modified for specific uses, it should become clear that the advantages of implantable drug delivery systems of this type far outweigh the disadvantages imposed by the need for surgical implantation and the difficulty in revising the implant once under the skin. Disorders for which current treatment is often inadequate include Parkinson's disease, for which site-specific infusion of dopamine or its analogs might prove beneficial, perhaps in response to electrical feedback indicating the delivery of too much or too little drug. Other potential applications of such devices include Alzheimer's disease, amyotrophic lateral sclerosis (ALS), and possibly some of the chronic and poorly understood psychiatric disorders such as schizophrenia.

A very large group of potential applications is in the general field of endocrinology. The infusion of insulin in diabetic patients is only the first of a relatively large group of disorders in which either basal rate or modulated hormone infusions might be of benefit. For example, it has been clear in recent years that several reproductive disorders can be successfully treated by basal or modulated infusions of gonadotropin-releasing hormone (GNRH) and its agonist or antagonist analogs. Examples of such disorders include precocious puberty, delayed or absent puberty, and hormonally responsive cancers such as those of the prostate and breast. Hormone infusions could also provide a means of contraception. Other disorders for which one or more hormones might be infused over the long-term with benefit include the treatment of short stature with growth hormone, the treatment of osteoporosis with parathyroid hormone or other therapeutic agents, and possibly even the delivery of anorectic agents into the systemic circulation or even the cerebrospinal fluid in the treatment of morbid obesity.

Finally, drugs may be delivered by this means to avoid some of the debilitating side effects seen with conventional therapy. As described above for heparin, the narrow range of plasma concentrations achieved with this type of drug infusion is actually safer and more effective than intermittent injections because of the risks of both over- and under-anticoagulation. Similar drawbacks are noted with the conventional administration of certain antiarrhythmic agents, as noted earlier. One potentially very useful drug is cyclosporine, which is a powerful immunosuppressant agent that prevents rejection of transplanted organs and which may eventually prove useful in the prevention of certain autoimmune disorders such as type I diabetes mellitus. Patients receiving cyclosporine therapy often develop characteristic renal toxicity, and a small percentage of patients appear at increased risk for the development of certain malignancies after long-term use. It seems possible that one or both of these severe toxicities might be prevented by maintaining the serum level in a narrow therapeutic range, one that is exactly appropriate for the rejection status of the transplanted organ or the current status of the ongoing autoimmune process.

[40] Implantable Infusion Pumps: Practical Aspects

By PERRY J. BLACKSHEAR, BRUCE D. WIGNESS,
ANNE M. ROUSSELL, and ALFRED M. COHEN

Introduction

As use of implantable drug delivery systems increases, both in experimental applications in animals and in clinical use in patients, it becomes important that descriptions of correct surgical and refill procedures be readily available. We have been involved in the development and use of one such device, the Infusaid implantable infusion pump (Infusaid Corporation, Norwood, MA) for nearly 14 years and during that time have evolved procedures for both human and animal surgical implantation and pump refilling which minimize the risk of infections and other complications and pump malfunction resulting from incorrect usage. Careful attention must be paid to practical aspects of the use of such devices, since they cannot be readily removed for servicing or adjustment. For details of the device and its current clinical usefulness, the reader is referred to [39].

Pump Implantation in Humans

As with the implantation of any foreign body, the development of an infection at the site of infusion pump implantation usually requires pump removal. Meticulous attention to surgical detail is of great help in avoiding such problems. The pump can be implanted in various areas of the body but is always placed subcutaneously, usually on the muscle fascia, to which it can be sutured at the time of implantation. This prevents excessive rotation or movement of the pump. The delivery cannula can be placed into the superior vena cava for systemic intravenous infusions or in the arterial circulation (most commonly the hepatic artery for regional chemotherapy of liver cancer) or in the carotid system for infusion therapy of head and neck cancer and brain tumors. The delivery cannula may also be placed in the epidural or intrathecal spaces for continuous infusion of opiates for analgesia or into the cerebral ventricles for chemotherapy of central nervous system tumors. The following sections discuss technical aspects of implanting pumps for infusion of drugs either intravenously or via the hepatic arterial route. The reader is referred to [39] for references describing cannula placement in other sites.

Pump Preparation

Before implantation, the appropriate pump model, preferred flow rate, and appropriate connectors needed (if any) should be selected. The pumps in current clinical use are between 1.5 and 2.5 cm thick and approximately 9 cm wide, with a reservoir size which permit percutaneous refilling every 2 weeks to 3 months. The cannula position will have an effect on eventual fluid flow rate, since the flow rate is directly proportional to the difference between the pump's internal pressure and the ambient pressure at the cannula tip (ΔP), which can vary from a low-pressure system such as the epidural space or the central venous system to a high-pressure system such as the arterial circulation. However, since the actual amount of drug delivered in any particular time period is a function of the concentration of that drug in the infusate, once the flow characteristics of the individual pump are determined *in vivo* over several weeks postimplantation, then the drug delivery can be calculated and is extremely accurate over a prolonged period of time.

The Model 100 Infusaid pump is the basic device, holds a functional volume of 45 ml, and is suitable for most intravenous and central nervous system uses. The Model 400 pump is identical to the Model 100 pump, but in addition it has an injection sideport attached to it. This allows direct injection of the catheter system distal to the pump reservoir, which permits direct injection of drugs or contrast media into the cannulated vessel. This pump is more appropriate for arterial chemotherapy. Thinner variations are available with 25- or 35-ml reservoirs—Model 500 and Model 200, respectively.

The pump's delivery cannula is relatively thick-walled, to minimize reflux of blood and prevent kinking and occlusion. The suture used to affix the delivery cannula in place should be tied firmly but not tightly enough to occlude the lumen. To prevent cannula dislodgement when placed in the arterial circulation, circumferential tying rings are bonded to the cannulae of Model 400 pumps. If direct surgical implantation of the catheter into the arterial circulation is not performed, then the distal portion of the delivery cannula should be cut off and the smooth portion of the silicone cannula placed intravascularly.

For percutaneous angiographic placement of the delivery cannula, e.g., into the hepatic artery, a number of stainless steel tubing connectors are available from the Infusaid Corporation. For example, a double-ended barbed friction adaptor is available to connect two pieces of tubing or to permit changing the pump while leaving the pump catheter in place. In addition, a single-ended friction adaptor is available to attach to a 5.3 French angiography catheter for integration of the pump with angiographic catheterization (see later). These steel adaptors have an external diameter larger than the internal diameter of the silicone delivery cannulas, a problem which can be easily resolved by soaking the silicone catheter ends to be connected for 5–10 min in a small amount of organic solvent (longer times may dissolve the rubber). We use a fluorocarbon skin degreaser [PreSurgical Skin Degreaser (Freon TF), Aeroceuticals, Southport, CT] that is routinely available in our operating room, but toluene or acetone or almost any other solvent can be used. The solvent softens the rubber so that the steel connector can readily be inserted into the catheter lumen; when the catheter is taken out of the solvent, the solvent evaporates and the silicone returns to its normal consistency and shrinks down around the connector. In general, with this technique it is not necessary to further tie the catheter around the adaptor, although a 2-0 silk tie can be used for further security.

The Infusaid pumps are supplied sterile and empty. However, it is possible to resterilize a pump by autoclaving using special instructions obtainable from Infusaid Corporation. The pump should be removed from its package only under sterile conditions. A separate table should be used on which to put the pump and the additional supplies used with the pump. It is imperative that no moisture be allowed to soak down through the drapes onto an unsterile table. For this reason, a waterproof barrier drape should be placed first on the table, with several layers of paper or cloth placed over it. Obviously, the model and serial number should be recorded and forwarded to the company, as well as placed in the patient's record.

Before implantation, the pump must be warmed to provide a positive internal pressure and to start fluid flow and must be filled with an appropriate placebo solution. The pump can be warmed to 40–50° by placing it in preheated water or saline or by using some type of electric heating pad or warming blanket. We have found that the simplest approach is to place the pump in warm water in a steel basin, avoiding temperatures higher than 50°. It is convenient to place sterile water or saline into the autoclave for a few minutes, then dilute it with room temperature water to approximately 45°. Until experience is gained with this temperature, a thermometer should actually be used to monitor the temperature. To avoid overfilling, each pump must be evacuated completely before it is filled with the maximum volume appropriate for each model. The pump should be filled only with special sidehole needles (22-gauge Huber needles) to avoid cutting a core from the refill septum with a standard needle; these are also available from Infusaid. After the empty pump has been in 45° water for about 5 min, a 22-gauge Huber needle is used to puncture the septum, at which point egress of a small amount of air or sterile water may occur. The pump should then be filled; the system may be flushed once or twice with 10–20 ml of infusate before instilling the final full amount. Further details of the refill maneuver after implantation are contained in a later section (see Pump Refills in Humans).

Once the pump has been filled, it is placed back in the warm water, maintained in the range of 45–50°. Depending upon the fixed flow rate of the device, it may take several minutes to hours before dripping is seen from the end of the catheter, which is necessary to establish proper pump function before implantation can proceed. If a pump with a sideport is used, the sideport should be flushed immediately prior to implantation to remove residual air from the catheter.

Patient Preparation

All efforts are directed toward avoiding an acute wound infection. If at all possible, the patient should not have his surgical skin area shaved the day prior to surgery. This is best done just prior to surgery to minimize the development of folliculitis. Prophylactic antibiotics are given to ensure high tissue levels of antibiotics at the time of the surgical incision. Staphylococcal wound infection is the most common, and an antibiotic such as oxacillin or a cephalosporin should be used. A parenteral dose of antibiotic is given on call to the operating room and immediately at the beginning of the operative procedure, and several more doses are administered postoperatively. Routine use of antibiotics for 5–7 days following implantation has not been necessary.

Pump Implantation

With Delivery Cannula in Central Venous System. The optimum position for central venous catheter placement is in the superior vena cava near the entrance to the right atrium; if the catheter slides into the right ventricle, it must be retracted. The access route to the superior vena cava is either through one of the subclavian veins, internal jugular vein, external jugular vein, or cephalic veins. The pump is placed on the pectoralis fascia in the infraclavicular fossa. In almost all cases, surgery can be done under local anesthesia using approximately 60 ml of 1% lidocaine infiltration anesthesia. On call to the operating room, patients receive analgesics and sedatives; we routinely prescribe diazepam, 10 mg by mouth, and morphine sulfate, 10–20 mg intramuscularly in adults.

When the surgery is done under local anesthesia, it is important that the patient wear a face mask during preparation of the skin to avoid potential contamination. The patient is placed supine on the operating table. The entire chest, neck to the chin, and both shoulders are prepared with soap and then povidone–iodine solution. Access to all major venous structures is afforded. The patient is then carefully draped including a large sheet across the midneck in front of the face; the face mask can then be removed if the patient feels somewhat claustrophobic. The standard approach is a left infraclavicular subclavian vein puncture using a needle, guidewire, and split sheath introducer (Cook Catheter Company Pacemaker Insertion Set, 11-French). Such a system obviates the need for a cephalic or external jugular vein cutdown. The left side is preferred because of the more suitable angulation of the left innominate vein. A C-arm fluoroscopy unit may be used if readily available in the operating room. However, such a unit increases the risk of contamination, and a simpler procedure is just to obtain a chest X ray at the completion of surgery. If the catheter ends up in an internal jugular vein or in the opposite subclavian vein, it can be pulled into the superior vena cava at a later time using an angiographic technique, preferably by a skilled angiographer. In this procedure, the radiologist inserts a pigtail catheter through a transfemoral venous route, hooks the catheter, and pulls it down into the superior vena cava. This is a simple procedure that has minimal risks.

After diffuse infiltration with 1% lidocaine, an 8-cm incision is made paralleling the clavicle 2–3 cm beneath the clavicle, lateral to the sternoclavicular junction. A 10-cm pocket is then made on the pectoralis fascia inferior to the incision, and the upper half of the incision is elevated to the clavicle. The pump should fit comfortably into the pocket. If the patient is extremely obese, a pocket can be made in the subcutaneous tissues rather than on the fascia. At this point, the pump is placed back in warm water, and the patient is placed in steep Trendelenberg position (approximately 45° head down). The left subclavian vein is then punctured, and the guidewire introduced. If the curved guidewire goes in practically to the hilt, one can be reasonably sure that the catheter is in the superior vena cava. If the guidewire meets resistance part way in, this almost always means it is going up into the internal jugular vein, and it should be pulled back and manipulated. This manipulation is somewhat easier with fluoroscopy, but this has not been routinely necessary after some experience with the technique has been gained. Once the guidewire is in position, it is important to immediately place the patient in the supine position, i.e., out of

Trendelenberg. The pump is then positioned on the chest wall, and a loop of catheter is kept free. The distance to the superior vena cava is estimated by placing the catheter on the chest wall from above the clavicle down to the junction of the angle of Louis at the sternomanubrial junction. The catheter tip is then cut off at this spot. The introducer and sheath are then passed over the guidewire, guidewire and introducer are removed, and the catheter slipped down the sheath. The catheter is firmly held with a pair of vascular forceps, and the sheath is split longitudinally and removed. The catheter is inspected to make sure that there are no kinks where it enters the muscular fascia. At this point, several silk sutures are used to immobilize the catheter but not to occlude the lumen. It is best to cut off a large excess of the catheter so that there is only one remaining loop of tubing in the pocket. The pump is then affixed to the chest wall fascia with several sutures of 2-0 silk or 2-0 Tevdek. The wound is irrigated with neomycin, hemostasis is confirmed, and a suction catheter placed. No matter how dry the pump pocket may seem, it is imperative to use a suction catheter. In this and all other cases, the suction catheter is kept in place for 2–3 days. The incision is then closed in two layers with catgut and fine nylon to the skin or subcuticular Dexon with Steristrips. A simple anteroposterior X ray in the operating room then confirms the appropriate position of the delivery catheter.

With Delivery Cannula in Hepatic Artery

DIRECT CANNULATION AT LAPAROTOMY. Hepatic artery infusion chemotherapy for both primary and metastatic liver cancer takes advantage of the fact that over 90% of the blood supply of the tumor comes from the hepatic artery, as compared to 30% for normal liver parenchyma. Because almost half of all patients have variants of the "standard" anatomy of the hepatic artery, all patients should have a hepatic artery arteriogram before pump implantation. It is important that the celiac axis and hepatic artery proper as well as the left gastric and the origin of the superior mesenteric artery be visualized. This is necessary to detect the two most common variants of the hepatic artery; namely, the left hepatic coming from the left gastric (10% of cases), and the right hepatic artery coming from the superior mesenteric artery (15% of cases). Angiography will also demonstrate the relative locations of the left and right hepatic arteries and the gastroduodenal artery. Since total liver perfusion is the aim, under certain circumstances two separate catheterizations and two pumps must be implanted, e.g., if one is dealing with the left hepatic or right hepatic variants described above. Occasionally, it may be more appropriate to ligate accessory hepatic arteries. If a pump is not available at the time of laparatomy (i.e., in the case of unsuspected metastases to the liver), a silicone rubber catheter can be placed in the hepatic artery and can be attached to the pump in the later procedure.

Laparatomy is performed under general anesthesia. A lengthy midline incision is made. Abdominal exploration is performed to rule out extrahepatic intraabdominal tumor. A liver biopsy is then performed to confirm the presence of cancer if this has not been previously done. A cholecystectomy is performed only in the presence of gallstones. The liver is then elevated superiorly, and the stomach and duodenum pulled inferiorly. All overlying tissue above the hepatic artery and gastroduodenal artery is then divided. It is imperative that any branches from the hepatic, gastroduodenal, or left or right hepatic arteries that run back to the antrum of the stomach be divided to prevent drug perfusion of the stomach. During this time, the pump will have been warmed and filled. A subcutaneous pocket on the rectus sheath is then formed. This can be done either through a second transverse incision on either side of the abdomen or simply by elevating the subcutaneous tissue from the midline incision. It is important that the pump pocket be placed in the lower abdomen, with the sideport angled so that neither the pump nor the sideport will hit the ribs when the patient bends over. The pump is then implanted, and the catheter is brought intraabdominally through a puncture through the posterior aspect of the pocket into the peritoneal cavity. If there is any suggestion of ascites, then a pursestring suture should be placed around the peritoneum and the anterior fascia to minimize ascitic leak. Despite this, ascitic leak from the abdomen into the pump pocket is a frequent occurrence. Excess catheter is then cut off so that approximately 5–6 mm of catheter remains past the most proximal of the circumferential rings. With proximal and distal control, a longitudinal arteriotomy is made, and the catheter slipped into the vessel. Silk ties (with 2-0 silk) proximal and distal to the ring are then used to affix the catheter in place. Since it is possible to tie these too tightly, an injection of 1 cm of heparinized saline through the pump sideport is done at this time to confirm that the system remains patent. Adequate perfusion of the liver can then be determined by the injection of 2 ml of fluorescein or 1 mCi technetium-labeled, macro-aggregated albumin through the pump sideport. In the first case, a Wood's lamp is used to visualize the liver; the second requires a portable gamma camera to image the liver. In either case, it is important to flush the sideport with heparinized water following injection. The incision is closed with running No. 1 nylon, and a Hemovac catheter is left in the pump pocket for 2–3 days.

Standard hepatic artery anatomy is present in 50–60% of cases. In this case, the access route to the hepatic artery is via the gastroduodenal artery. There must be at least 2 cm between the gastroduodenal artery and the bifurcation of the left and right hepatic arteries to ensure adequate mixing of the drug. Approximately 2 cm of the gastroduodenal artery is isolated. The artery is ligated distally, and the catheter is threaded up to the level of the hepatic artery proper. If at all possible, it is important not to have a significant amount of catheter in the bloodstream, but instead to keep the tip of the catheter near the sidewall of the hepatic artery. A common variant is a trifurcation which includes the gastroduodenal and left and right hepatic arteries. In this case, the easiest solution involves threading the catheter retrograde about 2 cm in the gastroduodenal artery. This procedure carries with it an increased risk of hepatic artery thrombosis. The other approach is to ligate the gastroduodenal artery and use the splenic artery as an access route as previously described. If the left hepatic takes off from the proper hepatic artery far proximal to the gastroduodenal, then ligation of the gastroduodenal artery, with access via the splenic artery, represents the best solution. In all of the above cases, ligation of any branches, particularly of the right gastric artery, perfusing the antrum of the stomach is crucial.

The variant in which the left hepatic artery comes from the left gastric can be approached by implantation of two catheters and two pumps. The right and middle hepatic arteries are perfused via the gastroduodenal artery, and the left hepatic is perfused via the left gastric just distal to the takeoff of the left hepatic, with ligation of the distal left gastric. Frequently, when there is a large middle hepatic artery, then the left hepatic from the left gastric is small and perfuses only the lateral segments on the left lobe. When the bulk of the tumor is in the right lobe of the liver, then it is safe to just ligate this artery.

The right hepatic artery branching from the superior mesenteric artery is the most complicated anomaly to deal with. When it is only an accessory right hepatic, then it may be ligated. However, it frequently represents the dominant blood supply to the entire right lobe and must be perfused. This can frequently be done with two separate pumps, one catheter threaded via the gastroduodenal to the left hepatic and the second pump with a catheter directly into the right hepatic. The problem with the right hepatic coming from the superior mesenteric artery (so-called "replaced right hepatic") is that there are no side branches from which to gain access. Technically, it is certainly possible to reimplant the right hepatic on the proper hepatic or the left hepatic arteries or even to perform a reverse saphenous vein bypass graft. It is not acceptable to make an arteriotomy and just to place the large silicone catheter directly into the right hepatic artery as this almost always will lead to thrombosis of the artery. The simplest approach to this problem is to gain access to the artery, using a 22-gauge Teflon-sheathed intravenous catheter, and to connect this to the pump's delivery cannula. First, the silicone rubber pump catheter is cut off, thus leaving approximately 2 mm of catheter past the most distal ring. The end of the catheter is soaked for 5 min in fluorocarbon to soften the rubber. Then, with proximal and distal control on the right hepatic artery, the artery is directly punctured at a highly acute angle with a 22-gauge Teflon-sheathed needle. The central needle is removed, and the Luer-lock plastic hub cut off of the Teflon sheath. The Teflon sheath is then threaded up inside the pump's silicone rubber catheter. A 5-0 Tevdek suture is then placed in the advertitia of the artery and tied proximal to the circumferential ring to prevent the system from being pulled free.

It is imperative postoperatively and every 2–3 months to check patency of the system. This can be done on outpatients by injecting through the sideport 3 ml of 30% Renografin, followed by digital subtraction angiography, or 1 mCi of technetium-labeled albumin microspheres for direct hepatic imaging. The technetium–albumin scans demonstrate not only hepatic artery patency but perfusion patterns within the liver parenchyma.

INDIRECT ANGIOGRAPHIC CANNULATION. For patients who are too weak to undergo abdominal surgery and for those with poor liver function, with ascites, or with considerable malignant hepatomegaly, angiographic catheterization of the hepatic artery may be used. Variants in hepatic artery anatomy are a limiting factor. At the time of preliminary angiography, it is imperative that the gastroduodenal artery be occluded. Although frequently an angiographic catheter can be placed in the hepatic artery distal to the gastroduodenal artery takeoff, it may retract proximal to that artery leading to severe duodenitis. Occlusion of the gastroduodenal artery with Gianturco coils solves this problem. Fortunately, the right gastric artery frequently comes off the proximal gastroduodenal artery and also will be occluded with this technique. However, angiographic cannula placement is precluded if there is a large, independent right gastric artery that is in the perfusion pattern and which cannot be separately occluded. In selected patients, it is safe to occlude right hepatic arteries from the superior mesenteric artery and the left hepatic from the left gastric, again with coils, to maximize perfusion of the liver. Obviously, it is preferable to implant the catheters surgically, but in many patients this is not appropriate, and the following technique suffices.

Surgery is performed in the angiographic suite with great care to preserve sterility. The patient's chest, axilla, and arm are shaved, prepared with soap and providone–iodine, and draped. The arm is circumferentially wrapped with sterile drapes to allow it to be raised over the head to facilitate axillary artery puncture. Surgery is done under 1% lidocaine infiltration anesthesia after sedation with intravenous diazepam and narcotics. Initially, a 3- to 4-cm traverse incision is made in the lower axillary fold just beneath the insertion of the pectoralis major. A pocket is elevated on the muscle fascia. The arm is then elevated, and a single-wall puncture of the axillary artery performed. Under fluoroscopic control, a 5.3 French H1 polyethylene angiography catheter (Cook Catheter Company) is then threaded retrograde from the axillary artery and antegrade down the aorta, out the celiac axis into the proper hepatic artery, and placed appropriately. This is kept patent with the heparinized saline flush. A 10-cm incision is then made along the edge of the pectoralis major, and a 10 × 12-cm pocket made in the infraclavicular area. The prewarmed and filled Model 400 pump is then used. The catheter is cut to approximately 20–25 cm in length. A Cohen friction adapter is then placed in the cut end as previously described (Infusaid Corporation, Norwood, MA). The pump is then implanted on the chest wall and attached to the pectoralis fascia. The sideport is placed vertically. The catheter is then placed subcutaneously and brought out through the separate arm incision. The polyethylene angiography catheter is then advanced so that the connector in the catheter is buried in the muscle fascia. The entire system is now flushed through the pump sideport using 3–4 ml of heparinized saline (400 units/ml). A 2-0 silk suture is then used through the muscle fascia to hold the sideport and the entire intravascular angiography catheter in place. Care is taken to make sure that the silicone catheter from the pump lies smoothly and is in a nice 180° curve in the upper arm to prevent kinking and occlusion. A suction catheter is used in the chest incision, and both skin incisions are closed in two layers with 2-0 catgut and fine nylon to the skin. Arm motion is minimized for the first few days, after which all but the most vigorous activities are permitted.

Pump Refills in Humans

The pump refill is a relatively simple procedure and can be performed on outpatients in 10–15 min. In scheduling this procedure, care should be taken to ensure that the pump reservoir has not been allowed to empty, resulting in flow interruption and possible clot formation at the catheter tip. It is also important to remember the factors influencing flow rate. First, the bacteriostatic water used in the operating room to prime the pump will cause an initial rise in the flow rate noticed on the first refill. Also, because patient temperature and elevation can increase flow rate, these should be taken into consideration when planning refill intervals. Pump volume divided by pump flow rate indicates the maximum interval in days between pump refills. Estimated return volume with each refill is determined by the flow rate times the number of days since last refill subtracted from total pump volume.

In preparation for a refill, the prescribed solution is removed from the refrigerator and warmed to 15–30°. For the Model 100 Infusaid pump, which has a 45-ml reservoir volume, a 50- or 60-ml syringe is filled with about 50 ml of the refill solution, appropriately diluted as needed. The filled syringe is then connected to a 30-cm female–male Luer-lock adult pressure monitoring line (North American Instruments Corporation), and the extra 5 ml solution in the syringe is used to flush this tubing. The filled syringe and tubing is then used for a hand-held injection, or when the refill solution is extremely viscous, loaded into a refill apparatus.

The patient assumes a reclining position with head elevated approximately 30°; this is done to make the top surface of other pump as horizontal as possible in all dimensions. The location of the refill septum is determined by palpating the perimeter of the pump. At this point, all personnel, including the patient, should don surgical masks. Wearing sterile gloves, the operator prepares the area with three iodine scrubs, applied in a circular fashion beginning at the center of the pump and extending widely beyond the pump periphery. This procedure is then repeated with 95% isopropyl alcohol. A sterile drape is then placed across the patient's lower chest and abdomen, just below the pump site.

After changing sterile gloves, the operator then anesthetizes the area immediately over the refill septum with a 1% lidocaine solution, first injecting intradermally and then subcutaneously. The special Huber point needle (Huber 22- or 20-gauge needle, 1 or 1½ in. (available from Infusaid Corporation) is then attached to the male port of a three-way stopcock (Pharmaseal, Inc.). Only Huber point needles should be used in order to ensure the integrity of the refill septum. The empty barrel of a 35-ml syringe is placed in the female port of the stopcock opposite the needle, so that the syringe barrel points vertically. All connections are checked for tightness, and the stopcock is turned "off" to the only remaining empty port.

By percutaneous stick at an angle perpendicular to the pump, the refill septum is pierced. When difficulty is encounted in locating the septum, there are several ways to troubleshoot. First, while holding the needle steady, gently move the pump itself in a clockwise motion. If this maneuver is unsuccessful, a template is available from Infusaid which can be positioned over the pump and will aid in septum location. Once the septum is pierced, allow the drug chamber to empty under its own power into the empty syringe barrel. The chamber should *never* be manually aspirated as this may cause blood to be drawn into the intravascular delivery catheter and result in occlusion. When there is no return volume, two possibilities exist: (1) the septum has not been penetrated, or (2) the pump is empty. To determine which one of these is the case, inject 5 ml bacteriostatic water into the empty port of the stopcock and release the plunger, allowing the fluid to return. If there is no return, try again to locate the septum and repeat the procedure. When the septum is pierced and the proposed return volume has been obtained, gently rotate the needle to ensure complete emptying of the reservoir.

While maintaining the position of the needle and sterile technique, the tubing leading from the syringe containing the new infusate is connected to the only remaining arm of the stopcock; this arm of the stopcock is then turned "on" to this apparatus and "off" to the empty syringe barrel. The solution is then injected slowly, in 5-ml increments. When resistance is encountered, injection speed should be reduced. The syringe should be completely emptied. Then, to prevent spillage of residual drug into the subcutaneous tissue, the stopcock handle should be turned halfway between both female ports, the special needle grasped at its hub, and the needle quickly pulled out in one straight motion. Apply pressure and a dressing.

Accurate refill records must be maintained. Data should include (1) date of refill; (2) time interval since last refill; (3) return volume (ml); (4) flow rate (ml/day); (5) infusate volume (ml); (6) drug concentration (with lot numbers); and (7) actual and predicted drug dosage (units/day).

If possible, as in the case of heparin or insulin infusions, it is advisable to check appropriate blood measurements, i.e., the PTT or blood glucose level, 1 and 4 hr after refill to indicate whether any of the drug inadvertently was injected subcutaneously.

Pump Implantation and Refill Procedures in Dogs

Successful, long-term pump implantation in dogs requires careful attention to three major differences from similar surgery in humans. First, wound healing is promoted and infection prevented by locating the pump where it is least likely to be licked or scratched by the dog and by forming the pump pocket at least 3–4 cm from the skin incision. Second, since the risk of infection is greatly increased by the implantation of the pump or other large foreign body, surgeons must pay scrupulous attention to aseptic technique, prophylactic, intraoperative, and postoperative antibiotic prescriptions, continuous microbiological screening of pocket seromas, and rotation of antibiotics in response to bacterial sensitivity. Finally, implant failure can result from pressure necrosis due to implantation of a device that is too large, heavy, or mobile for its anatomical site. In general, although any dog weighing 8 kg or more is a candidate for a pump implant, for implantation of Infusaid Model 100 pumps it is best to use dogs with short hair, thick hide, and weighing at least 15 kg. Breeds such as beagles or coon hounds seem to do better with implants than breeds such as collies or poodles.

Preoperative Preparation

In preparation for surgery, animals fasted for 24 hr are given an intramuscular injection of a freshly prepared mixture of atropine sulfate (0.03 mg/kg body weight) and acepromazine maleate (1.0 mg/kg body weight). Fifteen minutes later the skin over either the saphenous or cephalic vein is clipped and cleaned with isopropanol, and the vein is cannulated with a 14- to 20-gauge indwelling cannula. The animal is then anesthetized with intravenous sodium pentobarbital (17 mg/kg body weight). This is one-third the usual recommended dose and takes into account the potentiating effect of the preanesthetic sedative. The calculated dose is drawn into a syringe, and one-half of the total volume is given as a bolus so that Stages I and II of anesthesia are passed through quickly and the dog enters Plane I of surgical anesthesia. If initial anesthesia is given slowly, the dog will become excited and very difficult to handle. Symptoms of Plane I include relaxation of limbs, moderate pupillary dilatation, slow and regular respiration, lack of eye movement and loss of skin, swallowing, and pedal reflexes. After 1 min, reflexes are assessed and more anesthesia is given as needed, and an endotracheal tube is inserted.

A broad spectrum antibiotic intended for intravenous use (such as 500 mg of cephaloridine or ampicillin) is diluted to 6 ml in normal saline and given slowly over 10–15 min. This step establishes prophylactic antibiotic levels. The rest of the preoperative preparation is carried out parallel to this maneuver.

A generous surgical field is clipped, but shaving is not recommended. The field is scrubbed three times with an antiseptic microbicide such as Betadine (1% iodine) or Hibiclens (4% chlorhexidine gluconate in 4% 2-propanol). Between each scrub, the area is wiped dry with sterile gauze sponges. If the scrub procedure is attempted with insufficient soap lather, skin abrasions will invariably result, increasing the risk of postoperative infection. The animal is then positioned on a heating pad on the operating table, and a final coating of Betadine solution is liberally painted on the field.

Five hundred milliliters of lactated Ringer's solution containing 500 mg of cephaloridine or ampicillin is given intravenously at a rate of 30 drops/min during the operation.

The implant site may also be anesthetized with a series of intracuticular injections of 1% lidocaine. This maneuver will reduce the amount of additional general anesthesia needed as the surgery progresses.

Pump Implantation

After draping the animal, a 60-ml dilution of antibiotic is drawn up for use on the operating table as a rinse for the pump pocket and cannulation wound; we routinely use ampicillin (0.5 g in 60 ml normal saline).

There are two suitable locations for implanting the pump: the iliac fossa and the infraspinous fossa. The iliac fossa site is more likely to be worried by the dog and is best used as a backup site.

For implantation in the infraspinous fossa, a 10-cm incision is made parallel to the cranial edge of the infraspinous fossa, beginning at the dorsal boundary of the scapula. The pocket is achieved by sharp dissection of the plane between the cutaneous trunci and the latissimus dorsi muscles. After the pocket is made, it is rinsed periodically with the antibiotic dilution. The resulting pump pocket will apparently be located unnecessarily remote from the skin incision, thus making placement of the anchoring sutures difficult. The temptation to move the skin incision so that it is tangential to the eventual pocket should be avoided, however, since pocket closure complications due to animal interference are greatly reduced by the described arrangement. The difficulty in placing the anchoring sutures from within the pocket is easily overcome by placing them transcutaneously using a cutting edge needle and then retrieving the loose ends from within the pocket. Using this technique, three nonabsorbable sutures are placed through the muscle arising from the vertebrae. A second skin incision is made, this time over the cannulation site. The pump is installed by first forcing the catheter tip over the friction grips of a trochar and then passing the trochar from the pocket through the subcutaneous tissue and exiting through the cannulation incision. The catheter is then gently pulled through that track until the pump is pulled next to the pocket. At this point, the anchoring sutures are threaded through the pump suture eyelets, the pump is inserted into the pocket, a final antibiotic rinse is injected into the pocket, and the pump is tied into place. The pump pocket closure is accomplished in three layers. First, the cutaneous trunci and latissimus dorsi muscles are brought together employing an absorbable suture such as 0-Chromic or 0-Vicryl, a taper point needle, and continuous technique in such a way that the amount of space available for seroma accumulation is minimized. Second, the fascia layer is closed, again with a continuous absorbable suture. Finally, the skin is closed, this time using an interrupted series of nonabsorbable sutures such as 0-Ticron, silk, or Prolene with a 1/3 curve cutting edge needle. Drains and compression bandages are not recommended. At this point, the cannula is still protruding through the cannulation site skin incision.

Cannula Placement

The cannulation site is usually an experimental variable which may be a suitably large vessel, body cavity (e.g., peritoneum, brain sinuses, intrathecal space, and joints between bones) or tissue (e.g., subcutaneous layer or muscle). Surgical techniques for several possibilities will be reviewed.

Systemic intravenous infusions can be easily achieved by cannulating either the superior vena cava (SVC) via the jugular vein (JV) or the inferior vena cava (IVC) via the deep circumflex iliac vein (DCI).

In the case of the DCI–IVC route, the DCI is approached by making a 2-cm skin incision at the level of and about 5 cm cranial to the iliac crest. Following a branch vein, dissection proceeds to the main trunk of the DCI and continues to the point at which the DCI passes into the first muscle layer. The vessel is isolated with two 0-Ticron sutures and ligated distally with a third. A venotomy large enough to accommodate the catheter is made by nicking the vessel with a No. 11 surgical blade, and the catheter is inserted and advanced approximately 10 cm. Three ties are placed around the vein and catheter. The loose ends are sewn into the surrounding muscle and local fascia, thus creating a secure anchor. When this technique is used, there is approximately a 5% incidence of catheters turning retrograde in the vena cava, an event which invariably results in progressive slowing of pump flow rate.

If the JV–SVC route is selected, dissection and cannulation will be much easier since the vessel is large and superficial. However, a valuable site for blood sample collection will be lost since the JV will clot off within a few days, even if it is not tied off at the time of surgery. It is, therefore, important to advance the cannula past the junction of the JV and axillary veins since the JV will clot off to that point.

Portal vein cannulations can be performed directly, using a pursestring anchoring technique or via a splenic vein. Both of these methods involve extensive exposure, thus requiring a large midline or subcostal incision. A less obvious and less traumatic route involves making a small (5 cm) flank incision, retrieving a loop of small bowel through the incision, and threading the catheter through a mesenteric branch vein. The disadvantage of this technique compared to using either a splenic vein or direct access is that verification of the location of the catheter tip is very difficult if fluoroscopy is unavailable.

Postoperative Care

Usually the dog will recover from anesthesia within 2–3 hr and regain its appetite and preoperative vigor within a day or two. Therefore, postoperative considerations are generally focused upon prevention of infections secondary to normal seroma accumulation in the pocket surrounding the pump. It is crucial that an aggressive prophylactic antibiotic regimen be continued for 10–20 days following surgery. In addition, we recommend that pocket seromas be frequently screened for microbial contamination and that antibiotic sensitivities are determined on any positive cultures. The routine should include (1) daily systemic injections of a broad spectrum antibiotic for 10 days postoperatively; (2) daily aspiration of the seroma from the pocket for as long as it continues to accumulate (usually 4–8 days); and (3) following seroma aspiration, injection of a 5-ml dilution of a broad spectrum antibiotic. A surgical scrub of the skin over the implant must always precede puncturing the pocket. The skin closure sutures must be examined daily for signs of infection and swabbed with an antibiotic ointment (such as Furicin, Panalog, or Novalsan) and removed at 10–14 days postimplant.

Pump Refills

In general, dogs quickly learn to accept the pump refill procedure, and it is unnecessary ever to resort to anesthesia or tranquilizers. If one person calms and restrains the dog and another carries out the refill manipulations, this procedure can readily be carried out on conscious animals. The supplies needed for each refill are identical to those used in human refills plus a hair clipper and 20-gauge Huber point needles substituted for the thinner needles used in humans; the thicker needles are necessary to prevent bending when trying to insert them through tough dog hide.

Author Index

Numbers in parentheses are footnote reference numbers and indicate that an author's work is referred to although the name is not cited in the text.

Y

Z

Subject Index